Africa as a Living Laboratory

Africa as a Living Laboratory

Empire, Development, and the Problem
of Scientific Knowledge, 1870–1950

HELEN TILLEY

The University of Chicago Press
Chicago and London

The University of Chicago Press, Chicago 60637
The University of Chicago Press, Ltd., London
© 2011 by The University of Chicago
All rights reserved. Published 2011
Printed in the United States of America

25   24   23   22   21   20   19   8   9

ISBN-13: 978-0-226-80346-3 (cloth)
ISBN-13: 978-0-226-80347-0 (paper)
ISBN-10: 0-226-80346-5 (cloth)
ISBN-10: 0-226-80347-3 (paper)

Library of Congress Cataloging-in-Publication Data

Tilley, Helen, 1968–
    Africa as a living laboratory : empire, development, and the problem of
scientific knowledge, 1870–1950 / Helen Tilley.
        p.   cm.
    Includes bibliographical references.
    ISBN-13: 978-0-226-80346-3 (cloth : alk. paper)
    ISBN-13: 978-0-226-80347-0 (pbk. : alk. paper)
    ISBN-10: 0-226-80346-5 (cloth : alk. paper)
    ISBN-10: 0-226-80347-3 (pbk. : alk. paper)
    1. African Research Survey.   2. Research—Africa—History—20th
century.   3. Research—Africa—History—21st century.   4. Imperialism
and science—Africa—History—20th century.   5. Imperialism and
science—Africa—History—21st century.   6. Africa—Research—History—
20th century.   7. Africa—Research—History—21st century.   8. Africa—
Colonization—History—20th century.   9. Africa—Colonization—
History—21st century.   10. Great Britain—Colonies—Africa—History—
20th century.   11. Great Britain—Colonies—Africa—History—21st
century.   I. Title.
Q180.A5T55 2011
509.6'09034—dc22   2010023092

Scientific truth is always paradox, if judged by everyday experience, which catches only the delusive appearance of things.

KARL MARX, *Value, Price, and Profit*, part 6, 1865

All outdoor laboratories are of their own local kind.

WILLIAM MORRIS DAVIS, 1887

We live in a system of approximations.
Every end is prospective of some other end,
which is also temporary;
a round and final success nowhere.
We are encamped in nature, not domesticated.

RALPH WALDO EMERSON, *"Nature,"* 1844

A new power of self-criticism is to be secured by scientific knowledge of the facts and a human sympathy which can see the whole business from the African side.

J. W. C. DOUGALL, *Africa*, 1938

CONTENTS

ILLUSTRATIONS

TABLES

ACKNOWLEDGMENTS

The subject matter of this book occurred to me between 1990 and 1998, when I had the privilege of working with a wonderful cluster of environmental and social justice organizations. Several of my friends from this period will recognize in these pages answers to questions we posed in our own attempts at transnational networking and global analysis. For their camaraderie and shared interest in pedagogy and politics I remain grateful to Premesh Chandra, Red Constantino, Ann Doherty, Michael Dorsey, Steven Gan, Alec Guettel, Niclas Hällström, Danny Kennedy, Wagaki Mwangi, Naomi Swinton, and Miya Yoshitani, all formerly of A SEED. Some of what we did together was supported by members of the Third World Network in Malaysia, especially Martin Khor, Meena Ramen, and Chee Yoke Ling. On the U.S. front, I was blessed to work with Tom Swan, Karen Stults, Eli Lee, Jane McAlevey, Laurie Weahkee, and Taylor Root, all formerly of Youth Action in Albuquerque; Anthony Thigpenn, Deepak Pateriya, Ng'ethe Maina, and Adrianne Shropshire of AGENDA and the Economic and Environmental Justice Project in Los Angeles; Colin Greer and the board and staff of the New World Foundation in New York City; and Libero Della Piana of the Applied Research Center in Oakland. These individuals and the organizing they continue to do remain an inspiration to me.

In the more than ten years that it has taken me to see this project to completion, I have accrued profound debts. From my days as a graduate student at UC–Berkeley and Oxford University I would like to thank my two PhD supervisors, William Beinart and Nancy Leys Stepan, who were unfailingly supportive and thoughtful. For their feedback and intellectual exchanges I would also like to thank Megan Vaughan, David Hollinger, Carolyn Merchant, Ravi Rajan, Richard Drayton, Mark Harrison, Robert

Fox, John Darwin, Raufu Mustafa, Anthony Kirk-Greene, Cécile Fabre, Kakoli Ghosh, Damon Salesa, Zoë Laidlaw, Melanie Newton, Ingrid Yngstrom, Sloan Mahone, Chris Low, Lotte Hughes, Andrew Fairweather-Tall, Jessica Schafer, and Sara Rich Dorman. All of these individuals influenced my thinking in both direct and indirect ways.

For his incredible hospitality and willingness to answer questions, I thank the late Edgar Barton Worthington. I am equally indebted to the late Colin Trapnell, Kenneth Robinson, and the late Theodor Monod for agreeing to talk with me about their pasts. For assistance with my research trip to East and Central Africa in 1999, it is a pleasure to acknowledge Lyn Schumaker, Wagaki Mwangi, JoAnn McGregor, Mark Leopold, Chris Conte, and Mwanahamisi Mtengula. While in Mombasa, Kenya, I attended the fiftieth anniversary of the International Scientific Committee for Trypanosomiasis Research and Control (ISCTRC), where I first met a number of field and laboratory scientists working on the subject. For giving me copies of their publications and speaking with me about technical dimensions of the history of trypanosomiasis I owe a real debt to Stephen Leak, David Rogers, and especially Ian Maudlin. I would also like to make a special mention of Jean-Claude Vatin for graciously including me among the circle of residents at the Maison Française in Oxford and for cosponsoring two conferences that I organized.

Over the years, I have also managed to meet a range of scholars interested in some of the same historical and political questions who shared their ideas as well as their published and unpublished research freely. This includes Saul Dubow, Robert Gordon, Paul Richards, Christophe Bonneuil, Richard Grove, Steve Feierman, Luise White, Lyn Schumaker, Nancy Hunt, Shula Marks, Andrew Zimmerman, Fred Cooper, Vinh-Kim Nguyen, Michael Worboys, David Arnold, Sunil Amrith, Gabrielle Hecht, Suzanne Moon, Henrika Kuklick, Randall Packard, Jim McCann, Julie Livingston, Peter Pels, Patrick Harries, Benoît de L'Estoile, Peder Anker, Gregg Mitman, Warwick Anderson, Simon Schaffer, David Anderson, Tamara Giles-Vernick, Nancy Jacobs, Harry Marks, Angela Creager, Ruth Rogaski, Bob Tignor, Tony Grafton, Michael Gordin, Ben Elman, Gyan Prakash, Sue Naquin, Liz Lunbeck, Michael Mahoney, Emmanuel Kreike, Burt Singer, Anthony Appiah, Michael Laffan, Judy Laffan, James McDougall, Jane Murphy, Tania Munz, Suman Seth, Melissa Graboyes, Funke Sangodeyi, Hannah-Louise Clark, and Paul Ocobock. Although we never met, I also feel deep debts to James Scott, Melissa Leach, James Fairhead, Robin Mearns, and the late John Ford, whose ideas and arguments prompted me to sharpen my own analysis. Gyan Prakash,

Angela Creager, and Rob Gordon deserve special thanks for their support, friendship, and words of encouragement over the years.

During the 2004–05 academic year I was able to consolidate my ideas and do a considerable amount of new research while a visiting fellow in the School of Social Science at the Institute for Advanced Study in Princeton. The theme that year, appropriately enough, was "interdisciplinarity." For their insight and fellowship I thank Adam Ashforth, Joan Scott, Sarah Igo, John Mowitt, Duana Fullwiley, Patricia Clough, Kenda Mutongi, and the late Clifford Geertz. In 2006–07, I was a ghostly presence at Harvard University. For giving me the green light and arranging my affiliation on short notice my gratitude to Allan Brandt and Henry Louis Gates. I owe the first year of leave to the National Science Foundation through its Scholar's Award program (award number SES-0349928) and the second year of leave to Princeton University's Philip and Beulah Rollins Preceptorship.

For reading the entire draft manuscript and offering many constructive comments I will always be grateful to Steve Feierman, Eugenia Herbert, Simon Schaffer, and Grey Osterud. Any remaining flaws and errors in the book are of course my own. And to David Brent I offer my thanks for his responsiveness and encouragement as an editor.

Above all, the people who kept me going and gave me the courage of my convictions over the years were my mother, Susan Tilley, and my sisters and brother, Barbara, Terry, Kate, and Cary. In the final stages, I also had my daughter, Siyamo, to inspire me. My mother's acute mind and commitment to social justice have been paramount to my own intellectual development. It is to her and Siyamo that this book is dedicated.

Portions of this work appeared earlier, often in rather different forms. Parts of the introduction, chapter 2, and chapter 3 originally appeared in two different publications: "African Environments and Environmental Sciences: The African Research Survey, Ecological Paradigms, and British Colonial Development, 1920–1940," in *Social History and African Environments*, ed. William Beinart and JoAnn McGregor (Oxford: Heinemann/James Currey Press, 2003), 109–30; and "William Allan: The African Husbandman," new introduction to the reprint of William Allan, *The African Husbandman* (London: LIT Verlag and the International African Institute, 2004), xi–xxxix. Part of chapter 4 appeared in "Ecologies of Complexity: Tropical Environments, African Trypanosomiasis, and the Science of Disease Control in British Colonial Africa, 1900–1940," in *Landscapes of Exposure:*

Knowledge and Illness in Modern Environments, ed. Gregg Mitman, Michelle Murphy, and Christopher Sellers, *OSIRIS* 19 (2004): 21–38. Finally, part of chapter 5 appeared in "Ambiguities of Racial Science in Colonial Africa: The African Research Survey and the Fields of Eugenics, Social Anthropology, and Biomedicine, 1920–1940," in *Science across the European Empires, 1800–1950,* ed. Benedikt Stuchtey (Oxford: Oxford University Press, 2005), 245–87. They are all republished here by permission of James Currey Press, the International African Institute, the University of Chicago Press, and Oxford University Press, respectively.

Africa as a Living Laboratory

Africa's modern empire-builders had the habit of thinking in continental dimensions. The historian, by following them in this habit, may find the clue not only to their achievements, but also to their illusions and extravagances.

—Keith Hancock, *Survey of British Commonwealth Affairs*, 1937

In July 1929, the British and South African Associations for the Advancement of Science chose to hold their annual meetings as a joint assembly in South Africa. More than five hundred delegates came from Britain and several hundred more attended from South Africa, making it the largest scientific gathering the continent had yet seen. The delegates spent over six weeks in the country, dividing their time among Cape Town, Johannesburg, and Pretoria. Many took part in extended expeditions through southeastern Africa, stopping to examine the archaeological ruins in Zimbabwe and ending their overland journey in Kenya. At least a quarter of the participants remained in Pretoria to attend two international congresses scheduled to coincide with their visit: the Pan-African Congress on Agriculture and Veterinary Medicine and the International Geological Congress, to which the directors of agricultural departments and geological surveys in British Africa were invited. For many participants, the joint gathering generated considerable enthusiasm for future research across the African continent.[1]

During his welcoming address to the assembly, the president of the South African association, Jan Hofmeyr, a rising liberal politician, asked his audience, "What can Africa give to science? . . . What can science give to Africa?"[2] Examining questions relating to astronomy, geology, meteorology, and medicine, he concluded, "For these investigations the diversity of conditions prevailing in the various regions of the African continent make it a magnificent natural laboratory." Animals could be examined "not as stuffed museum species, but in the laboratories of their native environment." Research into tropical diseases, "of which [Africa] may well be described as the homeland," could provide "hope and healing for mankind."[3]

Africa was more than a natural or a medical laboratory for Hofmeyr, it was also a region in which the human sciences—anthropology, archaeology, philology, psychology, and even racial science—might flourish. Scientific inquiries in Africa, Hofmeyr contended, would address the origins of the human species, would unravel the mysteries of human language, and could even reveal the intricate "workings of the human mind" and of what he called "racial intelligence."[4] Above all, he argued, "in Africa, as nowhere else, the factors which constitute these problems can be studied both in isolation and in varying degrees of complexity and inter-relationship, that in Africa we have a great laboratory in which to-day there are going on before our eyes experiments which put to the test diverse social and political theories as to the relations between white and coloured races."[5] In sum, Hofmeyr believed that the study of Africa could influence and transform almost every domain of science.

For Hofmeyr, and for his compatriot Jan Christian Smuts, who carried the laboratory analogy to England three months later during his Rhodes Memorial Lectures at Oxford, scientific projects should benefit Africa as well. Within the framework defined and administered by the European powers, Hofmeyr argued, "science must harness the great resources of Africa" and overcome "the might of African barbarism and the defiant resistance of African nature." Hofmeyr linked scientific research to the dual aims of economic and social development. Concluding his address, he remarked, "The development of Science in Africa, of Africa by Science—that is the Promised Land that beckons."[6]

The Problem of Scientific Knowledge

Hofmeyr's speech marks a seminal moment in African imperial history when scientific knowledge and colonial development had begun to assume pride of place in the international arena, and it points to the two overlapping histories this book reconstructs. The first concerns the place of scientific expertise in the conquest and colonization of British Africa, focusing especially on environmental, medical, racial, and anthropological research. The second explores the execution and impact of a metropolitan project known as the African Research Survey, a decade-long (1929–39) endeavor designed to examine "the extent to which modern knowledge was being applied to African problems."[7] Historians have long interpreted the publication of the African Survey as a turning point in Britain's administration of its African territories, especially in terms of the impetus it gave to the Colonial Development and Welfare Act of 1940, but its preeminent purpose, to evaluate the kinds of scientific knowledge necessary for imperial administration, has often been misunderstood or overlooked. Both stories intersect with wider debates about empire and development and require us to assess the nature of the social and scientific "experiments" conducted in the name of British imperialism. In this book, three questions loom large. First, how did scientific research in and on British Africa impinge upon imperial ambitions? Second, what effects did the African Survey have on British approaches to science and development? Finally, how did studies conducted in the African laboratory influence conceptual and practical developments in other parts of the world and across different disciplines?

Although Hofmeyr may have been mistaken to feel such confidence in the "development of science in Africa and of Africa by science," he was prescient to recognize that this process would have momentous consequences. The colonization of Africa and its imbrications with field research generated lasting and profound epistemological and political struggles. Scientific knowledge was indeed a problem for a wide range of people beyond scientists themselves: for metropolitan and colonial administrators whose effectiveness depended upon reliable information; for technical officers who were charged with facilitating economic and social development; for settlers who were fighting their own battles of indigeneity and belonging; and for Africans who were forced to navigate the tumultuous waters of competing epistemologies. Whether people produced or used expert knowledge, or served as its object of study, the arena of science could be both controversial and subversive, revealing the fault lines and unfulfilled ambitions of empire. As scientific practitioners grappled with African realities via their respective disciplines, and as Africans were enlisted in greater numbers as translators, informants, and assistants, the illusion of colonialism's strong facade began to crumble.

The narrative of *Africa as a Living Laboratory* unfolds between the 1870s and the 1950s. These bookends represent the beginning of the "Scramble for Africa" on the one hand and the eve of decolonization on the other. The fulcrum of the story turns on events between World War I and World War II. During the 1870s, scientific societies across Europe, especially those with an interest in geography and anthropology, shifted their attention more explicitly to the African continent; during the 1920s and 1930s, scientific research was coupled more decisively to imperial policy making and colonial development. Not only did the technical services across tropical Africa begin to grow dramatically, but colonial states' research efforts also began to influence inter-territorial and inter-imperial coordination.

Science and the African Research Survey

No analysis of the production of scientific knowledge in British Africa would be complete without considering the African Research Survey, which owed its genesis, in part, to the 1929 South African conference. The project's entanglements with various traditions of field research and intelligence gathering form a key chapter in the history of science in colonial Africa. Coordinated by a network of academics, officials, and public intellectuals and forever associated with the names of Malcolm Hailey, its director, and Edgar Barton Worthington, his scientific "lieutenant," it was the African Survey that both reoriented the British government's policies toward colonial development in the 1930s and set in motion research trends that were felt across the African continent for at least the next twenty years.[8]

Indeed, the African Survey was a project that could easily be interpreted as furthering imperial hegemony.[9] Its initial aim was to coordinate and, if possible, standardize colonial policies in Africa across the British, French, Belgian, and Portuguese territories. The Survey's leaders, in their pursuit of more effective colonial control, embraced the application of scientific knowledge and its complement, scientific colonialism. They championed new forms of "aerial reconnaissance" to conduct inventories of Africa's vast natural resources. They sought to develop those resources through widespread experimentation, drawing on methods and practices learned in other parts of the British Empire. In private negotiations, they maneuvered to exclude and deemphasize more radical voices concerned with Africa's future. They recommended the use of new cultural technologies, such as film and radio, to infiltrate more efficiently the heterogeneous cultures of the continent. They accepted, almost without question, the need to modernize African ways of life, especially agricultural and medical practices, and to incorporate these territories into the international economy. They recognized the need for colonial administrators to learn Africa's many vernacular languages in order to control the populations. In sum, the African Survey was designed to master Africa's environments and its human inhabitants through scientific management and planning.

An interpretation of the African Research Survey from this perspective alone, however, fails to appreciate the complexity that appears in the historical record. Throughout the course of the project, from its inception in

1929 to the outbreak of World War II, a subtext of criticism, dissent, and debate flourished among the project's many advisors that at times challenged the very foundations of British colonial rule in Africa. The project had uneasy origins, particularly with respect to Colonial Office hostility, and a multiplicity of authors. An analysis that emphasized only those elements perceived to institutionalize colonial domination would ignore key shifts in colonial ideology—including the critical stance the African Survey's leaders took toward racial science—that the Survey embraced and helped to popularize. It would also overlook the fact that, following on the heels of the Great Depression, a range of alternative visions and models for colonial development was emerging in British Africa that paid a great deal of attention to local conditions and environments. Many of these efforts were deeply indebted to the burgeoning sciences of ecology and social anthropology. They were also the culmination of decades of local experience during which earlier assumptions regarding the economic possibilities in tropical Africa were sharply undermined. One of the central arguments of this book is that British efforts to coordinate scientific research in tropical Africa led to a self-consciously interdisciplinary approach that stressed the heterogeneity of Africa's environments and the interrelations among the various problems scientists studied. While their aim was unquestionably to transform and modernize Africa, they envisaged ways of doing so that stressed site specificity and even local knowledge.

From the mid-1930s onward, the African Survey played a decisive role in shaping research priorities in both Britain and colonial Africa. A close examination of the scientific controversies in which it became embroiled—such as debates about the significance of ecological science, the relevance of intelligence testing, the need for integrated public health services, the adequacy of witchcraft legislation, and the usefulness of social anthropology—exposes not only the values and preoccupations of its coordinators but also the wider moral and political climate of the period. When its results were finally published in 1938, Lord Hailey declared in the report's opening pages: "Africa presents itself as a *living laboratory*, in which the reward of study may prove to be not merely the satisfaction of an intellectual impulse, but an effective addition to the welfare of a people."[10] This book takes its title from that pronouncement.

Key Concepts: Science and Empire

An analysis of the history of the African Research Survey goes hand in glove with an analysis of the history of scientific activities in colonial Africa. *Sci-*

ence, like *nature*, has always been a multifaceted term. In this book, I use it to encompass a broad range of institutions and activities, including professional networks and systems of patronage, learned societies and research institutions, field sites and laboratories, and cognitive frameworks and disciplinary structures.[11] To avoid awkward repetition, I often assume that the term encompasses the fields of medicine and anthropology. Significantly, the historical actors featured in this book were themselves preoccupied with distinctions between science and nonscience, "pure" and "applied" science, and the boundaries separating old and emerging disciplines. This book explores the changing and contested meanings of *science* itself in the context of practitioners' engagement with Africans' ideas and knowledge.

Empire is an equally slippery concept entangled in notions of territorial sovereignty, juridical control, spheres of influence, and sociocultural interpenetration. To highlight the differing experiences of various regions of the world, scholars have devised a range of terms that help to classify types of empire: *formal* versus *informal*, *contiguous* versus *overseas*, *internal* versus *external*, *settler* versus *nonsettler*, *economic* versus *military*, and so on. Whether a region were a protectorate, a colony, a mandate, a condominium, a province, or a dominion; whether it were dependent or self-governing; whether it tried to accommodate permanent settlers—all these features influenced those who governed. These demarcations were not always sharp or stable because both legal definitions and territorial status could change, yet the distinctions themselves mattered.

The primary lens of empire for this study is juridical: in other words, formal political empire. This approach helps to frame which regions of British Africa receive the most attention. Territories such as South Africa, which amalgamated and became self-governing in 1910; Egypt, which achieved an attenuated form of political autonomy in 1922; and Zimbabwe (Southern Rhodesia), which was granted self-governing status in 1923 and was thenceforth overseen as a Dominion territory, are not the focal points of this book. Neither are the High Commission Territories—Swaziland, Lesotho (Basutoland), and Botswana (Bechuanaland)—that were administered by the British high commissioner in South Africa.[12] The regions at the core of this study are the dependencies ultimately overseen by Britain's Colonial Office, with greatest emphasis placed on Kenya, Tanzania (Tanganyika), Uganda, Zambia (Northern Rhodesia), Nigeria, and Ghana (Gold Coast). These were the *tropical* dependencies that played the most significant role in scientific research. This book offers an analysis of the ways developments in specific regions influenced imperial policy making more broadly.

Constituting National, Imperial, and International Infrastructures for Science

Africa as a Living Laboratory shows that national, imperial, and international scientific infrastructures were constituted simultaneously. Untangling these relationships and determining the changing dynamics of cause and effect, particularly with respect to centers and peripheries, can be quite complicated. Not only did individual practitioners often traverse actively from one bureaucratic or institutional level to another, but also institutions themselves were designed to serve multiple constituencies and purposes. The roles of the League of Nations' Health Organisation and the Rockefeller Foundation and its International Health Board have considerable bearing on medical developments in colonial Africa in the interwar period. Both the intergovernmental organization and the private philanthropy stressed cooperation and collaboration that moved beyond national boundaries. Both also highlight the importance of the United States in these international networks, since Rockefeller funding enabled the Health Organisation to achieve such autonomy from the League itself.[13]

International scientific congresses in such fields as soil science, botany, and anthropology connected members of Britain's colonial service with other scientists from around the world and crystallized particular research interests outside the framework of colonial needs. Their groups' activities and interests were cognizant of Europe's empires, but were not dictated by imperial considerations. Similarly, the International Institute of African Languages and Cultures brought together scholars, missionaries, and administrators from across Europe, North America, and Africa who shared an interest in investigating Africa's human inhabitants. This institution was founded with an international remit, but its field of operations was almost entirely colonial and its leadership structure was grounded in national representation.

Scientific Diasporas

At the same time, some scientific structures emerged almost entirely from the needs of colonial regimes and represent a transnational process that was deliberately demarcated under the authority of a single nation. The establishment of technical departments in African territories dealing with agriculture, forests, game, survey, and medicine provides a perfect example. Staffing these departments required a great migration of scientists, both amateur and professional, from Europe to the African continent, creating *scientific diasporas* complete with their own institutions, journals, societies, meeting places, and funding. These individuals were accountable both to the territories in which they served and to the European nations by which they were employed. Their studies, while frequently presented to international scientific communities, were first and foremost produced in the context of colonial governance for a more limited audience. Thus technical experts were forced to navigate implicit contradictions of colonial rule. Indeed, they operated simultaneously within international processes, which depended for their success on the free flow of ideas for the growth of disciplines and scientific theories, and national and imperial processes, which involved more tightly monitored funding patterns, structures of accountability, and government bureaucracies. Who would ultimately benefit from their studies? What was their primary aim? How could the needs of multiple audiences and constituencies be met?

Richard Drayton has recently contended that positions in the colonial technical services, "until perhaps as late as the Edwardian era . . . provided vital opportunities for those participating in emerging disciplines."[14] For British Africa this insight could arguably be extended to the 1920s and 1930s, when a good deal of groundbreaking research was undertaken. Echoing the laboratory motif, the evolutionary biologist Julian Huxley remarked upon what he saw as Africa's enormous research potential in his 1931 book *Africa View*. According to Huxley, the British colonial service had "a shortage of biologists of all kinds and an intense shortage of biologists of the first class." Yet, scientists' intellectual and professional opportunities were infinite:

> If a technical officer were engaged on research the problems that open out before him are so varied, so untouched, and so numerous that he has much greater chance [than "at home"] of arriving at results of striking practical importance. . . . In Africa the prosperity and indeed habitability of enormous areas hangs upon success or failure in research, and research along the broadest biological and medical lines. . . . To be a good . . . officer in Africa you must be a first-class biologist, and you must be a knowledgeable and sympathetic anthropologist as well.[15]

Huxley's observations were made at precisely the time Britain's Colonial Office was restructuring and expanding its technical services. These posts constituted roughly one-quarter of all appointments during both decades, so the research activities of these officers deserve closer scrutiny. Significantly,

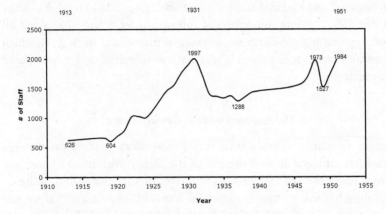

Figure 0.1. Technical services in British Africa, 1913–51. The territories were Kenya, Uganda, Tanzania, Zambia, Malawi, Nigeria, Ghana, Sierra Leone, and Gambia. Technical Services include agriculture, veterinary, medical, game, forestry, geology, and survey departments. Source: *Colonial Office Lists*, 1913–1951 (London: HMSO).

the dramatic increase in technical officers that took place in the 1920s, going from roughly six hundred in 1919 to approximately two thousand in 1931, generally precipitated a steady increase in original scientific research (figure 0.1 and the appendix). These developments coincided with and were reinforced by the "fact-gathering" activities of the African Survey.

A fundamental premise in the literature dealing with transnational issues in science is that national interests existed in dynamic tension with international cooperation. Elisabeth Crawford, Terry Shinn, and Sverker Sörlin have observed that "although there reigns an internationalist ethos among scientists, scientific work and careers are most often bound up with the nation. . . . Impetus towards international science was fuelled by national pride, the professional ambitions of a country's leading scientific personalities, and government policy."[16] According to Brigitte Schroeder-Gudehus, scientific internationalism "resides in the conviction that international agreement on theory and method and international communication and collaboration among scientists are inherent to scientific work and scientific progress."[17]

The layer of institutions established to meet the needs of the empire occupied an interstitial space that was neither national nor international. This tripartite structure of knowledge production—national, imperial, and international—has often been reduced in the scholarly literature to a simple dichotomy. Scholars concerned with national versus international science often omit imperial relations from their purview. Those focused on imperial and national science, while less myopic, have tended to leave to one side explicitly international infrastructures.[18] Only by taking all three institutional tiers into account can we understand the ways in which ideas and techniques moved across nations, empires, and international bureaucracies.

The Problems with "Colonial Science"

Among scholars concerned with empire and Africa, a more problematic trend has emerged around the use of the highly misleading phrase "colonial science."[19] Imperial historians have an understandable desire to explain what was specific to "colonial forms of knowledge."[20] If we can legitimately identify the "colonial state," which resembled, but was not identical to "modern states," can we not also legitimately speak of colonial science? The answer to this question is straightforward: no. While the designation colonial state remains debatable, particularly because there were many shades of gray between colonial and modern nation-states, there are enough durable features, especially in terms of their juridical constitution, to give the phrase analytic heft. The concept of colonial science, though, has been deviled by theoretical fallacies and ambiguous dualisms since its inception.[21] Built into its theoretical framework are several untenable assumptions. (Although not everyone who uses this phrase falls prey to these problems, they must be addressed.) First, that something called "Western science" developed in geographical isolation within Europe. Second, that colonial science existed as a distinct phenomenon separate from science proper. Third, that colonialism produced certain features that were pathological distortions of "real" or "good" science. Finally, that theoretically distinct bodies of knowledge, often called "indigenous knowledge," were disrupted or destroyed by colonialism. While all of these assumptions have had considerable intellectual purchase, none can stand in the face of close historical analysis.

Many of the conceptual difficulties are related to place.[22] States, colonial or not, are by definition territorially bounded, while science has circulated.[23] Disciplines and their supporting institutions are constrained by location; their epistemological existence may even depend on it, as is the case with both field and laboratory sciences. But they are almost never constructed to be immobile. Knowledge may be "situated," but it is designed to travel. There is too much circulation between metropole and colony, across colonies, and between colonies and nation-states to warrant the designation colonial science. If a concept or practice generated in a colony quickly became an accepted principle of metropolitan thinking, would it be aptly called colonial science? If principles and methods devised in metropolitan laboratories whose mission was to support the empire were picked up by scientists in other sovereign states, should we call this colonial science? Closer analysis usually reveals that differences between metropolitan and colonial contexts are relational: you cannot have one without the other; they are intimately intertwined.[24] To take the thorniest of the problems: even the codification of indigenous knowledge often required colonial relations to bring these patterns into focus, which is why the first studies to define indigenous knowledge were usually done by the colonizers.[25] To a great extent, theories of ethnoscience—folk, primitive, traditional, local, and indigenous knowledge—owe their existence to imperial structures and sociopolitical asymmetries.

A Living Laboratory: Epistemologies of Science and Empire

Studies of Africa's colonial past must come to terms with rapid and at times extreme social, environmental, and epidemiological changes. Western European nations' scramble for Africa occurred during a period when technologies of empire—steamships, railroads, machine guns, the telegraph, and quinine—were employed as part of ever-expanding networks of power.[26] Africa's partition also coincided with radical changes across scientific disciplines: as fields professionalized, new ideas and methods proliferated. Applied scientific institutions, such as experiment stations, technical departments, research institutes, and scientific societies, had already been established in many other regions of the world, often in support of colonial administration.[27] In tropical Africa, these capacities were a relatively late development being organized systematically only in the first decades of the twentieth century,[28] although the infrastructures set up earlier in both northern and southern Africa made each region its own "center of calculation."[29] The era of "interventionist" colonialism, which encompassed agriculture, public health, natural resource use, disease control, labor recruitment, conservation measures, and trade and investment, thus dovetailed with a set of trends and intellectual preoccupations peculiar, in large part, to the first half of the twentieth century. Among the most prominent developments of this period were: shifting geopolitical arrangements, including the rise of the United States as a dominant world power; increasing anticolonial sentiment emanating both from within and outside the British Empire; the growing recognition that colonial territories could not be expected to prosper socially or economically without sustained capital investments; far-reaching advances in a variety of scientific disciplines that fundamentally altered human understanding of themselves and their surroundings; and finally, a deepening concern for the "social relations of science" as a consequence of World War I, the economic depression, and the links between capitalist production and technology.[30]

Scholars have long likened colonial territories to laboratories, but the analogy was first used by those involved in empire building and had heterogeneous roots. During the last third of the nineteenth century, a number of individuals began explicitly to link the idea of improving the African continent with developing it through science. For them, science was not just a matter of theory; it was based on experimentation and practice in the field. The idea that geographical regions could yield important scientific insights was hardly new. Nor was the concept that colonies themselves were ideal locations to conduct tests of both a social and a natural kind.[31] Use of the word *laboratory* suggested a space of knowledge production touted for its rigorous and robust truth claims.[32] It also suggested a site in which manipulation might be manageable. Juxtaposed to the continent of Africa, however, its use drew tacit and explicit attention to the inherent limitations of the "laboratory revolution" in the sciences.[33]

The rise and increasing supremacy of laboratories for chemical, pharmaceutical, diagnostic, and experimental purposes developed in tandem with the ascendancy of field sciences whose domain of expertise was often colonial terrain.[34] Although laboratories encouraged many scientists to scale down their analyses, to go inside organisms, manipulate physical phenomena, isolate active ingredients, or even create experimental objects, they were also a catalyst for others to scale up their studies and develop new forms of surveillance and experimentation outside laboratory settings. Field scientists and their supporters recognized that some kinds of phenomena could not be investigated or controlled in a confined space.[35] While they evoked the authority of laboratory knowledge, they simultaneously challenged the physical boundaries and natural validity on which that authority was based. Studies in the field were not only posited as more real and true than laboratory analyses, but they were also organized around a range of concepts that resisted the tendency to examine parts separately from the whole: they highlighted interrelations, interdependence, and the "bird's-eye view."

Africa as a Living Laboratory studies the dynamic interplay between scientific fieldwork and research across metropolitan and colonial contexts and the ways this interplay helped to challenge and transform theories that had been prevalent during the scramble for Africa. It was largely field scientists—geographers, anthropologists, botanists, and specialists in medical geography—who took part in the partition of tropical Africa, and field research that informed many of the investigative projects that were part and parcel of colonial state building. Particularly during the interwar period, technical officers and scientific advisors began to advocate for cross-disciplinary methods that would bring to light connections among diverse phenomena in nature. By 1929 Jan Hofmeyr was well aware of these trends, which explains why he used the laboratory motif in pointing out that many problems in Africa "could be studied both in isolation and in varying degrees of complexity and inter-relationship."

This historical overview illustrates how certain imperial preoccupations regarding tropical Africa were largely overshadowed in the early twentieth century by new concepts and methods that intersected with shifting priorities in the realms of governance and economics. The moves from racial acclimatization to racial psychology; from miasmatic explanations of disease to germ theory, with its emphasis on vectors and infectious agents; from a belief in tropical fertility to concern over the infertility of the tropics; and from evolutionary theories of racial hierarchies to sociological theories of culture contact—all these transitions deserve greater scrutiny. They were significant not only for research paradigms but also for the underlying assumptions of colonial rule itself. The enduring importance of these debates becomes apparent when we realize that many contemporary discussions focused on the "problem of African development" continue to address the continent's environmental, epidemiological, and sociological conditions.

Power, Agency, and Commensurability

Readers today tend to be highly sensitive to any situation that suggests human experimentation by one social group over another, an image the laboratory motif inevitably conjures. These sensibilities are largely a product of twentieth-century experiences that exposed the extraordinary abuses committed in the name of scientific advancement. To understand the basis of this moral abhorrence, we need only think of the medical atrocities of Nazi Germany, which led to the Nuremberg tribunals and helped to transform international attitudes toward medical ethics.[36] Designed to ensure that experiments conformed to sound "moral, ethical, and legal concepts," the Nuremberg Code of 1947 placed great stress on the idea that humans had to give their consent to be subjected to experimentation.[37] In the colonies, whether in the realm of medicine, anthropology, psychology, agriculture, or even nutrition, consent was rarely an explicit concern.[38] The controversy surrounding the investigative journalist Edward Hooper's allegations of a relationship between the polio vaccine trials in the Belgian Congo in the 1950s and the origins of human HIV infections brought out just this point.[39] That at least some individuals in the colonial era viewed Africa as a literal, not just a metaphorical, laboratory suggests that they agreed with the idea that external manipulation and control were not merely acceptable but also desirable. Africans and their environments were in this sense captive subjects of such experiments.[40] This dimension of colonialism, which reveals the asymmetrical relationships between colonized and colonizers, has received considerable attention in the secondary literature on African imperial history.

Historians who have explored the relationship between science and empire in colonial Africa have recently gone to some lengths to demonstrate the importance of Africans as active agents in the production, application, and appropriation of scientific knowledge. Without losing sight of differences and inequalities, these studies have revealed intricate patterns of negotiation and exchange alongside those of domination and intervention. Richard Waller and Kathy Homewood have explored interactions among Maasai elders and colonial veterinary experts in Kenya and Tanganyika. Although they structure their account around a dichotomy between "Western medicine" and "indigenous" therapies, their evidence allows for a less rigid reading. While many Maasai chose to ignore or resist colonial veterinary interventions, some adopted certain measures they felt worked to their benefit. At the same time, not all veterinary experts rejected Maasai animal husbandry practices, and a handful even turned to their study and defense, illustrating a pattern that was more common than many critics presume.[41]

Megan Vaughan, Nancy Hunt, Julie Livingston, and Luise White have offered similar interpretations in the context of human medicine, underscoring how often Africans translated concepts and tools to suit their own purposes.[42] As Vaughan has argued, "in Africa at least, colonial medics were simply too thin on the ground and their instruments too blunt to be viewed either as agents of oppression or as liberators from disease, and studies of African demography confirm this view."[43] Hunt's emphasis on rituals of social reproduction and her interpretation of the ways Congolese leaders in Yakusu incorporated new symbols of wealth and fertility into their attempts to maintain healthy social groups offers a fine-grained analysis of Africans' efforts to amalgamate and synthesize different approaches to well-being. The growing literature on the history of prophet movements, social reactions to epidemics, and the adaptability of "traditional" healers not only illustrates the porous boundaries that could exist between introduced and endogenous knowledge systems but also draws attention to the often tragic powerlessness of these systems.[44] John Janzen observed more than two decades ago that in the face of the many upheavals caused by colonialism, "neither African nor European medicine was very effective."[45]

Struggles over the production of knowledge tell a comparable story. Lyn Schumaker, Jean-Hervé Jezequel, Sara Pugach, and Harry West, who have examined the role of research assistants in African ethnography, all stress the control they retained over information and ideas. As Africans took part in ethnographic studies and even became authors themselves, they demanded to be included among professional ranks and to receive tangible benefits from their contributions.[46] Bruce Berman, John Lonsdale, and Robert Gordon have examined the African anthropologists Jomo Kenyatta in Kenya and D. D. T. Jabavu in South Africa, who became credentialed scholars in their own right. For all these men—and most of the examples are indeed of men—producing ethnographic studies became a means to seize control of political terrain, both for personal aims and for purposes of collective representation. At the heart of African-authored ethnographies were political contests acted out on many different fronts. As colonialism recalibrated power relations on the ground, these interventions became a way for African elites to assert their own authority and expertise. None was exempt from struggles over how to interpret and react to people's cultural particularities, and all were making very deliberate choices about how to represent African societies. Scholars have recounted analogous stories regarding Africans who joined the medical profession as assistants, midwives, and doctors.[47] In the face of pejorative assumptions about their abilities and knowledge, these individuals often had ambivalent relationships to colonial ideologies and structures of rule. Their livelihoods usually depended on remaining in the good graces of their employers, but that did not stop them from expressing their own criticisms or from simply ignoring epistemological perspectives with which they disagreed. Nor did it stop them from forming opinions about their colleagues, both senior and junior, African and European, and taking issue with their points of view.

Interpretations of science and empire have thus concentrated on three central issues: the power colonialism conferred on science, the ways in which sciences were used as "tools of empire," and the agency of non-European peoples and places to reshape the sciences. For those who examine these complexities in historical contexts, one of the major theoretical and methodological questions hinges on the commensurability of divergent knowledge systems. When the terms of reference between two systems are far from equivalent and various groups have incompatible worldviews, it is

Africa (not including South Africa and Egypt), which amounted to more than 2 million square miles of new terrain that its officials sought to oversee.[54] The difficulties they faced trying to juggle state building at the level of a single territory with empire building across the African continent and the British Empire more broadly were formidable. There was no systematized body of scholarship on which they could draw to understand the peoples or regions that were now under their control. Nor was there any master plan for what to do with these territories beyond an amorphous vision of improvement and development, which some might call exploitation.[55] As many African historians have pointed out, underpinning these ostensibly benevolent motives was a penchant to use violence and coercion when colonizers deemed it necessary. The idiosyncratic and disjointed dynamics of territorial partition help to explain why it was not until 1906, after the transfer of Uganda, Kenya, and Nyasaland from the Foreign to the Colonial Office, that the Advisory Committee for the Survey of Tropical Africa felt able to announce that "there are now no 'hinterlands' in British Africa, the partition of the continent is complete, and the boundaries are, with few exceptions, well defined."[56]

British states in tropical Africa were grounded in an economic logic from their inception. Imperial officials viewed African populations less as prospective political actors and more as potential producers. Colonial states in this sense were designed to be development states: their telos, from the start, was focused on resources, revenue, and production rather than political participation.[57] If we accept this claim, it forces scholars to push back an analysis of colonial development to the first decades of the twentieth century and reconsider the prevalent assumption that development in tropical Africa only began after the Second World War. The term *development state* acknowledges that these were "a variant of the capitalist state," but it also draws attention to the constant friction that existed between production and social welfare, a friction that became most acute in the decades following the Great Depression.[58] The fact that many British African states were unable to be economically self-reliant and that basic infrastructure projects could not be supported from within posed constant challenges for imperial administrators.

A decade after the partition was declared complete, Britain's newly acquired tropical dependencies in Africa were still considerably weaker than those elsewhere, when defined in terms of their revenue, staff size, and bureaucratic infrastructure compared to the populations and geographies they were meant to rule. A quick review of the revenue base of the British Empire in this period demonstrates the relative poverty of most African states (table 0.1). Their financial foundations paled in comparison to other territories, falling well behind India, Canada, Australia, Egypt, South Africa, New Zealand, and even the Malay States. Although colonial dependencies were shored up economically and politically during the first sixty years of imperial rule (circa 1890 to 1950), their total staffing never managed to exceed 10,000 individuals (figure 0.2 and the appendix).[59]

Development and Empire Building as "Complex Problems"

A contentious debate among scholars today turns on why so many development initiatives failed to produce the benefits expected of them. One explanation focuses on the difficulties inherent in attempting to master humans and tame natural environments. Peter Taylor has referred to this ambitious philosophy as "technocratic optimism," which included a boundless faith in experts' ability to engineer better societies.[60] James Scott, in his book *Seeing like a State*, calls it "high modernism," which he argues emerged as a product of the rapid industrial and technological changes that occurred in Western Europe and North America between 1830 and 1914.

> At its center [high modernism] was a supreme self-confidence about continued linear progress, the development of scientific and technical knowledge, the expansion of production, the rational design of social order, the growing satisfaction of human needs, and, not least, an increasing control over nature (including human nature) commensurate with scientific understanding of natural laws. High modernism is thus a particularly sweeping vision of how the benefits of technical and scientific progress might be applied—usually through the state—in every field of human activity.[61]

The crux of many development problems for Scott and others who share this viewpoint is statesmen's desire to capture the intricacies of the natural world through the scientific methods of simplification, standardization, quantification, and abstraction. This approach, he argues, errs because it attempts to reduce highly complex and unpredictable systems to predictable and manageable ones. Scott illustrates his point with a quotation from Herbert Simon: "Administrative man recognizes that the world he perceives is a drastically simplified model of the buzzing, blooming confusion that constitutes the real world. He is content with the gross simplification because he believes that the real world is mostly empty—that most of the facts of the real world have no great relevance to any particular situation he is facing and that most significant chains of causes and consequences

Table 0.1 British territories in comparative perspective, 1916–17 (ordered by revenue)

Territory	Square miles	Population† (millions)	Revenue (£ millions)	Expenditure (£ millions)	Railway lines (miles)	Telegraph (miles)
India	1.8 million	315	96.8	90.8	36,286	87,480
Canada	3.7 million	8.4	47.8	30.5	37,434	66,055
Australia	2.97 million	4.9	34.0	87.5	23,774	57,221
Egypt	350,000	12.6	20.4	17.7	2,874	4,727
South Africa	473,100	5.97	18.7	19.3	9,419	16,219
New Zealand	104,751	1.1	18.0	14.1	2,989	13,687
Malay States*	52,592	2.9	9.4	6.1	876	2,810
Ceylon	25,332	4.5	4.4	3.7	706	2,024
West Indies	12,200	1.8	3.0	2.9	334	1,304
Nigeria	336,080	17.5	2.9	3.6	975	4,528
Sudan	1.01 million	3.4	1.9	1.8	1,500	4,710
Gold Coast	80,235	1.5	1.8	1.5	257	1,319
East Africa/ Kenya	246,800	4.04	1.5	1.2	618	1,305
N. & S. Rhodesia	440,000	1.7	.93	.99	2,472	7,956‡
Sierra Leone	31,000	1.4	.55	.533	331	985
Uganda	109,119	2.96	.32	.289	70	1,487
Nyasaland	39,573	1.1	.15	.128	129	824
Gambia	4,500	.146	.10	.83	NA	NA

Source: Sir John Scott Keltie and M. Epstein, eds., *The Statesman's Year-Book for the Year 1918* (London: Macmillan, 1918), xvi–xix.

* The Malay States include the Federated Malay States, the Straits Settlement, and the Unfederated Malay States.

† Population estimates must be regarded with skepticism, but they are included here for rough comparative purposes.

‡ This figure is somewhat higher than it ought to be because it includes lines owned by the African Transcontinental Telegraph Company feeding to other African territories.

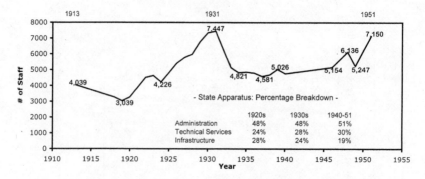

Figure 0.2. British African tropical dependencies: Colonial Service employment, 1913–51. The territories were Kenya, Uganda, Tanzania, Zambia, Malawi, Nigeria, Ghana, Sierra Leone, and Gambia. Administration includes district and provincial officers, the executive branch, the judiciary, and the police. Technical Services include agriculture, veterinary, medical, game, forestry, geology, and survey departments. Infrastructure includes railway, telegraph, harbor, and road staff, including most engineers. Source: *Colonial Office Lists*, 1913–1951 (London: HMSO).

are short and simple."[62] In Simon's and Scott's scenarios, the coupling of an oversimplified technical knowledge with centralized state planning is to blame for development failures. This system is generally unable to account for heterogeneity, diversity, or change; more to the point, it cannot accommodate what Scott calls "mētis," or practical and experimental "forms of knowledge embedded in local experience."[63] Advocates of approaches that might serve as alternatives to "high modernism" occupy peripheral positions: "A critique of such hegemonic ideas comes, if it comes at all, not from within, but typically from the margins, where the intellectual point of departure and operating assumptions . . . are substantially different."[64]

For all its strengths, Scott's analysis still takes inadequate account of the history of European empire building, especially in tropical Africa, and of the rise of scientific disciplines that considered complexity and interrelations their key problematics. These significant trends occurred during precisely the period in which Scott is most interested. Scott's arguments, however, remain valid: modern states and administrative personnel have indeed exhibited many of the tendencies he describes, and deliberate efforts "to improve the human condition" on a large scale have often failed. Yet important questions remain. Did colonial states, and empires more broadly, exhibit the same features of high modernism? Is it possible that certain scientific disciplines included foundational concepts that took into account the importance of place, interactions among phenomena, and even local knowledge?

Extending Scott's analysis, this book argues that imperial management and control, particularly in the wake of the empire's dramatic expansion in Africa, forced British officials and other interested parties to grapple actively with transnational and inter-territorial trends. Ideas of heterogeneity and diversity, far from being absent, were ever present in their discourse and became integral to the logic of empire building. *Thinking like an empire* meant drawing attention to competing interests and pointing out the ways in which one issue or problem was nested within another. It also meant avoiding watertight compartments of knowledge and drawing upon burgeoning disciplines, such as ecology and anthropology, that emphasized social and natural interrelations. Indeed, commentators often likened the empire to a "living organism" whose very "complexity" required not only new kinds of institutions but also a new "machinery of knowledge."[65]

To take an example from the interwar period, consider the report of the scientific subcommittee of the first Colonial Office Conference held in 1927. The committee was charged with exploring how Britain might augment scientific research across the colonial dependencies and contribute to their economic and social development. Its members seized upon the idea that cross-disciplinary and multiregional research was the best way to address economic and epistemological challenges simultaneously. As they framed the issue in their final report, the "Colonial Empire presents, in its diversity, many complex problems, which provide a fruitful and largely untouched field for investigation. . . . This may entail the simultaneous co-operation of institutions and workers in more than one science and in more than one part of the world."[66] In recommending new measures to assist scientific research they wished to assuage any fears that individual colonies or "local problems" might receive short shrift: "Our aim is to provide the body with a brain, nerves, and arteries, not to clothe it in a strait-jacket."[67] A decade later Charles Jeffries, the assistant secretary of state in the Colonial Office, summarizing his experiences trying to manage the growth of the administrative and technical services, invoked a similar vision.

> The difficulty of the task lies in the fact that we have to deal with no cut-and-dried system, but with a living growth. The Colonial Service is not a machine. It is an organism which must adapt itself to an almost infinite variety of conditions. There is no sealed pattern of official organisation which can be imposed upon each Colony irrespective of local circumstances. The

members of the Service are not a set of chessmen moved by invisible hands in Downing Street, nor is the Colonial Empire a board upon which pieces can be disposed at will.[68]

The Need for New Knowledge

The challenges that colonial administrators confronted were widely understood in Britain as well as across the British Empire. Those at the helm of imperial management were acutely aware of the complexities involved in holding the empire together in the face of both white settler nationalism and indigenous anticolonial resistance. Even constructing state bureaucracies entailed constant problems, given the paucity of individuals and resources at their disposal.[69] When officials called upon scientists to help them and created new structures to facilitate coordination, they were not just trying to depoliticize the problems, as critics of colonial experts sometimes suggest; rather, they were admitting that existing institutions were inadequate to the tasks at hand. To them, new knowledge seemed essential.

The first half of the twentieth century witnessed the creation and consolidation of African and empirewide bureaus and advisory committees designed to produce and disseminate this knowledge. Regional, pan-African, and imperial coordinating conferences functioned to settle policies and establish research priorities.[70] These official and semiofficial initiatives were complemented by special gatherings at scientific congresses and annual meetings of scientific societies. In 1905, for instance, the British Association for the Advancement of Science held its first joint gathering in South Africa. Between 1902 and 1914, the Royal Society, working in tandem with the Foreign and Colonial Offices, sent special commissions to investigate the sleeping sickness epidemics in Uganda, Nyasaland, Southern Rhodesia, and the Sudan.[71] In the 1920s and 1930s technical officers from the African territories convened special sessions at international congresses relating to the continent's soils, flora, fauna, ecology, and anthropology. These interterritorial and transnational endeavors achieved their greatest scope as a consequence of state building in sub-Saharan Africa.[72] The African continent was not just swept up in this process, but helped to catalyze and constitute it.

A preeminent example of thinking like an empire is the African Research Survey, which took these principles to new heights. It not only helped to popularize the idea that the problems of British Africa were inherently complex, but it also encouraged officials to sharpen their transnational and comparative perspectives, situating sub-Saharan territories in the wider context of the British Empire as a whole. What makes the story of African empire building so noteworthy in this respect is that ideas of complexity in the imperial realm were melded with ideas of complexity in the sciences. Those statesmen and scientists most interested in deepening connections between empire and expertise were responsible for characterizing Africa as a "*living* laboratory."

Subversive and Vernacular Sciences

Recently, historians have amply illustrated how scientists have given credence to pejorative, inaccurate, and sometimes patently false theories. Even those committed to the description of objective reality and suspicious of the view that science is socially constructed have come to acknowledge that all truth claims are produced within a specific political economy and social milieu that shape their content. These insights form an essential part of this book's analytical framework. Yet, what scholars sometimes miss in their analyses of scientific research in colonial Africa is that historical actors' attempts to get "full, precise, and accurate facts"—to quote the missionary anthropologist Edwin Smith—often undermined and transformed the assumptions that brought them to the continent in the first place.[73] Experts' collective emphasis on accuracy, especially when they had the opportunity to encounter their subjects of study firsthand in the field, could call into question not only the limits of the ways they pursued knowledge but also the very content of that knowledge.

Scientists and Auto-Critique

An important, hitherto unappreciated dimension to the story of the African Survey and of the scientific research conducted in Britain's dependent empire during this period is the pervasiveness of criticisms generated from within by experts active in the field. The interwar generation's appeal for more and better "facts" is so fascinating because it often stemmed from a recognition of past errors. Their concerns were not premised simply on naive assumptions regarding the objectivity or universal authority of science per se, but were based on an admission that existing knowledge could be markedly improved upon. The objects of their criticism were twofold. First, they centered on discrepancies between scientific representations and ontological realities. Practitioners in fields as diverse as ecology, soil science, epidemiology, agriculture, anthropology, and psychology repeatedly drew attention to the inadequacies of existing theories and assumptions.

At stake for them was not simply the truth of their claims but the kinds of interventions these claims facilitated. Second, criticisms concentrated on perceived failures or shortcomings of particular forms of colonial governance, the flaws and contradictions in the management of particular territories. Careful attention to the epistemological and political critiques made by scientists reveals that while at times their views were expressed as loyal opposition and were intended to bolster colonial control, they also represented an active challenge to existing imperial practices and ideologies.

Cedric Dover, an important voice in debates during the interwar period on the subject of miscegenation and racial hybridity, wrote a tongue-in-cheek poem in which he declared, "The reely scientific view, you know, / Is palsy-walsy with the status quo."[74] In many respects, this book offers conclusions that undermine Dover's declaration. Indeed, it consciously foregrounds a dimension of the interactions among power, political economy, and knowledge production that has as yet received little scholarly attention: the subversive relationship that could exist between science and empire, particularly in the era of late European colonialism. The south Asian historian David Arnold has suggested that biomedicine in British India eventually enabled the "emergence of anti-colonial discourse and an aspirant counter-hegemony."[75] Christopher Bayly has used Michel Foucault's concept of "insurgent knowledges" to signal "the weaknesses of power and legitimacy" in colonial knowledge production.[76] Both are referring to the "margins of power" in colonial India. In contrast, this book demonstrates that processes of knowledge production subversive of the status quo could emanate directly from epicenters of colonial power in Britain and tropical Africa. It offers not so much a view from below as a careful examination of the ways in which scientific research began to decolonize Africa by challenging stereotypes, destabilizing Eurocentric perspectives, and considering African topics on their own terms.[77]

We need not lose sight of the exploitative nature of colonialism or endorse previous scientific perspectives to see that their interactions played a role in undermining colonialism. During the 1920s and 1930s many certainties regarding industrialization, agriculture, labor, natural resources, and human identities were destabilized. With the rise of cultural relativism and the simultaneous critiques of both "science" and "Western civilization," these new viewpoints found expression in scientific theories and practices. We find considerable evidence of epistemological innovation across the empire, generated in dialogue with colonial power structures and colonized peoples and regions.

When the ecologist Paul Sears first proposed the idea of a "subversive" science in the context of his own discipline in 1964, his main argument was that this relatively young field served to challenge fundamental economic and cultural assumptions about the modern world. Ecology was subversive, he believed, because it drew attention to the complex web of relations among all living things and demonstrated that problems of the nonhuman world were inextricably bound to human activities. He chose the word *subversive* because he believed that, intentionally and unintentionally, the insights of ecological science were being used to confront hegemonic socioeconomic and political values. "By its very nature ecology affords a continuing critique of man's operations within the ecosystem."[78]

Sears's assertions have been challenged implicitly by several histories of ecological science in which its practitioners are shown to have held quite varied intellectual and political loyalties.[79] Yet Sears was right about his field's potential to call into question, or at least place on display, contradictory and competing human ambitions. Efforts to coordinate colonial policies and to understand African populations and environments had a similar effect: they drew attention to the numerous failures, mistakes, and misunderstandings that permeated African colonialism. Ecologists played an important role in this process, but so did social anthropologists. The field's emphasis on people rather than resources and profits cut against the grain of empire. That anthropologists were relatively few and that the effects of their efforts were visible largely behind the scenes has meant that this feature of the history of science in tropical Africa has been underappreciated.

The interpretation of the African Survey's history and its underlying scientific networks that this book presents does not negate the fact that external imperial control in British colonial Africa was autocratic and bound to an ideology of European superiority. Nor does it deny that efforts to mobilize scientific advisors had the simultaneous effect of drawing these individuals and their institutions into the colonial apparatus. But it complicates that picture by exploring the heterogeneous ideas and proposals that emerged from this process, many of them bearing a striking resemblance to alternative agendas and critiques that emerged in Europe and North America in the second half of the twentieth century.[80] Should these intellectual traditions be characterized, in the words of Fiona Mackenzie, as a "counter-colonial ecology," or would it be more accurate to refer to them as social ecologies of colonialism?[81] By misapprehending, mislabeling, and facilitating new forms of control, emerging sciences had the potential to coerce. Yet, by introducing new concepts, new ways of knowing, and new methods for understanding, these disciplines also had the potential to liberate. In the end, they did both.

Vernacular Knowledge and Vernacular Science

A related ambition of *Africa as a Living Laboratory* is to examine how particular scientists dealt with Africans' knowledge. Colonizers in all parts of
the world were forced to consider what indigenous inhabitants knew, how
they knew it, and how reliable vernacular knowledge really was. Whether
administrators wished to admit it or not, understanding why colonial subjects lived and behaved as they did could determine the success or failure
of specific colonial interventions. A central question that this book seeks
to answer is precisely how studying African ideas and practices gradually
became a separate scholarly venture for Europeans and Americans, fore-
shadowing contemporary investigations into African indigenous knowledge systems.

Much of the emphasis on Africans' systems of thought owed its origins
to anthropological research, which influenced the conceptualization of
the issue in other fields. Another source for such investigations, however,
arose out of the modernizing project itself. If Africa were to be brought
into the modern world, then modernity had to be defined against African
"traditions." While many administrators and scientists believed most cultural traditions should be abandoned, some argued that they contained
important insights that could be useful to colonial states and to the empire
more broadly. This book aims to explain how scientific research in colonial
Africa contributed to the construction of various ethnosciences, whether
these related to agricultural, epidemiological, or therapeutic knowledge. I
refer to this research as *vernacular science* to call attention to the fact that
scientists and technical officers were often responsible for constructing the
very categories that helped make vernacular knowledge visible and communicated it beyond the locality within which it arose.[82]

Through most of the colonial period, Africans had few opportunities to
join the ranks of scientific professionals. Not only were the institutions of
higher education that existed in the tropical colonies inadequate for scientific training, but also the racial underpinnings of colonial states meant
that Africans' entry into scientific and medical professions was either deliberately blocked or slowed, for the simple reason that officials wished
to avoid shaking up the social order too quickly. Having little direct access
to the production of knowledge, Africans were often shunted into indirect
roles as assistants, informants, translators, and ethnographic subjects. Ironically, this subordinate relationship bolstered colonial states' ethnographic
interests and dovetailed with anthropologists' concerns to investigate the
content of Africans' natural and medical knowledge. Had European-trained

scientists existed in greater numbers in tropical Africa, it might have been easier to stamp out or ignore Africans' cosmologies, but their relatively thin distribution and the bureaucratic weaknesses of colonial states unintentionally abetted scientific interest in indigenous knowledge.

Unpacking the Laboratory

Colonial Africa was indeed a laboratory for scientific research, development experiments, and social engineering. It would be wrong, however, to stop there: it was also a laboratory for social criticism, interdisciplinary and transnational methods, the study of interrelated phenomena, and the codification of new areas of ethnoscientific and vernacular research. This book examines all these facets of imperial administration, exploring how the continent became a laboratory in the first place and how the production of scientific knowledge helped to challenge the very foundations of empire. Unless we consider all these dynamics in a single frame, we risk doing a serious disservice to the historical record.

Chapter one considers the multifaceted preoccupations of geographical societies and the field expeditions they sponsored in tropical Africa. Had these societies not proliferated so rapidly across Europe at the end of the nineteenth century and had their membership not superseded those of other scientific societies, it seems unlikely that the juridical partition of tropical Africa would have begun when (and where) it did. Geographical interests and expeditions became a pan-European and transnational phenomenon, turning the African continent simultaneously into an object of scientific study and an arena of empire. Before this process was complete, however, scientific societies had already begun to take their interests in other directions. The chapter briefly explores the topics that came to the fore during the decades when politicians continued to oversee the "paper partition" of the continent. These debates revolved around questions of human and nonhuman, physical and cultural survival. This analysis reveals the longevity of concern for the development of Africa and the integral part scientists played in shaping its epistemological content.

Getting accurate and up-to-date knowledge about Britain's various territories in tropical Africa proved to be a lasting problem. Chapter two tells the story of the origins and fortunes of the African Research Survey, which was designed to remedy the dearth of information in Britain about its African colonial dependencies. Its main aim was to "ascertain with reference to an entire continent to what extent the resources of modern science are being utilised by governments to assist them in the exercise of an intelligent control and wise direction in the course of development."[83] The chapter explores why scientific experts played such an important role in this project and explains how they shaped its ultimate ambitions.

Chapters three through six, which cover environmental, medical, racial, and anthropological subjects, examine intersections between disciplinary and epistemological developments on the one hand and colonial state building and imperial policy making on the other. The laboratory motif draws attention to the production of *new* knowledge. Africanists are well aware of the lively debates over the continent's relative richness and fertility, topics taken up in chapter three. Long before the scramble for Africa, naturalists and political economists had speculated about the region's productivity, and some even urged its enclosure.[84] It took a century before European governments acted on these aims, driven in part by geographers' and botanists' positive assessments of territorial resources. In the late nineteenth century, Africa was called the *"tropical* continent par excellence," since 90 percent of sub-Saharan Africa falls between the Tropic of Cancer and the Tropic of Capricorn.[85] In the twentieth century it became the *colonial* continent par excellence. This process necessarily called attention to Africa's natural economy—its flora, fauna, climate, and minerals, as well as its labor reserves—and prompted investigations and inventories of its wealth.

Chapter three examines a pivotal transition that took place during the colonial period, facilitated by agricultural departments and environmental scientists. At the start of the twentieth century, most colonial rulers considered African agricultural practices fundamentally wasteful and unsound, yet by midcentury a number of field scientists had shown that many African techniques "could hardly be bettered by science."[86] This shift was accompanied by a transition away from a view that tropical environments were inherently fertile toward an understanding that they were fragile and susceptible to permanent damage. As officials began to reposition African agriculturalists in their development schemes, they tended to express explicit critiques of the capitalist foundations of colonial states. These critiques were never powerful enough to challenge capitalism, but they were sufficiently pervasive to take some agricultural development strategies in new directions. Acknowledging that certain African cultivators possessed better knowledge of their environments than many colonizers did could implicitly call into question the wisdom of large-scale, profit-driven revenue schemes.

Colonization and epidemics have long been intricately linked: not only has imperialism created new conditions for disease pandemics, but also epidemics have served to justify empire. This destructive reinforcing loop is the subject of chapter four. In much the same way that the intensified colonization of India in the first half of the nineteenth century led to outbreaks of cholera, an infectious disease that was then little known, the scramble for Africa touched off epidemics of sleeping sickness (trypanosomiasis), the etiology of which was still barely understood. Europeans' fear that the disease might spread beyond the bounds of the continent and their recognition that its animal variants in cattle and game prevented much of equatorial Africa from being fully developed made African trypanosomiasis a cause célèbre at the turn of the twentieth century. Chapter four considers research into infectious diseases, the so-called vertical interests of colonial states, and its intersections with research into nutrition and rural health care, often labeled horizontal concerns. In the course of this transition from vertical to horizontal preoccupations, directors of Africa's medical services began to express their own critique of the political economy of health, forcing colonial rulers to consider Africans' standard of living and the "almost derisory" level of funding allocated to improve conditions.[87] Perhaps more significant in the long run, technical officers' field research on the history of infectious diseases in tropical Africa encouraged them to produce a vernacular science that incorporated insights from Africans themselves about the best ways to manage disease environments.

Neither the British government nor its colonial authorities in tropical Africa ever spent much money to interrogate racial differences or capacities explicitly, despite periodic demands to do so. Since African colonial states were racial states, we must explain why deliberate efforts to make racial science a research priority ultimately failed.[88] It is not enough to suggest that colonial officials did not need to interrogate race because their minds were made up about racial hierarchies and difference. The evidence demonstrates the opposite: real fissures about race existed both within colonial states and among scientific specialists. This had as much to do with the goals of colonial practices and policies, which focused on liberal and economic transformations in the name of development, as it did with changing assumptions within scientific disciplines and societies. Nor can we say that almost all scientific efforts were imbued with racial assumptions and leave it at that. In fact, by the early twentieth century scientists and politicians involved in imperial affairs were beginning to challenge racial categories more systematically, with policy makers expressing concerns about racial bias, antipathy, hatred, and prejudice and scientists challenging race as a conceptual system for differentiating among human groups. Chapter five examines these dual dynamics, exploring the mixed fortunes of racial politics and racial science, and illustrates just how ambivalent imperial authorities were about privileging race as a primary prism for settling colonial questions. The instability of racial categories and their susceptibility to criticism had the potential to unsettle the racial foundations of empire.

Of all the disciplines brought to bear on colonial Africa, none has been more controversial than anthropology, the focus of chapter six. Anthropologists were by far the least numerous field experts in tropical Africa during the entire colonial period. They also tended to be the most outspoken critics of colonialism's imperfections and inadequacies. Indeed, anthropologists and their administrative allies were often the thin edge of the wedge in terms of critiques of empire. Their attempts to use the African continent to professionalize paradoxically allowed them to illuminate the Eurocentric and contradictory assumptions contained within colonialism. Anthropologists rarely set out to dismantle the empire; some were actually strong believers in paternalistic overrule. But their commentary and observations often inadvertently drew attention to its injustices. Even as they helped to reify ethnic and tribal identities and codify assumptions about authentic versus invalid patterns of social organization, they brought questions of Africans' agency, autonomy, and knowledge to the forefront. More than any other field science, anthropological research helped to challenge both cultural and cognitive absolutes, leaving the door open for pluralistic perspectives of all kinds.

These epistemological legacies were not always welcome or benign, but African states have been forced to confront them repeatedly since achieving political independence. Various regions within Africa still provide many counterexamples to ostensibly modern norms, whether in the context of land use, intellectual property, therapeutic pluralism, or even law and politics, to give but a few examples.[89] Many of the philosophical struggles that began during the colonial era and helped to bring about the empire's demise remain salient to this day. Indeed, in the postcolonial period a key question circulating around debates relating to African knowledge systems, including therapeutics, is whether *official* recognition means states are explicitly endorsing epistemological relativism. Is it possible, in other words, to accommodate a variety of epistemic frameworks, otherwise known as epistemic pluralism, without becoming mired in the challenges associated with incommensurability? The seeds of these questions, as we shall see, were planted by experts, administrators, and social critics in the colonial era who were themselves grappling with the problem of scientific knowledge.

that in the decades leading up to the conference, geographical expeditions and scientific knowledge had played a significant role in shaping imperial ambitions. During that period, Green explained, "a new world was revealed rich in treasures desired by Europe. In Europe, meanwhile, a new idea had grown—the idea of world dominion, now made possible for the first time by the development of science and mechanical arts which had increased a thousand-fold the conception of human power." Green was struck by the sheer magnitude of the imperial undertaking and the speed with which it was completed. "The history of Africa is without parallel. . . . [The partition] was the first event in the world on that stupendous scale—the beginning of a new age. The revelation of Africa, came at the moment when the notion of World Power swept through Europe. The great experiment was opened on the new continent."[3]

In his provocative history of European explorations of tropical Africa during the nineteenth century, Johannes Fabian has echoed Green's analysis and called attention to the intimate connections between knowledge production and territorial conquest. "Geographic space to be discovered and explored [was] turned into a laboratory in which scientific assumptions were to be tested, as well as into a territory that was to be occupied." "And yet," Fabian concedes when examining explorers' efforts at "making knowledge," "during the first encounters they had no frame for their relations. There was no colonial regime, no colony to simulate a laboratory, a place where objects held still and variables could be controlled."[4] The questions implicit in Fabian's and Green's remarks deserve careful examination: How did an entire continent become a laboratory and an empire almost simultaneously? In what ways did the ideas and activities of scientific and learned societies influence the process of projecting imperial power? Which scientific disciplines were involved and what were their driving questions and benchmark ideas? Most centrally, how did concerns about development dovetail with scientific debates and efforts at colonial state building?

During what historians call the "Scramble for Africa" and its ensuing conquest—the period from about 1870 to the outbreak of world war in 1914, when the major European powers partitioned the continent between themselves, set up the machinery of trade and rule, and strove to establish effective dominion over Africans—scientific societies' efforts to promote research on Africa were actually uneven and episodic.[5] Before the Berlin Conference, the Royal Geographical Society was the primary institution that took the lead in sponsoring fieldwork and cartographic projects, yet the research fund devoted specifically to Africa that it established in 1877 proved to be short-lived.[6] From the 1850s onward, the British Association for the Advancement of Science included papers on African geography and anthropology at its annual gatherings, but only in 1891 did it create an interdisciplinary standing committee to investigate the "climatological and hydrological conditions of Tropical Africa."[7] Although the Royal Society had sponsored small-scale studies of malaria and disease-carrying tsetse flies in the 1890s, its Sleeping Sickness Commissions launched between 1902 and 1914 represented its first large-scale undertaking within the continent itself. Even the Royal Botanic Gardens at Kew, which had published its first volume on the *Flora of Tropical Africa* in 1868 and had joined forces with the British Association in 1883 to send Harry Johnston to Mount Kilimanjaro to study "migrations and modifications of species," had difficulty finding the funds for later volumes.[8] Likewise, the leadership of the Royal Anthropological Institute and the British Museum expressed some interest in African ethnography prior to the 1890s, but it took an outsider to the profession, Mary Kingsley, to popularize the idea that the sociocultural aspects of African societies deserved greater attention. Her desire to encourage "West African Studies" inspired a handful of social reformers to found the Royal African Society in 1901, which they hoped would "create an interest in Africa, [as] a field for investigation."[9]

Scientific societies' desire to gather and produce new knowledge about Africa, especially its vast interior, was thus a relatively recent development that had shallow roots within the continent itself. During a presidential address before Britain's Statistical Society in 1872, the epidemiologist, Sir William Farr, reviewed existing knowledge about the different regions of the world and announced: "Of Africa, statistics knows little or nothing *certain* . . . as yet all Africa is for science a great desert."[10] While Farr was embracing the familiar motif of Africa as a terra incognita, which justified the European imperative of exploration and discovery, he was also drawing attention to the challenges of ignorance and uncertainty. For those who valued accurate and exact knowledge, admitting that important topics remained unexplored and vital questions unanswered carried with it an implicit critique of the geographic limitations of contemporary science. Even more significantly, it encouraged experts to think of Africa as an object of scientific study in its own right.

The Scramble for Africa and Its Explanations

There is no single explanation for the multiple territorial seizures that occurred across Africa between 1870 and 1914. To paraphrase Paul Kennedy, each annexation must be understood within a far-reaching web of causes.[11]

Still, most interpretations of the scramble for Africa tend to privilege certain developments over others, emphasizing economic, diplomatic, technological, military, or ideological factors. From this vast and contentious body of scholarship a consensus has emerged on several key points. Until the 1880s, for instance, many political leaders in Great Britain, France, and Germany were reluctant to pursue formal empires in tropical Africa. Establishing legal political control over territories would require substantial treasury expenditures as well as an extended military commitment; both were difficult for leaders to embrace or promote. The British, in particular, relied on their naval superiority to ward off other powers and enjoyed virtually uncontested dominance around much of the African coastline. Prussia's victory over France in 1871 and the ensuing unification of Germany and Italy altered diplomatic and economic relations across Europe and stimulated challenges to Britain's near monopoly in Africa. The British Empire remained the largest economy, but as its continental counterparts developed their own industrial capacities, the economic center of gravity began to shift.[12] Many countries simultaneously reconfigured their own borders and began to redefine their sense of national and imperial identity.[13] The possibility of new bilateral and multinational alliances generated both international competition and imperial momentum.

Before the establishment of effective control of African colonies, trade with most regions of Africa yielded negligible financial returns. Capitalist economies in Europe did not need tropical Africa to sustain them, even during the industrial slowdown that affected many parts of Europe between 1873 and 1896.[14] The most lucrative regions of the continent were at its northern and southern tips—Egypt, Algeria, the Cape, and Natal—which were almost entirely under the control of the Ottomans, French, and British before the center of the continent had been partitioned. While the idea of finding new products and markets in tropical Africa was seductive, in actual fact the bulk of the continent had not yet generated substantial financial returns. "European economies," Trevor Lloyd has observed, "would have gone on perfectly happily even if Africa had never existed."[15]

Only in a few areas, such as Ethiopia and Liberia, did Africans possess sufficient organization as well as firearms to deflect military incursions and land confiscations.[16] In most other regions, African polities were unable to prevent or halt the process of conquest, although they had the power to disrupt and transform it. Uprisings, rebellions, and prophet movements with political implications occurred during this period in almost every region of sub-Saharan Africa. Prominent among those that attracted significant attention in metropolitan centers were the Asante wars in the Gold

Coast (1874–76 and 1896–1900), the Zulu war in southern Africa (1879), the Mahdist revolt in the Sudan (1881–98), the Ndebele and Shona resistance in Rhodesia (1896–97), the Maji Maji uprising in German East Africa (1905–07), and the Herero movement and massacre in German Southwest Africa (1901–06).[17] Terence Ranger, however, has cautioned against seeing these actions as hostile solely to European invasions, since some also expressed internal critiques of indigenous elites. Prophet and religious movements in particular were "the idiom of profound and perhaps 'revolutionary' challenge to the official orthodoxies of African societies themselves."[18] This period was marked by dramatic realignments of social power across tropical Africa, which diminished Africans' abilities to unite against European conquest.

Diplomats and politicians, with few exceptions, paid little heed to the idea that African polities ought to have legal standing on the world stage. Indeed, the notion that non-European peoples in the less developed regions of the world were entitled to national self-determination was not part of their worldview. In the words of one of Britain's leading cartographers, European governments had acted "for the most part, without the knowledge or irrespective of the wishes of the native chiefs and rulers whose lands have been thus apportioned."[19] The absence of African leaders from the Berlin Conference was taken for granted by the participants, and the language of the Berlin Act that followed "irreversibly exclud[ed] any pretensions to sovereignty that indigenous communities might have entertained."[20] The "New Imperialism" of this period was justified as part of a wider "civilizing mission" and was rooted in interpretations of international law that relegated Africans to the position of objects acted upon by Europeans. Paradoxically, the only recognized agency they could exercise was the "right" to give up their sovereignty.[21]

European technologies ranging from destructive weaponry to preventive and curative medicine also played key roles in enabling imperial conquest. Philip Curtin and Daniel Headrick have shown that biomedical technologies functioned as "tools of empire"; sanitation measures and quinine, which was used both as cure and as prophylactic, enabled Europeans to survive long enough to subdue Africans.[22] In the mid-nineteenth century, Europeans' death rates from infectious diseases were so high as to prohibit military occupation, let alone long-term settlement. Survival rates for European soldiers gradually improved from the late nineteenth to the early twentieth centuries.[23] While the improvements were initially slight, they were still sufficiently visible to make political leaders more willing to bring troops and officials to the tropics.

Firearms, steam technologies, and new modes of communication and transport also helped Europeans overcome Africans' resistance to conquest.[24] The Maxim gun was introduced just in time to quell the Ndebele rebellion in the 1890s. Steam engines, railways, and telegraphy made it feasible to administer vast territories and establish outposts of rule in Africa's tropical interior, which had been considered too remote. New technologies assumed even greater significance in Europeans' imaginations, bolstering their belief in human mastery of the natural world and catalyzing utopian visions that were as grandiose as they were beyond reach.[25]

"An International Battlefield"

On the eve of the Berlin Conference a correspondent for the London *Times* offered a chilling prophecy: "It is evident that Africa is in danger of becoming a sort of international battlefield. All the other continents have been, more or less, appropriated by races of European origin; Africa cannot much longer escape the same fate."[26] The dramatic partition of the continent's tropical interior differed substantially from the much longer histories of European contact and conquest along Africa's coasts and in its northern and southern regions. The very act of creating vast territorial units de novo, beginning with the Congo Free State and rapidly constituting dozens of others, radically transformed Africa's political landscape and spawned renewed interest in studying the continent's distinctive attributes.

In order to understand the role of science, broadly construed, in Africa's partition and early decades of state building, we must consider how learned societies were caught up in this process. Scientists and geographers had long been interested in tropical Africa, but only in the 1870s did that interest gather momentum and coalesce into a genuinely pan-European effort. At the forefront of these activities were geographical societies. In the second half of the nineteenth century, too, bioscientists began self-consciously to scale up their objects of study and refine their methods of fieldwork.[27] The African continent served as an important venue for both endeavors. Significant research had previously been undertaken in tropical Africa during the seventeenth and eighteenth centuries, particularly in geology, botany, medicine, and zoology, and a few individuals had asked similar theoretical questions about Africa. Yet these antecedents differed from developments at the end of the nineteenth century in both their scope and scale. The rapid growth of geographical societies across Europe between 1870 and 1890 and, in particular, their scientific preoccupations, field expeditions, and entanglements with commercial and juridical concerns

Table 1.1   Number of geographical societies by region

Location	1870	1890
Europe	12	99
Asia	1	3
North Africa	0	3
North America	1	3
Central and South America	2	5
Australia	0	2
Total	16	115

Note: The societies in North Africa included one in Egypt founded by the Khédive Isma'il (1875) and two in Algeria, founded in Oran (1878) and Algiers (1879). A South African Geographical Society was not founded until 1917.

affected the pace and timing of partition in tropical Africa. Geographical societies made an essential contribution to the conditions that precipitated the scramble for Africa, acting not in isolation but precisely through their intricate connections with economic, diplomatic, and military forces and their myriad intersections with demographic and cultural concerns. Without them, it is unlikely that Africa would have become a "sort of international battlefield."

When the Suez Canal officially opened in 1869, there were just twelve geographical societies spread throughout Europe and only four in the rest of the world: two in Latin America (Mexico City and Rio de Janeiro), one in Asia (Bombay), and one in the United States (New York). The Paris, Berlin, and London societies were the oldest, founded in the late 1820s and early 1830s. By 1890, European geographical societies had multiplied from twelve to ninety-nine, while those in the rest of the world went from four to only sixteen (table 1.1).[28]

In 1881, these groups had an estimated membership of 30,000 individuals, with France, Germany, and Britain accounting for 60 percent.[29] While these societies ran the gamut in terms of their size, influence, and topical interests, their uniting concern was the earth itself and the myriad insights, both useful and theoretical, that its study could reveal. Francis Galton, one of the doyens of fieldwork in Victorian Britain, expressed the hope during an 1872 address before the British Association for the Advancement of Science (BAAS) that geographical science would become more popular, "for its problems are as numerous, as interesting, and as intricate as those of any other [science]. The configuration of every land, its soil, its vegetable covering, its rivers, its climate, its animal and human inhabitants act and react upon one another. It is the highest problem of Geography to analyze

Table 1.2 International geographical congresses

Date	Location	Number of participants
1871	Antwerp	600
1875	Paris	1,477
1881	Venice	1,049
1889	Paris	618
1891	Berne	556
1895	London	1,529
1899	Berlin	1,238

Source: Totals for all the congresses held between 1871 and 1968 from
George Kish, "The Participants," in *Geography through a Century of
International Congresses* (Paris: UNESCO, 1972), 35–49.

their correlations, and to sift the casual from the essential."[30] Lest anyone
in his audience forget why this approach mattered, Galton reminded them
that knowledge was central to "the power of man over the phenomena of
nature. He is not always a mere looker-on and a passive recipient of her
favours and slights; but he has power, in some degree, to control her processes, even when they are working on the largest scale."[31]

A key impetus for these societies' growth was the first International Geographical Congress held in Antwerp in 1871, which attracted six hundred
participants from twenty different countries across Europe. Four years later,
the second congress, held in Paris, attracted more than twice as many participants, with nearly 1,500 registrants from thirty-four countries (table 1.2).
The largest group of participants at these and later congresses was that from
the host country: in 1871 nearly half the participants were from Belgium,
and in 1875 almost two-thirds were from France. In this respect, international geographical gatherings helped to reinforce national and regional
pride. Yet, more significantly, they allowed participants to share ideas and
information across national borders.

A widely publicized discussion that took place during the 1875 congress focused on "Colonisation, Emigration, and Manual Labour in Tropical Countries." The organizers were especially keen to consider the "best
systems of colonisation" as well as the resiliency of "indigenous races" and
the ability of Europeans to acclimate to hot climates. This panel was followed by a more focused debate about "the best paths to follow and the
most favorable points of departure for expeditions attempting to fill the
gaps in knowledge of the interior of Africa."[32] Most of the speakers, who
had already undertaken extensive expeditions, described the difficulties
they faced during their journeys.[33] Illnesses affecting humans and animals
and a lack of suitable "beasts of burden" made travel arduous and slow.

Europeans' dependence on local porters, guides, and domestic help made trips expensive and politically vulnerable. The German explorer, Gerard Rohlfs, explained that he thought it was virtually "impossible, as a single individual, to make meteorological, zoological, and mineralogical observations alone."[34] He and his compatriots, Georg Schweinfurth and Gustav Nachtigal, agreed that detailed scientific studies in tropical Africa were best pursued only after a region had been explored. Even instruments warranted some discussion, since measuring distances and determining latitude and longitude were difficult in certain environments.[35] The delegates sought to find ways of overcoming the daunting obstacles these expeditions encountered in order to achieve their aims.

Just a few months earlier, Schweinfurth had framed the challenges faced by geographers in terms of an intellectual imperative. He had moved to Cairo for health reasons and had quickly developed a close relationship with the Khédive Isma'il. A modernizing hereditary ruler, Isma'il was committed to ensuring that Egypt played its part in scientific research. In his view, Egypt "effectively became the gate by which science was going to make its entry into Africa."[36] In 1872 Isma'il wrote to Samuel Baker, a British geographer and explorer who was then working on his behalf in the Sudan: "The idea of opening the centre of Africa to science, commerce and progress is a great idea, which has gained hold of me to such a high degree."[37] He established a learned society that he hoped could match the "mother societies" in Europe and chose Schweinfurth to serve as its first president. When Schweinfurth delivered his inaugural address to the newly founded Egyptian Geographical Society in the summer of 1875, he clearly articulated its wider mission: "No question has acquired more general significance for science today than the exploration of Africa."[38] Although scientists in Europe might have considered this claim an exaggeration, Schweinfurth was correct to suggest that as geographical societies proliferated they heightened interest in Africa's interior.

King Léopold of Belgium recognized that these developments had a direct bearing on his country's international position. As early as the 1860s, Léopold had been asking his advisors for their thoughts on what strategic international or imperial adventures Belgium might undertake. After a number of false starts, including an overture to buy the Philippines from Spain, he turned his attention to Africa. Léopold was already a member of the Paris Geographical Society and an honorary Fellow of the Royal Geographical Society, so he sent a representative to the second International Geographical Congress.[39] From this gathering, he learned—if he did not know already—that other European powers were fielding exploring expeditions: Germany had recently founded a Society for Equatorial African Research (1873) and had launched an expedition to the West African region surrounding Loanga (in present-day Gabon); the Paris Geographical Society was preparing to send a new expedition to Central Africa led by Pierre Savorgnan de Brazza; and Britain was awaiting news from an expedition in Central Africa led by Verney Lovett Cameron, who took up the task after David Livingstone's death in 1873.[40] Léopold's interest was particularly piqued by the congress's debates about how best to carry out and coordinate African exploration. A few days afterward, he wrote to a colleague, "I intend to find out discreetly whether there is anything to be done in Africa."[41]

Belgium was among the few western European nations without a geographical society, so Léopold had directed his patronage toward London and Paris.[42] In an effort to galvanize interest within Belgium, he planned an international gathering to consider African geographical questions. In late May and early June of 1876, he traveled to England and Scotland, paying visits to his cousin, Queen Victoria, as well as the Prince of Wales, and meeting with key leaders of the Royal Geographical Society (RGS), including Cameron, who had by then returned from his equatorial trip, and Sir Bartle Frere, who had recently stepped down as the RGS's president.[43] By the middle of June he was drawing up plans for the conference itself. He wrote to a Belgian politician, "I desire that the main effort to open Africa permanently to civilisation has its founding date in Belgium."[44] At that time, this remark might well have sounded absurd; Belgium had no navy to speak of and was a relatively minor political and economic power. In retrospect, however, it seems both honest and shrewd.

"What Is the Present State of Our Knowledge of Africa?"

In contrast to the international geographical congresses that brought together large numbers of participants from across Europe, King Léopold's 1876 Geographical Conference brought together the scientific elite. Thirty-seven delegates from seven European countries took part, including the presidents or vice-presidents of five geographical societies and eight of Europe's leading African explorers.[45] Their task was fourfold: to begin a "general exposition of the geographic, scientific, and ethnographic knowledge that is possessed today on central Africa";[46] to determine the best means to explore the interior of the continent more fully; to define the functions and possible location of field stations across the equatorial belt; and to establish the structure and mission of an international association that would guide their work.[47] Britain's was the largest foreign delegation, with ten members; Germany and Austria-Hungary followed with four each; France had two, with two additional members joining on the final day; and Italy and Russia each had one.[48] Members of the Belgian delegation, which consisted of thirteen individuals reflecting a mixture of scholarly, political, and commercial interests, had convened twice before the conference itself in order to discuss their own perspectives and strategies. They insisted on starting with a review of existing ethnographic and scientific knowledge in order "to establish the lacunae still to fill, [and] the obscure points in need of clarification."[49]

In his summary of the conference proceedings, one of Léopold's advisors, Emile Banning, highlighted the breadth of questions delegates raised during the course of their discussions and focused on fundamentals. He began by asking, "What is the present state of our knowledge of Africa? What do we know of the general aspect of its soil, of the elevations of its surface, of the distribution of its waters? What are its climate and resources? To what families do the races [peuples] which inhabit it belong? To what degree of civilisation have they raised themselves? Do there exist reasons for believing that they may reach higher?"[50]

These were questions that many geographical societies across Europe were trying to answer.[51] In a series of articles written earlier in 1876, Banning summarized the state of African exploration, including the recent reports by Verney Lovett Cameron.[52] Over two and a half years (1873–76), Cameron and a cadre of assistants not only surveyed Lake Tanganyika and the river systems comprising the Congo basin but also managed to traverse the entire equatorial region from east to west, a distance of nearly 3,000 miles. Complying with Bartle Frere's instructions, Cameron kept systematic records of his locations during his travels.[53] He made more than 3,900 measurements regarding altitude, averaging three per day, and more than 1,000 observations of longitude and latitude verified by nearly 600 lunar observations.[54] According to the president of the RGS, previous observations of longitude in Africa's tropical interior had been based on only a single lunar observation taken at Ujiji on Lake Tanganyika.[55] Cameron's achievements prompted the RGS's "computer" at Greenwich— the person responsible for making mathematical calculations at the Royal Observatory, who verified his data—to remark that the number of his observations was "greater possibly than any traveller has before, in an unknown country, accumulated in a similar short space of time."[56] This level of detail "astonished the scientific officers" and was considered a major contribution to the "advancement of pure, substantive, scientific Geography."[57] In his letters and reports to the RGS, Cameron described an "experience of rough fieldwork" and "privations and hardships so great that now looking back it is a matter of wonder to myself that any registers or journals were kept at all."[58]

When the conference participants turned to the question of how best to explore the continent's equatorial interior, they deferred to the experts in their midst. Expedition leaders had already established three different axes for their activities in Africa: one following a north–south orientation, taking in the territories between the Cape and Cairo; another a north-to-west trajectory from Algeria to Senegambia; and a third an east–west route from Zanzibar to Angola. All three routes were shaped by the continent's rivers: the Nile and the Congo were increasingly seen as its premier waterways, while the Zambezi, the Niger, and the Gambia occupied significant subsidiary positions. The delegates had good reason to avoid any entry from the north or south, since they were well aware that the balance of power in northern Africa was fragile as the Egyptian Khédive, the Ottoman Empire, and France and Britain contended for supremacy, and that the Portuguese and the British largely controlled the southern coasts. The route that remained open to multinational exploration was the east–west path Cameron had recently traversed. A map produced by the RGS's African Exploration Fund a few months later showed just how few expeditions had traversed the region surrounding the Congo River (see plate 1).

To determine the functions and locations of the field stations, conference participants divided into two groups: the Austrian, Russian, and German delegates in the first and the British, French, and Italian in the second. The Belgians had decided beforehand that they would play the role of observers, emphasizing their nation's neutrality and the more extensive experience of the invited guests.[59] When the groups reconvened, they discovered that the "two sections were not entirely in agreement for the reason that one side [Austrian, Russian, and German] had favored an exclusively scientific point of view while the other [British, French, and Italian] placed more emphasis on an economic and practical point of view."[60] The main differences were in the geographical scope each group designated for exploration and the density of stations they proposed. The second group put forward a more ambitious scheme that involved "a continual line of communication between the eastern and western coasts of the continent," including primary and subsidiary roads that would link the lakes region on the east with the Nile basin to the north, the Congo to the west, and the Zambezi region to the south.[61]

In order to reconcile the two proposals, the delegates agreed that their first priority should be to establish "*scientific* stations" since, in the words of Gustav Nachtigal, anything broader would "suppose by necessity the intervention of Governments. . . . It is necessary to take action slowly [and let] the work progress naturally. The scientific stations must not precede exploration, they must follow it."[62] The majority of participants felt that a more modest undertaking, which established two or three bases in the interior and two sites on the coasts near Zanzibar and Loanda, would be sufficient to give scientific explorers from European nations jumping-off points for their expeditions.

Their vision of creating scientific stations corresponded with ongoing efforts to establish Christian missionary and trading stations near lakes and rivers in tropical Africa. "The scientific mission of the station should consist so far as possible in astronomical and meteorological observations, collecting specimens of geology, botany and zoology, mapping the surrounding country, preparing a vocabulary and grammar of the languages of the natives, making ethnological observations, collecting and reporting the accounts of indigenous travelers in unknown regions, and in keeping a journal of all events and observations worthy of note."[63] The staff would include a director who was both "a man of action and a man of science . . . a medical naturalist, perhaps a physical astronomer, [and] five or six artisans, skilled in various handicrafts."[64] The delegates were clear that the stations would serve no explicit military function. The negotiation of legally binding "conventions" would provide their point of entry.

The next step was to decide upon the form and function of the "international commission" that would oversee and coordinate their work. Participants sought to strike a balance between international coordination and national autonomy by allowing individual nations to establish their own committees for the "exploration of Africa" but still receive benefits from a supranational group. According to Bartle Frere's letter to Britain's foreign secretary, establishing a pan-European commission could "be useful in avoiding some chancey diplomatic friction in prosecuting African exploration." Its affiliates would "form a sort of executive exploring association, for work which the R.G.S. does at present only by fits and starts when assisting travellers like Livingstone or Cameron."[65] Ultimately, the delegates decided that the leadership structure would consist, first, of the presidents of geographical societies (beginning with those present at the conference and adding those who wished to join later) and, second, of two members chosen from each of the national committees. Frere proposed that King Léopold assume the presidency of the international commission; he, in turn, was nominated to represent Britain on the executive committee. (After being appointed governor of the Cape Colony a few months later, Frere resigned and was replaced by an American delegate.) The other two members were Nachtigal for Germany and the eminent naturalist and anthropologist Armand de Quatrefages, a professor of anthropology at the Musée National d'Histoire Naturelle, for France.[66]

After less than eight hours of formal meetings, the delegates had managed to canvass the entire continent, agree upon the best strategies for its exploration, and determine the means by which they would move forward. What they discussed during the many hours they spent together after the daily sessions adjourned at 1:30 p.m. was not recorded, but participants had ample opportunities to compare experiences and evaluate new ways of solving common problems, as well as to size up the competition. Most were convinced that this undertaking was both momentous and necessary. Rutherford Alcock, president of the RGS, felt he was speaking for all the participants when he reported that "the Congress . . . was a graceful recognition of the claims of science, and more especially, perhaps, of the Geographical Societies so largely represented on that occasion and so directly concerned in showing the way into the heart of Africa."[67]

The experts who attended the Brussels conference were united in their belief that naturalists and scientists of all sorts would be central players in the "opening up" of Africa. They also seemed aware that any effort to "civilize" Africans through science would inevitably encroach upon the activities and powers of indigenous governments. Although delegates attempted to sidestep this question during the meeting, it proved impossible to avoid for long. Just a few months after the conference, Cameron told his readers in Across Africa that "in any scheme for forming stations in Central Africa"—"be they for missionary, scientific, or trading purposes—the fact that those in charge would soon have to exercise magisterial powers must not be lost sight of. . . . Many people may say that the rights of native chiefs to govern their countries must not be interfered with. I doubt whether there is a country in Central Africa where the people would not soon welcome and rally round a settled form of government."[68]

Cameron's assessment of Africans was belied by rebellions and resistance in the following decades, but he was right about one thing: proposals to establish scientific stations carried with them an implicit legal mandate. It would prove far more difficult than the conference delegates had imagined to separate scientific from "commercial and practical" activities. Field operations and research at permanent stations would require a kind of governmental or magisterial authority that individuals operating under the auspices of learned institutions seldom possessed. Before the mid-1870s, geographical societies had helped to generate interest in tropical Africa and had even trumpeted its wealth of resources, but they had not been given the power to negotiate territorial rights. That position was about to change.

Few leaders of geographical societies intended to touch off a multinational land grab. Many of their proposed plans, whether for collective research or large-scale infrastructure projects, ended in disappointment. Yet expeditions themselves were sufficiently influential, both directly and indirectly, to bring about political action. Although knowledge does not automatically confer power, the drive for knowledge, when pursued on a massive scale and by multiple parties simultaneously, could force the hands of power and produce acute diplomatic friction. These developments form just one part of a complete picture of the scramble for Africa, but they go some way toward explaining why interest in the continent's tropical interior gained such momentum during the 1870s.

Geographical Societies and Geopolitical Negotiations

Few scholars have paid much attention to the scientific questions raised by Léopold's Geographical Conference, although many historians of empire note its significance. Those who say anything about Léopold's interest in geography suggest that this preoccupation served as a mere smoke screen for his imperial ambitions. Geoffrey de Courcel has remarked that "Leopold II, who for a long time had been trying to get reluctant Belgium interested in colonization, initially pretended, with great cunning, that his reasons for wanting to operate in Central Africa were purely scientific and humanitarian." H. Gründer has argued that Léopold "initially succeeded in concealing his colonial ambitions behind the mask of an enquiring geographer."[69] These interpretations underestimate the significance of geographical societies, relegating them to a subsidiary rather than a substantive role in the scramble for Africa. The fact that twelve different countries, including the United States, felt compelled to establish special committees for African exploration after 1876 suggests that we must evaluate the conference and its effects in terms of institutions and not just individuals (table 1.3).

Whatever Léopold's private intentions for his own future, his conference served to concentrate geographical societies' attention on tropical Africa as a primary locus for their work. In the period between 1877 and 1884, a veritable rush of expeditions traversed the continent's equatorial belt.[70] Indeed, the blurring of boundaries between scientific fieldwork and legal jurisdiction generated great geopolitical anxiety in the months and years following the gathering. According to Frere, Portugal's representatives had

Table 1.3. African exploration: Committees and societies founded between 1876 and 1877

	Country	President of society or committee
1.	Belgium* (IAA & Belgian Cmte)	King Léopold II & Comte de Flandre
2.	Austria-Hungary	Archduke Rudolph, heir apparent
3.	France	Count Ferdinand de Lesseps
4.	Italy	Prince of Piedmont, heir apparent
5.	Spain	King of Spain
6.	Russia	Grand Duke Constantine Nikoleyvich
7.	Holland	Prince of Orange
8.	Portugal	Duke of San Januario
9.	Switzerland	President M. Boathelier de Beaumont
10.	German African Society*,†	Prince von Reuss
11.	British African Exploration Fund	Prince of Wales, heir apparent
12.	USA	John Latrobe§

Sources: Rutherford Alcock, "African Exploration," *Times (London)*, July 18, 1877, p. 10, col. E; "Work of the German African Association," *PRGS* (1882): 678–85; Lysle Meyer, "Henry S. Sanford and the Congo: A Reassessment," *African Historical Studies* 4 (1971): 19–39, on p. 23.
* These societies were vested right away with the authority to establish field stations, which were to undertake scientific research and "serve as bases of operations for travellers, and partly as centres for the spread of civilisation and commerce." France's committee adopted the same practice in 1879.
† The German African Society (Deutsche Afrikanische Gesellschaft) ultimately amalgamated with the Society for Equatorial African Research (founded in 1873); these are not to be confused with the company established by Carl Peters in 1885, the German East Africa Society (Deutsch Osta-Afrikanischen Gesellschaft).
§ Latrobe was also president of the American Colonization Society, with strong connections to Liberia.

been excluded from the 1876 conference "on account of her bad reputation in Africa."[71] Leaders of the Sociedade de Geografia de Lisboa (founded in 1875) responded quickly to the perceived slight by petitioning their king to undertake additional expeditions in order to "safeguard" Portugal's "territorial rights."[72] Adopting an assertive rather than defensive stance, they stressed the "urgent necessity and imperious duty imposed on Portugal by her traditions, by her situation as the second colonial power in Europe, and by her economic and political interests beyond the seas, to enter definitively into an expansionist movement."[73]

In the aftermath of the conference, British statesmen and naturalists were equally concerned that fieldwork might segue into land claims, but in contrast to the Portuguese they responded by drawing even sharper lines between geographical and political operations. After Bartle Frere had apprised the foreign secretary, Lord Carnarvan, of the proposed plans for an International African Association (IAA) and suggested that the Prince of Wales become Britain's patron of these activities, Carnarvan candidly expressed his skepticism about any country other than Britain taking the lead on such a project. He replied to Frere:

> I should still watch all the proceedings with a careful eye, and should keep the international exploration well within the geographical limits which I understand to be proposed. I should not like anyone to come too near us either on the South towards the Transvaal, which *must* be ours; or on the north too near to Egypt and the country which belongs to Egypt. In fact, when I speak of geographical limits I am not expressing my real opinion. We cannot admit rivals in the East or even the central parts of Africa: and I do not see why, looking to the experience that we have now of English life within the tropics—the Zambesi should be considered to be without the range of our Colonisation. To a considerable extent, if not entirely, we must be prepared to apply a sort of Munro [sic] doctrine to much of Africa. Bearing however all this in mind the Prince of Wales may perhaps not disadvantageously connect himself with this international and scientific project.[74]

The close connection between exploration and territorial appropriation was clearly evident as early as 1876 and, in Carnarvan's mind, international competition superseded scientific cooperation.

Carnarvan's cautiously watchful attitude toward the IAA was soon shattered by reports from Henry Thring, Britain's parliamentary counsel and a specialist in international law, as well as from his subordinates in the Foreign Office and from two of Britain's most respected field agents in tropical Africa, John Kirk, in Zanzibar, and Frederic Elton, the consul general in Mozambique.[75] All of these officials pointed to two significant problems with the International Association's proposed structure and function: it could potentially incite, rather than prevent, international tensions; and its scientific activities would likely raise a range of political and economic questions as well.

Thring went to the heart of the matter of juridical control. In his briefing to the Foreign Office, he wrote that despite the fact that "at the first glance," the International Association included "features of great attraction, especially to scientific and philanthropic men . . . closer inspection displays the difficulties which must arise in applying to nations principles applicable only to individuals of the same nation." In such a multinational endeavor, "the national elements must necessarily be brought into conflict with the international; for international rights of possession and profits cannot practically exist, and who is to decide as to their distribution be-

tween the constituent nations; or how are such nations to be satisfied with
the portions allotted to them by a private society, in which the largest nation and the smallest are on a par, and in which a combination of two
or more nations may exclude the others from participating in benefits to
which they deem themselves entitled."[76] If rights of possession could not
be held internationally, using these stations for any purpose beyond scientific research had political implications: "the establishment of stations
throughout Africa is in effect the establishment of factories such as laid
the foundation of the British power in India, and must of necessity involve grave commercial questions and interference in disputes incident to
all trading transactions in uncivilized countries."[77] Britain would do well,
Thring concluded, to avoid these potential conflicts by opting out of the
project in the first place.

Complicating the inherent tensions between multinational and national rights was the problem of how to set limits on geographical expeditions and scientific activities. In Kirk's view, "it is not easy to conceive that
a mutual understanding would long subsist between the rival interests represented [in the IAA], unless the Association [were] restricted to scientific
objects." But who would decide where science stopped and commercial
and political activities began? The whole endeavor was fraught with problems because it incorporated infrastructure projects that fell under the purview of states and political leaders. Kirk argued that "once trade-roads and
resting-houses are constructed in Central Africa, there are many questions
that must arise in regard to their defense, and the pecuniary rights of the
promoters, that will of necessity open questions to be dealt with by each
national section of the Association from its own commercial and political
situation and the interests of its own people."[78]

An official in the Foreign Office had already articulated these concerns
in a briefing produced for Carnarvan in December 1876. The problem, in
his view, was not scientific research per se, but all the other activities that
attended it.

> Among the objects of the Society are stated the *acquisition of stations* in vari-
> ous parts of Africa, notably upon the principal routes of travel, which are to
> form centres of civilization. These are to be connected by roads to be formed
> by the Society. The Society is to conduct scientific research but there is noth-
> ing to prevent it acquiring any amount of land or entering upon any com-
> mercial undertaking, on the contrary so far as I could gather these objects
> would be rather encouraged than not under the guise of introducing human-
> izing and civilizing agents among the Natives.[79]

The far-reaching functions of the field stations seemed destined to lead to trouble.

As a result of these detailed and extended consultations, Britain's Foreign Office became convinced that the Royal Geographical Society should have no formal relationship with the IAA. The Prince of Wales was persuaded to write to King Léopold to explain that he could not accept any position in a geographical committee that would be subordinate to the international association. Steering clear of all transnational coordinating efforts, the RGS decided that it would move forward with its own plans for African research, and the Prince of Wales agreed to serve as its patron. To do anything else would be tantamount to relinquishing not only a certain degree of political power but also their nation's perceived status as the leading sponsor of geographical expeditions. Frederic Elton was especially insistent on this point. Why, he asked Carnarvan, would England join forces with other nations to collaborate in exploring efforts, or even antislavery activities, when its own endeavors had already been in the vanguard in tropical Africa?

> At its best—its very best—such co-operation can never be otherwise than disjointed, and embarrassed by unforeseen eventualities that may arise, by jealousies, and by want of that perfect concert vitally necessary for the accomplishment of great works. Our single-handed work has borne some fruit—why should it be hampered and stinted in future yield by binding rules, regulations, and—continental conventionalities—in a word why should we be forced to work under a "rule of thumb"?[80]

Concerns about Britain's ability to act autonomously in tropical Africa reinforced concerns about the volatile mixing of science with political power.

Acutely aware of the problems of political jurisdiction, Bartle Frere asked Henry Thring to draft the RGS's mission statement for its proposed African Exploration Fund. This way, he felt, the RGS could avoid any breach of international law or the possibility of provoking international tensions. Thring's language intentionally took into account officials' multiple anxieties, but also managed to assure those who chose to ally themselves with Léopold that as long as they pursued their collaboration privately it would not jeopardize their reputation. This compromise made "more palatable to the King of the Belgians the refusal of the Prince of Wales to act as President of the National Committee."[81] But on one issue there could be no room for compromise: future RGS expeditions under British auspices would carry with them no juridical mandate.

While the final circular announcing the African Exploration Fund's launch publicly endorsed "the enlightened efforts of the King of the Belgians to give a new impulse . . . for the systematic and continuous exploration of Africa," it took pains to point out that the RGS's council was "of [the] opinion that exploration is, from its very nature, a subject fitted for national enterprise, rather than for the complicated agencies of an international association, which could not be carried out in practice without giving rise to embarrassing questions of a political and international character." The African Exploration Fund would support "the scientific exploration of Africa, (especially the central part of that continent,) in a systematic and organised manner, with a view to the attainment of accurate information as to climate, the character of the inhabitants, the best routes of access, the resources of the country, and all such other matters as may be instrumental in preparing the way for opening up Africa by peaceful means to the influences of religion, civilization, and commerce."[82]

The close parallels between the research questions initially posed by the founders of the IAA and those selected by the RGS for its Exploration Fund indicate that despite political tensions, there already existed a consensus on the topics that needed investigation. The international conferences sponsored by European geographical societies between 1871 and 1876 had helped to establish the need for more comprehensive baseline inquiries. In the process, these gatherings had focused the spotlight on the legal and political controversies that could arise if fieldwork within tropical Africa were conducted under multinational auspices. When the foreign secretary reviewed Thring's circular for the RGS's new fund, he had those tensions in mind. With palpable relief, Carnarvan told Thring approvingly that his document was "not only very skillful—as any composition of yours must be—but quite safe."[83] For the largest and most powerful empire on the planet, safety was no small thing. Avoiding conflicts with other nations and problems with constituents was, for British officials, a valuable way to ensure the empire's stability. Other European countries had less reason to be so cautious.

National Jealousies, Scientific Study, and Territorial Partition

What the Royal Geographical Society and the Foreign and Colonial Offices failed to appreciate in 1877 was that their reservations and consequent withdrawal from the International Association opened the door for other nations to enter the fray by undertaking geographical explorations of their own. Britain's insistence that expeditions ought to be pursued within a national framework prompted the IAA's leadership to allow its affiliates to establish field stations under national rather than multinational auspices. At the same time, its headquarters in Belgium continued to operate as a transnational entity. Between 1877 and 1884 the IAA never resolved the question of its international status, but oscillated ambiguously between an intergovernmental and nongovernmental body. The RGS's decision to limit its African Exploration Fund to supporting "pure Geography and general survey," rather than commercial geography, which was then gaining popularity in France and Germany, led them to rebuff an expedition proposal submitted in 1877 by Henry Morton Stanley. They chose instead a proposal for East African research prepared by Keith Johnston, who planned to travel with geologist Joseph Thomson in order to do research on "physical geography." Geographical studies, in Johnston's view, ought to be about more than description or classification; they should also examine "mutual relations," which would allow "causes and laws of action [to] come to light": "for every branch of knowledge touches upon geography on some side."[84] Stanley already had a checkered history with the RGS, in part because several of its fellows believed his field operations were too violent and insufficiently focused on scientific research. So he turned to the only other patron then seeking his services, King Léopold himself.[85] Stanley worked on behalf of the IAA and its offshoots, the Association Internationale du Congo and the Association du Congo d'Hautes Études, for the next seven years.[86]

Britain's defensive moves to keep political and geographical undertakings separate had the unintended effect of encouraging other nations, especially those less concerned with policing their existing boundaries, to initiate new projects. G. N. Sanderson has aptly observed that there could have been no "general multi-power partition of Africa if the British had continued successfully to deter other powers from encroaching upon their informal African empire."[87] Among the range of interrelated factors that played into this political recalibration is the simple fact that while British officials were putting on the brakes with respect to geographical expeditions in tropical Africa, statesmen in other countries were beginning to entertain the possibility of more extended undertakings. Not only did Britain see its political and economic influence diminish during this period, but it also witnessed the gradual eclipse of its geographical supremacy. The irony is that British nationals themselves made this shift possible by insisting that RGS affiliates take no part in making territorial claims.

A significant turning point in the synergy between expeditions and formal empire came in 1879, when French officials gave the "scientific explorer" Savorgnan de Brazza permission to make treaties along the Congo River in the name of France's chapter of the IAA, directly challenging the treaties Stanley was then collecting for King Léopold and the IAA head-quarters.[88] The lead explorers made little attempt to distinguish their geo-graphical from their territorial aims, but persisted in speaking of their missions as scientific. While British officials monitored these developments, they remained relatively inactive. Other countries responded differently, calling immediate attention to possible political conflicts over different territories. Portuguese officials were alarmed at what they perceived as encroachments on their coastal sphere of influence. The secretary of Portugal's Geographical Society went so far as to write to the IAA inquiring whether "Messrs. Stanley and Savorgnan de Brazza [are] to be considered as the explorers of the International African Association, and as such to be quite subordinate to the purely scientific and humanitarian intentions of the said association, excluding absolutely all individual ideas and all political mission and authority."[89] The IAA's secretary replied that de Brazza and Stanley were working independently, one for France and the other for King Léopold, and stated that Stanley's primary mission was "to found scientific and [resting] stations on the Congo, and also to furnish it with any elements of study likely to further enterprise in that country." Portuguese officials received a more definitive answer in the autumn of 1882 when France ratified a treaty between de Brazza and King Makoko that claimed a strategic piece of territory near Stanley Pool on the Congo River. Looking back on this period, Stanley told the London Chamber of Commerce in 1884 that de Brazza's presence forced him to rearrange the treaties he was gathering: "In the parliamentary papers you will find copies of some treaties wherein we have obtained all rights to trade, to mine, to plant, sow, build houses—in short, do all things. The reason . . . is simple enough. We have left the natives nothing to give to anyone else."[90] Between 1880 and 1884, Stanley gathered more than three hundred treaty agreements in the region, far surpassing those of any other field explorer, and also helped to found thirty-nine field stations along the Congo and its tributaries under the auspices of the IAA.[91] Whatever their usefulness for scientific study, they were more important as markers of territorial possession.

The national jealousies and the impetus to seize lands that British officials had feared soon materialized. While imperial historians such as Ronald Robinson and John Gallagher have played down these moves and the rivalries they generated, dismissing de Brazza's and Stanley's "skimpiest of credentials," this focus on the insubstantial character of their professional identities seems to miss the point.[92] European geographical societies had been active within tropical Africa for decades and had sporadically suggested formal colonization. Yet it required government sanction to turn

such appeals into reality. Only after African exploring committees proliferated widely and brought questions about their scientific and geopolitical
remit to the fore did political leaders confront the issue of whether they
would authorize territorial claims in the tropical interior. To the litany of
explanations for both the origins and the rapidity of late nineteenth-century
high imperialism must be added the catalyst provided by geographical expeditions and their coordinating conferences.

Although historians tend to think of the scramble for Africa in terms of
those nations that ultimately secured territories in the continent itself—
Germany, France, Britain, Portugal, Belgium, Spain, and Italy—this view is
too narrow. If we include those nations that established special societies to
"explore and civilise" Africa (to quote from the Swiss national committee),
we realize that this endeavor was a pan-European and even a trans-Atlantic
phenomenon.[93] After the RGS's leadership stepped away from encouraging
treaty negotiations, other scientific explorers such as Savorgnan de Brazza
(acting for France), Stanley (acting for King Léopold and the IAA), Gustav
Nachtigal and Gerhard Rohlfs (acting for Germany), and Serpa Pinto (acting for Portugal) set in motion a more intensive exploration of tropical Africa that owed much of its momentum to the Geographical Conference of
1876. If these men had been operating in isolation without the support of
nationally oriented geographical societies, then it is unlikely that their ad
hoc treaties along the Congo and Niger rivers, on the east coast and in the
interior, and across the western and southern territories would have been
greeted favorably by metropolitan legislators. Geographical interests laid
the groundwork for geopolitical negotiations, while geopolitics, in turn,
circumscribed not only *who* would be included in international alliances
but also *how* these would take place and *what* would be considered. This
segue from science to politics helps to explain why the Berlin West Africa
Conference (1884–85) took on such significance. Not every representative
at the table wanted to pursue territorial rights, but all fourteen countries in
attendance hoped to play a role in constructing the rules of engagement.[94]

Just before the Berlin Conference convened, the Royal Geographical Society prepared a map on the "partition of the coast of Africa," indicating
which European nations held sway along the continent's long coastline.
Tellingly, next to the European "colonizing powers" the legend made room
to include "independent kingdoms" and "native possessions," several of
which still existed (see plate 2).[95] Despite the circumscribed geographical
focus of the Berlin Conference, members of geographical societies and officials concerned with African affairs understood that much more was at
stake. In a London *Times* article that popularized the phrase "the Scram-

ble for Africa," an astute observer remarked that "the coast itself is of no value, except as giving access to the interior. What each of the annexers will do with their lots so far as the development of resources of the interior is concerned remains to be seen."[96] By the end of the Berlin Conference, a massive territory had been carved out in the center of the continent, comprising the region open to free trade and much of what would become the Congo Free State. This move was highly provocative, even though what kind of state the Congo would be remained uncertain. Regardless of the Congo's legal status or the political intent behind its foundation, the symbolic impact was dramatic: the creation of the Congo Free State seemed to indicate the beginning of a full-scale partition, especially given delegates' emphasis on the need for each European power to back all claims to new territory with "effective occupation."[97] Henceforth, any of the signatories of the Berlin Act that wished its possessions to be recognized by the other signatories "had to insure the establishment of authority in the regions occupied by them."[98]

The origins of the Congo Free State and the rapid creation of numerous other "paper protectorates" in the "heart of Africa" must be explained in a broad context that includes diplomatic, economic, and political exigencies. Yet, the geographical quests that preceded and coincided with partition also shaped the process in significant ways. Indeed, the delegates gathered in Berlin recognized the importance of learned societies' activities. Stanley, the only African explorer who was present in any official capacity (as a member of the U.S. delegation), did his best to promote the interests of Léopold's International African Association. During his speech before the conference, Stanley spoke in glowing terms of "the natural wealth and attractiveness of the Congo region, where Europeans could live and thrive as well as in their own temperate zone if only they adapted their habits to the African climate."[99] Speculation was rife in Berlin as to whether Europeans could permanently settle in the African tropics. The region had earned a reputation as deadly to travelers, but little information was available on either the causes of ill health or the role climate played in producing high European mortality rates. The head of the Russian delegation, Count Knapist, told the conference that "precise data on the climate of Africa are absolutely wanting, whereas the [International] Meteorological Committee have already gathered them in every other part of the world." It would be a tremendous service to science if the conference might "facilitate the establishment of a meteorological station in the upper regions of the Congo."[100] The lead Italian negotiator, Count de Launay, said during one of the first sessions that he hoped to see some provision in the final protocols to protect scientific travelers working in the continent. "I think that a mention of this kind will produce the best effect amongst the men of science, the explorers, and the numerous geographical societies, so well represented by the Special Delegates, and other most competent persons assembled at Berlin on account of the Conference. . . . It is to scientific men and explorers . . . that we owe the marvelous discoveries made during these latter years in Africa. . . . It is our duty to encourage them, to protect them all, in their researches and expeditions, both present and future.[101] The gathering approved this provision and incorporated it into Article 6 of the final Berlin Act. Scientific research had come full circle: previously it had helped to pave the way for empire, and now empire was opening new avenues for science.

Taking Stock: Resources and Races

The Berlin West Africa Conference marked the beginning of an exceedingly rapid and politically fraught legal process. As Lord Salisbury, Britain's prime minister, noted in 1889, "Africa is the subject which occupies the Foreign Office more than any other." To manage his new responsibilities, Salisbury had had the maps of Central Asia and the Balkans taken down from his rooms at the Foreign Office and replaced with an "aggressive profusion" of African maps.[102] Between 1885 and 1895, European diplomats steadily oversaw the completion of the "paper partition." Edward Hertslet's three-volume study on the *Map of Africa by Treaty*, produced for the Foreign Office in 1894–96, illustrates just how busy they were: at the turn of the twentieth century, only a small fraction of the African continent remained free of European control.[103] The French had largely preserved their sphere of influence in the north, securing a large bloc of territory from Algeria to Senegal. Thanks to the efforts of de Brazza, they had persuaded the other European powers to accept French claims to a vast tract of land north of the Congo River. Britain had succeeded in protecting its influence in Egypt and South Africa, the areas Lord Carnarvan had said "must be ours," and had consolidated its control of key waterways: the Nile, the Niger, the Zambesi, and the Gambia. With the support of the Belgian legislature, King Léopold established his sovereign control over the Congo. The colonial stalwarts in Africa, the Portuguese, had managed to hold onto two huge swaths of land on the eastern and western coasts of southern Africa. The latecomers to empire building, Germany and Italy, had also justified their rights to extensive territories. The Germans appropriated a portion of East Africa that had long been coveted by the British (see plates 3 and 4). For some fellows of the Royal Geographical Society and for colonial statesmen such as Cecil Rhodes, the German presence was a particularly bitter pill to swallow, as it broke up the contiguous empire they had envisioned stretching all the way from Cairo to the Cape.

The biggest losers in this game of geopolitical chess were the "independent kingdoms" and "native" leaders "to whom," as John Kirk wryly observed at the time, "are not accorded the usual rights prescribed by international law."[104] The partition also dealt a fatal blow to the Ottoman Empire and the Zanzibar Sultanate, which had previously exerted influence in northern and eastern Africa respectively. Formal partition might have been "good for the map-makers," lamented the British missionary and political activist Horace Waller, but "Africa will have something to say about all this before she is done with 'colonization.'"[105] The scale and speed of the scramble for Africa and the growing impression that it reflected the basest kind of "earth-greed" even caused members of the RGS's executive council to raise questions about the "ethics of African geographical explory."[106] Although the RGS had helped to prompt a new era of "geography militant," to borrow Joseph Conrad's apt phrase, geographers were deeply divided about the land rush.[107] Verney Lovett Cameron, who was staunchly committed to imperial expansion, was still taken aback that statesmen at Berlin would try "to dispose of" the Congo Basin "without consulting the wishes of the inhabitants, and in entire ignorance of a very large portion" of the region.[108]

The Berlin Conference also ushered in a new era of scientific assessment and debate stimulated by what Joseph Thomson called the "mania" for Africa that swept across Europe. A geologist by training, Thomson had been selected to join the African Exploration Fund's first expedition to East Africa in 1878 and, like Cameron before him, had taken his research instructions quite seriously, collecting mineral and botanical samples, reporting on meteorological conditions, offering ethnographic commentary, and considering variations in vegetation and natural resources. Upon his return to Britain, he expressed skepticism toward unrealistic expectations about tropical Africa's economic prospects. He wrote sarcastically after the Berlin Conference

[t]hat nothing might be wanting to make Africa more attractive to the public imagination, even its climate was rehabilitated. It was no longer spoken of as the white man's grave, but as a happy colonizing ground for emigrants. . . . A load was taken off the mind of many of weak faith who saw in the immediate future a struggle for existence in an overstocked world. Through the

medium of Africa a crisis in operations of Nature's law of the survival of the fittest was indefinitely postponed by this new discovery of fertile and healthy regions for our surplus population.[109]

Thomson's take on the situation was correct: optimism about the possibility that large numbers of Europeans might relocate to tropical Africa reached a climax at the Berlin Conference. Population growth within Europe was a growing concern for both national and colonial policies: unemployment, crime, and deviant behavior appeared to be on the rise. The specter of "racial degeneracy"—the fear that human evolution was being reversed and people were sliding backward from civilization toward barbarism, especially in urban centers—haunted public officials. The continuing economic depression exacerbated the problem. Propagandists for colonialism, including many geographers, played on these concerns during the Berlin gathering to bolster enthusiasm for imperial expansion.[110] They were largely responsible for the idea that tropical Africa was a new "El Dorado" or a "second India," "from which the fruits and produce of every climate may be flung into the lap of busy Europe . . . fields yet unfertilised, forests yet unutilised, mines yet unexplored,—all offer a full cornucopia, a marvellous exuberance of tropical and subtropical wealth."[111]

These positive projections, Thomson argued, were deeply flawed. Much of tropical Africa was neither fertile nor particularly healthy, yet it was difficult to be heard above the din made by Africa's boosters. Anyone sharing his opinions, he claimed, was deemed "weak-minded, irresolute, and senile . . . put down as having much to learn."[112] Thomson was not opposed to colonial rule in principle, and he had already helped to secure the Nigerian hinterlands for the Royal Niger Company, but he had serious doubts about whether any area in tropical Africa could sustain permanent European residents. Invoking a pervasive concern with "vanishing races," he also worried that, rather than elevating Africans, colonialism might lead to a "downward path . . . to extinction, though the Negro, if he die out at all, will die hard, for, unlike the aborigines of New Zealand or North America, he is in Equatorial Africa the fittest to survive."[113]

Others shared Thomson's reservations about the colonization of Africa. In the 1890s, pessimism vied with more sanguine characterizations of the continent's climate and disease environments. The litmus test for pessimists and optimists alike was European survival. No one disputed the fact that European mortality rates in tropical Africa were high, but some experts began to contend that this unhealthiness could be overcome. What Thomson shared with these optimists was a keen faith in empirical study, especially when conducted on the continent. However misleading "the African craze" had been, "it would be wrong . . . to convey the impression that [it] was barren of results. Far from it. In no part of this century has there been a richer harvest of scientific facts."[114] The way forward was to investigate Africa further.

Experts and Empire

Geographers were always ready to pronounce their opinions and offer their expertise about Africa, so they remained at the forefront of these debates in Britain. Their influence remained strong through the 1890s; in the first decade of the twentieth century, as we shall see in chapter three, they were also responsible for opening the door to another nascent field science, ecology. Significantly, it was geographers who insisted that scientific specialists take the lead in shaping ideas about "the development of Africa." "Surely . . . the time has come," commented the secretary of Scotland's geographical society in 1890, "when we should take into our councils those whose special knowledge fits them to advise or control?"[115] Professionals had advanced this argument to justify their discipline's presence within British universities. One advocate of geographical education wrote, "Many a generalisation would have been avoided, had political thinkers been trained in knowledge of the earth they live in, and of the influence which . . . [it] exert[s] on the varying political tendencies and institutions of peoples who part its empire between them."[116] Thanks to the RGS's extended campaign, the first formal positions in geography were established in Oxford and Cambridge in 1887 and 1888 respectively.[117]

Medical specialists were steadily increasing their involvement in African affairs at the same time. Some approached their subject as medical geography, while others began to describe it as the study of tropical diseases. Britain's Foreign Office acknowledged the increasing importance of this subspecialty in 1898, when it was considering how to support the recently proposed Schools of Tropical Medicine in London and Liverpool: "the most pressing case is that of Africa," "the scheme is mainly intended for the benefit of Africa."[118] Patrick Manson, the advisor to both the Colonial and Foreign Offices on medical matters, was even more blunt: "Africa is not only undergoing political and social changes, but is undergoing an enormous pathological, shall I say, revolution. The introduction of European ways, of methods, of travel, of rapid communication, is upsetting the whole pathological arrangement of Africa."[119] Imperial administrators, he argued, now had an obligation to address the problems their compatriots had created.

A third group that participated in debates about tropical Africa's future was anthropologists, although their professional identity was even less established than that of medical researchers or geographers. Indeed, thanks in part to the publication of the British Association's *Notes and Queries on Anthropology* (1874), a joint endeavor between geographers and anthropologists, it was expedition leaders and field officers rather than self-professed anthropologists who first published specialized studies on African people and cultures. Frustrated leaders of Britain's Anthropological Institute (AI) remarked as early as 1876, following Verney Lovett Cameron's report on his cultural research, that in Africa it was "quite right that geography should take the lead, and anthropology should follow afterwards, but anthropology cannot afford to be far behind."[120] Like their counterparts in geography, anthropologists hoped to make their usefulness felt within university and government circles. As Francis Galton put it in his presidential address before the Colonial and Indian Exhibition of 1886, "Anthropology teaches us to sympathise with other races, and to regard them as kinsmen rather than aliens. In this aspect it may be looked upon as a pursuit of no small political value."[121]

Although professorships in anthropology were established at Oxford and Cambridge in 1884 and 1900 respectively, the Anthropological Institute continued to face serious obstacles in its efforts to convince the British government to sponsor research on ethnographic and ethnological subjects alongside geographical investigations in the field. When an Imperial Institute was first proposed in 1887, Galton himself declared that it "cannot fail incidentally to become an important centre of anthropological intelligence."[122] But fail it did, leading successive AI presidents to observe that it was "little short of a national disgrace that in the largest empire in the world . . . there is no Imperial Department whose function should be to collect and classify the facts of the physical, psychical, and ethical histories of our fellow-subjects."[123] In the late 1890s, African anthropology received a significant boost in Britain, initially through the private efforts of Mary Kingsley and her supporters and then through the work of missionaries, linguists, and administrators who sought to publish in the institute's journal and present their findings before the British Association. Indeed, it was Kingsley who urged anthropologists to turn away from measuring physical attributes—"for no self-respecting person black or white likes that sort of thing from the hands of an utter stranger, and if you attempt it you'll get yourself disliked in . . . Africa"[124]—and focus instead on African systems of religion, law, and "fetish beliefs," as she put it. Kingsley was actually the first fellow of the Anthropological Institute to equate "witchdoctors" with "consulting physicians" and to analyze the philosophical, linguistic, and practical elements of "African therapeutics."[125]

At stake for many of the specialists who turned their attention to Africa during the 1890s was a single question: how should European rulers approach the continent? While geographers and physicians thought about Africa's natural resources and prevalent diseases, those interested in the burgeoning field of anthropology were more focused on Africa's peoples. All were well aware that the scientific evidence was patchy at best. As the RGS's lead cartographer put it in 1891, "Of Africa we know next to nothing."[126] While the British government in London and its official representatives in Africa were absorbed with the turbulence of conquest and state building, the Fellows of scientific societies began to hone in on questions concerning the continent's climate, physical geography, epidemiology, racial groups, and cultures. These issues had sporadically attracted scholarly attention in previous decades, but with new evidence and reports surfacing from a variety of regions, many were now able to revisit these subjects more systematically.[127]

Between 1890 and 1900, geographers, biomedical scientists, and anthropologists were responsible for producing continent-wide maps of the features that were central to the architects of empire: these included not only the "value of African lands," but also its vegetation, temperature, rainfall, geology, altitudes, commercial products, population densities, languages, and religions.[128] Robert Felkin, a specialist in tropical diseases in Scotland, published in 1893 the first continental scale map on the "distribution of disease" that was based on direct experience in East and Central Africa (see plate 5). In his discussion of his methods and findings he explained that he wished to indicate both the "amount of disease in any area," and "the comparative salubrity or unhealthiness of a district . . . [which would] be useful, especially as the rush to Africa still continues, and extensive schemes of colonisation are in the air."[129] Information about geography, demography, epidemiology, and sociocultural anthropology was correlated spatially, in order to reveal patterns and causal relations that related directly to the future imperial development of the continent.

"The Conquest and Development of African Lands"

At the forefront of these efforts were John Scott Keltie and Arthur Silva White, the librarian and secretary, respectively, of the Royal and Scottish geographical societies, who also served as secretaries for the British Association's Geography Section between 1889 and 1892. In 1890, White published the first English-language monograph on *The Development of Africa*. That same year, Keltie published the first textbook on *Applied Geography*, taking "the geography of Africa and its bearing on the development of the continent" as his main case study. These geographers shared concerns about the territorial rapaciousness of the European powers, which they thought might worsen rather than improve African conditions. "European nations," according to Keltie, "have somewhat lost their heads during the last five years. They have been grabbing blindly at whatever land remains unannexed, apparently regardless of their adaptability, and as if anxious only to add as many square miles as possible to the statistics of their foreign possessions."[130] Their motives for writing their books were largely pragmatic and instrumental: they sought to ensure that officials and traders had the knowledge they would need to pursue more far-sighted policies. "The African Question is in the main a geographical problem," asserted Silva White in his book's preface. "In its initial state—the conquest and development of African lands—we have to deal not so much with political as with geographical conditions. It is only after the latter are understood that we can effectually control the former."[131]

In approaching their subject matter, both Keltie and Silva White assumed that tropical Africa already lagged behind most of the other continents of the world, a situation Keltie summed up as "tardy development." Some measures of these lags were, in their eyes, indisputable: most of Africa lacked communication and transportation infrastructures of the sort that were proliferating across Europe, North America, and South Asia.[132] With the exception of Islamic strongholds and areas of northern and southern Africa that were dominated by white settlers, most regions lacked a critical mass of schools and libraries, which they believed were important to the growth of trade and industry. Africa also seemed to have fewer natural harbors along its coastline and navigable rivers in its interior. Its climate was thought to be hotter and more dangerous than that of any other continent except perhaps Australia. Keltie remarked: "Had it not been for its latitude, Africa would have been overrun by European colonization and European commerce long ago."[133] Silva White reasoned along similar lines: "Nature, whilst lavishing on her the most bounteous gifts, has, at the same time, imposed certain barriers and restrictions to their enjoyment which hamper no other continent."[134]

While both geographers adopted popular racial assumptions in their analysis, including the belief that European cultures were inherently superior to all others, they were more likely to blame tropical Africa's "backward" state on its environmental conditions than on the capacities of its indigenous peoples. Decades of experience had already shown that Africans could be "skilled diplomatists" and "astute traders": "That the race, and far more the individual, is capable of development—of what we call progress—has been demonstrated over and over again."[135] Although Silva White occasionally referred to "the African" as an "untrained child of Nature," he challenged the "popular idea of the Negro" represented in "the travesty one sees of him on tobacco-labels or on the stage."[136] Keltie, too, expressed a paternalistic attitude and criticized racial stereotypes. He asserted that Africans were "on the whole a fine race. . . . We are often told that the Negro is a lazy being. . . . But as a universal statement facts belie that assertion."[137] In their view, "popular dogma" about Africans was not helpful when it came to thinking about future development because it glossed over useful details and obstructed cosmopolitan interchange. However "unsophisticated" Africans might seem to Europeans, Silva White cautioned, "we Europeans appear . . . a most immoral and wicked people [to Africans]; their judgment of us is, in fact, crippled by the same limitation that prejudices our judgment of them."[138]

What is most unusual about White's and Keltie's approach is that while they wished to foreground the determining power of natural factors such as physical geography, climate, natural resources, and the inherent attributes of its "indigenous populations," they also connected these to an assessment of factors under social control, including external trade, "indigenous political conditions," and foreign influences. White argued, "It is only by understanding the interaction between physical and political phenomena that we can hope to lay the foundation for a rational policy in Africa: for violation of Nature's laws brings the inevitable Nemesis."[139] Those who could reveal these laws and mediate between political and natural realms were experts with specialized knowledge based on scientific research. As Keltie put it, "whether we call it geography or not, our Consuls abroad are continually telling us that ignorance of a country, ignorance of its people and their peculiar ways, ignorance of their language, ignorance of the wants special to the country, is constantly placing [us] at a disadvantage."[140] Keltie's dictum was simple: "Superior knowledge, in the end, must win the race."[141]

These points were not entirely original, yet in advancing their arguments in relation to tropical Africa, Keltie and Silva White shaped the contours of debate. First, their works encouraged readers to recognize that much of what people thought was known about Africa was actually quite indeterminate. As Silva White put it, many "phenomena" were "so manifold and complex, and our *data* in Africa are, moreover, so limited and imperfect, that we can refer to them here only in very general terms."[142] Far more extensive research was needed. Both implicitly and explicitly, the two geographers reinforced the view that Africa required external European intervention. Second, they each played an important role in strengthening links between geographical thinking and ideas about development that turned on assumptions about nature, race, and politics. A key supposition of their arguments was that the most effective approach to African development was to work with, rather than against, the existing attributes of the continent. This idea became a kind of catchphrase in Silva White's work: "I have endeavoured to arrive at the general principles that underlie the development of the continent along what may be regarded as its natural lines." Third, their efforts were part of a wider behind-the-scenes campaign to ensure that "earth-knowledge" was considered foundational to political and commercial activities in tropical Africa: the better the laws of nature were understood, the more effective the European powers would be in administering their overseas possessions. In Silva White's and Keltie's eyes, geography would serve as the overarching discipline for syncretic analyses since it allowed scholars to fuse insights across various "branches of knowledge." In an 1893 article, Silva White urged his readers to consider geography not merely as a descriptive science but as a whole "body of thought" that could help "man" manage his relations to diverse environments (figure 1.1). By

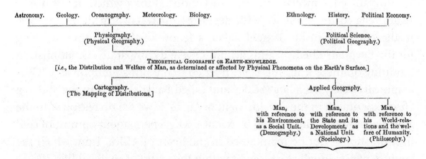

Figure 1.1. The position of geography among the sciences, according to White. The original legend reads, "Geography is shown in the above Table to draw its facts from eight elemental or derivative sciences, and to group these facts under two divisions, from which general principles are deduced. This forms the Theoretical side of Geography. Its Practical side is sub-divided under two heads: (1) Cartography, in itself a distinct science, involving Mathematics; and (2) Applied Geography, which takes account of Man in his three characteristic aspects—towards his Home, towards the State, and towards the World— and considers them in conjunction, their more precise definition being left to the three other sciences for separate treatment." Source: Arthur Silva White, "The Position of Geography in the Cycle of the Sciences," *Geographical Journal* 2 (1893): 178.

extension, this knowledge would help Europeans administer and control African lands that were very different from their own.

"The Development of Africa by Science"

Historians of British colonial development and its intersections with scientific expertise usually point in this period to the career of Joseph Chamberlain, a member of Parliament from Birmingham who became colonial secretary in mid-1895.[143] A liberal reformer, Chamberlain helped to inaugurate a new policy of "Constructive Imperialism."[144] In widely publicized speeches, he compared Britain's colonies to great "undeveloped estates" that the nation, like a landlord, had a duty to improve. He promoted infrastructure and disease control projects in West Africa and agricultural development in the Caribbean. The Colonial Office embraced tropical medicine at his initiative.[145] Yet science, empire, and development had long been linked in British thought. As early as 1858, the Parliamentary Select Committee exploring the possibility of European settlement in India argued that its tropical climate and the absence of constitutional government would likely deter large numbers of Britain's "labouring classes" from choosing to settle there. The committee concluded that colonization should "clearly be limited to a class of superior settlers who may, by their enterprise, capital, and science, set in motion the labour, and develope the resources of India."[146] Similar ideas circulated regarding tropical Africa. In 1869, the RGS organized a special gathering in London during which the Prince of Wales regaled the audience with details of his recent trip down the Nile in the company of Sir Samuel Baker, a fellow of the RGS then working for the Egyptian Khédive, and Richard Owen, a comparative anatomist. In reporting on these feats, the *Times* referred to the British Empire as a "Geographical Society on a large scale" and called Baker's efforts in the Sudan a "first experiment in the development of Africa": "we see no reason with the advantages now being secured by science, why a permanent outpost of civilization should not be maintained in the heart of Africa, from which the country may gradually be brought within the realm of civilized life."[147]

Situated in this longer historical context, Silva White's and Keltie's monographs on African development represented the culmination of at least two decades of thinking on this subject and prompted new research projects designed to investigate Africa's economic and social prospects. Less than a year after Silva White's and Keltie's books were published, while they were still secretaries of the Geographical Section, the British Association appointed a special committee to study over a ten-year period the "climatological and hydrological conditions of Tropical Africa."[148] "A continuous series of such observations is greatly wanted," the committee reported, for "until we possess them we have no means of coming to any precise practical conclusions as to the possibility of European acclimatization."[149] The chair was Ernst Ravenstein, a cartographer and statistician who in 1890 presented a paper to a joint session of the Geography and Economic Science sections offering an analysis of the "lands of the globe available for European settlement."[150] In his talk, Ravenstein provided estimates of world population growth, population densities by continent, and the limits placed on demographic growth by the availability of arable land. His statistical calculations were a first attempt to determine what would later be called the globe's "carrying capacity," a pursuit that grew out of European anxieties about supplies of food, fuel, and the diminishing possibilities for overseas expansion. Although Africa fared fairly well in his evaluations, with half of its territory categorized as "fertile land," Ravenstein admitted that trustworthy data were hard to come by.[151] Taking a conservative approach, he estimated Africa's population at 127 million, a figure much lower than other authorities suggested, and calculated its population density at 11 per square mile, compared to 101 in Europe and 57 in Asia. These figures alone made Africa worth a closer look.

In order to remedy various gaps in knowledge, Ravenstein's committee drew up a pamphlet in early 1892 offering "Hints to Meteorological Observers in Tropical Africa," which it distributed, sometimes along with measuring equipment, to dozens of European residents and scientific travelers who were asked to take at least one and preferably three measurements per day of barometric pressure, temperature, and wind speed and, where relevant, of rainfall and lake levels as well.[152] By 1895, the committee was receiving annual reports from at least forty-three sites: sixteen coastal stations, eleven inland stations at altitudes below 3,000 feet, and sixteen inland stations at higher altitudes.[153] These data allowed Ravenstein to map "Africa into a series of hydro-thermal regions," which would be useful not just to plan human settlements but also to think through agricultural development.

Shortly before Chamberlain made his own momentous speeches on Britain's great estates, the RGS raised these issues for public debate during a special panel it organized for the 1895 International Geographical Congress in London. This session brought together a group of renowned naturalists, geographers, and administrators, including John Kirk, Arthur Silva White, Henry Morton Stanley, Frederick Lugard, and Emin Pasha, to discuss "the extent to which tropical Africa may be developed by the white races." Among their chief concerns were the issue of whether European races could acclimate to tropical climates and the question of whether Britain should establish a managerial or a settler empire in its tropical territories. (India was the model they had in mind for a managerial empire while South Africa stood in as an example of a settler colony.) Given the celebrity status of many of the speakers and the widespread interest in the subject, the session attracted the largest audience of any panel at the congress and led observers to title the discussion "The Development of Africa by Science."[154]

After the formal presentations, Ravenstein reported to the congress that his committee had "found that localities likely to suit European constitutions were limited in area, and also, as a rule, difficult of access and poor in their resources."[155] His conclusion was in accord with the other speakers, all of whom expressed grave doubts about tropical Africa's fitness for intergenerational European colonization. As Silva White put it during his own address: "the heart of the continent remains, and is likely to remain, the true home of the Black race. . . . Nature has, in short, marked off Tropical Africa as the abiding home of the Negro and indigenous tribes."[156] If white settlement were to be tried anywhere, John Kirk suggested, it should be undertaken very slowly, with small numbers, and only at altitudes approaching 5,000 feet. The two areas of British Africa Kirk was familiar with that fit his criteria were "Matabeleland" (named after the people now called Ndebele and located in present-day Zimbabwe) and "Maasailand" (named for the Maasai people, in present-day Kenya and northern Tanzania).

The transition from the view that diseases arose from poisonous vapors in the environment, or miasmas, to the view that they stemmed instead from specific parasites or germs, each of which caused its own disease etiology, occurred with relative rapidity at the end of the nineteenth century. This conceptual change not only allowed medical specialists to overturn long-standing assumptions about African epidemiology, but it also forced them to think differently about relationships among climate, disease, and human physiology. For instance, after the revelation that sleeping sickness was caused by a parasite carried by the tsetse fly, Cuthburt Christy, a member of the first Royal Society commission to Uganda, announced in an article published in the *Journal of the African Society*, "We realise at last that there is nothing mysterious shrouding the diseases, of the once 'Dark Continent.' The word acclimatisation has now no more significance than the word miracle." The commission's research in effect undermined so many of the assumptions then prevalent in acclimatization debates that Christy felt it had become an almost useless concept. "There is every reason to believe that careful research, on the spot, by men trained in research methods, cannot fail to bring to light the causes of disease." This new knowledge, he argued, would "increase the welfare not only of the white man in our Colonies, but of the negro subjects of the Empire."[157] These investigators did not so much "rehabilitate" Africa's climate, in the words of Joseph Thomson, as point their finger at other factors, such as insects, parasites, and indigenous drugs that deserved investigation.

As physicians and geographers sharpened distinctions between temperate and tropical climates, however, the temperate environments won the day. British migration to South Africa doubled between the 1880s and the 1890s, going from 76,000 to 160,000. Even after the South African war, migration continued to climb, reaching almost 270,000 for the first decade of the twentieth century.[158] Algeria, the main settler colony in the north, had approximately 130,000 resident Europeans by the 1850s and a white settler population that surpassed the half million mark by 1900.[159] By contrast, migration patterns to tropical Africa were negligible and barely made an impact on the statistics charting the great exodus from the British Isles. It was in large part indirect migration, from South Africa northward, that initially made up the European populations settling in the British territories closer to the equator.[160]

Although Chamberlain consolidated an interest in coupling science and development during his tenure as colonial secretary between 1895 and 1903, a range of other scientists and statesmen were equally invested in the subject and often brought a deeper knowledge of scientific debates and African affairs to the table. During these years, a significant part of British Africa was still overseen by the Foreign Office rather than the Colonial Office, placing it as much within the purview of the prime minister and his foreign secretary as under the colonial secretary. Official and public debates about tropical Africa's future transcended any single bureaucracy and were typically conducted on the assumption that these questions were shared by all the European powers.

Even in these early years, advocates of scientific investigations acknowledged their potential to transform and challenge existing assumptions about the continent and its peoples. In a series of tributes to Mary Kingsley and Britain's newly founded African Society, the public intellectual and statesman, Edward Blyden, writing from Sierra Leone, stressed Kingsley's efforts to draw attention to the "deplorable ignorance" and "misinformed conceptions" that existed "in Europe . . . of the true condition of the natives" and of "African law, religion, and customs." Quoting from her letters to him, he endorsed her view that "England is dealing in West Africa with a purely imaginary African. . . . They only know the African from a lot of unscientific reports." The African Society's objective to investigate "indigenous" culture and institutions would thus help to "disperse many errors and dispel many illusions." "The studies, discoveries, and publications of the African Society will make it possible in the future not to dispose of native questions by an *a priori* logic, but conclusions will be arrived at with a full knowledge of the facts."[161] As we shall see, the call for more and better facts, especially by those monitoring the fate of Britain's African empire, had only just begun.

CHAPTER TWO

A Development Laboratory: The African Research Survey, the Machinery of Knowledge, and Imperial Coordination

Because the Empire includes so many diverse peoples differing in physical and mental attributes, and because it yields such a variety of natural products, it offers problems political, social, and industrial of peculiar interest and complexity. . . . When action is taken [governments of the empire] have the largest outdoor laboratory in the world: a field for action which allows in some part or parts the investigation of problems of almost every description.

—Sir Robert Greig, *British Association meeting in South Africa*, 1929

The [African Research] Survey is the first attempt that has been made to ascertain with reference to an entire continent to what extent the resources of modern science are being utilised by governments to assist them in the exercise of an intelligent control and wise direction in the course of development. Such a survey may be expected to provide a basis for experiments in economic planning which would not only benefit Africa, but prove of wider and more general interest.

—Description of the African Survey to Rockefeller Foundation, 1935[1]

In 1928, Sir George Schuster, the economic and financial advisor to the secretary of state for the colonies and a former financial secretary of the government of Sudan, remarked during a brief tour of Northern Rhodesia that "it is astounding to find each little Government in each of these detached countries working out, on its own, problems which are common to all, without any knowledge of what its neighbors are doing and without any direction on main lines of policy from the Colonial Office."[2] Among those concerned with colonial affairs, Schuster's critique of the lack of imperial coordination was widely shared. Since at least 1910, a small but influential group of men known as "Milner's Kindergarten," named for their role as civil servants in South Africa under the High Commissioner Lord Milner (1897–1905), had been actively campaigning to promote more systematic studies of the empire's constituent parts.[3] One of their first efforts was to found a new British journal, the *Round Table*, which served as a repository of imperial information. Philip Kerr (later Lord Lothian) observed in the inaugural editorial, "No one can travel through the Empire without being profoundly impressed by the ignorance which prevails in every part, not only with the affairs of the other parts, but about the fortunes of the whole."[4]

Following the 1919 Paris Peace Conference, Kerr's closest collaborator, Lionel Curtis, and others from Milner's circle established a new institution in London, the Royal Institute of International Affairs, to address the "lack of adequate knowledge" in world politics, "the remedy for which is continuous and organised study."[5] By 1924 affiliates of Chatham House (as it was known colloquially, after the stately building in St. James's Square that serves as its headquarters), had created a special "Africa group" for the "disinterested study of African affairs."[6] For at least two of its members, Joseph Oldham and Frederick Lugard, these efforts did not go far enough.[7] They were "dissatisf[ied] with the prevailing order of things" in colonial administration, and particularly with the "negative policy of drift" they perceived concerning the "burning problems in Africa at the time."[8] By 1925 they were circulating a proposal for another institution to "promote a better understanding of the distinctive character and contribution of African peoples."[9] The aim of this new body, the International Institute of African Languages and Cultures, was to "bring scientific knowledge and research to the solution of the practical problems which presented themselves in" any effort to administer "primitive races."[10]

Maintaining the British Empire and comprehending its different elements was, these men realized, far more difficult than extending its reach had been. Whether they cared more about Africa or about empire, they agreed that the production and dissemination of knowledge held the key to effective governance. Early in 1929, several of these individuals began to circulate plans for another think tank, based in Oxford and connected with the Rhodes Trust, of which Philip Kerr was now secretary. The proposed "Institute of Government" would undertake a "systematic and scientific study" of the British Empire, "the most interesting experiment in government in the history of the world."[11] In their view, the empire provided a "unique laboratory" in which to examine problems relating to "democratic states and their international association," in conjunction with "the good government of the coloured peoples and the harmony of their relations with the white peoples." Unfortunately, there was still "no adequate equipment for the study of those matters anywhere in Britain. . . . Nor was any adequate attempt being made to compare British methods with those of other colonial administrations."[12] An Institute of Government established within Rhodes House, the trust's new headquarters, would help to remedy this tendency toward imperial neglect. Its first task would be "to explore the whole problem of Africa."[13]

A Project of "Exceptional Fertility"

Frederick Cooper has argued that "post–[World War II] imperialism was the imperialism of knowledge."[14] D. A. Low and John Lonsdale have characterized the experience in Africa at this time as "the second colonial occupation" during which "technical experts" armed with new funds for "development" virtually invaded the territories and usurped the role of emerging African professional classes.[15] The implication of each claim is that knowledge and expertise became prime vehicles through which colonial power was exercised. To trace the roots of these phenomena, and particularly of the ways in which science and development became so intertwined in British Africa, we must examine the history of the African Research Survey, the most important intelligence-gathering project of the interwar period.

The conception for the African Survey emerged from a series of conferences held at Oxford University in the autumn of 1929 to coincide with the Rhodes Memorial Lectures delivered by General Jan Smuts and intended to provide definition to the proposed Institute of Government. Ultimately, this gathering broadened the institute's mission beyond political science to other fields of knowledge, including the natural, medical, and human sciences. It also brought to the fore questions about the nature of economic versus social development in British Africa. Smuts, an Afrikaner educated at Cambridge, not only was a leading statesman, having served as prime minister of the Union of South Africa and aided in the formation of the League of Nations, but also played an active role in scientific societies, having participated in botanic expeditions, been elected president of the South African Association for the Advancement of Science in 1925, and later served as the British Association's president during its centenary in 1931.[16] Smuts carried South African preoccupations to Britain during his term at Oxford, echoing his compatriot Jan Hofmeyr's laboratory analogy during his first lecture on European settlement in tropical Africa's highlands.[17] The BBC broadcast his lectures nationwide, and the press devoted considerable attention to them. While the public debated his views, a group of thirty-one men—national politicians, Oxford dons, natural scientists, and anthropologists—gathered behind closed doors to discuss colonial Africa's future. Like the gathering of geographers convened by King Léopold in 1876, the 1929 conference had ripple effects across the African continent for decades to come.

The African Survey was planned and directed by men who were prominent in a number of fields: in African affairs, Frederick Lugard and Joseph Oldham; in imperial affairs, Malcolm Hailey, Reginald Coupland, William Ormsby-Gore, and Malcolm MacDonald;[18] in international politics, Philip Kerr (who in 1931 succeeded to the title of Lord Lothian), Lionel Curtis, and Arthur Salter; and in science, Richard Gregory, Julian Huxley, John Boyd Orr, and Bronislaw Malinowski.[19] This group was obsessed with gathering scientific and other empirical data on the African empire. Studying the continent in comparative terms, they argued, would enable colonial states to pursue "African development" less erratically and more successfully. Lord Lothian, who would eventually chair the African Survey's general and executive committees, emphasized in 1931 the project's "exceptional fertility."

> I don't think that any concrete scheme has ever yet been put forward for preparing a survey of a great human problem by bringing under review through qualified expert persons the four main fields of knowledge, namely political science, natural science, economics, and anthropology, using these words in their widest significance. Hitherto, research has been to a considerable extent conducted in water-tight compartments leaving the synthesis to be effected by individual writers. . . . The essence of the African proposal is to experiment with a new and more comprehensive form of research than, I think, has ever previously been attempted.[20]

Lothian was correct to see the African Survey as an unparalleled undertaking: the entire continent became the object of transdisciplinary study.

Proposals for the African Survey changed as its proponents sought an institutional location and financial support. Throughout these transitions, however, the project retained its focus on scientific expertise and its intersections with African development. The idea began as an Institute of Government at Oxford. By early 1930, its promoters had begun to describe it as an Institute of African Studies. In mid-1931, after it became clear that neither British nor American philanthropies would provide funding for the full project, it was scaled down to a "Survey of African Research." At that point, the Royal Institute of International Affairs at Chatham House offered to serve as its home and the Carnegie Corporation provided a $75,000 grant for the first phase of the project. The organizers hoped that once the Survey issued its reports, they could attract more substantial funds to found a Bureau for African Research. Indeed, the African Survey was a linchpin in the process that led to the passage of the 1940 Colonial Development and Welfare Act, which not only shifted British policy away from the expectation that each colony ought to generate its own revenue but also stressed the role of research and scientific expertise as necessary components of any development plan. The African Survey, in this sense, played a central role in consolidating British Africa's status as a development laboratory.

The term *development* has had myriad definitions.[21] Its use between 1860 and 1960, while typically associated with harnessing nature's bounties and generating wealth, was also embedded in more complex cultural and moral meanings. Trusteeship, the dual mandate, and indirect rule—the catchwords of the first half of the twentieth century—were intimately associated with nineteenth-century ideas of development and harbored within them vestiges of an older notion of a European civilizing mission.[22] There were, in fact, clear continuities between the colonial secretary, Joseph Chamberlain's turn-of-the-century imperative to develop Britain's "Great Estates," and Frederick Lugard's popularization of the "Dual Mandate" in which Britain was positioned as trustee of the people and of the resources of its African empire.

Underlying these continuities, however, were important transformations both in development policy and ideology. These were actually reflected in the African Survey's use of a "living laboratory" motif to describe the continent's future possibilities. While, on the one hand, the phrase signaled the immense scope for studying social and natural phenomena *in action*, on the other hand, it also anticipated the moral implications and consequences of colonial interventions. The leadership of the African Survey was increasingly aware of the tenuous nature of their nation's responsibilities. Speaking in metaphorical terms, where there was life there was also uncertainty and risk. "[H]istory," the Survey's executive committee concluded in 1938, "looking back in retrospect on the part played by Imperial Powers in Africa, will be more concerned with the nature of the contribution which the European occupation will have made to the future of the African peoples, than with the profit or loss which the African connexion may have brought to Europe."[23] An emphasis only on economic factors in development would overlook the human factor.

The Survey produced three official publications amounting to over three thousand pages of text: *An African Survey*, credited to Lord Hailey; *Science in Africa*, by Edgar Barton Worthington; and *Capital Investment in*

Africa, by Sally Herbert Frankel, all published by Oxford University Press in late 1938.[24] Natural science, medicine, and anthropology were central to the project from its origins in 1929–32 through its execution in 1933–39. Indeed, scientific research far outstripped all other areas of interest to the African Survey, including economics and native administration: it warranted its own subcommittee, was allocated the largest share of the budget, had at least three times the number of advisors as other matters, and had the longest-serving coordinator, E. B. Worthington. By contrast, Frankel's volume, Robert Kuczynski's book on *The Cameroons and Togoland: A Demographic Study*, and Margery Perham's analysis, *Native Administration in Nigeria*, which were initially conceived as preparatory studies for Lord Hailey, were more conventional academic monographs, undertaken with little guidance or assistance from the committee.[25] Only Worthington's volume represented commissioned research. *Science in Africa* was unprecedented. At the inauguration of the study in 1933, the coordinators acknowledged that the "scientific side of the Survey constituted a peculiar problem" because no other field would require "preliminary research on this scale." This inquiry, they claimed, was "almost a separate task."[26]

Nearly three hundred scientists served as advisors during the 1930s, including many who had firsthand research experience in tropical Africa. These individuals spanned a wide range of disciplines: meteorology, geology, soil science, botany, forestry, zoology, human and veterinary medicine, nutrition, psychology, and anthropology. Although several historians have discussed the African Survey in relation to colonial development and native administration, few have considered its scientific dimensions, despite the fact that one of its official volumes was an encyclopedic summary of environmental, medical, and anthropological research.[27] The Survey's three greatest legacies were also directly related to scientific knowledge: the establishment of the Colonial Research Fund in 1940 (and the Colonial Research Committee in 1942), which dramatically increased British support for research both in and on its colonies;[28] the creation of a set of inter-territorial research institutes in British East Africa following the Second World War, all of which were supported by colonial development funds;[29] and the founding in 1950 of the Scientific Council for Africa South of the Sahara, an inter-imperial coordinating body led by scientific experts.[30] These legacies are all the more noteworthy when we recall that the project began as an independent effort, initiated outside formal government circles, and was received in its early years with indifference and hostility by Britain's Colonial Office.

In order to understand the Survey's significance it must be evaluated not just in terms of its long-term consequences but also in terms of its original aims. This approach allows us to appreciate key dimensions of the project that have hitherto been examined only in a piecemeal fashion, if at all: its interest in ecological science, soil fertility, "native" agriculture, infectious diseases, rural health care, demographic patterns, eugenics and intelligence testing, witchcraft legislation, and social anthropology. In the late 1920s, at the Survey's genesis, the question of further European settlement in colonial Africa was still open for debate. By the time it issued its reports, in 1938, the Survey's authors no longer considered this question relevant for British officials. The project traced first the rise, and then fall, of technical departments in the territories in the wake of the world economic depression and lamented the effects retrenchment had on the pursuit of original research. Its advisors advocated a shift away from an ideology of economic self-sufficiency in the colonies toward one of human welfare and external aid. Finally, they recognized the need to analyze past failures, accept the complexity of problems, and pursue multifaceted and comparative research. The late 1930s did indeed seem to the coordinators of the African Survey as "the most formative period of African history." "The task of guiding the social and material development of Africa," in their eyes, "gives rise to problems which cannot be solved by the application of routine knowledge; they require a special knowledge, which can only be gained by an intensive study of the unusual conditions."[31]

Over the course of the 1930s the African Survey moved from the margins of the British government to its highest echelons of power, ultimately superseding the advisory capacity of the Colonial Office. Two of its earliest leaders, William Ormsby-Gore and Malcolm MacDonald, were appointed colonial secretary. By 1935, the Survey had become Britain's gatekeeper for African research; by the late 1930s, it was helping to transform colonial policies. When the results of the Survey were published, the Colonial Office and the African governments distributed over 2,000 copies to their respective territories; 5,500 more went to individuals in Britain and to the French, Belgian, and South African governments. By comparison, Lord Lugard's *The Dual Mandate in Tropical Africa*, into its fourth edition in 1937, had sold only 2,242 copies.[32] In 1957, the president of the Carnegie Corporation wrote, "If at any time in the past twenty years you had asked an expert on Africa what you should read as a serious introduction to that continent, he probably would have placed *An African Survey* by Lord Hailey at the top of the list. . . . It has been *the* standard reference work on Africa—a comprehensive, scholarly, readable, 'one volume education on Africa.'"[33]

The African Survey's controversial origins and its official triumphs afford a lens through which to observe the internal tensions and fault lines of Britain's African empire. Its history provides a microcosmic view of the political, moral, and epistemological debates of the period. The project linked the fortunes of research in specific African territories with wider metropolitan and transnational trends. A closer examination of the Survey's genesis and behind-the-scenes debates reveals not just how it achieved widespread influence with the British government and its colonial dependencies, but also how it went from a project involving only a handful of investigators to one that incorporated the research and views of hundreds. Unless its inception is understood and unless the sheer collectivity of the Survey is made visible, it becomes exceedingly difficult to explain why, as John Hargreaves asserts, the final published *Survey* "not only provided signposts and stimuli to enquiry and problem-oriented research in all sorts of directions, but exercised a substantial liberalizing effect on official policy."[34] As the following sections reveal, the African Survey's emphasis on science and its desire for colonial reform were the common threads that held the project together during its many transformations. To appreciate this point, we ought to turn to its genesis in late 1929.

Letting "Science Speak": The Conference on African Problems

On the afternoon of November 9, 1929, a veritable who's who of Britain's imperial and academic elite met at Oxford University for the start of two days of private discussions about colonial Africa. The gathering followed the second of three public lectures delivered by Jan Smuts and convened at the recently opened Rhodes House, a building the trustees hoped would become a beacon for imperial studies. In his opening remarks, the historian H. A. L. Fisher called on the participants to bridge expansive visions with practical results. The occasion provided a unique opportunity "for considering what help Oxford might give to the solving of scientific, economic, political, ethnological, and other problems of Africa—a continent which as General Smuts has said, must be regarded as a unity. . . . The time has come to substitute fundamental thinking for aimlessness and drift in the management of the Empire."[35]

The prime movers behind the gathering were four individuals: Philip Kerr, acting as secretary of the Rhodes Trust and one of the principal architects of the proposed Institute of Government; Fisher himself, the warden of New College, a Rhodes trustee, and an advocate of educational reform; the imperial historian, Reginald Coupland, who helped found, with Lionel Curtis, Oxford's Raleigh Club, an affiliate of the *Round Table*; and the missionary-statesman, Joseph Oldham, who had spent several years advocating for scientific research in colonial Africa and had recently served with George Schuster on the Hilton Young Commission for Closer Union in East Africa. Fisher and Coupland secured the participation of five heads of Oxford's colleges, four professors in the natural sciences, and four professors in the social sciences, as well as the directors of the Imperial Forestry Institute (Robert Troup) and the Agricultural Economics Research Institute (Charles Orwin).[36] Oldham's connections to British Africa and the International Institute of African Languages and Cultures drew in the German linguist and ethnologist, Dietrich Westermann, and the commissioner of labor for Tanganyika, Major Orde-Browne. The Manchester economist, Henry Clay, who had recently returned from the British Association's gathering in South Africa, also attended.

Among the best-known politicians who participated in the meeting was Leopold Amery, a conservative member of Parliament who had that June completed a five-year term as Secretary of State for the Colonies and Dominions. Amery had overseen the creation of the Civil Research Committee (CRC) and the Empire Marketing Board (EMB) and had been a driving force behind the passage of the 1929 Colonial Development Act (CDA) (table 2.1).[37] He had a long-standing friendship with Smuts, had been closely associated with Milner's Kindergarten and the *Round Table*, and was a Rhodes trustee. Along with his undersecretary of state, William Ormsby-Gore (who had been invited, but was unable to attend the conference), Amery ardently advocated that scientific insights be brought to bear on colonial advancement. "The development of the empire," he declared in a speech in 1925, "was not only a great political and administrative problem, but a great economic problem. It was also, no less, a great scientific problem—a problem of applying scientific research to the practical task of making the most of our immense resources."[38] For Amery, as for many of Britain's educated elite, development was then synonymous with the task of exploiting natural resources.

To address matters of administration and finance, the conference organizers invited the secretaries of the Empire Marketing Board, Stephen Tallents, who was authorized to make direct grants for research, and of the Civil Research Committee, Thomas Jones, who helped to coordinate small-scale scientific projects in metropolitan and African centers. Both men had close connections to the Colonial Development Advisory Committee, established that summer, which would oversee new grants to British Africa. Along with Amery and Smuts, they brought questions of science and development to the fore. Tallents, who spoke first, stressed the connections.

Table 2.1 Key events in Imperial and African affairs during the 1920s

1920	British (later Royal) Institute of International Affairs founded
	League of Nations Permanent Mandates Commission approved
1922	Egypt declared by British Parliament to be a "sovereign state"
	The Dual Mandate in British Tropical Africa published by Sir Frederick Lugard
	Education in Africa published by the Phelps-Stokes Fund under director Thomas Jesse Jones
1923	Advisory Committee on Native Education in Tropical Africa (June) (expanded to all colonies in 1929)
	Kenya White Paper on the "paramountcy" of native interests issued (July)
	Southern Rhodesia granted "responsible" government (October)
1924	*Education in East Africa* published by Phelps-Stokes Fund under director Thomas Jesse Jones
	Northern Rhodesia comes under auspices of Colonial Office (April)
1925	Report of the East Africa Commission issued (April)
	Committee on Civil Research founded (June)
1926	International Institute of African Languages and Cultures founded
	Empire Marketing Board founded
1927	Colonial Office Conference, London
1928	*The Native Problem in Africa* published by Raymond Leslie Buell, Harvard University
1929	Report of the Royal Commission on Closer Union of Eastern and Central Africa issued (January)
	Passage of the Colonial Development Act (July)

"The economic demands of the world" were placing an "immense and increasing strain . . . upon Africa." To avert "disasters," he believed, "the best brains in the Empire and elsewhere, the best activities of White and Black in Africa, must be mobilised. . . . Great reserves of man power might be made available by training and by more effective prevention of epidemics and the scientific improvement of native diet. There was also a wide scope for the mechanising of agriculture." To develop any of these areas produced an "immense demand for scientific research and its application in the fields of agriculture, entomology, forestry, and irrigation." The Empire Marketing Board was doing much "in the encouragement of research and the collection of data," but more could be done.[39] Smuts, who spoke next, wholeheartedly agreed with this sentiment and declared that Rhodes House should become "*the* venue for coming to grips with Africa's problems." "A whole continent was lying at the feet of the British Empire"; the principal question to be addressed was how "African development" would proceed. Oxford's purpose

should be to get the African problem out of the political atmosphere and away from the sentimentalists, and to let science speak. . . . There must be a "General Staff" to think out African problems as one co-ordinated whole. . . . If Oxford would, as it were, turn from the Greeks to the Negroes, it would help Africa as nothing else could. If we applied the scientific method, if we obtained the sinews of war from the EMB, we might found an institution which might in fifty years be one of the greatest things in Oxford.[40]

Beyond the grand visions that inspired the meeting, participants disagreed about fundamental aspects of imperial policy. When they were originally planning the agenda, Oldham had suggested, and Kerr and Coupland had concurred, that their primary objective should be "getting Smuts to state his case" regarding Africa "and letting the Conference criticize it."[41] They were referring largely to Smuts's racial politics. Oldham's main point of contention with Smuts was the Afrikaner's pejorative attitude toward Africans. Oldham was inspired by Christian idealism, which compelled him "to stand unequivocally for justice and brotherhood in international and inter-racial relations."[42] Throughout the 1920s, Oldham strongly supported the rights of Africans and resisted the idea of white settler self-government in the Kenya Colony, a position he had been able to uphold while a member of the Hilton Young Commission.[43] After his first meeting with Smuts in 1926, Oldham had written with disappointment that "for [Smuts] the problems of South Africa were essentially those of the development of White civilization, in the working out of which the Native could, for practical purposes, largely be left out of account."[44] Following Smuts's lectures in Oxford, Oldham was even more disparaging: Smuts's "conception of the native expressed in private conversation and hinted at in his Rhodes Lecture [on African Settlement] is that he is, in Aristotle's sense, by nature a slave, and that his function is to be the servant of the white man."[45]

Conference participants considered the question of Africans' political and social standing through the debate then raging in Britain over the future of Kenya. For nearly two decades, Kenya had been a lightning rod of imperial controversy. Once Britain managed to secure responsibility for the German territory of Tanganyika following the Great War, what to do about the East African territories became a matter of heated struggles. New territories placed in sharp relief old questions regarding European settlement, voting rights, "race relations," and territorial governance. Kenya's future became a litmus test for Britain's African empire. If the settlers triumphed, critics believed that Kenya would follow the segregationist policies of South Africa and neglect the needs of Africans; if African interests were held to be "paramount," as a landmark 1923 government white paper declared, they hoped that Britain might fulfill its trustee obligations.[46] At the root of the controversy was an issue that surfaced again and again throughout the interwar period: the moral responsibilities implicit in trusteeship and often contradicted in practice by colonial administration.

To get to the heart of these issues, Coupland suggested a dinner debate that would allow conference participants to consider what was at stake.[47] Oldham made the case against settler self-governance in Kenya, and Amery spoke for those who were in favor of increasing settlers' numbers and power. At the time, Kenya's European population was still comparatively small, estimated at only 12,530. Amery believed Smuts's first lecture had offered "a very challenging affirmation of the white settlement doctrine" and asked the group to join him in promoting a "vision of 100,000 white people in Kenya."[48] Oldham argued that it was unlikely that the number of settlers could be increased by an order of magnitude in the foreseeable future; even if it were possible, the suggestion was dangerous because it would further polarize East African politics and reinforce Africans' subordinate position.

None of the organizers intended for the debate to resolve the question of white settlement or self-governance definitively; they sought only to air the key issues. Yet Oldham was convinced that the "Oxford representatives were all on my side. . . . Oxford has a mind of its own."[49] The Greek specialist Gilbert Murray, who was an active participant in the League of Nations Union and a longtime mentor and ally of Norman Leys (a leading critic of the Kenya government), decried "the economic exploitation of the helpless territories and nations by the strong ones."[50] Likewise, Coupland believed fervently that the British Empire's purpose was to benefit "natives" and not necessarily settlers. Oldham had invited Westermann and Orde-Browne partly because they could present a "different side of the picture to that painted by Smuts." Speaking about his experiences in Tanganyika, Orde-Browne tried "to introduce some facts into the discussion," to the amusement of some participants and the irritation of Amery, by observing that settlers' activities were often at odds with the needs of "the overwhelming majority of the inhabitants" in the region.[51]

In spite of Oldham's perception that he had bested Amery, Amery felt confident that he had many allies in the room and recorded in his diary that he thought he got "a good deal the best of the discussion."[52] Philip Kerr, for one, was sympathetic to Amery's position. Two years earlier he had written a long article on the African Highlands for the Rhodes trustees in which he explored connections between South African and Kenyan politics: "The idea of self-determination is in the air [for Kenya]. As soon as a moderate number of white people settle in any area it seems to be impossible to resist their demand for self-governing institutions." Yet, unlike

Amery or Smuts, Kerr tended to equivocate when it came to governance because, in his view, the integrity of the empire could only be maintained if
the British government retained direct control of its colonial dependencies.
"The foundations of [Britain's] authority as against local white opinion [in
Kenya] are fundamentally weak. Yet how can it surrender authority over an
area of 2,000 miles by 500 containing perhaps 10,000,000 natives to the
control of a few thousand white settlers? On the other hand, how is it to resist the demand?"[53] This last question had taken on vital importance in the
late 1920s as those for and against increased settler control ratcheted up
their lobbying efforts. As most of the conference participants would have
known, the prime minister and his cabinet would discuss the issue of governance structures in Kenya and Tanganyika the following week.

Controversies involving racial politics were compounded by disciplinary politics. While almost all the speakers over the course of the first two
days stressed the need for integrated research and "cross-linking" between
fields, a noticeable tension appeared between support for investigations
that dealt directly with natural resources and those that concentrated on
"the native as a human being." This friction was tied directly to ambiguities
in participants' understanding of development. Amery and Tallents tended
to think primarily in terms of economic development, which in their eyes
placed greatest emphasis on exploiting natural resources, improving Africans' qualities as laborers, and generating financial returns. Others, including Westermann, Oldham, Orde-Browne, and Murray, thought there was
a danger in limiting studies to resources alone since they tended to relegate to the background the social aspects of development. As Murray put
it, "over-rapid economic development" often had "devastating effects upon
primitive races." If Britain were serious about its role as colonial trustee,
Murray believed it was necessary to give "constant prominence to the non-
economic aspects of African questions."

Oldham and Westermann were among a handful of individuals spearheading the International Institute of African Languages and Cultures
(IIALC), founded just three years earlier. On the morning of the conference's second day, Westermann was invited to describe the IIALC's work
and chose to return the discussion to the "human element." The institute
existed, he said, "to study—in particular—the changes entailed for the native by the presence of the European." It was important to recognize that
"whatever might be [Europeans'] future policy, they would always in Africa have to rely on the work of the African." Taking Africans into account
strictly in economic terms "could not be in itself enough"; Africans had to
be considered in wider sociological terms. To this end, "the mental cul-

ture and the languages of the native" and their "inner life" would need to be better understood. Like Westermann, Oldham agreed that the "development of Africa's natural resources required Western science, capital and enterprise," but he admonished conference participants that we "have to make up our minds whether we are to regard the peoples of Africa as instruments of our advantage or as ends in themselves."[54] "The [investigative] work initiated by Mr. Amery and Mr. Ormsby-Gore had to be carried further. The sub-human factors, e.g. the insect, the bacillus, and the germ formed only one part of the problem. Disinterested study must equally be devoted to the human factor."

The representatives of the Colonial Office and governmental funding agencies took the opportunity to respond. Thomas Jones explained that the Civil Research Committee was primarily an advisory body; while it could shape research priorities, it had "neither money nor executive powers." It was the EMB that was "endowed with funds, from which investigation on the 'sub-human' level had been freely financed. Their slowness in taking up Mr. Oldham's suggestions had been due partly to the inherent difficulty of human studies, whether in Africa or England, and partly to the want of qualified researchers."[55] Alexander Lindsay, the Master of Balliol College, concurred with both points. It was difficult to pursue human studies in a "scientific spirit" because they invariably gave rise to "passionate controversy." "There was moreover a temptation for the specialist to give his mind a complete holiday when outside his field: to be altogether unscientific on economic and political questions and not wholly to withstand the improper influence of their economic and political ideas upon the production and presentation of their scientific results." The Civil Research Committee was not averse to encouraging the EMB to support studies that considered humans as biological agents in terms of nutrition or disease, as it did with the East African Dietetics Committee and the Tsetse Fly Sub-Committee, which had recently launched projects in Kenya and Tanganyika respectively.[56] The difficulties arose in investigations that considered humans as social and cultural agents, topics that fell, not to biology or medicine, but to sociology, psychology, and increasingly anthropology.

Oldham himself, ironically, had been initially mistrustful of anthropology, worrying that anthropologists were "more interested in theoretic aspects of the subject than its practical application."[57] Even when he testified in early 1927 to the Civil Research Committee in favor of a "Native Welfare Research Organization" for Kenya, he commented that he thought it was "doubtful whether anthropology could at present give as much assistance as it might." In his view, "anthropological science" needed to be brought "more into touch with the realities of administration and education."[58] When appointed to the Hilton Young Commission later that year, Oldham began collecting observations concerning the place of scientific research in various African administrations. Through this process he came to work more closely with Bronislaw Malinowski, then at the London School of Economics, who was already convinced of the need to make anthropology a science more relevant to colonial affairs.[59] Just as Philip Kerr and Lionel Curtis began to lobby colleagues on their proposed Institute of Government at Rhodes House in early 1929, Oldham and Malinowski were undertaking a concerted effort to focus the IIALC's research agenda toward "practical anthropology."[60] They justified this move on the grounds that it would enable policy makers to consider colonial interventions "from the native point of view, rather than [trying] to force the natives into a European mould."[61] Oldham hoped the new Oxford Institute would take a similar perspective.

Bringing Together the "Different Phases of Science"

Participants in the Oxford conferences developed their resolutions in two stages: when the first set was approved on the afternoon of November 10, they decided to continue the meeting for a third day, which meant that the second set of resolutions was approved only on the afternoon of November 16. At the close of the first two days, the conference called for "the establishment of a Centre of African studies in association with the University of Oxford." On the 16th, a committee was formed to undertake negotiations with the university, and a special "committee of the men of science" was charged to develop a "report on the researches which it would be desirable to undertake." Neither Westermann, who had to return to Berlin, nor Oldham, who had previous commitments, was able to attend this final day of meetings.[62] This meant, crucially, that key representatives of the IIALC did not play an active role in shaping the final proposals.

The last day of the Oxford "Conference on African Problems" followed Smuts's third lecture, which focused on "Native Policy." During the previous sessions, human, economic, and political questions had received the bulk of the attention. During the final session, the speakers turned to agriculture, botany, forestry, and veterinary and medical research, the subjects Oldham thought were sometimes given too much prominence. Even so, speaker after speaker declared that in their fields much more could be done. Oxford's professor of botany, Arthur Tansley, said that agricultural and botanical surveys were in their infancy, although "nowhere was there a wider field for such studies than in Africa." Robert Troup, director of the Imperial Forestry Institute, deemed "discreditable" that "while good work had recently been done by a Belgian and by two Americans, no comprehensive study of African vegetation had been made by any one from this country." Oxford itself appeared to be lagging behind other institutions, such as Cambridge, and displayed a "regrettable weakness . . . in some of [its] scientific departments." The physiologist J. S. Haldane lamented that "nowhere in England was there a better organized veterinary laboratory than at Onderstepoort [in South Africa], [or] at Johannesburg [where] there was an important institute for medical research. Oxford paid no attention at all to those questions" in the context of either human or animal health.[63]

Most of the natural scientists at the conference stressed the need for interdisciplinary approaches to topics they believed would become increasingly important with respect to African development. Specific problems such as desiccation (now often called desertification), soil erosion, diet, health, or land use, they argued, were "bound up" in wider social and political issues. Troup concluded: "As Mr. Amery has pointed out, the great need was for collaboration between scientists working on different phases of science; the study of vegetation alone, without parallel studies of an anthropological, geological, and by no means least important, a climatological character, would be of little profit. . . . Such problems [as desiccation] needed not only the forester, but the meteorologist, the anthropologist and the historian."[64] Reiterating William Ormsby-Gore's appeal "for more candidates [in the Colonial Service] with a good general biological training and a wide ecological outlook," Tansley stressed that he was trying to "reorientate the work of his Department as to provide the specialized training [they] would need." So impressed was Fisher with the discussion, which underscored "the lacunae in our African knowledge," that he encouraged Arthur Tansley to develop a committee "to systematize their desiderata and make a report."

The second series of resolutions called not only for the formation of an Institute of African Studies at Rhodes House, which would concentrate on "scientific, economic, human, and governmental" research, but also for funds to support "a permanent Director of African Studies." As Coupland had remarked during the previous session, "It was almost ridiculous that there was nobody in England who was giving his whole time to the study of these problems." A small interim committee was formed, and Kerr and Coupland were charged with writing up a general statement to be sent to the Rhodes trustees, the Empire Marketing Board, the Colonial Development Fund, the Colonial Office (for the Advisory Committee on Native Education), the prime minister (as the chair of the Civil Research Committee), and the Committee of the Privy Council for Industrial and Scientific Research. Although several of these institutions were potential sources of funding, Kerr and Coupland would also explore "financial backing in America."

As the conference came to an end, several participants expressed the conviction that a collaborative project of this sort would not only be important for Oxford but would also "prove of value to Africa."[65] The main challenge, Fisher thought, was "to find the best method of approaching the University," which might become "a rather delicate problem." Kerr, like Fisher, was concerned that the "tradition of individual work is so entrenched in Oxford that it is almost impossible to get into the heads of the older dons that there is scope for a certain amount of collective thinking about the subjects" in which they specialized.[66] Amery "felt sure that anything on the economic side or in the ecological field would gain the interest of the [Empire Marketing] Board." Tallents pointed to the support the EMB had recently provided to the London School of Economics when it established a professorship in imperial economic relations. Smuts offered the most optimistic closing remarks: "so good was the cause that he did not doubt the prospects of its success." Their consideration "of African problems was one of the most remarkable discussions in his experience." Combining the expertise of scientists and statesmen offered a unique perspective on Africa: "no such survey . . . had ever before been made."

"The Million Dollar Interlocking Scheme for African Research": The Transition from an Institute to a Multiyear Survey

Participants left the Oxford conference impressed and excited. Oldham called the weekend "one of the most interesting I have ever spent."[67] Westermann wrote that it had been "a great experience" and that it would be "a great pleasure to me to give any help in the formation of such an Institute."[68] Coupland and Kerr seized upon what they believed was the driving vision: to create "a 'Fellowship' of all men *co-operating in building up a new body of knowledge* and inspiring research in all directions."[69] Thanks to the conference's geographical emphasis and the multiple fields represented, what had been conceived as an Institute of Government Studies was quickly being transformed into a transdisciplinary Institute of African Studies. As Oldham put it to Lord Lugard, "The main object was to discuss the establishment at Oxford of a centre of African studies which the people connected with Rhodes House have had in mind for some time. Smuts has given it a push."[70]

Between December 1929 and April 1930, the contours of the Oxford project were fleshed out as advocates pursued institutional and financial support. In late December, Lothian and Smuts journeyed to the United States (where Smuts attended the annual meeting of the League of Nations Union) to meet with the Rockefeller Foundation staff to see if they could secure the necessary funds. Upon his return, Lothian, with Coupland, prepared a follow-up memorandum to conference participants and wrote the proposals and resolutions required to secure full university sponsorship. Meanwhile, Tansley convened a meeting of the scientific committee to consolidate its statement on the role and function of the natural and human sciences in the proposed center. Equally significant for the future of the project, four individuals who had missed the conference but who shared a keen interest in Africa—Lord Lugard, Lionel Curtis, Julian Huxley, and Bronislaw Malinowski—became actively involved in the planning. Even before the science committee presented its conclusions, Huxley, an evolutionary biologist at Kings College, London, eagerly put forward his own proposal to establish a Department of Ecology in the institute. As Coupland put it to Lothian, Huxley "is immensely bitten with Africa and enthusiastic about our scheme."[71]

Most of the anthropologists and environmental scientists in Oxford were already involved with imperial affairs. In 1924, the university had been selected as the site for a new Imperial Forestry Institute, which by 1929 was coordinating research with field officers in West Africa. From 1926 on, Troup, Tansley, and the anthropologist Robert Marett had been contributing to the Colonial Office's annual training programs for officers who enlisted in the Colonial Service. Several of these men had recently taken part in the British Association's annual gathering in South Africa, which intensified their desire to offer their expertise to the British government and its African territories.

When Tansley's scientific committee issued its report, its members had already reviewed existing efforts, both official and unofficial, and found them to be fragmented and inadequate.[72] "The various problems of Africa are closely interwoven, and . . . they all in common demand a series of general surveys as a basis for special work."[73] They envisaged the new bureau providing "some central and continuous policy of research" aimed at a "complete or synthesized survey of the field." In their view, a comprehensive overview was a matter of "urgent importance." It was essential to move beyond narrow "specialist Departments" and coordinate data, not only through technical measures such as the compilation of bibliographies and new research monographs but also through cross-disciplinary approaches and "the study of problems which can only be investigated on the spot." A concerted effort, they believed, could provide a new approach to the many unanswered questions regarding the continent, including "the nature and distribution of African soils," "the type of agriculture carried out by individual native tribes," "the five to fifty year pulsations of the African climate," "the question of desiccation in certain parts of Africa," "the botanical composition of the forests of Tropical Africa," "the density and distribution of population," "the importance of environmental influences . . . [on] human migration," "medical ecology in relation to disease," and "the economic . . . religious . . . [and] social organization of the different peoples." Answering these enormously intricate questions would require extensive, long-term research. Yet, so central were these problems to development efforts that significant returns would come from studying them "on a broadly ecological basis."

Julian Huxley made many of these same points in the proposal he submitted to Fisher and Coupland shortly after he returned from East Africa. His tour was sponsored by Britain's Committee on Native Education in Tropical Africa, which asked him to consider how biological science and knowledge of the natural world might be integrated into general educational efforts.[74] Huxley suggested that the proposed African studies center adopt an ecological framework. "At the present moment, it is clear that many if not most problems of applied biology can only be satisfactorily solved by reference to a background of ecological ideas, by whose aid the interrelations of different branches of biological science can be studied." Several specialist institutions dealing with the biosciences already existed in Britain and Africa; "what is needed is rather a coordinating centre which can both receive help from and give help to such bodies. . . . General ecology, in the sense of a broad correlating science, should be its main aim; only so will it yield its maximum return."[75] Like the members of the science committee, Huxley thought it important to avoid duplicating current efforts. Indeed, he believed that existing departmental and disciplinary divisions got in the way of dealing with the larger questions relating to "African ecology." Huxley was actively promoting a relatively new field, ecological science, in much the same way that Malinowski was promoting "social anthropology" within the International Institute on African Languages and Cultures (IIALC). Huxley claimed that "it is often possible for the ecologist to point out to this or that specialist new lines of approach to his particular

problem—disease of man or of domestic animals may prove to be correlated with a cycle of abundance and scarcity in some wild animal . . . game
migrations or . . . climatic cycles or variations in mineral content of food-
plants." Achieving effective results, Huxley insisted, depended on a "close
liaison between the Department of Ecology and any Anthropological work
prosecuted in the School of African Studies, and with medical work bearing on Africa." This triumvirate—ecology, anthropology, and medicine—
was central to colonial Africa's economic and social development. Lothian
was in full agreement. Bringing Lionel Curtis up to date on these events, he
concluded, "I don't think there is any doubt that if we can get an Institute
of this kind really going in Oxford with four or five first-class men at its
head, it would . . . revolutionise the future of Africa."[76]

During the winter of 1930, Lothian undertook liaison work with other
institutions across Britain. In order to attract substantial funds from an
American philanthropy, the Oxford center would have to establish "har-
monious relations" with all organizations "operating in the African
field."[77] Oldham warned Lothian about the challenges he might encounter
on this front: "One problem that will need a good deal of thought is the
relation between the proposed institute and the International Institute of
African Languages and Cultures." Oldham had left the Oxford conference
enthusiastic about the general outlines of the project, but in the intervening months he had left the planning to others while he wrote a "critical examination" of Smuts's Rhodes lectures, focusing on the questions involved
in the governance of East Africa, which was being considered by a parliamentary committee.[78] Only after he received Lothian's revised proposal did
he begin to realize that the new center's embrace of anthropology might
be regarded as threatening.[79] The IIALC was still in its own developmen-
tal phase and had only recently begun to receive small contributions from
"the British colonies in Africa, Southern Rhodesia and also the French and
Belgian Colonies." Oldham worried "that the International Institute may
feel that the new proposal rather cuts the ground from under their feet."
He promised to help in whatever way he could to alleviate tensions, "but
they are real and will need careful attention."[80]

Both groups were planning to submit grant proposals to the Rockefeller
Foundation, so Lothian and Oldham recognized that the best solution
would be to develop complementary research programs. To do this, Lothian had to win over Lord Lugard and Bronislaw Malinowski. Lugard had
spent that autumn in Geneva, where he served as Britain's representative
on the League of Nations Mandates Commission, and upon his return was

briefed by Oldham about the Oxford project. As chair of the IIALC's Executive Council, he was keen to enter into the discussions, but was concerned about overlap. Somewhat precipitously he decided to write to Sir Samuel Wilson in the Colonial Office, asking him to ensure that Lord Passfield, the colonial secretary, "not commit himself [to Oxford] too far for the moment." He justified a cautious approach on the grounds that the proposed "Bureau of African Studies" would "undertake practically identical work" as the IIALC, although he added that there was "no suggestion of rivalry or hostility" between the two groups.[81]

Lord Passfield and his staff at the Colonial Office, who learned of the proposal in January, were ambivalent about it right from the start. They were concerned that some of the project's promoters, such as Smuts and Curtis, were reputed to hold positions on African affairs that were overly favorable to European settlers and overly critical of Britain's imperial administration. Poor communication allowed unfounded rumors and misinformation about the Oxford project to circulate in the office without correction, creating an atmosphere of suspicion and distrust regarding the institute's ultimate objectives even before its full scope had been clearly defined. After a meeting with Lothian in early March, Passfield reported, "He did not ask me for anything beyond my 'moral support', or at least my refraining from opposition. I made no promises either way."[82]

Just a few weeks after sending the cautionary letter to the Colonial Office, Lugard became convinced of the need for two research organizations focused on Africa: one with an international leadership structure, the IIALC, and the other under British auspices at Rhodes House. At a planning meeting Lugard hosted at his home in early March, Oldham, Lothian, Malinowski, and Westermann clarified the division of labor between the two groups. Afterward, Lugard expressed his understanding in a memorandum: "The Rhodes Institute at Oxford backed by men of great academic weight and interlocked with our Institute must exert a powerful influence on the training of cadets [bound for the colonial service], on the standing of Anthropology in the University, and on the attitude of Government to African problems." If the two centers coordinated research, Oxford could do what the London-based IIALC was constitutionally barred from doing: directly influence the government. Drawing on data that the IIALC would gather from across the different African territories, the Oxford institute could formulate informed policy recommendations that "cannot be ignored."[83] Later that month the two groups submitted applications to the Rockefeller Foundation for £10,000 per year for ten years. Among a small cohort of supporters, which included Lugard, Oldham, Malinowski, Lothian, Coupland, Curtis, Fisher, and Huxley, this joint effort became known as the "Million Dollar Interlocking Scheme for African Research."[84]

Once the IIALC and the proponents of the Oxford project had worked out a synergistic relationship between the two initiatives, they pursed major funding enthusiastically. The historian George Stocking has characterized the negotiations between the IIALC and the proposed Institute of African Studies as a matter of expediency, arguing that the IIALC felt "forced to coordinate planning" given the high profile of the Rhodes House initiative, but maintaining that they privately "made it clear [to the Rockefeller Foundation] that their own plans had priority."[85] While the Oxford proposal quickened the pace of the IIALC's request to the foundation, once the fear of competition had been overcome the IIALC's endorsement of Oxford was "genuine," even adamant. The IIALC's main goal was to promote research and fieldwork in the human sciences. While its definition of anthropology was quite broad, it did not concern itself with other disciplines. The research envisaged by those promoting the institute at Oxford was comprehensive: it would undertake a "survey of the problem of Africa as a whole," which might "permanently change the thought of the world about Africa." It would shed light on the "different and often conflicting policies [being] adopted in the different parts of Africa" and help lay a foundation for the European powers "to formulate a common policy."[86] The Oxford institute would have an anthropologist among its four specialists, but it would also involve research in areas that had been identified by Tansley's committee of scientists as well as by Julian Huxley.[87] These documents emphasized ecological science and environmental disciplines: forestry, agriculture, entomology, geography, and botany, as well as "medical ecology in relation to disease."[88]

As negotiations with the Rockefeller Foundation unfolded over a following year, until a formal decision was made in April 1931, the two groups, led by Oldham, Lugard, and Malinowski for the IIALC and by Lothian and Coupland for Rhodes House, shared information regarding the foundation's staff and its funding priorities. At crucial moments when rivalry or competition might have risen to the surface, the evidence suggests instead a sincere effort at mutual assistance and fair play. For example, John Van Sickle of Rockefeller asked Malinowski to comment in confidence on the quality of the Oxford proposal (he offered a positive assessment), and Oldham, Coupland, and Lothian discussed the features of the Rhodes House institute that they needed to emphasize in order to gain the foundation's favor.[89] Oldham and Lothian worked together especially closely.

Why then did the Rockefeller Foundation decide to fund one project, the IIALC, and not the other? Why did the promoters of the Oxford institute not pursue funding more actively in Britain? What was at the root of the Colonial Office's enduring opposition to the project, and what light does its stance shed on British imperial dynamics with respect to scientific knowledge? To find answers, we must delve into the negotiations and explore both the organizational landscape in Britain and the sequence of events that led the advocates at Oxford to shift from the idea of a permanent institute in favor of a two- or three-year survey. The institute's redirection had as much to do with "watertight" compartments of knowledge as it did with institutional and personal politics.

"There Is Not One Science for Africa and Another for Asia"

One of the dimensions of the Oxford initiative that appealed to its supporters but unnerved the staff at the Colonial Office was the university's independence from government. The academic organizers viewed their project as a way to push Oxford to develop along the lines of research universities in Germany and the United States, which allowed scholars to address current affairs and influence official policies but remain to a considerable extent free from government intervention. During a trip to Germany just before the Oxford conference, Lothian summarized their approach approvingly: "When you find one or more people who clearly have zeal and ability to research in a field, give them money and buildings and then give them freedom."[90] As Lionel Curtis worked to reform research at Oxford, he became quite pointed in his criticisms: "a University like Harvard has money to . . . [send] out men like Buell to Africa . . . whose results, sound or otherwise, are in fact influencing public policy; Oxford is doing nothing of the kind."[91]

Raymond Buell was actually one of the few political scientists in the United States who had devoted considerable attention to colonial Africa. His two-volume study, *The Native Problem in Africa*, was widely acclaimed; Oldham praised it as the "most comprehensive and best documented study of the subject yet published."[92] As Curtis's comments suggest, the prestige and quality of the work reflected poorly on Britain's own research institutions and served as an important stimulus to the reform efforts at Oxford. Buell's work was sponsored by the Bureau of International Research at Harvard, which was funded by the Rockefeller Foundation and in many respects served as a template for the early proposals for Oxford's Institute of Government.[93] Curtis and others were convinced that Oxford had the potential to play a significant role in world affairs, yet were concerned that it was surrendering that role to other universities by failing to equip itself adequately. Using dramatic rhetoric, Curtis warned that the course of affairs in the British Commonwealth might be shaped more by experts from other nations than by British researchers and policy makers.[94]

This desire for political relevance sparked anxiety and concern on the part of the Colonial Office when it first learned of the proposal for an Institute of African Studies. Oxford, along with Cambridge, trained cadets in the Colonial Service and was among a handful of locations where active recruitment took place. If an institute were established there that criticized imperial policies toward African peoples or attempted to influence cadets' outlook, it could "cause considerable difficulty."[95] Officials were particularly perturbed by a rumor that General Smuts or Lionel Curtis would be its first director. Even after this rumor was dispelled, the Colonial Office seemed wary of the project's ulterior motives and political biases.[96] "I am . . . quite certain that an active Institute meddling in politics would prove a thorn in the side of the CO," wrote one official. "Lionel Curtis, for example, is no friend to us and I can see endless trouble if he were associated with the body."[97] Another cautioned, "The whole scheme wants watching carefully, not to say suspiciously."[98] Some took a less hostile attitude: the recruiter Ralph Furse remarked that the Colonial Office should "be sympathetic, but watchful." The sentiment that the Oxford center would be "yet another Institute for Teaching Colonial Governments Their Business" prevailed well into 1933.[99]

Officials' other objections to the project proved equally enduring. They wondered whether the "various enquiries" proposed by the institute were really "necessary or desirable."[100] The fear of "meddling in politics" aside, staff at the Colonial Office remained skeptical that the institute would meet any genuine intellectual needs, even when the idea of a systematic survey was substituted for a permanent institute. Since "the ground has very largely been covered before,"[101] what new knowledge could the project generate? The assistant secretary of state for West Africa remarked that "apparently, they cannot get away from the idea of research as a sort of general thing altogether distinct from the particular subjects or Departments in which the research is to be carried out. The idea that one investigator with one or more assistants can really go into all these varied questions seems to me frankly absurd."[102]

The scientific dimensions of the proposal elicited one of the more virulent objections. R. V. Vernon, an assistant secretary of state, thought the project's emphasis on Africa was misguided and "unsound": "science has no geographical boundaries and there is not one science for Africa and another for Asia."[103] The Colonial Office staff believed it would be better to organize all research along imperial rather than African lines. The British government had been working to expand its scientific services across all of its colonial dependencies since 1919, when Lord Milner, then colonial secretary, had prompted a review of staffing in medicine, agriculture, and veterinary science. Concerns with the lack of well-trained specialists continued, and committees were appointed in 1925 and 1927 to address the agricultural and veterinary services' capacity to undertake research as well as the establishment of a Colonial Office scientific and research service.[104] Some of the committees' recommendations were similar to those put forward by the Oxford conference. They also faced similar bureaucratic and fiscal challenges: the British government tended to be a "reluctant patron" of scientific research.[105] Still, a number of steps were taken: the appointment of Medical and Agricultural Advisors to the Secretary of State for the Colonies (in 1927 and 1929, respectively), the formation of an Advisory Committee on Agriculture and Animal Health (in 1929), and increased coordination between the Colonial Office and the various institutions responsible for particular pieces of recruitment, training, and research, such as the London and Liverpool Schools of Tropical Medicine, the Imperial College of Tropical Agriculture in Trinidad (established in 1921), the Imperial Forestry Institute in Oxford, and the network of Imperial Agricultural Bureaus (founded in 1927).[106]

The Colonial Office staff was somewhat defensive about the Oxford institute's suggestions that official efforts were insufficient and that analysts outside imperial administration might be better able to devise strategies for scientific research. As assistant secretary of state R. V. Vernon put it: "Professor Julian Huxley's proposal for a Bureau of African Ecology at Oxford seems to me a typical example of the way this scheme is being rushed with various 'bright ideas' without looking round to see what the existing organisations are. It is perfectly obvious to me that if there is to be anything in the nature of a Bureau of Ecology, it ought not to limit its sphere to Africa [and] that it will have to be brought into close relation with the existing Agricultural Bureaux."[107] Vernon was unwilling to concede that Huxley and members of the Oxford science committee were well informed about the existing scientific infrastructure, both in Britain and in Africa. Huxley was likely more familiar with the strengths and weaknesses of these research efforts than many of the individuals in the Colonial Office. Acknowledging the need for a comparative approach, he urged that staff at his Bureau of African Ecology be "free to undertake research upon any general ecological question, without reference to its immediate application to African problems; for experience shows this is in the long run the surest way to achieve progress in fundamentals." The geographical emphasis was important, however, for specific locations posed particular scientific problems. Huxley argued that a multidisciplinary study of Africa would help illuminate the "ecological nexus" in which many problems were embedded. A coordinating bureau would not replace research on the ground, but would help make sense of it: "the *special* African problems" that needed resolution "can best be undertaken by men working in Africa."[108] Huxley made many of these points publicly in his testimony for the Joint Committee on Closer Union in East Africa: "Co-ordination is perhaps more needed in science and medicine. The multiplicity of independent departments—Medical, Agricultural, Game, Fisheries, Geological, Survey, Tsetse Research, Forestry, Veterinary, etc.—is so great, and yet the way in which their problems interlock is so intricate, that without some central organisation, considerable overlap, waste of energy, and lack of co-ordination is almost inevitable."[109] His memorandum concluded, "Close co-operation with any similar bodies at home which might be concerned with co-ordinating scientific ideas and policy for Africa and the Empire is also desirable."

The Colonial Office's apprehensions about the scope of the institute's work were exacerbated by the project's decentralized and shifting leadership. At various stages in the negotiations Lothian, Coupland, and Oldham all "spoke for" the Oxford project, often with different officials, and the information they provided was on occasion contradictory. The CO staff knew that "investigations into African problems cannot even get underway without the goodwill of Colonial Governments and therefore of the Colonial Office. But the . . . Oxford authorities were not alive to this obvious fact."[110] Several staff members pointed out that if the project wished to secure government funding, it should report its results directly to the prime minister and the secretary of state for the colonies.[111] Reading the proposal as a deliberate effort to sidestep the chain of command, Cecil Bottomley, assistant undersecretary for the African colonies, remarked: "Certainly not much tact, but the tactics are ingenious. The C[olonial] D[evelopment] F[und] C[ommittee], the Education Committee, and the EMB, can each, if care is not observed, make it difficult for the S[ecretary] of S[tate] to refuse Colonial contributions which, if asked for by a frontal approach, would I hope have been denied."[112]

Yet the Oxford project's initial letter to the CDFC in December 1929 had been gently rebuffed. The problem, according to Basil Blackett, the committee's secretary, lay in the disciplinary parameters of the Colonial Development Act. The scientific research that could be supported was defined in much narrower terms. Blackett wrote, "We are taking as wide a view as possible of the meaning of Colonial Development, so I do not wish to state at this stage that a grant for the purposes mentioned would be quite impossible. It may, however, be advantageous to you to know the difficulties at the outset."[113] Blackett's interpretation of the limits of his mandate was right on the money. He had, as yet, little authority to fund projects that were not framed explicitly in terms of economic development or natural resources.

As the secretary of a private philanthropy, Lothian was aware of the differences in scale between American and British funding patterns. Over the course of the 1920s, Rockefeller Foundation support had virtually transformed social science research in Britain. The London School of Economics was one of the primary beneficiaries: between 1923 and 1928 it had received £249,000 ($1,245,000) from the Laura Spelman Rockefeller Memorial Fund.[114] In 1929, the two branches of the Rockefeller family funds were amalgamated, making it the largest single philanthropy in the world.[115] By 1931, the Rockefeller Foundation was allocating nearly £3.8 million annually (approximately $19 million) in international public health, the natural and social sciences, the humanities, and medical research.[116] By comparison, between 1929 and 1939, the Colonial Development Advisory Committee approved grants for research amounting to only £554,000, or roughly £55,000 ($277,000) per year.[117] The Empire Marketing Board was somewhat more generous, allocating £1.65 million in scientific research grants between 1926 and 1932, or roughly £236,000 ($1.2 million) per year.[118] Neither could come close to the general support grants being offered by U.S. philanthropies for higher education and scientific research. In both the United States and Britain, Rockefeller support proved pivotal to the direction in which research universities and institutes were developing.[119] The large sums of money at the Rockefeller Foundation's disposal and its emphasis on building "centres of research excellence" appealed to the advocates of both the IIALC and the Oxford institute. The economist Henry Clay, who had attended the Oxford conferences and later became a member of the African Survey's general committee, told Selskar Gunn, the foundation's senior staff person for European activities, that the dearth of funds in Britain would be an enduring obstacle to both projects. Gunn wrote in his report that "Clay hopes that the Foundation will aid the Oxford scheme. [He] doubts seriously if financial aid can be obtained in England at the present time . . . [and] is of the opinion that the Oxford Institute should be started on a basis of at least £10,000 a year."[120]

"Saving the Scientific Core of the Scheme"

When the Rockefeller Foundation's staff and trustees convened on April 15, 1931, the only funding requests relating directly to Africa came from the IIALC and the Oxford institute. After brief deliberations, the trustees decided to provide a grant of £50,000 ($250,000) to the IIALC and decline the request from Oxford.[121] Their official reasons for rejecting the proposal were institutional: they felt the university's research capacity was insufficient for the "specialized program" the institute proposed to undertake. They also expressed concern that "there has been no suggestion of financial support from any source other than the Rockefeller Foundation," which they thought could have damaging political ramifications if any controversies were to arise in Britain regarding the institute's activities.[122] The IIALC, by contrast, had already received preliminary support from the foundation and secured small contributions from various territorial governments. The trustees believed its proposed multiyear program of fieldwork in colonial Africa would produce new publications concerning Africa's peoples and train a new generation of specialists in anthropology. The London School of Economics, Malinowski's home institution, had already been building up its research program in the social sciences. Hearing news of the outcome, the Colonial Office declared that in the "competition" between the two groups, the IIALC "won."[123] The full story, however, is far more complicated and interesting.

Tracing the history of the Oxford institute's fortunes within the Rockefeller Foundation reveals an important reason for the proposal's failure that remained largely invisible to the staff: the project's transdisciplinary research program simply did not fit into the foundation's main areas of funding. Because the proposal seemed heavily oriented toward scientific research, it was first given to the director of the Division of the Natural Sciences, Max Mason, who after six months decided that "the problems presented do not fall within the present program" of his division.[124] It was then handed to the director of the Division of the Social Sciences, Edmund Day, who was "much more interested in the development of the Social Sciences at Oxford than in the African project."[125] When the Division of International Health was asked to weigh in, its staff acknowledged that the proposal contained interesting research measures related to the control of "the pests of Africa," but these elements were not enough to induce them to support the larger project.[126] Each division was concerned only with those elements of the proposal that fell within its mandate. The project's overall objectives seem to have been overlooked or undervalued. Echoing comments made by Day, Selskar Gunn concluded in his memorandum to the trustees that "it would be more significant to aid Oxford in . . . the development of its general program of research in the Social Sciences than in connection with the proposed Institute of African Studies."[127]

Well before the trustees made their formal decision, the proponents of the Oxford proposal were given indications of its general line of thinking. On hearing from Oldham that the foundation preferred social science research,[128] Curtis wrote in frustration to Lothian that it had been a "strategic blunder" to follow "Smuts' influence in narrowing the project from a School of Government to a School of African Affairs."[129] Lothian disagreed, contending that the emphasis on Africa and the "collective" method gave the plan much more focus than a nebulous project promoting the social sciences. "I am not sure that the Rockefeller people have ever really grasped what we were really driving at," he told Curtis.[130] When Lothian realized in late March that the proposal would almost certainly be rejected, he wrote directly to Edmund Day urging him "not to lose sight" of the "exceptional fertility" of the project.[131] Lothian's letter failed to convince the foundation to change its mind, but it did encourage them to "keep the door open" to future proposals concerning a new Institute of African Studies.[132]

Meanwhile, efforts to find an institutional home for the African research center were underway. Julian Huxley had valiantly pursued, as part of the Oxford institute, his proposal for a Bureau of African Ecology throughout 1930 and 1931.[133] One of his collaborators was a former student, Charles Elton, author of the landmark text, *Animal Ecology*, and a lecturer in Oxford's Department of Zoology. In March 1931, Coupland wrote to Lothian that he was working with Huxley and an Oxford agriculturalist, C. T. Morison, on "saving the scientific core of the scheme."[134] After the Rockefeller request had formally been denied, Huxley drafted a revised memorandum that was sent to the most active advocates of the Oxford institute, who were invited to "an informal meeting at Rhodes House . . . to discuss African Research."[135] Those who attended the meeting included Huxley, Coupland, Lothian, Oldham, Lugard, Curtis, Morison, Amery, Blackett, George Tomlinson of the Colonial Office, and Philip Mitchell, then secretary of native affairs in Tanganyika.[136] The discussions proved quite fruitful.

The group began by evaluating why the proposal to the Rockefeller Foundation had failed and decided that the best course of action would be for Oxford "to build up its Econ and Political Faculty without any particular reference to Africa."[137] They considered ways to begin the research work envisaged for the Oxford institute and arrived at the idea that it would be best to find someone to conduct a two- or three-year survey of "African research" culminating in a report that would identify "what is most worth doing and how it can best be done."[138] Their intent was to determine which topics should be given priority for research both in the United Kingdom and in African territories and how "a better correlation of existing research" could be achieved. The idea of an institute was not abandoned, but before its location could be determined, its functions would need more careful elaboration.

The discussion went far enough to include suggestions of personnel for the survey. The most prominent was Richard Feetham, a member of Milner's Kindergarten then serving as a judge in South Africa. Given the survey's strong emphasis on the natural and human sciences, the director would need a "scientific lieutenant," and Huxley proposed that Charles Elton be brought on to conduct a six-month "survey of the ecological field" in advance of Feetham.[139] No matter who was selected to lead the project, they believed that the

> person undertaking the enquiry would have in the first instance to make himself familiar with the present scientific organization of the Empire. It is in the field of medical, agricultural and veterinary research that most progress has been made in recent years, and it might be found, consequently, that the most urgent need is for corresponding advance in other fields. . . . It would be necessary to examine what is at present being done and what ought to be done in regard to the study of such questions as population, economic resources, communications, land, labour, anthropology (including native religion, law, economics, and social institutions) and administration.[140]

The discussion then turned to funding. A survey would take from two to three years and require at least £5,000 ($25,000) each year. The president of the Carnegie Corporation, Frederick Keppel, had planned a trip to the United Kingdom that spring to discuss how the corporation could best disburse grants through its "British Dominions and Colonies Fund," which had approximately £100,000 ($500,000) to allocate annually. Keppel was staying with Lionel Curtis that weekend at All Souls College, and Curtis and Oldham had planned a luncheon for him at Chatham House on May 21. According to Keppel, the "Trustees in New York had thought they would feel a little more comfortable about voting money from this Fund if from time to time they got in touch with a group of men in London who knew about the Empire as a whole."[141] The meeting in May 1931 was the first of a series of annual luncheons that Chatham House hosted throughout the decade to help Keppel guide the Carnegie Corporation's funding strategies.

At that meeting, Keppel was lobbied to provide the funds for the "African Survey of Research."[142]

Like the Rockefeller Foundation, the Carnegie philanthropy was established "to promote the advancement and diffusion of knowledge."[143] Although Carnegie made his fortune in the United States, he had been born and raised in Scotland, so he wanted to ensure that some of his profits went to Britain and its empire. The Carnegie philanthropy was much more flexible than the Rockefeller Foundation. Keppel, its president from 1923 to 1941, had considerable latitude in determining the corporation's priorities and was able to act quickly. The Carnegie endowment was smaller than the Rockefeller Foundation's, and the two philanthropies had different organizational cultures. For Rockefeller staff, issues of funding criteria and strategy were of great importance, while Keppel pursued a more entrepreneurial and eclectic approach.

When Keppel was asked to support a "preliminary survey" of "African research" in late spring 1931, he approved it almost at once. This project could be Carnegie's contribution to a more far-reaching Anglo-American strategy for tropical Africa.[144] His only concern was to ensure that the Rockefeller Foundation would accept a two-phase funding scheme in which Carnegie awarded the first grant and then the Rockefeller Foundation stepped in to fund whatever projects were recommended in the final report. After discussing the idea with three of the foundation's senior officials, Keppel felt assured that this plan was viable.[145] He then wrote to Lionel Curtis, his liaison for the project, that the trustees were prepared to consider the proposal. With the grant now all but guaranteed, in Keppel's view, the next task was to identify a suitable director.[146] By October 1931, Carnegie had formally approved a sum of £15,000 ($75,000) and the Royal Institute of International Affairs had agreed to be the project's fiscal sponsor.[147]

In November, Lothian invited those who formed the core of a committee to oversee the work of the survey's director to a luncheon at his residence in London.[148] Although no chair for this "informal committee" was appointed until the summer of 1933, Lothian tacitly assumed that role on the recommendation of Coupland and Oldham.[149] The transition from an Institute of African Studies to a "preliminary survey of African research" was complete. Although the contents of the survey continued to be debated between 1931 and 1936, its defining features remained largely unchanged. It would be a comparative study including not just British, but also Belgian, French, Portuguese, and South African territories. It would concentrate both on research and on "the machinery of knowledge," with considerable attention to empirical "fact-gathering." In addition to administrative, economic, and legal issues, it would emphasize the natural sciences and anthropology, using ecological science as a bridge between the human and nonhuman subjects. Such a synthesis, the organizers agreed, would provide the most useful analysis of the factors that affected colonial Africa's social and economic development. Finally, it would consider the issues on which a more permanent Institute of African Research should concentrate. Beyond this broad outline nothing more could be decided. As Keppel put it, "everything will depend primarily on the man who can be persuaded to take up the direction of the work."[150]

From the Margins to the Center: The Execution of the Survey of African Research

In searching for a director, the survey's committee had several criteria clearly in mind. The director's political reputation had to be beyond reproach in government circles; any doubts on this front might jeopardize the project's ability to gain access to sensitive information and secure the trust of colonial officials. The director would have to possess the intellectual credibility and authority for his final report to carry sufficient weight with policy makers and be able to distill insights from large bodies of knowledge in ways that were both accurate and compelling. He would need to demonstrate what they called "the necessary balance" between "sympathy for the under dog," Africans, and "understand[ing of] the white man's point of view as settler in Africa." Perhaps most startling in the context of colonial politics at the time, the committee agreed that he would have to be willing to see "self-government as the only possible goal, however remote . . . for the African people."[151] Prospective candidates were often evaluated in terms of their attitudes not only to the Union of South Africa and to Kenya, where social tensions were increasing, but also to India, which was seen as a political forerunner to what might be expected for British tropical Africa. Nearly two years passed and more than thirty suggested candidates were vetted before Malcolm Hailey agreed to direct the project.[152] A colonial administrator who had spent more than twenty years in India, eventually as governor of the United Provinces, Hailey brought both experience and a fresh perspective: "The subject matter with which he has to deal is in the highest degree controversial, but he is not identified with any of the conflicting views, he can really bring an unbiased and therefore a judicial mind to bear on the proposed Survey. . . . Hailey has this supreme qualification that neither he nor anyone else knows at this moment what attitude he will take on these burning questions and his views, when he forms and states them, will at any rate be entitled to be regarded as unprejudiced."[153]

Although the intellectual historian Paul Rich has argued that Hailey embraced and sanctioned "a common pattern of Western control over separate African Territories, as had been produced in the Indian Raj,"[154] this was not a foregone conclusion for Africa. While the Survey's leadership wished to find a political moderate who had no objections to the goal of systematizing colonial relations, they also hoped the director would approach this subject with a critical eye. Hailey was far more likely to entertain policies and ideas that challenged the status quo in Africa than other candidates who had been considered, such as Edward Grigg, the former governor of Kenya.[155] Hailey's participation in the Survey has been carefully explored by John Cell.[156] He accepted the post in June 1933 and took part in the first two general committee meetings in July and October of that year. He had expected to begin work on the project full time upon his retirement in September 1934, but the transfer of responsibilities to his successor and his preoccupation with the Government of India Bill took longer than expected. He was not able to begin work on the project full time until June 1935.[157] The intervening years were extremely productive thanks to the coordinating efforts of Hilda Matheson, who had been brought on as the project's secretary, and the executive committee, which commissioned a range of specialist studies in advance of Hailey's return to the United Kingdom.[158] These delays, coupled with Hailey's breakdown in late 1937, account for the lag time between the project's launch and the release of its official publications. Had the Survey gone according to plan, the "preliminary results" would have been completed in two or three years and would have been published in 1935 or 1936 (table 2.2).

The project's extension to an almost decade-long undertaking made it a far more diffuse and collective endeavor than originally intended. While the core of the Survey's leadership remained relatively stable over the course of the 1930s, the network's reach grew progressively wider, taking in hundreds of individuals from all tiers of the imperial and scholarly elite. Paradoxically, the Survey's amorphous and fluctuating leadership structure and its entanglements with other institutions, including the IIALC, Chatham House, the League of Nations Mandates Commission, the Rhodes Trust, the London School of Economics, Oxford University, and, eventually, the Colonial Office, only served to heighten its influence. Hailey's eight-month tour of sub-Saharan Africa (1935–36), which was conducted in the company of Donald Malcolm, a former district officer from Tanganyika, and

Table 2.2 African Research Survey: A chronology of select events

1933	Hailey agrees to direct (June)
	1st general committee meeting (July 15–16)
	Edgar Barton Worthington secured to "Buellize" the scientific material (October)
	2nd general committee meeting (October 31)
	African governors first notified of the project (December 11)
1934	Meeting between executive committee (including Worthington) and the British Association Committee on the Human Geography of Tropical Africa (January 2)
	Meeting of executive committee (including Worthington & Perham) and anthropologists (March 13)
	Preliminary chapters of "Science in Africa" issued (October)
1935	3rd general committee meeting (March 9–10)
	Scientific subcommittee created, 1st meeting held (April 30), John Boyd Orr and Richard Gregory added to General Committee
	Hailey begins full-time work (June)
	Malcolm MacDonald resigns from general committee to take up post as Colonial Secretary (June)
	4th general committee meeting (July 20–21)
	Hailey and Donald Malcolm depart for sub-Saharan tour (August to June 1936)
	2nd draft of "Science in Africa" chapters distributed (October)
	2nd Scientific subcommittee meeting (December 11)
1936	Hailey flies back to London (January) and returns to African tour (late February)
	Worthington joins sub-Saharan tour (April to June)
	William Ormsby-Gore resigns from general committee to take up post as Colonial Secretary (May)
	Hailey appointed Lugard's successor on League of Nations Permanent Mandates Commission (September)
1937	5th general committee meeting (June 29)
	Hailey experiences "nervous" and physical breakdown in Geneva and Italy (October 1937 to early April 1938)
1938	An African Survey published (November)
	6th general committee meeting (November 24)
	Science in Africa and Capital Investment in Africa published (December)
1939	Interim African Research Survey Committee constituted (March 22, 1939)
	Lord Hailey approved to join IIALC's Executive Council (June)
	IIALC Executive Committee votes to continue work of African Survey under its auspices and approves a name change to International African Institute (June)

E. B. Worthington, the project's "scientific lieutenant," gave both colonial administrators and technical officers a stake in the outcome.[159] As the executive committee recorded in its progress report to the Carnegie Corporation at the close of 1936, "The importance with which the Survey has come to be regarded, both by British and foreign governments, has been greatly, almost embarrassingly, increased since [Hailey's] tour of Africa, and his conversations with Governors and officials in both British and foreign territories." This experience reinforced the executive committee's commitment to produce a report that would be a "model of comparative study"; only this, they believed, could "live up to the expectations aroused."[160]

By the time various chapters of the final report were being drafted in 1937, dozens of authors, including Margery Perham, Lucy Mair, and a junior staff member from the Colonial Office, Frederick Pedler, had been enlisted to digest copious amounts of material that the coordinators had received from various experts and advisors. Hailey himself had right of final approval and played a substantial role in the editing process, but the end result was the product of a group effort. Nonetheless, in 1938 the Survey's executive committee decided to attribute the nearly two-thousand-page volume, *An African Survey*, solely to Lord Hailey because it would give the report greater weight and allow Hailey to become an official spokesman for their shared vision. Indeed, they were right. Within weeks of its publication, Hailey was *the* central figure in Britain on African affairs. His preeminence as a government advisor rapidly eclipsed the contributions made by the large body of experts who had been assembled to develop that perspective.

The end run around opponents within the government that Colonial Office staff had initially feared actually came to pass. While doubts about the African Survey's substance and necessity lingered among the staff for much of the 1930s, and relations with certain officials in England and in British Africa were at times strained, the colonial secretary, Philip Cunliffe-Lister, reported to his subordinates in 1933 that "the whole scheme has been placed on a new footing owing to Sir M. Hailey's being brought into it." The Survey's coordinators were assured that they would now "receive every possible assistance from the Colonial Office."[161] By the mid-1930s, when controversial topics for African research came to the office, such as a proposal from the Kenya Colony to study Africans' "mental capacity" and "backwardness," or for new research institutes, such as the plan to establish an institute in Northern Rhodesia for ecological and anthropological research, the prime minister, Ramsey MacDonald, and the various colonial secretaries relied less on their technical advisors and more on the coordinators of the African Survey itself.[162] MacDonald's son, Malcolm, had been an active member of the Survey's general committee since 1931, which undoubtedly played a role in this early realignment of power. Most significant was the perception, both within and beyond official circles, that the African Survey was going to provide a more thorough overview of research priorities than any government body would be able to produce. When critics of colonial policies came forward, especially those with decades of official experience in specific African territories, they found a more sympathetic ear with the African Survey's staff than they did with the Colonial Office, which had been relatively successful in keeping individual critics at bay through a labyrinth of bureaucracy. Neither the technical advisors in the Colonial Office nor the advisory committees became irrelevant, but where African matters were concerned, the African Research Survey came to be seen as the principal adjudicator.

What Might Be Done to "Buellize" the Scientific Material

An enduring challenge for the coordinators of the African Survey was how they would integrate the four fields of knowledge—politics, economics, anthropology, and natural sciences—that they proposed to encompass. By 1931, the scientific experts affiliated with the project had put forward a more coherent and detailed plan of research than the experts in economics and politics. Both the scientists and nonscientists agreed that if the project were to incorporate issues relating to "modern science" successfully into its wider framework, the scientists had to determine this agenda. When the general committee held its first meeting to discuss its aims and objectives in July 1933, Julian Huxley reminded them of the need to undertake "a full consideration of scientific research being carried out in different parts of Africa." He insisted "it was essential to secure a man with a broad ecological outlook" who could "find out on the spot what was being undertaken and suggest methods of its better coordination."[163]

The scope and scale of the project Huxley had in mind seems initially to have eluded the nonscientific members of the committee, especially Hailey. At an informal discussion from which Huxley was absent, the group decided that the scientific work was simply a "routine compilation" that could be undertaken by a "competent" person "without any considerable difficulty." Most of the evidence required, thought Joseph Oldham, was already known and could be found in Britain and Europe.[164] When Hilda Matheson began to canvass scientists and scientific administrators, she learned that this undertaking was not so simple; she reported to Lothian that they "showed me the very wide range which required to be covered, the amount of personal travelling involved, and the highly technical nature of much of the investigation." Huxley and Tallents "took the view that only a trained ecologist, or someone familiar with current scientific work, could possibly extract the information from different specialists, and know how to relate it and how to estimate its value."[165]

At an informal meeting in late September among Huxley, Matheson, Oldham, and Hailey, Huxley explained the thinking behind the proposed scientific survey. Nothing of this sort had been prepared before for colonial Africa, or even British India. Since technical knowledge was increasingly seen as indispensable to state building and development, they could provide a great service to all the European powers if they took the time and trouble to prepare an accurate summary. Huxley suggested that this survey would require at least six to eight months of full-time work and would likely involve a research trip to Africa as well. It had already been settled that Hailey would undertake an extended tour of Africa. The executive committee agreed that the person brought on to "Buellize the scientific material" might accompany him on this trip as a "specialist assistant." To reflect this new consensus, they recorded in the minutes that they now recognized that "the scientific side of the Survey constituted a peculiar problem, and that no other preliminary research on this scale was likely to be required in relation to other questions."[166] Although Huxley had initially hoped that Charles Elton would be able to take the post, by the time the Survey was fully underway Elton had assumed new responsibilities at Oxford as director of its Bureau of Animal Population and as founding editor of the *Journal of Animal Ecology*. In his stead, Huxley quickly put forward a second candidate, Edgar Barton Worthington, whose ecological research in East Africa Huxley felt would serve as excellent preparation for the overview.

When Malcolm Hailey and the African Survey committee made the formal offer of employment to Worthington, they had already defined his overarching tasks. He was to consider the "better coordination of scientific work for its own sake and as regards its bearing on general human and administrative problems."[167] He would conduct a careful review of the "eight or nine chief scientific fields" and cover "all Tropical and Equatorial Africa, British and foreign, but a skeleton sketch only of the Union of South Africa."[168] He was to explore not just the technical content of scientific disciplines but also their "degree of correlation, the interrelation of parallel or kindred research, the availability of printed results, and the lacunae, in geographical areas and in subject matter, waiting to be filled." In sum, the Survey was an invitation to discuss the relevance of existing research "from the point of view of public welfare" and its "adequacy or inadequacy [in terms] of resources in men, money, training, and . . . results."[169]

Of the two candidates considered for the scientific work, Elton was the more original and theoretically inclined. Yet Worthington's field experience and his exposure to questions of development made him strategic to the project. Worthington had completed a Bachelors of Science in 1926 in Zoology at Cambridge, where he had taken part in an interdisciplinary group known as the Biological Tea Club. By a fortunate accident, an expedition to the Great Barrier Reef of Australia on which he was to serve was postponed for a year, making him available for the 1927–28 "Fishing

Survey of Lake Victoria," which was designed to investigate why yields in the lake were declining. Sponsored by the governments of Kenya, Uganda, and Tanganyika, this survey provided Worthington with an entrée to a region of the world that became a mainstay of his scientific and administrative activities for the next twenty-five years.

Between 1927 and 1932 Worthington spent twenty-four months in the field with a team of specialists conducting research on food chains, species interactions, and the relationship between organic and inorganic elements of lake environments.[170] They were asked to consider how best to conserve particular fish species while simultaneously increasing their yield. To do this required placing each of the fisheries in its widest biological context. He and his wife, Stella, wrote: "Results could not be expected from an examination of the *Tilapia* fishery alone, but a survey had to be made of the entire fauna and flora, the chemistry and depth of the water, the native population and so forth. Only when all such information had been collected and analysed, and the general life processes of the water had been understood, could recommendations be made."[171] When Worthington presented one of his first papers on the results of this research in 1932 to the Royal Geographical Society, one of the project's instigators, G. Northcote, underscored the value of a comprehensive approach. "Dr. Worthington's account of his party's . . . studies . . . showed the lakes and rivers of Uganda, the Sudan, and Kenya as closely inter-related parts of a single whole; one can only hope that other scientists are following [him] along other . . . zoological, biological, and anthropological lines so as to further complete the picture of Africa as one entity and in relation to the rest of the world."[172]

While Charles Darwin and Alfred Russel Wallace were the British trailblazers for the biosciences, Worthington was one of its foot soldiers, joining the small but growing cadre of researchers who spent a good portion of their careers outside their homelands. Although he was not in the colonial technical services, his background and initial experiences resembled many of theirs. Members of these scientific "diasporas" tended to be male, in their early twenties, and from relatively privileged middle-class backgrounds. By the early 1930s they often held degrees, usually from Oxford, Cambridge, London, or Edinburgh, and had already begun to publish in their respective fields. Once situated in a particular territory they could work for long periods in extreme disciplinary isolation, having little contact with anyone knowledgeable in their fields or research methods. At the same time, they might have brief, but significant access to individuals from Britain or elsewhere who were touring their territories, or they might participate in territorial and regional conferences intended to aid in the coordination of research. These encounters could have important consequences for their personal relationships and intellectual development. Worthington met Joseph Oldham and the other members of the Hilton Young Commission in 1928 while in the field; he also interacted with Jack Driberg, an ethnographer and former administrator in Kenya, Geoffrey Hale Carpenter, a physician and medical entomologist, and Captain Charles Pitman, the game warden in Uganda.

Many of the technical officers and field researchers were immediately, and sometimes self-consciously, dependent upon local informants and translators for their knowledge of specific environments. In several scientific subjects, officers worked with trained African assistants; in medicine and health care, they worked with African assistant surgeons and nurses. These interactions could yield new understandings that were often incorporated into scientific publications and experimental practices.[173] In his memoirs, Worthington listed the members of the scientific expeditions to the East African lakes: "Most important were the complement of African sailors, firemen, and local fishermen who, on balance, probably taught us more about the fish and fisheries of Lake Victoria than we were able to teach them."[174] They first pointed out to him that he was examining not one distinct species, but two. "This episode stimulated us to question local fisherman all round the lake about their names for fish, as well as their fishing methods and they revealed a good deal of ecological understanding. The fishermen told us that *mbiru* breed near the shore and *ngege* [tilapia] do so in deeper water, which also helped our research."[175] Worthington and his counterparts in other fields worked with local informants in a vein that I call "vernacular science."

Producing "Science in Africa"

When Worthington accepted the position as scientific director of the Survey in October 1933, his first step was to go through the previous ten years of *Nature*, then considered the "*Times* of the scientific community," extracting anything relevant to Africa.[176] With the help of Hilda Matheson and Lord Lugard he began to establish contact with individuals in Europe and the United Kingdom he needed to consult. Between December and April, he traveled to London, Brussels, and Paris to meet scientists at the centers of research for zoology, agriculture, forestry, entomology, and botany. These included Kew Gardens (London), the British Museum of Natural History (London), the Musée National d'Histoire Naturelle (Paris), the Jardin des Plantes (Paris), and the Jardin Botanique de l'État (Brussels).[177] He also met with colonial ministers and scientific advisors, including Frank Stockdale and Thomas Stanton at the Colonial Office.[178] In the early winter months he visited almost all of the Imperial Agricultural Bureaus and helped to arrange a meeting between the directors of the Committee of Human Geography in Tropical Africa and the Survey's executive committee.[179] By the spring he had turned his attention to anthropology, initiating a dinner consultation among a group of anthropologists and African Survey staff and advisors and attending Malinowski's seminars on "functional anthropology."[180] By October 1934, he had submitted to the executive committee 250 typescript pages of a "Preliminary Report on Science in Africa," which Matheson sent to each of the experts he had consulted.[181]

Worthington's report took considerably longer to complete than expected, but the executive committee received his drafts "with enthusiasm."[182] Matheson reported, "So impressed is Dr. Oldham that he has actually asked him to add a further section dealing with the almost sacred subject of anthropology."[183] Another chapter dealing with "pure science" had been recommended by several experts, although ultimately its content was folded into different sections.[184] When Hailey finally was able to turn his attention to Worthington's work in March 1935 it made him realize the "peculiar importance of the scientific side of the report."[185] To both the general and executive committees he "emphasized the need for ensuring the authoritative character of anything which might be published in respect of any of the sciences covered." Distributing Worthington's chapters to a large body of scientists "in order to elicit criticism, comment or approval" would "enable the names of leading specialists in the various subjects to be associated with the Survey in its final form."[186]

Hailey's desire to solicit endorsements stemmed from the reaction Worthington's drafts provoked in the Colonial Office and the Economic Advisory Council (EAC). Stanton and Stockdale, as well as Francis Hemming, the EAC's secretary, considered Worthington's first drafts "very poor stuff."[187] Their "misgivings" were sufficient to prompt Cecil Bottomley and George Tomlinson, the Colonial Office liaison to the Survey, to hold a joint meeting on the subject.[188] As Hemming later emphasized to Hailey, Worthington's project "was an impossible task for any one man to be set; . . . noone [sic] could be expected to produce reasonably reliable memoranda covering so wide a field."[189] They were concerned not only with errors and omissions regarding particular scientific debates but also with matters of policy. If his report were incorporated into the "Hailey Survey" uncorrected, it might give undue authority to "wrong conclusions" and "misleading statements."[190] Besides communicating their views to Lord Lothian, Tomlinson met with William Ormsby-Gore, who seemed "much more alive to the danger of Sir M. Hailey's being misled."[191] These criticisms stimulated the Survey's general committee to devote much of its March 1935 meeting to a discussion of the "scientific authoritativeness" of Worthington's work. They also led Ormsby-Gore to recommend that other senior scientists, including John Boyd Orr and Richard Gregory, be invited to join the general committee.

Both Boyd Orr and Gregory had firsthand experience in Africa, although only Boyd Orr had actually conducted research there.[192] As editor of *Nature* since 1919, Gregory was one of the best informed generalists in the scientific community: "no single individual was then more closely in touch with the spectrum of activities pursued by Britain's natural scientists."[193] Boyd Orr was a highly respected medical doctor with expertise in human and animal nutrition. A harsh critic of conditions of poverty both at home and abroad, he popularized the notion of a "marriage of health and agriculture" to meet "human needs." With this aim in mind, he served as an advisor to the League of Nations Health Organization, the Empire Marketing Board, and the Colonial Office Advisory Council on Agriculture and Animal Health. His primary research was conducted at the Rowett Research Institute in Scotland, which was also the headquarters for the Imperial Bureau of Animal Nutrition; he directed both.[194]

After Boyd Orr and Gregory agreed to advise the African Survey they were invited to a special meeting in April 1935 to discuss Worthington's report. This group, which included Hailey, Lothian, Huxley, Gregory, Boyd Orr, Worthington, and Matheson, decided to establish a "Scientific Sub-Committee" to help Worthington revise and edit his drafts so they were scientifically "beyond reproach."[195] Huxley and Gregory agreed to draw up an extensive list of experts to whom the chapters could be distributed; they would also assist Worthington in writing a compelling introduction in which the "urgency of the various problems dealt with in the Report" was underscored.[196]

The creation of the Scientific Sub-Committee formalized what had already been a two-tiered structure in the African Survey's organization. Although Hailey would draw upon Worthington's work for his own report, the scientists recognized that Worthington's material was itself a "valuable example of the Survey method" and held "intrinsic importance to scientists and to the general public."[197] Not only would the monograph "sell on its own merits," as would a volume on economic issues, but it might help to reorient research priorities.[198] *Science in Africa*, a book of nearly 750 pages, was published in December of 1938 after going through three drafts.[199]

When the second draft was produced in the late summer of 1935, it was distributed to Huxley and Gregory's "panel of experts." Of these, "133 were British or Colonial authorities, 18 Dominion authorities, 66 Foreign authorities and 4 international authorities. In addition 20 complete copies were sent to the Colonial Office to be sent in duplicate to each of the Colonial Governments in Africa."[200] Malcolm MacDonald, a general committee member, had been appointed colonial secretary that summer. He took responsibility for sending an "urgent" dispatch to each territory instructing them that they were to distribute the chapters to the "Heads of the various [technical] Departments."[201] Comments and criticisms should be prepared in time for the tour of Africa that Hailey, Donald Malcolm, and Worthington would undertake.[202]

The colonial African governments took MacDonald's request seriously. From the mandate territory of Tanganyika, to take just one example, the game warden and the Department of Lands and Mines each sent back four pages of comments, the Agricultural Department submitted seven pages, the director of medical and sanitary services offered ten, and the Forestry Department wrote eleven.[203] In addition to the comments received on the ground in Africa, which amounted to hundreds of pages, another eighty-five individuals responded directly to the Survey's main office in London, including, to name just a handful, the technical advisors to the British Colonial Office and the directors of the International Institute of Agriculture (Rome), the Indian Research Fund, the Ross Institute of Tropical Hygiene, the Imperial Institute of Entomology, the Imperial Bureau of Soil Science, the Imperial Bureau of Animal Nutrition, and the Amani Agricultural Research Station, as well as numerous professors and staff at research institutes across the United Kingdom.[204]

Hailey took the opportunity of his voyage to South Africa to read Worthington's text "more carefully than before." He was impressed with the "excellent" material in the drafts and felt it was time to consider the format for publication; since the book would be "read largely by scientists," he thought the volume could benefit from "fuller documentation."[205] Worthington and the Scientific Sub-Committee agreed.[206] The letters written in reaction to Worthington's second drafts were largely positive, and many offered valuable suggestions for revision. A handful, however, continued to be quite critical, sometimes with respect to the way Worthington portrayed the content of a discipline and other times to the way developments in specific territories were described. In part to "disarm criticism" and in part to familiarize himself with regions of Africa he had never visited, Worthington joined Hailey and Malcolm for the final three months of the West

African leg of the tour.[207] Hailey described his arrival as "a great benefit to me," but also to Worthington who "has heard what local officers have to say about his draft Report."[208]

The tour of sub-Saharan Africa and the process by which the drafts were reviewed had significant effects on Worthington's viewpoint and the contents of his volume. Prior to his departure, he had developed and deepened contacts among a wide array of specialists. He had been recruited to take part in an interdisciplinary committee on nutrition in Africa—with the anthropologists Raymond Firth and Audrey Richards, and the pathologists and biochemists Robert McCance and Elsie Widdowson—under the auspices of the International Institute of African Languages and Cultures.[209] His drafts reflected his growing commitment to describe the "interrelations" among various disciplines and demonstrated his understanding of the importance of regional variation across the continent. However, Worthington had not returned to Africa since his last lakes expedition in 1931, creating a gap between his field experiences and his conceptual framework. The trip bridged this gap and helped him to crystallize his overarching analysis; it also drove home the importance of research being done on site. Later, in evaluating the significance of his West African tour, he stated that "it brought to my attention the enormous way in which all the different sciences could help in the colonial development of the colonies."[210]

Even more important than the direct impact of the tour was the body of evidence Worthington began to compile from local research officers, whose work had no central or collective forum. Either their efforts were buried in annual reports, where they remained essentially "useless," or they were published in separate specialist journals where their relevance was often overlooked. After a "long talk" with Nigeria's "dietetics expert," Dr. Zaria Turner, and one of the medical health officers "about the possibility of intensive work on one tribe from many angles," Worthington wrote in his diary that they "want [a] centre where research workers from Africa could go and chat in England about their work and get in touch with any experts."[211] *Science in Africa* came to be seen, by Worthington, Hailey, and the African Survey committee, as a precursor to just this possibility.[212]

Over 275 individuals were ultimately involved in reading, commenting on, and contributing to Worthington's book. Of this figure, eighty-three, or 30 percent, came from "government departments" in Africa. While the Colonial Technical Services were weighted heavily toward medical officers, Worthington relied equally on environmental scientists. Although in numerical terms anthropologists made up the smallest group of advisors, they were disproportionately represented when compared with their very small numbers in African colonial governments. Their influence on the substance of both *Science in Africa* and *An African Survey* was significant and far-reaching. The same was true, as we shall see in the following chapter, for ecologists.

Conclusion

Throughout its execution, the African Survey's leadership adhered to their original objective of creating a "permanent scheme of research" that could contribute to the study and management of Africa's imperial problems. The official volumes produced by the Survey were intended to be both *"a model of comparative study"* as well as "a prelude to long-term research."[213] By 1938, the vision for the project had thus come full circle from the early proposal to create an Institute of African Studies at the University of Oxford. The anthropologist Daryll Forde traced this history most succinctly in a confidential memorandum to the International African Institute's new leadership in 1944. Shortly after *An African Survey* and its complementary volumes were published, he reported, the IIALC and the Survey's leadership engaged in a series of conversations about amalgamating the functions of the two institutions.

> This would not have involved an entirely new departure from the [IIALC's] point of view since it had already provided a considerable information service in the more strictly anthropological and linguistic fields; but a reorganisation was contemplated which would have provided a comprehensive service on scientific, cultural and sociological matters in Africa, together with the publication of an Information Bulletin in addition to the journal "Africa". It was also hoped to establish branches of the Institute in other national capitals to serve as centres both for information and the promotion of research activities. Finally it was proposed to organise over a period of years a series of international conferences concerned with various aspects of the research field.

Although IIALC's executive council approved these proposals in 1939 and Lord Hailey joined its executive council, "[t]he outbreak of war abruptly terminated efforts to implement these plans."[214]

While the "unofficial" dimension of the Survey's history was curtailed, its official trajectory was able to move forward, albeit still in the context of a world war. When the Survey's proposals were considered in the Colonial Office in 1938 and 1939, the question arose whether it was best to address

Africa alone or extend the plan to include the entire dependent empire. The British government recommended the latter course of action, particularly in light of the Royal Commission on the West Indies, which drew attention to a similar set of problems in the Caribbean.[215] And so it was that a colonial, rather than an African, research committee was established, and a Colonial Development and Welfare Act was passed. Before we consider postwar developments in chapter 7, we ought to turn now to the scientific debates and topics in which the Survey became embroiled and that ultimately shaped its recommendations for the future.

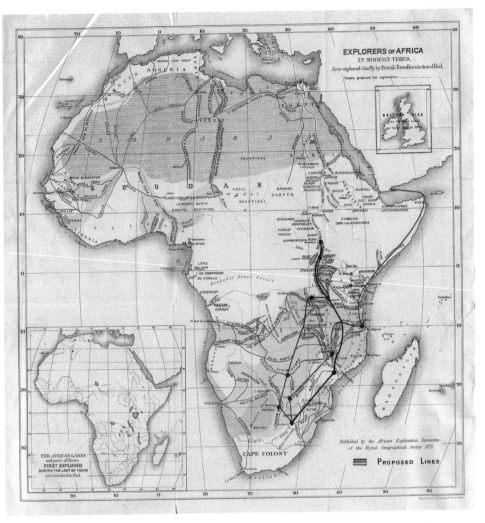

Plate 1. Royal Geographical Society's African Exploration Fund, 1877. Presented to Colonial and Foreign Offices. Map of explorers' routes and proposed telegraph lines. *Shaded areas,* British exploration. Source: CO 879/15, "Correspondence Respecting the Projected Telegraphs to South Africa, 1879," BNA.

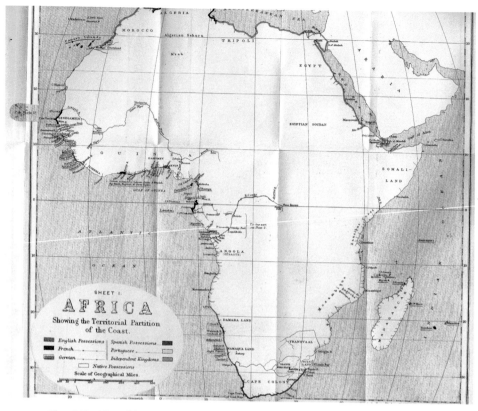

Plate 2. Partition of the coast of Africa in 1884. Source: Rawson W. Rawson, "The Territorial Partition of the Coast of Africa," *PRGS*, 6 n.s. (November 1884): 615–31.

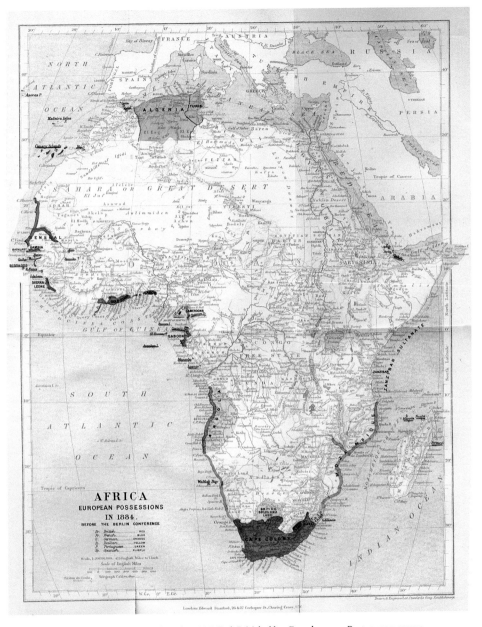

Plate 3. European possessions in 1884. *Red*, British; *blue*, French; *green*, Portuguese; *orange*, German. Source: J. Scott Keltie, *The Partition of Africa*, 2nd ed. (London: Edward Stanford, 1895).

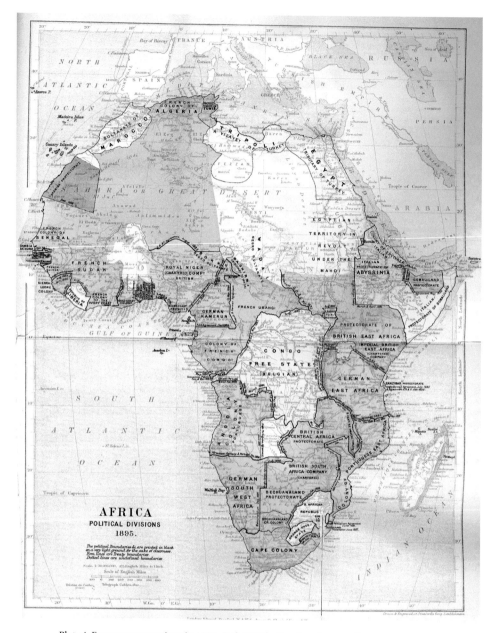

Plate 4. European possessions in 1895. *Pink*, British; *blue*, French; *green*, Portuguese; *orange*, German; *purple*, Spanish. Source: J. Scott Keltie, *The Partition of Africa*, 2nd ed. (London: Edward Stanford, 1895).

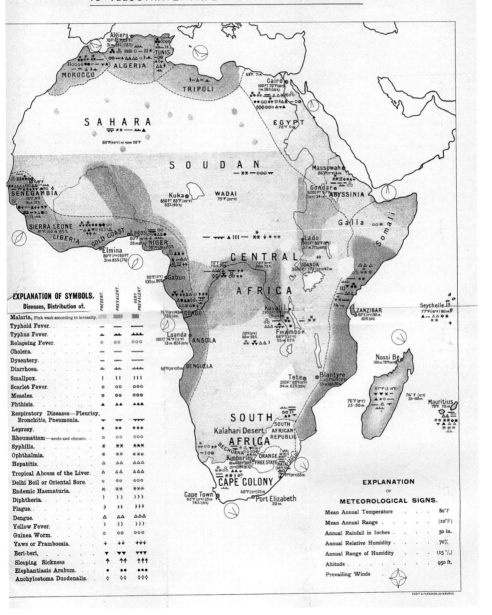

Plate 5. "Distribution of Disease in Africa." *Shaded areas*, malaria. Source: Robert Felkin, "On the Geographical Distribution of Tropical Diseases in Africa." *Proceedings of the Royal Physical Society* (Edinburgh) 12 (1894): 415–87.

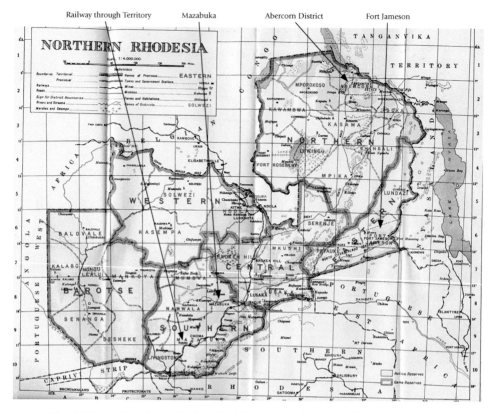

Railway through Territory   Mazabuka   Abercorn District   Fort Jameson

Plate 6. Northern Rhodesia, administrative divisions, 1938. *Green areas,* "native reserves." The administrative divisions totaled 291,000 square miles; the native reserves, 111,000 square miles (71 million acres).

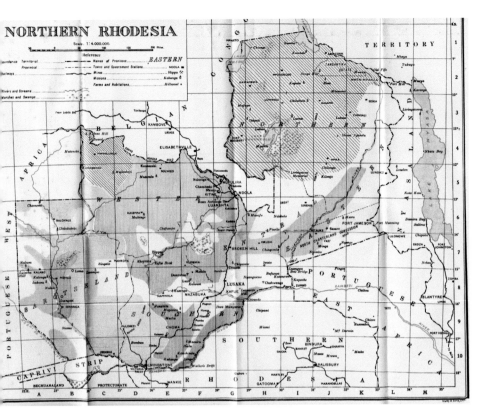

NORTHERN RHODESIA

Plate 7. Northern Rhodesia, staple crop distribution, circa 1938.

Maize

Sorghum

Bulrush Millet

Finger Millet

Cassava

Maize and Sorghum

Sorghum and Bulrush Millet

Maize and Bulrush Millet

Cassava and Finger Millet

Cassava and Bulrush Millet

Mwanza   Shinyanga   Railway Lines

Plate 8. Tanganyika provinces and districts, circa 1930. Source: *Handbook of Tanganyika*
(London: Macmillan, 1930).

CHAPTER THREE

An Environmental Laboratory: "Native" Agriculture, Tropical Infertility, and Ecological Models of Development

The picture really presented by Africa is one of movement, all branches of physical, biological and human activity reacting on each other, to produce what biologists would refer to as an ecological complex.

—E. B. Worthington, *Science in Africa*, 1938

Our agricultural officers . . . would do well to make a careful study of native methods of agriculture before they arrogantly assume that the methods of the European are far superior to those of the African. Unfortunately this is the assumption generally made, and much that is valuable in native crop raising has probably been lost.

—Archibald Church, *East Africa, a New Dominion*, 1927

In his 1983 memoir, Edgar Barton Worthington remarked that the African Research Survey "initiated a good deal of ecological thinking and the application of ecological principles. . . . related not only to the biological sciences, but also to the development of natural resources and to the affairs of mankind."[1] Worthington's understanding of the ecological perspective as a comprehensive view that linked science to society in the service of development was exemplified in his own career. Shortly after he completed his work for the African Research Survey, Lord Hailey invited him to serve as the scientific secretary to the Colonial Research Committee. Following the Second World War, he joined the East Africa High Commission, where he oversaw the establishment of its inter-territorial research institutes and wrote Uganda's first Ten-Year Development Plan. In 1950, he was recruited as the first secretary-general of the Scientific Council for Africa South of the Sahara, which coordinated research efforts across the colonial dependencies as well as South Africa and Southern Rhodesia. All of these projects and institutions emphasized strategies of colonial development that drew upon multidisciplinary research and reinforced "one of the main morals of Lord Hailey's Report": "the close interdependence of all scientific effort."[2]

Yet neither ecological thinking nor ecological research originated with the African Survey. Indeed, the Survey consolidated and propagated a range of approaches that were already being tested across a number of different territories in tropical Africa. Before Worthington became involved in the project, a group of statesmen and scientists, including Leopold Amery, William Ormsby-Gore, Julian Huxley, Arthur Tansley, Stephen Tallents, and Charles Elton, had called attention to the need for ecological studies in British Africa. The ecological perspective itself had emerged a generation earlier and found strong advocates among those scientists concerned with the colonies. By the time the British Ecological Society was founded in 1913, British-based specialists in botany, parasitology, geography, and anthropology were folding its insights and vocabulary into their own disciplines. Alfred Haddon stressed the importance of an ecological approach during his presidential address to the Anthropology section at the British Association's first meeting in South Africa in 1905: when studying "any given people in the field, materials are available for tracing the interaction between life and environment and between organisms themselves, to which the term oecology is now frequently applied, but we still need to have this interdependence more recognized."[3] In 1906, during a discussion of the empire's natural resources at the Royal Geographical Society, its president, the Africanist George Taubman Goldie, declared that "the oecologists will be supported by this Society" because of their talents for thinking about the natural world "as a whole . . . as a living entity."[4] A few months later, he exhorted the Royal and Scottish Geographical societies that "mankind has hitherto dealt with the surface of Mother Earth in a haphazard, a hand to mouth fashion, without much scientific study of the varying ecological conditions in different localities." Taking diverse environments into account in a systematic fashion might hold the key to the "future welfare and progress of mankind."[5]

During the interwar period, ecology was coming to be seen as more than a concept or a discipline. Many of its adherents positioned it as a tool of environmental management and a way of organizing knowledge about complex phenomena. British Africa became a major field for experimenting with these ecological approaches. Some initiatives were carefully planned, while other ecological studies developed serendipitously. The African Survey ensured that their insights and methods were given pride of place not only in the final reports but also in the metropolitan and colonial institutions that stemmed from the project. To understand why and how this came about, we must consider what these advocates thought ecology would contribute that other disciplines could not. How did they define "ecological thinking" and "ecological principles"? What place did ethnography hold in ecological fieldwork, and what epistemological status was granted—or denied—to Africans' own techniques of natural resource management?

This chapter explores these questions by examining the history of agricultural departments and research institutes, which were second only to medical departments in the number of bioscientists and technical officers they employed (see Appendix).[6] Focusing on agriculture draws our attention to several interconnected themes. Because agriculture has its roots in regenerative nature, both scientists and imperial officials pointed out its long-term value. Unlike mining, which produces a nonrenewable product, agriculture could potentially be sustained indefinitely. All societies engage in agriculture, so its significance extended beyond concerns with food production to cultural anthropology. By the 1920s, the human side of agriculture in tropical Africa had begun to attract more focused scientific attention. Any efforts to develop agriculture also involved botanical, zoological, and meteorological sciences concerned more broadly with flora, fauna, and climate. The science increasingly enlisted to make sense of these interacting fields in tropical Africa was ecology. In practical projects, too, agriculture was closely connected with the work of other technical departments, such as medicine, forestry, and game. Indeed, throughout decades of debate about how best to develop the British territories, one fundamental consensus endured: agriculture was, and would remain, at the center of African economies. The Colonial Office Conference of 1927 declared: "particularly in Tropical Africa, agriculture, including stock rearing, is the principal and the historic occupation of the native inhabitants, and the main source of their wealth. . . . It is, then, on agriculture that the wealth of the Colonies and all schemes for the amelioration of the lot of their inhabitants must depend."[7]

Genealogies of African Environmental History

Looking closely at agricultural research in tropical Africa provides a significant corrective to several recent interpretations of science in colonial contexts. Many historians and anthropologists, reacting against older, imperialist views featuring "backward" and resistant natives and "progressive," enlightened colonial officials, have emphasized colonialists' "misreadings of the African landscape."[8] They argue that scientists helped to *construct* inaccurate "environmental orthodoxies" that became "received wisdom," and which had negative consequences for Africans that often extended into the postcolonial period. Among the earliest proponents of this view was the ecologist, John Ford, who was himself employed for over twenty-five years, between 1938 and 1965, by various colonial states and research institutes in British Africa. In 1971, with funding from the Wellcome Trust, Ford published *The Role of the Trypanosomiases in African Ecology: A Study of the Tsetse Fly Problem*, a historical account that quickly and deservedly became a classic in African studies.[9]

Ford's book appeared just as a new wave of historical research documenting colonialism's adverse effects came to prominence. Imperial control was still being challenged actively in Zimbabwe, Namibia, and Mozambique, while apartheid in South Africa showed no signs of waning. Although he was not the first to focus on environmental or epidemiological issues, Ford certainly offered the most ambitious and far-reaching account to date.[10] He also introduced key vocabulary and angles of analysis that were picked up and developed by scholars working in social, medical, and environmental history. Ford explained in the book's preface that he had written his study because, "Many years ago I began to have doubts about *official colonial doctrine* concerning the importance of the tsetse in tropical Africa."[11] Given his desire to offer a critical account of colonialism, Ford pointed the finger at his own colleagues: "It is a curious comment to make upon the efforts of *colonial scientists*," he wrote, "that because of *preconceived notions* about Africa and Africans . . . basic circumstantial epidemiology has often been wrongly described."[12] According to Ford, these workers' principal failures were fourfold. First, "Western science continues to make a piecemeal approach to disease whenever and wherever it breaks out." Without recognizing the need for an "ecological" framework and "for joint investigation by zoologists, ecologists, and parasitologists," he argued, "the colonial governments in Africa . . . left to the new countries a legacy of ideas that had little relevance to the biological processes with which they had unwittingly interfered." Second, he believed officers had "almost entirely overlooked the very considerable achievements of the indigenous peoples in overcoming the obstacle of trypanosomiasis to tame and exploit the natural ecosystem of tropical Africa by cultural and physiological adjustment both in themselves and their domestic animals." Third, officials made "erroneous" assumptions about their own effects and, as a consequence, pursued "misdirected" efforts to control the disease. "Unfortunately, with very few exceptions, it was psychologically impossible for men and women concerned in imperial expansion in Africa to believe that their own actions were more often than not responsible for the manifold disasters in which they found themselves caught up. The scientists they called in to help them were as ignorant as they of the problems they had to tackle." And finally, he asserted that officials pursued an overconfident and misguided quest to "eradicate" tsetse flies and find a "permanent cure" for trypanosomiasis. "The real problem is not how to get rid of tsetse flies or how to cure sleeping sickness and the infections of domestic animals, but how to apply techniques of control in such a way that an expanding economy is not hampered by their mishandling."[13]

Ford was implicitly arguing in his 1971 book that new states in tropical Africa ought to pay attention to what today might be called *subaltern* strategies of disease control. He spelled this out a few years later in a volume edited by anthropologist Paul Richards: "If there is any lesson to be learned from the results obtained by applied science in tropical Africa in the twentieth century, it is that vast projects designed to solve artificially isolated problems do not work except with the willing collaboration of the people they are intended to help."[14] It was no accident that Ford stressed the inadequacies of the laboratory and its methods in his critique. He had been trained as a field scientist at Oxford University, circa 1929 to 1938, under the tutelage of Charles Elton, among others, and it was only by working in the field that he had an opportunity to speak to "the local people" and hear what evidence they had to offer on the history of trypanosomiasis in the areas where they lived.[15]

What Ford left out of his narrative, however, was any discussion of the terms of his own employment. Had he explored this, he would have had to acknowledge that his book was in many respects an outgrowth of what he had been hired to do. As E. B. Worthington reported to the Colonial Office in his 1948 "progress report" on the East African "research and scientific services," one of the main studies being done in the realm of tsetse and trypanosomiasis research was an examination of the "relationship of Tsetse to its Environment, including Man, Animals, Vegetation and Climate. This, the broad ecological approach, is the special task of Mr. Ford."[16] Worthington also drew the office's attention to "an important paper" that Ford and two of his colleagues had recently published, as an example of cross-disciplinary research, in which they "gave a general account of some of the main problems involved [in tsetse and trypanosomiasis control], and detail[ed] the approach to these problems (a) through the trypanosomes, (b) through the tsetse flies, and (c) through broad ecological relationships." This research method seemed one of the more promising lines of work, *but*, stressed Worthington, "It is desirable to re-emphasize that, in spite of long years of hard work by many investigators, there is still a great deal of knowledge lacking on the innumerable questions of how tsetse flies and trypanosomes live in relation to their environments."[17]

Ford's arguments have been perceived by many to be a radical departure from colonial thinking. I would suggest instead that key elements of Ford's work were less a rupture with the colonial era and more an expression of continuity. After all, he derived his evidence from his own extensive field-work and from a close reading of the published accounts of technical officers who had been employed in the colonial service. It was they who had also done many of the oral interviews and field studies on which Ford's analysis rested.[18] Ford was not so much rejecting as he was embracing "official colonial doctrine" when he took up these research questions.

Ford's influence among a handful of social scientists in the 1970s—Helge Kjekshus, Paul Richards, and Leroy Vail, for instance—and a wider group of medical and environmental historians in the 1980s, including Maryinez Lyons and James Giblin, was profound and helped to establish a research agenda that reexamined colonial history along new lines.[19] No longer was the aim of research to help colonial states or the imperial project; now scholars undertook studies that focused on Africans' agency and, often, on the negative effects colonialism had on people's lives. Kjekshus, for instance, was one of the first historians to examine at length the demographic effects of the rinderpest cattle epidemic that spread across eastern and southern Africa in the late nineteenth century.[20] Using a vocabulary and periodization very similar to Ford's, he argued that colonial conquest had touched off a demographic and "ecological collapse," largely because officials had "fail[ed] to consider the African positively at the centre of administration and development." Quoting Julius Nyerere, the first president of independent Tanzania, Kjekshus concluded that "development means the development of people." Kjekshus believed that because African states were no longer under the yoke of colonialism that they could now "question older assumptions about the unadaptive nature" of "'traditional' economies" and acknowledge that Africans' precolonial systems of environmental control had been based on "change and adaptation."[21] Like Ford, Kjekshus held out an implicit hope that subaltern populations might become central to development strategies.

Paul Richards, in turn, took up the question of Africans' environmental knowledge, calling it "folk ecology" and "people's science," and wrote one of the first (postindependence) anthropological studies of West African "indigenous agriculture."[22] Richards was well aware that he was following in the footsteps of certain technical officers and scientists during the colonial period; in fact he took a great deal of interest in these precedents, but his examination of the colonial past was more utilitarian than comprehensive. His central preoccupation was to explore "the relationship between science and development—more especially the relationship between environmental science and prospects for increased food production in West Africa [today]."[23] To do this he framed his argument in terms of the differences between "scientific universals and ecological particularism." "Following John Ford, I have argued that West African farmers are especially good i) at solving ecological problems of the kind that arise when human and 'wilderness' ecosystems intersect, and ii) at exploiting the risk-spreading possibilities of ecological boundaries and landscape sequences."[24] The goal, in other words, was to ensure that people who lived on the land and generated their livelihoods from it received support for the tools and ideas they had developed across the generations. This was necessary not just to redress the balance of power but also because these methods worked. Richards's historical analysis, however, led him almost incidentally to offer a very interesting insight about "scientific" studies of "local practices" during the colonial period: they sometimes resulted in the authors "discovering" the value of these practices and "reinvent[ing] traditional agriculture!"[25]

A body of work that has built on these themes emerged in the mid-1990s and has straddled the fields of African environmental history, social anthropology, and development studies. According to Melissa Leach, James Fairhead, and Robin Mearns, among others, "the origins and persistence of received wisdom about environmental change in Africa lie in the substance of science, on the one hand, and in its social and historical context on the other, including the effects it has through development in practice."[26] Subscribing to many of Ford's initial criticisms, they argue that "colonial scientists" "had little evidence to support their hypotheses about African environments . . . [which] [n]evertheless . . . became institutionalized in the colonial agricultural, forestry, livestock and wildlife departments, forming the rationale for intervention."[27] Labeling this knowledge environmental "orthodoxy," they conclude that it often rested on "bad" science and on "racialist, pejorative views" of Africans and prevented more locally sensitive policies from gaining sway.[28] When competing perspectives did exist, those who "challenge[d] the prevailing orthodoxy found their views either suppressed or unable to influence higher levels in the institutional hierarchy."[29]

Fairhead, Leach, and Mearns aim, above all, to highlight how dynamics of race, power, and empire served to marginalize most Africans not only from the production and reproduction of knowledge, but also from its application. This has meant, they argue, that Africans' "local agro-ecological management strategies" were either ignored or disregarded as "ignorance and wanton destructiveness."[30] In fact, they believe existing African knowledge could become a foundation for more sensitive environmental policies, "which would be more appropriate to [Africa's] ecological and social conditions."[31] By reframing debates about African environments, they wish to reposition African actors and privilege an "ecological pluralism" that avoids "the fundamental dualism in western thought which conceptually separates people from 'nature.'"[32]

Rather than embrace these critiques wholesale, this chapter examines instead the ways that scientists working in British Africa helped to *deconstruct* late nineteenth-century ideas about the tropics and early twentieth-century assumptions about the unproductive and wasteful nature of indigenous agricultural practices. Indeed, it was research officers associated with agricultural departments who first began to undermine widespread assumptions of tropical Africa's natural fertility and drew attention to the relative infertility of the soils. Along with a handful of field ecologists, they also began to speak in positive rather than pejorative terms about the techniques associated with African agricultural practices. The fact that these ideas originated during the colonial period requires us to reinterpret the interplay between science and empire in new ways.

As British Africa was subjected to scientific scrutiny, the politics of knowledge increasingly took center stage. The agricultural research described in this chapter repeatedly pushed back against the wholesale transfer of European norms to African environments and offered a wide range of different approaches to modernization and development. As part of this process, Africans' subaltern, or orally transmitted, knowledge became an object of scientific study, a pattern I refer to as *vernacular science*. While none of the British scientists involved in these activities was moved to call for the end of empire, many were keen to criticize how it worked in practice. They also often inadvertently opened up a space to reconsider and even redefine state and imperial aims. The nature and scale of the problems research officers and anthropologists were forced to confront in the African "field" led them to adopt conceptual and methodological tools that attempted to bridge previously distinct areas of disciplinary expertise and cross the boundaries that separated one kind of problem from another. Although these perspectives did not always fundamentally reshape the views of policy makers in the metropole or redirect all the scientific and technical work within colonial states, they were enormously significant in shifting evaluations of African expertise from the negative to the positive.

The Political Economy of Bioscientific Research

Africa's environments were fundamental to the colonial project for one reason: resources could be extracted or utilized to yield revenue for the colonial state, which was essential to the maintenance of the empire. As British officials soon discovered, however, turning resources into revenue was a difficult business. Tensions between social needs and commercial uses, conservation and exploitation, and differing systems of knowledge complicated the equation. Colonial states sought to make tropical Africa's natural economy more profitable; they also aimed, in principle if not always in practice, to enhance the well-being of their inhabitants. The dual aims of wealth and welfare, along with the structural asymmetries inherent in the imperial project, forced colonial administrators to confront a series of questions that were politically charged and epistemologically fraught. What were the best uses of different kinds of land? Which productive systems would be the most lucrative, and which the most beneficial? What were the best measures of agricultural success? Implicit in these questions was a default assumption that officials would have the means to implement the findings of scientific studies and be able to control the outcome of these experiments. In much the same way that colonizers' attempts to rule indigenous peoples often backfired, leading to social resistance, wars, and weakened states, their efforts to manage the natural world sometimes exploded in their faces, turning what many thought would be a relatively easy extraction of resources and profits into an intricate courtship dance with nature itself. As Worthington observed in his memoirs, "The ecology of land use could only be learned on the spot and the environment had a way of kicking back at mistakes."[33]

From the late nineteenth century on, geographical descriptions of tropical Africa typically included inventories of potential crops, mineral deposits, and pharmaceutical specimens. Colonies were regarded as "sources of supply of foodstuffs and of raw materials for the industrial enterprises of the mother land."[34] In Africa, the British articulated their imperialist vision in unambiguous and unapologetic terms. During a 1907 tour of East Africa, for example, Winston Churchill described tropical Africa as Europe's future breadbasket and claimed that it would soon "play a most important part in the economic development of the whole world."[35] Nature's apparent abundance lay open to exploitation, and environmental scientists were brought in to find the best ways to make it both pliable and portable.

Agricultural Departments and the Place of the "Native"

Colonial administrators were expected to create a viable "export trade" in their territories and to extract wealth from Africans by means of economic coercion, including land appropriation, forced labor, and requirements that cultivators grow specific crops.[36] The severity of these interventions varied from territory to territory. The atrocities of King Léopold's regime in the Congo attracted the greatest attention in Europe, but British policies and practices were socially disruptive too. By the turn of the twentieth century, Britain's tropical African colonies had begun to undergo an "export boom" in agricultural and mineral products.[37] Many of these commodities—palm oil and kernels, rubber, cotton, groundnuts, timber, coffee, cocoa, and metals—were fed directly into industrial production in Europe and elsewhere: oils for food, soaps, and manufacturing; cocoa and coffee for the chocolate and drinks industry; cotton for clothing and fabric; and rubber for vehicle tires and for belts to transmit motive power to machinery in factories. Tropical Africa's trade statistics were still much lower than many other regions of the world, but exports from this region had already surpassed those from the British-controlled islands in the Caribbean,[38] and this rapid growth seemed to imperialists a positive sign of things to come.[39]

Agricultural departments were founded between 1890 and 1914, usually in tandem with forestry, meteorological, and geological research. Botanical stations were central to these efforts, and from the early 1900s the Colonial and Foreign Offices promoted "agricultural and forestry surveys" (table 3.1).[40] Initially, governors faced serious obstacles in finding the funds to support territory-wide research. When Harry Johnston was appointed High Commissioner of British Central Africa in 1891, he founded a "scientific department" for the study of its plant, animal, and mineral resources, a project that addressed geographical and anthropological questions as well as economic ones.[41] The naturalist-turned-administrator was responsible for establishing East Africa's first botanical stations, in Nyasaland and Uganda, and for recruiting Alexander Whyte, a botanist with experience in Asia, to oversee their crop experiments.[42] Funding for science was not forthcoming, so Johnston had to pay for many of the scientific instruments himself, and he secured financial support from both the Royal

Table 3.1   British botanical stations in tropical Africa, 1887–98

Location	Founded
Lagos (Nigeria)	1887
Gold Coast (Ghana)	1890
Niger Coast Protectorate/Royal Niger Company	1891
Gambia	1894
Sierra Leone	1896
British Central Africa (Nyasaland/Malawi)	1891
East Africa Protectorate (station in Uganda)	1898

Sources: "Botanical Station at Lagos," *Kew Bulletin of Miscellaneous Information* (hereafter *KB*) 18 (1888): 149–56; "Gold Coast Botanical Station," *KB* 55 (1891): 169–75; [on Niger Protectorate dates] "Botanical Enterprise in West Africa," *KB* 130 (1897): 329–33; "Gambia Botanic Station," *KB* 135 (1898): 35–43; "Botanic Station—Sierra Leone (with plan)," *KB* 130 (1897): 303–17; "Botanical Enterprise in British Central Africa," *KB* 104 (1895): 186–91; and "List of the Staffs of the Royal Gardens, Kew, and of Botanical Departments and Establishments at Home, and in India and the Colonies," *KB* (1898): appendix III, 55–62; also see "West African Botanic Stations," *KB* 84 (1893): 364–66. Kew listed the Ugandan station under East Africa Protectorate, but it was always in Uganda.

Geographical Society and the Royal Society.[43] State building was an expensive business, and scientific research was not always the top priority.

Lack of funds also plagued Kew Gardens' studies in British Africa. The same year Johnston moved to Nyasaland, Britain's prime minister, Lord Salisbury, wrote privately to Kew's director urging him to consolidate the information its correspondents had collected over the years: "A proper knowledge of the Flora of Tropical Africa would do much to aid the development of the territories over which this country has recently acquired an influence."[44] By 1900, however, the treasury had yet to approve any grant for this effort, and the second volume of the *Flora of Tropical Africa*, which had been completed in 1898, was "still unpublished." In the eyes of Kew's director, William Thiselton-Dyer, these delays raised serious questions about Britain's commitment to development. "It is, of course, a waste of time tendering to Her Majesty's Government such advice as is contained in [these reports], unless there is a real desire to deal with our immense responsibilities in Tropical Africa in a practical spirit."[45] Thiselton-Dyer argued that colonial governors should be provided with funds to undertake field surveys. Metropolitan bodies like Kew, he stressed to Joseph Chamberlain in 1900, could only offer advice "of a vague and general kind. A man of trained experience brought face to face with the problem on the spot is in a very different and more effective position. . . . The appropriate methods [of agriculture and forestry] vary in different cases, and it requires a skilled expert to work them out locally."[46]

At the same time, Harry Johnston was trying to drive home the significance of systematic scientific fieldwork to the imperial project. Soon after his appointment as Special Commissioner to Uganda in 1899, he asked his technical officers: "What resources does the country at present possess for the development of a profitable trade?"[47] In a special dispatch to the Foreign Office in 1901, just a few months before he returned to England to attend the inaugural meeting of the African Society, he contended that research should be coordinated at the imperial level and asked the Foreign Office to discourage scientific exploration "at the hands of persons not connected to the Administration." Johnston said he would welcome and support "experts" appointed by the Foreign Office to conduct field research, but he would guard against "persons who wish to acquire knowledge for purely selfish purposes, or who wish to enrich museums outside the British Empire . . . or merely to drive a lucrative trade in specimens." In addition to worrying about their profit motive and affiliation with competing imperial powers, Johnston charged that "nearly all these individuals give untrue descriptions of the country, because their presence in it has been brief and their capacity for forming a correct opinion naturally poor."[48] In his eyes, long experience in the field was an important prerequisite for producing accurate and true knowledge.

For all his reputation as a proponent of "constructive imperialism," Chamberlain initially seemed to undervalue the kind of scientific fieldwork Thiselton-Dyer and Johnston were advocating. For two years after his appointment as commissioner, Frederick Lugard lobbied Chamberlain to recognize "the necessity of hiring a small scientific staff" for the purpose of "development and conservation" in Northern Nigeria,[49] but each year the funds he set aside in the budget for a "Department of Scientific Research" were "over-ruled" by the Colonial Office.[50] Chamberlain justified this decision by remarking that "the expenditure of a small sum would probably effect little or no good." Lugard, however, finally succeeded in getting Chamberlain to approve a preliminary field survey. The natural inventory, conducted along the lines that Kew's officials had suggested, would focus not just on the size and location of forests and types of soils but also on the "crops . . . actually grown at the present time by the natives in different parts, and [their] methods of cultivation."[51] Until the survey had been completed, Chamberlain and Lugard agreed, it was inadvisable to start new agricultural projects.

Officials at the Colonial Office and Kew Gardens were already reluctant to promote plantation-based agriculture in the West African territories, since they thought it unlikely that large numbers of white settlers would ever live there and the region lacked the transportation networks that a profitable export trade required. As early as 1888, Alfred Maloney, the governor of Lagos and an avid botanist, sought to persuade the Colonial Office that agricultural diversification was the best strategy. "Blind adherence to one industry only means commercial ruin, as was proved, to the cost of many, in some of our colonies. 'Eggs-in-one-basket' policy has proved disastrous."[52] Pursuing a policy of diversification would require information about which crops might thrive in specific regions.

As they received field reports from the different territories, imperial officials and advisors began to articulate general principles that were supposed to guide the development of agricultural industries. Northern Nigeria played a key role in this process because its reports conveyed the unexpected news that the region's "inhabitants are as a rule good agriculturalists" who achieved "splendid" results with both local staples and newly introduced cash crops, including cocoa and coffee.[53] The social reformer Edmund Morel made a similar observation after touring this region: "In the northern part of Zaria and Kano the science of agriculture has attained remarkable development. There is little we can teach the Kano farmer. There is much we can learn from him."[54] Chamberlain's successor at the Colonial Office, Alfred Lyttelton, announced that "the only sound plan" of agricultural development "is to begin by carefully taking stock of *existing* industries, and to aim, in the first instance, at improving and developing these (where they are capable of development) before proceeding with attempts to introduce new products."[55] Colonial states, he continued, should also take into account "the character of the inhabitants and their existing agricultural development."[56] This approach was designed to enable frail colonial states to avoid provoking social unrest. Equally important, it would allow Britain to keep pace scientifically with its European competitors in the region, especially Germany.[57]

Other European colonial powers were also moving toward the study of "native agriculture" at the turn of the twentieth century. As Christophe Bonneuil and Mina Kleiche have shown, French administrators had begun to point out the practical problems associated with ignorance of "native" practices.[58] In 1906, Yves Henry, an agricultural director for French West Africa, criticized experiment stations in the region for failing to account for "the environment in which [Africans] live and the procedures they use in order to make the land productive (*mettre en valeur*)." "The three elements of our public riches are: the producer, or in other words, the *indigène*, the productive milieu, or rather the landscapes and climate, and finally the product, its nature and means for improvement. Through the system of experiment stations, it is possible to know only some part of the third element, while one gains very limited knowledge of the second and virtually nothing of the first."[59] Henry and his colleagues were convinced that uninformed interventions were likely to fail. In one of the first articles to address African agriculture, the botanist Auguste Chevalier suggested that "judicious observations of the cultural procedures that [Africans] employ, the [crop] rotations they practice, the yields they obtain, and the distinctions between the varieties that they cultivate . . . will often help the colonial to avoid useless groping and will guard against applying procedures associated with European culture prematurely."[60]

Before the First World War, however, no European power incorporated this kind of research into a systematic strategy of agricultural development. In British Africa, neither the agricultural departments nor the Colonial Office adopted a consistent or coherent approach. Agricultural activities proceeded in an ad hoc fashion, and only a few scientific studies were supported through botanical stations and field surveys. As part of the state-building process, all territories tended to encourage their subjects to cultivate cash crops well before administrators had an adequate understanding of land tenure or subsistence agriculture. By default and by design, scientific concerns were subordinated to colonial states' economic priorities.

Fieldwork, Interdisciplinary Collaborations, and Emergent Critiques

Even during the interwar period, field research was not given the high priority that many scientists and some officials in Africa believed it should be accorded. The proposal to create an integrated scientific and research service within the Colonial Office was never adopted,[61] neither were a series of recommendations, endorsed by the participants at the 1927 Imperial Agricultural Research Conference, to establish a chain of research institutes within the colonial territories.[62] Yet the projects that were implemented often had significant effects on imperial thinking. Research completed during these decades reached beyond the boundaries of specific territories and shaped both metropolitan policies and inter-territorial practices. Importantly, scientific studies drew attention to contradictory elements and fault lines within existing programs in colonial agriculture. Most apparent in these reports was the tension between promoting commodity crops for

Table 3.2   Research officers: Agriculture, veterinary,
forestry, and entomology, 1929 and 1938

Territory	1929	1938
Nigeria	57	70
Gold Coast	34	38
Sierra Leone	13	6
Gambia	4	2
Anglo-Egyptian Sudan	28	27
Tanganyika	47	43
Kenya	34	36
Uganda	21	17
Nyasaland	15	17
Northern Rhodesia	11	11
Southern Rhodesia	27	32
Zanzibar	8	10
Total	299	309
Other Colonial States:		
India	171	439
South Africa	217	364

Source: List of Agricultural Research Workers in the British
Empire, 1929 (London: HMSO, 1930); List of Research
Workers, Agriculture and Forestry, in the British Empire, 1938
(London: HMSO, 1938).

export and supporting "native agriculture." The alternative approaches that
field scientists and specialists promoted might be described as "bottom-up
development schemes," although the contemporary ethos of participatory
development that aims to allow producers a role in decision making was
absent from their recommendations.

The emergence of this perspective was made possible by an interesting
confluence of factors that expanded opportunities for scientific fieldwork
in diverse African environments. Above all, this included the metropolitan push to augment the technical services themselves. Between 1918 and
1931, employment in the agricultural, medical, survey, forestry, and game
departments more than tripled across British Africa, a growth unparalleled in any other decade during the colonial period (see figure 0.1). A
growing number of officers were assigned to the production of knowledge,
rather than restricted to imperial management tasks (table 3.2). Although
their numbers remained small compared to their counterparts elsewhere
in the British Empire, their work had a disproportionate impact on imperial thinking and policy making. Staff members who remained in the
same area for years became intimately acquainted with the place and its indigenous inhabitants. Staff retrenchments during the 1930s made research
officers' results more prominent, since their numbers remained relatively

stable compared to other officials, and their studies came to be seen as benchmark examples.

Acting in concert, statesmen and scientists began to push for research institutes and scientific investigations within British Africa that could serve more than one territory. Not only were technical officers forced to agree on shared problems, a process that involved controversies over definitions, but also they became more aware of the need to account for local differences. When the directors of agriculture from Tanganyika, Kenya, Uganda, and Zanzibar met in 1921 to discuss the Amani Agricultural Station, which Britain inherited from the Germans, one of their biggest concerns was that its investigations might fail to address the "considerable variety in natural conditions" throughout "these vast, new countries."[63] While they endorsed the idea of increasing the amount of research done in the region, they doubted that a "centralized" approach would yield the sort of knowledge they needed to do their jobs well. The drive to localize knowledge was just as strong an imperial imperative as the desire to standardize and systematize it.

At the same time, because individual territories had so little funding at their disposal, state leaders encouraged interdepartmental collaborations in research, which they referred to as scientific "teamwork." A handful of metropolitan advisors were beginning to promote field sciences, which stressed multidisciplinary investigations that made connections between natural and social phenomena. Upon his appointment as director of Amani in 1927, William Nowell spent five months touring the six East African territories and learning about the distinct research needs of their agriculture departments. Besides short-term crop studies, which were well established, he stressed the urgency of embarking on long-term investigations, especially "a description of East Africa in which the inter-relations between geological structure, soil, climate, vegetation, animals, and man and his industries would be worked out."[64] He told a gathering of naturalists in London that "while this proceeds, valuable information may be obtained by a shorter route through ecological studies."[65] By the late 1920s, plans were circulating in Britain and its African dependencies for three multiyear research projects featuring teamwork, ecological methods, and ethnographic sensitivities based at the Amani Agricultural Station, the Tsetse Research Department in Shinyanga, Tanganyika, and the Ecological Survey in Northern Rhodesia. The histories of these projects and the fortunes of the African Survey were closely intertwined. All three projects received grants through various metropolitan funding schemes, including the East African loan of 1926, the Empire Marketing Board, and the Colonial Development Fund.

Table 3.3   Interwar League of Nations and international conferences concentrating on Africa

1925, 1928	International Conference on Sleeping Sickness in Africa, League of Nations (London, Paris)
1932	International Conference of Representatives of Health Services of African Territories and British India, League of Nations (South Africa)
1933, 1938	International Conference for the Protection of Fauna and Flora of Africa (London)
1935	Pan-African Health Conference, League of Nations (South Africa)*

* A third pan-African health conference had been proposed by the East African governments for March 1940 and would have been held in Nairobi, Kenya, had the war not intervened.

Table 3.4   Interwar Imperial coordinating conferences

Empire Forestry Conferences	1920, 1923, 1928, 1935
Imperial Entomological Conferences	1920, 1925, 1930, 1935
Imperial Botanical Conference	1924
Imperial Mycological Conferences	1924, 1929, 1934
Imperial Agricultural Research Conference	1927
Empire Meteorologists Conference	1929
Colonial Directors of Agriculture Conferences	1931, 1938
Empire Survey Officers Conferences	1931, 1935
Commonwealth Scientific Conference	1936

Note: With the exception of the 1923, 1928, and 1935 Empire Forestry Conferences, which were held in Canada, Australia, and South Africa respectively, all other events took place in Britain.

These initiatives became hallmarks of bioscientific research in colonial Africa during the interwar period.

Both the British government and the League of Nations increasingly supported scientific coordinating conferences in specific territories and metropolitan centers (tables 3.3 and 3.4). At these events, technical officers were able to pool information, consider common problems, and share ideas and strategies. Inadvertently, these conferences created a forum for individuals to criticize the status quo. The most prevalent and enduring criticism was the lack of scientific knowledge required to inform policy and practice. In a memorandum on agricultural research prepared for the 1927 Colonial Office conference, William Ormsby-Gore declared: "Our ignorance about the soils of the tropical Empire is profound. The only book on the soils of Tropical Africa is the work of an American." Britain must rapidly increase the scale and pace of its "acquisition of scientific knowledge of our resources," he contended.[66] As Ormsby-Gore and others found, the new knowledge they sought could take Agricultural departments in unexpected directions.

Those who worked in tropical Africa were occasionally nervous about the lead taken by South Africa in agricultural research, arguing that the divergent natural and social conditions facing colonial states required different research priorities and policies. In 1929, after South Africa hosted the fifth Pan-African Agricultural and Veterinary Conference, the South African government proposed to Britain's Colonial Office that it coordinate botanical and agricultural research for the entire continent. South African officials argued that this plan would take some of the financial burden off cash-strapped colonial dependencies and would put at these territories' disposal the far more numerous research workers based in South Africa (see table 3.2). Several directors of agriculture in East and West Africa objected to this move. Tanganyika's director explained to the Colonial Secretary that "there is at present a sharp divergence of outlook between the centre and the South of Africa." The biggest problem, in his view, was South Africa's overemphasis on white settlers' efforts to the detriment of African agricultural development. "I feel our delegates would be more at home, and could better express themselves at conferences where native work and European work are given equal prominence. . . . We should preferably more strongly develop our own conference" system.[67] No further pan-African agricultural conferences were organized in South Africa in the 1930s, while the East African coordinating conferences flourished and had significant effects on both agricultural and epidemiological research (tables 3.5 and 3.6). Conflicts still arose among the East African states because of Kenya's settler politics and the overwhelming support it provided to white farmers, but these were easier to adjudicate since colonial dependencies were more on a par, economically and politically, with one another than with South Africa.

The final factor that affected both scientific research and agricultural development was the Depression. Charlotte Leubeuscher, an advisor to the African Survey, assessed the situation in 1934: "One of the paramount facts by which one is faced in Africa is the enormous development that has occurred in almost every province of economic and social life, as compared with pre-war times, a development that has been pushed forward impetuously in the years from 1920–29 and was brought to a sudden, in some cases catastrophic, stop in 1930 by the economic world crisis."[68] The Depression had a direct and almost immediate effect on each territory's export revenue. In British West Africa, export income was cut in half between 1929 and 1931. The situation was just as dire in East Africa, which already had lower levels of revenue. Only Central Africa (limited to Northern Rhodesia and Nyasaland) experienced a slight increase in exports, mostly from Northern Rhodesia's emerging copper mining industry.[69]

Table 3.5   South African and pan-African coordinating conferences, 1900–1940

1904	Pan-African Veterinary Conference, Bloemfontein
1905	British Association for the Advancement of Science joint session with South African Association for the Advancement of Science, South Africa
1907	Pan-African Veterinary Conference, Cape Town
1909	Pan-African Veterinary Conference, Pretoria
1923	Pan-African Veterinary Conference, Nairobi, Kenya
1929	Pan-African Agricultural & Veterinary Conference, South Africa
	British Association for the Advancement of Science joint session with the South African Association for the Advancement of Science, South Africa
1934	Inter-State Locust Conference, Pretoria*

* This event was initiated by South Africa and included representatives from Angola, Basutoland, Bechuanaland, the Belgian Congo, Northern Rhodesia, Southern Rhodesia, Portuguese East Africa, Swaziland, Kenya, Uganda, Tanganyika, and South-West Africa; see B. P. Uvarov, *Locust Research and Control, 1929–1950* (London: HMSO, 1951), 15.

Table 3.6   Interwar East Africa research coordinating conferences

1931	Agricultural Research Conference at Amani, Tanganyika
1932	East African Soil Chemists at Amani, Tanganyika
1933	Coordination of Tsetse & Tryps Research (Animal & Human) at Entebbe, Uganda
	Coordination of General Medical Research at Entebbe, Uganda
1934	Coordination of Agricultural Research and Plant Protection at Amani
	Coordination of Veterinary Research at Kabete, Kenya
	East African Soil Chemists at Zanzibar
1936	Coordination of Agricultural Research at Amani
	Coordination of Tsetse & Tryps (animal & human) Research at Entebbe
	Coordination of General Medical & Veterinary Research in Nairobi, Kenya

Note: Sources for these events include published conference proceedings, library records, and E. B. Worthington's *Science in Africa*. There were earlier conferences on these subjects in Tanganyika, but none was designated an inter-territorial research conference.

As the economic historians Michael Havinden and David Meredith have pointed out, the Depression had particularly acute consequences for government finances in British Africa. By 1936, territorial governments were spending between one-fourth and two-fifths of their budgets on serving their debts, a far higher proportion than the British territories in the Caribbean and Asia.[70] The dramatic declines in colonial treasuries affected Africa's white settlers and indigenous inhabitants unevenly.[71] Recent historical research disaggregating African and European producers has shown that at some times and places the economic crisis conferred greater power on African producers and rural communities within local economies. When subsistence agriculture, noncash exchange, mixed cropping, and flexible production were all part of the equation, the economic downturn could mean opportunity rather than calamity.[72] For the research officers and administrators who were responsible for studying and attempting to remedy the situation, dire circumstances could justify strategies that might have been dismissed in territories with more capital.

Putting Farmers First? "Native Agriculture" and the Question of Soil Fertility in the Tropics

Agricultural departments' research agendas had begun to undergo a subtle shift away from European and toward indigenous practices even before the Depression. This trend was intimately connected to burgeoning studies of soils, nutrition, and land tenure. In many parts of the world, agricultural and ecological sciences were coming together to form the new field of agro-ecology, but only a handful of specialists who advocated this approach took the trouble to include rural cultivators in their studies.[73] Even fewer drew upon the emerging discipline of social anthropology. Indeed, colonial Africa was the main region of the world in which studies of "native agriculture" were connected explicitly to concurrent developments in ecological and anthropological research.[74] Scholars may lament that these approaches failed to protect subsistence agriculturalists, alter capitalist modes of production, or transform hierarchical power structures and economic policies that constrained African producers. But the attention that colonial officials and scientists paid to local agricultural techniques and vernacular knowledge was still unprecedented. The roots of agroecology and a "farmer first" mindset can be found in British colonial Africa.[75]

Shortly after the First World War, the American ecologist, Homer Shantz, urged administrators in Britain to study "native agriculture" and soil conditions in tropical Africa. Shantz had been a student of the ecologist Frederick Clements and was employed in the U.S. Department of Agriculture (USDA) when the British government's expert committee for the Paris Peace Conference requested an overview of African land and soil types.[76] A small group compiled a preliminary report under Shantz's leadership. Shantz was then invited to join the Smithsonian Institution's 1919–20 African expedition. He and the museum's field naturalist, Henry Raven, spent nearly a full year traveling overland from the Cape to Cairo. Shantz collected 2,600 botanical and soil specimens, while Raven gathered more than 1,000 mammals, birds, reptiles, and fish.[77] Upon his return, Shantz enlisted a USDA soil scientist, Curtis Marbut, to help him analyze the results. Together they produced a continental map classifying "the lands of Africa according to their potential productivity," which was intended to facilitate "development."[78]

During his travels, Shantz made extensive notes on the agricultural practices of the people he encountered, although only a tiny fraction of this information was included in his 1923 book.[79] When he was asked to join the Phelps-Stokes Fund's Commission on East African Education, he had an opportunity to extend his field studies.[80] The project had the consent of the Colonial Office, and its activities in the field dovetailed with those of the Parliamentary East Africa Commission on which Ormsby-Gore served. Historians have frequently noted that the parliamentary commission raised the profile of scientific research in relation to imperial development, but Shantz's study more closely reflected emerging trends in the agricultural departments across British Africa. The British commissioners argued with great conviction that "the potential riches of East Africa lie in the cultivation of the soil. There is a vast area of the most wonderful land, adequately watered and capable of yielding economic crops of almost all tropical, sub-tropical, and temperate varieties."[81] Shantz criticized both their assumption that the region's soils were highly fertile and their almost total omission of African cultivators. His contribution to the East African Education commission's report addressed these issues at length.

Shantz began his analysis with the amount of land under cultivation and the number of livestock held by Europeans and indigenous Africans in nine East African territories. Although European-owned farms produced the bulk of the crops sold for export, Africans carried out their own agricultural production on "about 23 million acres, or over 20 times the amount of land under European cultivation."[82] Shantz found similar disparities within the livestock industry: Europeans owned many fewer domesticated animals than indigenous peoples did.[83] Approximately 95 percent of Africans in the region continued to pursue some form of subsistence agriculture, which was "much diversified" and, as often as not, was conducted under the leadership of women.[84] Shantz argued that it made more sense for "the agricultural development of the country" to rely heavily on Africans' own initiative rather than anything imposed by Europeans. He boldly declared:

> From the agricultural point of view . . . the wellbeing of the farmer is the first consideration. The interests of the farmer, his family and community come first. . . . The agricultural methods of the Natives in Africa have often been condemned as shiftless, wasteful and destined to decrease the productivity of the country. Again, one meets continually the statement that the Native knows nothing about crop production. These statements, in a way, reflect the attitude of the European toward the Native, the assumption being that since he does not follow our methods and our practices he must be essentially wrong. But there are many testimonies in the literature to the effect that the

Native is an excellent agriculturalist. It is well to bear in mind at the start that very little serious attention has been given to his methods and practices and that there is no adequate scientific study of Native agriculture on which to base sound conclusions.[85]

During his two trips to Africa, Shantz had not carried out systematic surveys of the sort he deemed necessary to evaluate Africans' agricultural methods, but he offered some hypotheses based on his observations. He was struck by the wide range of environments he had traversed, from the temperate highlands of Ethiopia and Kenya to the drought-prone savannas in Tanganyika and Uganda, and from the lightly settled forested areas of the Rhodesias to the densely populated regions around Mount Kilimanjaro and Mount Kenya. Adapting agriculture to these settings required a high degree of knowledge and skill. Shantz speculated that Africans associated particular crops with different soils and types of natural vegetation. Plants that needed more water were grown in locations closer to streams, where he observed "irrigation aqueducts" of considerable "engineering skill," while those that could withstand drier soils were grown elsewhere.[86] On hillsides, around tree stumps and anthills, and near small depressions that collected flood water, Shantz observed methods of cultivation that showed "wonderful ingenuity."[87] Happily, most indigenous cultivators in these areas appeared to be free "from the tyranny of the plow"; "if they used plows, it would not always be easy to choose the best land."[88] He also praised pastoralists for developing "the highest type of agriculture possible" in "semi-desert and desert lands." Shantz reminded his readers that in these areas, just producing a subsistence was a real achievement.[89] "Modern agriculture" was not necessarily more effective; indeed, its methods often caused soil erosion, decreased yields resulting from monocropping, and damage from pests and diseases. What anthropologists now call shifting cultivation might well be more productive, Shantz concluded. "Natives, by their method of abandoning the land and taking a new piece, accomplish what the European, with all his staff of scientifically trained men, has not yet satisfactorily accomplished. They escape the problems of soil fertility and physical condition, and the question to a very great extent of plant diseases. They allow the land to lie fallow, to go back to grasses and brush, and return to it only when they are sure they can secure a good crop."[90]

Above all, Shantz sought to guard against "dogmas," whether new or old, that might hinder successful agricultural development in the region. "Modern agriculture," he asserted, incorporated "an unusually large amount of dogmatic teaching said to be based on [a] scientific foundation." Plowing, fallows, and fertilizers "all . . . have proved failures under certain conditions, although they are desirable under others. We have swung from one extreme to the other, and the 'slogans' of agriculture must be questioned very seriously before they are allowed to interfere with the well-established customs of the Native."[91] Since women managed food consumption as well as a substantial share of its production, he recommended that their practices be studied as well. "To interfere with the Native's diet would seem even more serious than to interfere with his agricultural methods, until we are sure that the changes proposed are not to his detriment."[92]

Shantz's unorthodox analysis deepened interest among a handful of reform-minded imperial statesmen and environmental scientists in the fertility of African soils and moved some of them to take Africans' cultivation techniques and knowledge more seriously. The reproof Ormsby-Gore delivered to his British colleagues for their ignorance of Africa's tropical soils was backhanded praise of Shantz's work. Joseph Oldham thought highly of his studies as did Ormsby-Gore's compatriot on the East Africa Commission, the MP Archibald Church, who acknowledged Shantz's views in his book on the "experiment of tropical development" in East Africa: "Dr. Shantz . . . pleads strongly for a careful study of native methods, for which he expresses no little admiration."[93] As Shantz himself knew, colonial states' agriculture departments had already begun to explore these questions for themselves, albeit sporadically and with an eye to potential state revenue.

Odin Faulkner put the study of African methods of cultivation high on the agenda of Nigeria's agriculture department when he became its director in 1922. In an inaugural article on the aims and objects of his department, he insisted that technical officers should obtain "a thorough knowledge of the minutiae of local farming, such as can only be gained by one who has actually engaged in farming in the country." That would require "experiments and investigations . . . which aim . . . at the gradual acquisition of a detailed and accurate understanding of the local systems and methods of agriculture, and the scientific and economic facts which have led to their evolution. Such knowledge is needed because, without it, the department's activities may be ill-guided; and without such knowledge the department will command no-one's confidence."[94]

Agriculture departments were charged with promoting "improvement," not merely with supporting subsistence agriculture and small-scale production. Defining what constituted agricultural progress and what ends it was to serve was fraught with contradictions: African farmers, white settlers, colonial states, and imperial authorities all had distinct and potentially conflicting ideas and interests. Before we can understand the broad influence of agroecology and vernacular science on inter-imperial development policies during the interwar period, we need to consider the ways territorial and metropolitan aims were bridged. The trajectory of studies of agriculture in Northern Rhodesia is a prime example of the ways scientists, colonial officials, and technical specialists wrestled with these issues. Their work demonstrated the value of new ecological models, disproved previous theories about the lack of productivity of indigenous agricultural methods, and pointed to new directions that influenced imperial thinking as well as the practices of colonial states.

The Roots of Agroecology: Field Studies of African Agriculture in Northern Rhodesia

In 1924, shortly after the white settlers who controlled Southern Rhodesia were granted responsible government, political authority over Northern Rhodesia (now Zambia) was ceded to the Crown by the British South Africa Company, and the region became a protectorate under the direct control of the Colonial Office. The new territorial dependency had approximately 5,500 Europeans (including government officials), most of whom were concentrated in the booming Copperbelt, and roughly 1.1 million Africans across its 290,000 square miles.[95] Accordingly, colonial authorities adopted a policy of indirect rule. Conditions in Northern Rhodesia were particularly well suited to the approaches Faulkner and Shantz advocated; agricultural research there eventually came to be focused on indigenous cultivators. Indeed, Northern Rhodesia served as a laboratory where new models of "ecological development" were tested.

In late 1927, the newly founded Agriculture Department in Northern Rhodesia initiated two field studies of African cultivation that had profound effects both on the policies of the colonial state and on metropolitan points of view. The colonial administration was faced with the question of how to "improve" agricultural pursuits both for Africans and for white settlers, most of whom had moved there when the British South Africa Company was in charge. White farmers had tried two export crops, maize and tobacco, but the landlocked country's lack of transportation networks meant that markets were relatively inaccessible. The acting director of agriculture, John Smith, worried that a dip in international prices would spell ruin for anyone without "alternative crops."[96] Recent attempts to grow cotton had also failed, he reported; farmers had rushed into it too quickly and "sacrificed good principles of agriculture" in the process, causing "considerable loss to not a few settlers."[97]

In order to ascertain better ways to cultivate these and other crops, including cereals, grains, and oil seeds, the department set up a 265-acre research station at Mazabuka situated in the Southern Province near the railway line, which ran from Victoria Falls through the middle of the territory and into the Belgian Congo (see plates 6 and 7). The Mazabuka district had the largest number of cattle in the territory (over 160,000 head), many of which were African breeds owned by the Plateau Tonga.[98] Mazabuka soon became the headquarters for the veterinary and forestry departments as well. In addition to "field and plot experimentation," the station was equipped for "laboratory investigation," and veterinary and agricultural officers were each provided with their own "research wing."[99]

Right from the start, officers in the agriculture department were well aware that they needed to conduct "field experiments" and investigations across the entire territory. In 1927, Smith wrote to his superiors: "Information is desired which can only be obtained from an ecological survey of the land available for agriculture" that would classify "the chief types of vegetation in relation to climate, rainfall and soil types of their habitats."[100] The territory lacked the funds for such a study, so Smith and his senior research officer drew up reports of Northern Rhodesia's scientific priorities that Smith submitted to the first Imperial Agricultural Research Conference slated to convene that October in London.[101] Unwilling to wait for a metropolitan response, Smith decided to proceed on a smaller scale and arranged for a coordinated research project to begin the following year. This plan linked crop trials at Mazabuka to field experiments in the Abercorn District (now Mbala) in the northernmost reaches of the territory.[102]

These two field studies were designed to complement each other. The work in Abercorn was carried out by Unwin Moffat, one of the few research officers who had been born in Northern Rhodesia and was conversant in local languages, including Bemba, which was spoken by large numbers of people in the area. Moffat had been appointed to the department with a partial fellowship from the Empire Cotton Growing Corporation specifically to study "native methods" of cultivation on the Tanganyika Plateau, which bordered the Belgian Congo (see plate 6).[103] Conducted in the midst of village settlements, his work explored two forms of shifting cultivation that colonial administrators and British South Africa Company officials characterized as "slash and burn" agriculture and considered inefficient and wasteful. Moffat and his colleagues were concerned that these methods exacerbated deforestation and soil erosion, especially in the limited space of "native reserves," regions that had been set aside solely for indigenous inhabitants and which were one way the colonial state tried to control populations (see plate 6). Moffat's field research was meant to ascertain these practices' relative productivity and effects on the land and to explore ways of improving upon their techniques.

The second study was coordinated by T. C. Moore, an instructional officer who later worked with the Ecological Survey, and was carried out at the Mazabuka Research Station. Moore identified a range of "indigenous crops" that were cultivated across the territory, including peanuts, beans, peas, finger millet, cassava, and sorghum. By this time the agriculture department employed eleven Africans as subordinate staff, so Moore and a group of "European and native collectors" began their work by gathering seed samples from across the territory and interviewing African cultivators, taking down "all possible information regarding the vernacular names, areas where grown, season when grown, climate conditions, types of soil, . . . how planted, reputed yields, resistance to attack by animals, insects or other pests, the popularity of the variety compared to other local varieties, its keeping qualities and any other facts of possible economic importance."[104]

While Moffat examined agricultural practices, offering detailed observations of the advantages and limitations of different agricultural methods, Moore spent his time evaluating crops and seeds. Moore was interested in trying to balance the aim of increasing yields with a desire to understand cultural preferences for particular crops. Both Moore and Moffat pursued their studies "with a view to future improvement and development." At the inception of his study, however, Moffat acknowledged that Africans did not share this imperial aspiration: "There is no doubt about the keenness of natives to take up agriculture, but at present, they do not see any reason why they should change their methods of growing their foodstuffs, because they are able to produce all the food they require."[105]

Northern Rhodesia's colonial officials sought to expand crop production for both internal and external markets. They anticipated that the mining industry in the Copperbelt, like that in the adjacent province of Katanga in the Belgian Congo, would raise the demand for food supplies. They were also aware that industrial development in Southern Rhodesia and South Africa had prompted "a great exodus of men from the villages" to urban and mining centers.[106] In the late 1920s, officials were increasingly anxious about the effects of labor migration on social dynamics in the territory, especially on the women and children who were left behind and remained responsible for food production. Agriculture department staff knew that women took the lead in cultivation, but they worried that, in the absence of men, the women might lack the labor power to move to new sites, which could lead to soil depletion. Scientific advisors, too, expressed concern that recent economic developments had negative effects on many people's lives.[107] H. C. Sampson, a botanist at Kew Gardens who spent time in Northern Rhodesia in 1924, reported to the Colonial Office that in Fort Jameson conflict had erupted between village chiefs and women whose husbands had departed for Southern Rhodesia. Agricultural development programs that enabled people to "earn money in their own country" by "growing crops, the produce of which can be marketed," might mitigate social instability.[108] Unwin Moffat's and T. C. Moore's agricultural studies were the first attempts to ascertain which crops and production methods the colonial state might try to develop.

Moffat spent four years conducting scientific and ethnographic research on alternative methods of cultivation. When he presented his findings, he explained that "native methods of agriculture are usually looked at from the point of view of the highly specialised European agriculturalist." This approach was bound to lead officials astray, because "little account is taken of the local conditions . . . [or of] the temperament of the native farmers." "In trying to improve native agriculture," he insisted, "it seems essential that the people themselves should be considered. The best and most natural way for introducing any improved methods should be by using their own established methods as a basis on which to build." In Abercorn, Moffat not only observed these methods as Africans conducted them but also set up experimental plots that replicated local practices, including planting and harvesting times, sowing and weeding patterns, and crop rotations.[109] He enlisted a range of individuals from the region to help conduct these field experiments.[110]

"Citemene" was the name Bemba-speaking peoples gave to a method of shifting cultivation that relied heavily on cutting and burning trees to prepare circular sites for planting (figure 3.1). The 1,200 square mile area he studied included at least "1,346 'chitemene' gardens." According to his observations, "over 21 square miles of forest was cut in order to make these gardens. This is an annual operation." Mambwe-speaking peoples pursued a form of production that relied on "mound-cropping" and a system of green manuring and fallows. Each garden in this system was "about one and a half acres in extent," and each household cultivated at least one garden. Since multiple crops were grown during the three- or four-year cycle, the fallow period could range from months to a full year depending on the pattern of crop rotation.[111] One of the main crops in both systems was *Eleusine coracana*, finger millet (figure 3.2).[112] When Moffat started his research, he had little information on the relative yields of citemene and mound cropping. To ascertain whether indigenous methods were as

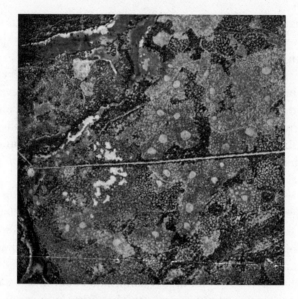

Figure 3.1. Aerial survey of citemene production, Northern
Rhodesia, circa 1930s. The original legend reads, "Incursion of
Northern Chitemene 'large circle' gardens in an area in south
Mpika District formerly cultivated by Southern Chitemene
methods. Large circles with surrounding cut areas have been
cleared out of regenerating woodland still showing sites of
former small circles. Small white areas to the left are subsidiary
gardens." Source: C. G. Trapnell, *The Soils, Vegetation, and Agriculture of North-Eastern Rhodesia: Report of the Ecological Survey*
(Lusaka: Government Printer, 1943).

productive as conventional European methods under African conditions,
he added a third set of plots in which he followed the "normal" practice of
straight-line hoeing.

These experiments demonstrated that the prevalent assumptions fa-
voring European over African methods of cultivation were, as Moffat had
suspected, entirely misplaced. The millet yields for the European test plots
were "markedly inferior" to those for citemene, producing only 440 pounds
per acre while citemene produced 1,200 pounds per acre. "In 1928–29,
the hoed plots were so inferior that the crop, after a very uneven germina-
tion . . . was not worth reaping." Mound cropping was more productive
than straight-line hoeing, but its yields were outstripped by the citemene
plots by an average of 500 pounds per acre.[113] Moffat deemed these results
encouraging, since he thought a system of fallows and rotations might
eventually be substituted for citemene, but citemene was clearly the most
productive of the three techniques.

Moffat and his colleague, the soil chemist H. B. Stent, conducted several tests comparing artificial and organic fertilizers and analyzing the effects of citemene burning on soil fertility. Animal manures were a complete failure: "the weeds could not be kept under control" and "the yield was depressed." Phosphate- and potassium-based fertilizers had little effect. "Nitrogenous" fertilizers stimulated plant growth, but made them susceptible to diseases and had only a "trifling" effect on yields. Citemene practices, by contrast, rid the plots of weed seeds and produced changes in soil composition that had a positive effect on crop yields. When Stent tested the mineral content and pH level of soils in and near citemene plots, he found that burning "appreciably enhanced" the quantities of phosphate and potash in the soil. "The freshly burnt soil is more highly saturated with calcium, which accounts for its improved physical condition so noticeable in the field. . . . The ash . . . add[s] large quantities of mineral plant food" and has a "beneficial effect which shows itself in an increased mineral content in the grain and consequently in the improved dietetic properties in the meal, which forms the staple food of the people."[114]

These striking and robust results had a ripple effect throughout the territory. C. J. Lewin, formerly a research officer in Nigeria's agriculture department, had been appointed chief agriculturalist in Northern Rhodesia

Figure 3.2. Citemene garden, Northern Rhodesia, circa 1930s. The original legend reads, "Millet ash culture. A newly prepared Northern Chitemene garden, showing the deep ash layer still hot after the burning. After planting catch-crops the burnt soil will later be thrown over the millet seed with the hoe. Trees filled; the well-known lopping practice is not an invariable characteristic of Chitemene." Source: C. G. Trapnell, *The Soils, Vegetation, and Agriculture of North-Eastern Rhodesia: Report of the Ecological Survey* (Lusaka: Government Printer, 1943).

in 1930 and assumed the directorship in 1932, the same year Moffat and Moore submitted their findings.[115] In an editorial written to accompany their studies, Lewin announced the department's new position toward African agricultural practices.

> Were a history of attempts to improve native agriculture written, it is certain that many passages would substantiate the truth of the proverb concerning the impetuosity of fools and the timidity of angels. Recognition of the inherent soundness, under natural conditions, of native agricultural practice has only become general in recent years. Practices apparently contrary to the accepted principles of good farming, usually prove on investigation to be the best possible in the circumstances under which the native cultivator works. Shifting cultivation is frequently condemned, yet what more easy method of raising food could be devised where cultivation is done by hand and abundance of land is available? . . . It behooves an agricultural department to investigate local practices with the utmost care before presuming to attempt to improve them.[116]

Lewin's editorial signaled a marked departure from the position expressed by Frank Stockdale, the Colonial Office's newly appointed agricultural advisor in 1929. Stockdale had informed his colleagues: "the indigenous population of Northern Rhodesia have not evolved any system of agriculture worthy of the name and, consequently, further extended development by them or by settlers must be based upon a system of trial and error."[117]

The reorientation of agricultural research that took place in Northern Rhodesia between 1929 and 1932 coincided with and accelerated a shift in thinking that took place in the metropole. The confluence of these developments led to the design and funding of the Ecological Survey, which was initially conceived as a two-year project to assess the territory's agricultural and forestry resources.

In 1929, Northern Rhodesia submitted its proposal for an ecological survey to the Colonial Office after enlisting the support of both Oxford University's Forestry Institute and Kew Gardens. Two years before, the Forestry Institute sent Ray Bourne "to report to the local government on forests."[118] Working closely with Thomas McEwan, the senior research officer, and John Smith, the department director, Bourne seized the opportunity to demonstrate the value of employing "aerial survey in the economic development of new countries." In his final report, which was sent on to the Colonial Office, he endorsed McEwan and Smith's plan to survey the territory's landscape and natural resources both by air and on foot.[119] Reconnaissance surveys with a comprehensive design were essential because "the distribution of vegetation is subject to the varied influences of climate, geology, soil, disease, and man. . . . The special study of the changes taking place is the subject of the comparatively young science of Ecology."[120] Bourne recommended that "each field party be representative of several sciences." The surveyors should be accompanied by "an agriculturalist and a forester, both having special knowledge of the local flora and of the principles of ecology and soil science," as well as "a geologist, experienced in the application of geophysical methods of survey." He argued that approaching land-use problems in Northern Rhodesia "from an ecological point of view" would enable officials and scientists to account for the myriad of interacting variables that affected productivity and help them make better decisions about development.[121] Northern Rhodesia's proposal was also sent to H. C. Sampson at Kew, who expressed great enthusiasm for comprehensive field studies. An ecological survey supported with metropolitan funds for colonial development could set a new direction for the future, he suggested. "This is I believe the first time that a vegetation survey has ever been proposed in any of our colonies before it was too late to avert the damage which has been caused by both promiscuous and by extensive settlement or by the development of a new phase of native agriculture."[122] Thanks to these specialists' positive reviews, Stockdale and the advisory committee for the Colonial Development Fund (CDF) were convinced to support the plan.

The CDF had just been established, and in the first six months of its existence it made three grants for scientific research and agricultural development across tropical Africa—all to Northern Rhodesia.[123] As the Colonial Office official, Acheson, explained, "There is now an opportunity to start the intensive development of the territory on scientific lines. If our knowledge of African conditions is to be useful, it cannot be restricted to one territory alone, but must form a connected whole, so that the work done in each district [i.e. dependency] takes its place in an organised scheme of enquiry." Not only would the territory-wide Ecological Survey "contribute to our fundamental knowledge of African flora," but it would also "link up . . . with the ecology of Nyasaland," with the research work at Amani, and with "the tsetse research in Tanganyika."

> Instead of assuming that a certain amount of knowledge concerning the [territory] . . . has already been attained by day to day experience, the ecological survey proposes to acquire the necessary first experience in a scientific man-

ner. *This has never been attempted in the course of any colonisation up to the present day.* It is possible that many mistakes might have been avoided in the development of various colonies had their flora and fauna been scientifically studied before intensive development began. The time for such a study of Northern Rhodesia would appear to be the present.[124]

What John Smith had envisaged as a two-year survey expanded into a fifteen-year endeavor that encompassed nearly the entire territory, making it one of the most extensive and thorough bioscientific projects in the history of British Africa.[125] The final soil-vegetation maps and the field notebooks kept by its primary coordinator, Colin Trapnell, have recently been reproduced by Kew Gardens, "because no comparable study in terms of scope has superseded it."[126]

Consolidating Vernacular Science: The Ecological Survey of Northern Rhodesia

At the time of its conception in 1927–28, Northern Rhodesia's Ecological Survey was designed to serve a range of development-related concerns, including an assessment of white settlers' agricultural practices and the inadequate quality of the grazing land for European cattle. By the time of its inauguration in the summer of 1932, although some attention was paid to Europeans' needs, the fieldwork had shifted toward two main aims: an investigation of local soils and vegetation, and a detailed account of the various types of "native agriculture." Moffat's and Moore's studies, published just as the survey was launched, drove home the importance of including sociological analyses alongside ecological investigations and the present and potential productivity of indigenous agricultural methods. In a 1933 review of the state of agriculture in British Africa, Hugh Wyndham, a member of Milner's Kindergarten and an affiliate of Chatham House, emphasized that: "Instead of, as formerly, regarding Africa as a potential supplier of the tropical raw materials required by the home country, modern African Agriculture Departments [now] take Native agriculture as it has been practised in the past as their starting-point, and concentrate on trying to increase its output from this point of view. It is a development of the theory of trusteeship. The process has demonstrated that the African Native is not, as is so frequently stated, specially unprogressive."[127]

The project's research priorities were also influenced by the 1932 report of the Agricultural Survey Commission, which addressed major issues related to land policies.[128] The commission's undertaking, however, was more administrative than scientific. Approximately 3 million acres of the territory's "best arable land" had already been appropriated by Europeans, while another 5.5 million acres remained under the jurisdiction of the British South Africa Company and the copper-mining industry, which from the mid-1920s on served as the main engine for the territory's economic growth.[129] The agricultural commission studied those areas not yet taken up by white farmers and determined their suitability for European-style agriculture.[130] With these questions addressed, at least to the satisfaction of the Legislative Council, the Ecological Survey focused on exploring the 71 million acres of "native reserves," the 11 million acres of forest and game reserves, and the nearly 94 million acres still unallocated. The project's concern with Africans' agricultural practices became so central to agriculture officials that eventually European settlers protested "the entire lack of interest shown by the Department towards the white farmer."[131] Over the course of the 1930s, the director, C. J. Lewin, often found himself "doing battle with the settlers" because he made research into "native agriculture" a top priority.[132]

The researcher whose work became most important to the Ecological Survey was a relative neophyte who was keen on newly emerging ecological perspectives. Colin Trapnell was only twenty-two when the Colonial Office approved his application for employment in Northern Rhodesia. While studying classics at Oxford, he cultivated his interest in ecology through the university's Exploration Club, which he cofounded with the ornithologist, Max Nicholson, during his third year. They invited Charles Elton to serve as its faculty advisor; later he was joined by Arthur Tansley and Julian Huxley.[133] Club members were taught by its faculty sponsors that its purpose was "ecology, scientific nature study, [with a] minimum of collection." Their travels were designed less to serve museums and the field of systematics, which addressed the "what" of nature, and more to investigate the "how and why": the "ecology and aesthetics" of the regions they visited.[134]

Arthur Tansley shepherded Trapnell through the application process, putting him in touch with other senior researchers and setting up a meeting with the Colonial Office recruiter, Ralph Furse.[135] Initially, he was offered a position in systematics at the Botanical Garden in Ceylon, but Trapnell preferred ecological fieldwork. During the interview, an official recalled receiving Bourne's report and explained that Northern Rhodesia had just such an opening. Before Trapnell could take up the post, he needed more formal scientific training. So, like many recruits to the colonial technical services, he spent a year (1929–30) at the Royal College of Science taking courses on soils, chemistry, botany, mycology, and entomology. As part of his fellowship, he visited research facilities with any bearing on "ecological methods," including the Imperial Forestry Institute, where he met Bourne, Kew Gardens, where he worked briefly with Sir Arthur Hill, and several of the Imperial Agricultural Bureaus. He completed his training in South Africa, spending two months with a field study under I. Pole Evans in Pretoria and then another month meeting with research officers responsible for botany, ecology, and agriculture. By the time he arrived in Northern Rhodesia in December 1931, so much time had passed that officials at the agriculture department had almost forgotten he had been hired.

One of the first hypotheses tested by the Ecological Survey in 1932 was whether African cultivators were conscious of relationships between soil and vegetation types, on the one hand, and "agricultural practice and potentialities," on the other. Trapnell and his collaborator during the first three years, J. N. Clothier, a research officer fresh from training at the Imperial College of Agriculture in Trinidad, tested this hypothesis in two different sites: the Kafue River Basin, a region of approximately 13,000 square miles that bordered many European farms and was bisected by the railway line, and a 27,000 square mile portion of Barotseland, the westernmost province that extended to the Angolan border.[136] Echoing Moffatt, they reported to their supervisors that "Barotse agriculture would call for no intervention but for the fact that there is urgent need for the development of a cash crop or of some product capable of export."[137]

During their safaris, Trapnell and Clothier traversed an average of eighteen miles each day, almost always on foot, which was roughly the distance between villages, and employed one to two dozen porters and at least one translator.[138] Typically, in each village they met with the elders and "asked a series of routine questions, [including] methods of land selection, processes of clearing land, planting, duration of cultivation, rest periods."[139] They supplemented their field data with information from administrators and agricultural officers, including Unwin Moffat, C. J. Lewin, and William Allan, who were enduring allies of the project, as well as specialists from the Amani Agricultural Research Station in Tanganyika, which Trapnell visited in 1935.[140] The amazing heterogeneity of the agricultural techniques they encountered is recorded in Trapnell's diaries and notebooks, which fill more than 1,200 printed pages.

In their first full report for 1933, Trapnell and Clothier claimed that their hypothesis of a "direct correlation between vegetation types and agricultural practice . . . has by now been established by abundant evidence."[141] As Trapnell later reported in the Kew Gardens Bulletin, "It was found that [the native] recognised, and employed in the selection of his cultivation sites, the same types of bush and plant indicators as those which the survey would employ, and had a definite . . . conception of differences in fertility which they indicated."[142] Referring to this knowledge as "intuitive ecology," they praised its "astonishing" "precision" and deemed it "infallible"—"so long as the native is in his traditional environment" and retains control over patterns of migration within the region.[143] There was considerable "scope for improvement," however, since some groups were more success-ful than others in avoiding famines and producing sufficient subsistence crops; they observed that "poor land and backward villages go together." But, they concluded, "native agricultural systems are normally admirably adapted to their environment [and] any criticism of them must for this reason be made with the greatest caution."[144]

The Ecological Survey's emphasis on sociological dimensions of agricul-ture was strengthened over time by the presence of Audrey Richards, a so-cial anthropologist, and by the Rhodes-Livingstone Institute (RLI), founded in 1937 to focus on applied anthropology.[145] Richards had completed her doctoral research on nutrition, which she called "a biological process more fundamental than sex," using published accounts of "patrilineal, cattle-loving Bantu" societies in southern Africa.[146] In 1929, she sought the op-portunity to do fieldwork, shifting her focus to a matrilineal society where she could explore "the life of the native women—with a view to increasing our knowledge of their economic status and the possibilities of their future education."[147] Thanks to the backing of Bronislaw Malinowski, Reginald Coupland, Frederick Lugard, and Gilbert Murray, Richards received a small grant from the Rhodes Trust in Oxford. The trust initially declined to fund any anthropological research, but, as Coupland had pointed out when he asked them to reconsider, "Miss Richards does not propose to work at the old-fashioned anthropology—which concerns itself with much that is of no practical value—but at the new social anthropology which studies the present-day life of the native in relation to his environment."[148] This expla-nation not only persuaded the Rhodes Trust to fund Richards's application but also helped its director, Philip Kerr (Lord Lothian), to see the signifi-cance of anthropology for the multidisciplinary research program he envis-aged for the African Survey.

The International Institute of African Languages and Culture's interest in understanding the effects of labor migration on land tenure and agricul-ture and Richards's own knowledge of various regions of British Africa led her to study Bemba societies in Northeastern Rhodesia.[149] Her fieldwork sites overlapped with those of Unwin Moffat as well as Lorna Gore Browne, an "honorary agricultural officer" who oversaw the Shiwa Experiment Station in Northeastern Rhodesia from the mid-1930s onward. Richards located her 1939 book, *Land, Labour, and Diet in Northern Rhodesia*, which drew upon Moffat's and Trapnell's field notes, on "the border-line between two different sciences, biological and social."[150] While her work painted a richer picture of the social side, the Ecological Survey's reports provided a broader understanding of the environmental. What they shared was a concern to explore "indigenous" knowledge, which their research made increasingly visible and tangible.

Richards, like Trapnell and Clothier, set out to test a hypothesis about agricultural practice: "Our first problem is therefore to discover whether some traditional body of knowledge guides the Bemba in the choice [of land to cultivate], and if so, what measure of success or failure their method of land selection gives."[151] Although she pointed out religious and political motives for site selection and instances in which sites with poor soils were chosen, Richards concluded that the Bemba used plant species as a gauge of soil fertility: "Some Bemba told me, in answer to questions, that certain trees and grasses indicated suitable soil for the making of gardens, and I collected from different, mostly elderly informants, a list of ten such trees and four grasses."[152] Not all her informants agreed, however, and she realized that many Bemba employed an explicit experimental method called *ukupansa* (to try the soil) and *ukueshya pambilibili* (to try at random), recognizing that failure was sometimes necessary to eventual success.[153] Richards was convinced that "a certain body of knowledge" guided land selection. Like Trapnell and Clothier, she saw this wisdom as a product of time and experience and worried that it would be unable to withstand the rapid changes resulting from male labor migration.

In their historical study, Henrietta Moore and Megan Vaughan have pointed out that in the published accounts of these field-workers, "agricultural practices, cropping patterns, and even such physical structures as soils seem constantly to be in danger of escaping from the typologies created to contain them."[154] These typologies, they rightly argue, were prompted as much by a desire to control as by a need to understand. The representations generated by the Ecological Survey emerged from a series of underlying imperatives concerning colonial intervention: production for markets as well as subsistence, a preference for settled rather than shifting cultivation, anxiety about environmental deterioration, and ambivalence about existing divisions of labor between men and women in agriculture. In light of all these considerations, what seems most surprising is that the field research succeeded to such a large extent in capturing the constant variation, movement, and change that was inherent in indigenous peoples' methods of cultivation. The "unfathomable intricacy of the classification systems designed by" Trapnell and Clothier was indeed, as these scientists put it, a new "ecological approach" to African agriculture.[155]

> In its broadest application it [ecology] may be defined as a study of plant, animal, or human life in the light of the control exercised upon living things by the external factors of the environment and by their own inter-actions. Agriculture, as a human activity concerned with living things, admits of such an approach. Instituted with a view to making an ecological investigation of the vegetation of the country, and thereby of its natural resources, *the Survey has extended the application of ecology to native crops and agricultural customs and finally to native agricultural development*. It may be said to have provided a new approach to native agriculture whereby native practices can be investigated as the product of their environment, and their development guided by comparative study and scientific knowledge of the country.[156]

Trapnell and Clothier did not "essentialize" Africans' knowledge of the nonhuman world, as some contemporary critics might charge; they themselves cautioned against this notion, as they did against static conceptions of the environment or a rigid separation between nature and culture. The fundamental distinction they tried to make was between an appropriate development of agriculture, based on a comprehensive understanding of complex socioecological patterns and an inappropriate one that placed too high a premium on cash crops to the detriment of indigenous people's livelihoods.

Significantly, Trapnell and Clothier used the word "system" throughout their study. At its core, the concept was a nuanced attempt to engage empirically with interwoven social and biological variations. Arthur Tansley, one of Trapnell's mentors, had introduced the term "ecosystem" just a few years earlier.[157] According to Tansley, an ecosystem was both ontologically real and a unit of analysis that was "isolate[d] mentally" and therefore "partly artificial."[158] By combining the idea of a "physical system" with the concept of ecology, Tansley argued, scientists could describe patterns that were inherently complex. Ecosystems consisted of "components that are themselves more or less unstable—climate, soil, organisms. . . . The relative instability of the ecosystem, due to the imperfections of its equilibrium, is of all degrees of magnitude, and our means of appreciating and measuring it are still very rudimentary."[159] Humans were dynamic agents within ecosystems: "human activity finds its proper place in ecology . . . [when]

regarded as an exceptionally powerful biotic factor which increasingly upsets the equilibrium of preexisting ecosystems and eventually destroys them, [but] at the same time forms new ones of a very different nature."[160] Tansley's idea of "dynamic equilibrium," a system constantly evolving in response to changed conditions, was captured in the concept of a "complex system."[161]

Trapnell, Clothier, and Richards' investigations of indigenous people's environmental knowledge helped to cement Northern Rhodesia's agriculture department's policy of investigating Africans' "principles of land selection" and, more broadly, of "native" norms and practices regarding land tenure before undertaking "improvement schemes" or forcibly moving portions of the African population from one region to another.[162] Northern Rhodesia was becoming a model for a new kind of integrated research and "ecological development," thanks both to the sociological emphasis of the Ecological Survey and to the investigations of land rights, political economy, and social organization conducted by social scientists affiliated with the Rhodes-Livingstone Institute.[163] The Ecological Survey's published reports, which appeared in 1937 and 1943, helped to dispel previous assumptions that African agriculturalists in Northern Rhodesia were "backward" and to persuade policy makers that agricultural improvement must be based on comprehensive ecological research. An educational officer

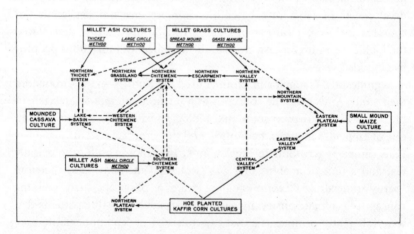

Figure 3.3. "Probable Origins and Development of Agricultural Systems in Northeastern Rhodesia." The original legend ends with this explanation: "The main influences from various quarters which seem to have contributed to the make-up of the systems are shown by arrows with solid lines and lesser influences by arrows with broken lines." Source: C. G. Trapnell, *The Soils, Vegetation, and Agriculture of North-Eastern Rhodesia: Report of the Ecological Survey* (Lusaka: Government Printer, 1943).

with a personal interest in African ecology remarked in his review of the second report: "Complicated diagrams express the inter-relations of these [agricultural] systems and their influence on one another. The popular idea that African agriculture is a crude and inadaptable affair is completely contradicted by the Survey's findings. . . . African agricultural practice is very closely adapted to local conditions, and changes with them"[164] (figure 3.3). Clement Gillman, one of Tanganyika's engineers and a self-professed "human geographer," declared that the Ecological Survey's reports provided "those ecological fundamentals without which any talk or action about 'development' of African native communities is, at its best, pious and sterile lip-service, or, at its worst, disastrous guess-work and blundering."[165]

"Indirect Agriculture": Inter-African Trends Arising from Localized Research

The history of British efforts to improve agriculture in its African colonies is filled with the kind of "guesswork and blundering" Gillman deplored. Significantly, it was technical officers and field scientists who first pointed out these problems, often in retrospective analyses of their own work. After spending more than a decade in Nigeria's agriculture department, Odin Faulkner and his assistant director, James Mackie, published *West African Agriculture* in 1933 in the hope of teaching best practices to incoming "candidates for Government service . . . in the Administrative and Agricultural Departments." Their introductory chapter roundly criticized Nigeria's administrators for their misplaced goals and mistaken assumptions, which were especially serious before the First World War. Their mandate "to achieve . . . quick [financial] returns left [the agricultural staff] little opportunity for . . . a study of local farming and local conditions as alone could show what improvements were really feasible and likely to commend themselves to the native farmer. In the past, Europeans have gone much further than merely to suggest, and have recommended, advised, persuaded, almost forced, the farmer to adopt their proposals, often without having first attempted to ascertain whether they were acceptable to him." Many of these efforts were "a complete failure" and caused the "native farmer" to have a "justifiable suspicion of all new ideas."[166] "This is not because the native farmer is inherently conservative; indeed, he seems generally to be less conservative than the European farmer."[167] "The prevalent idea that the native farmer is excessively conservative is largely due to the mistakes of Europeans in the past." Not only should schemes for improvement be thoroughly tested in the field, but also social factors should be taken into account. Africans' "agricultural methods are very intimately bound up with tribal traditions, with land tenure, and with the economics of village life," and farmers had to consider all these factors when they contemplated any change in their cultivation techniques.[168]

G. Howard Jones, a research officer in Nigeria from 1924 to 1929, devoted an entire book to this kind of auto-critique in order to help Europeans understand "the human point of view, the social anthropology, of the native farmers who actually work the system."[169] Jones's official responsibilities included the study of cotton diseases, but his long-term interest was in an examination of "local farming practices," since these challenged so many of his own assumptions. A European might dismiss the farms on Africa's west coast as "laughable and ridiculous," he wrote, because at first glance they were "apparently without any order or arrangement."

> Yet if one looks at it more closely there seems a reason for everything. The plants are not growing at random, but have been planted at proper distances on hillocks of soil arranged in such a way that when rain falls it does not waterlog the plants, nor does it pour off the surface and wash away the fine soil: the stumps of bushes and trees are left for the yams to climb upon and the oil palms are left standing because they yield valuable fruit: and although several kinds of plants are growing together, they were not sown at the same time nor will they be reaped together: they are rather successive crops planted in such a way that the soil is always occupied and is neither dried up by the sun nor leached out by the rain, as it would be if it were left bare at any time. This is but one of the many examples that might be given that should warn us to be very cautious and thorough before we pass judgment upon native agriculture.[170]

These methods, Jones concluded, were highly effective in preserving the soil's precious—but precarious—fertility.

Jones was surprised by what he discovered about soil conditions in tropical Africa. Nineteenth-century "explorers spoke of the intensely fertile virgin soils of West Africa which they implied were the cause of the prolific vegetable growth that they saw everywhere towering in masses around them." But the luxurious foliage rested on fragile foundations. "Researches have certainly shown that tropical soils, as assessed by modern methods, are not in themselves remarkable for their richness."[171] Instead, indigenous cultivators had found methods to conserve and enhance soil fertility. Following the guidelines set by his department, Jones made an analogy between the philosophy of indirect rule and the agricultural strategies colonial states should pursue that suggested "an alternative solution of the problem of how tropical countries are to be . . . developed." Rather than promoting "large estates" and plantations, as had been done in the Caribbean, he advocated an approach he called "indirect agriculture," which built upon "the strong points of peasant farming," especially its "great resistance to adverse conditions."[172] Urging restraint and humility in imperial interventions, he admitted that "as one gets to know the social anthropology of the West Coast peoples, one often wonders whether their ideas may not sometimes be wiser than ours."[173]

Recent scholarship on the colonial state's use of the threat of soil erosion to intervene in the lives of its African subjects has alerted us to the ways in which coercive conservation efforts planted the seeds for anticolonial and protonationalist movements, especially in East Africa.[174] Less is known about research into African soil conditions by European soil scientists. Fiona Mackenzie, in her important study of gender and agriculture in the Murang'a district of Kenya, calls the research results of the soil scientist, V. A. Beckley, a "counter-colonial ecology" because, unlike other agricultural officers in Kenya, he focused on "local crops and agricultural methods" and was less insistent on pushing the idea of uniform agriculture with an emphasis on exports and high yields.[175] Beckley was part of a group of soil chemists in East Africa who made research into indigenous agricultural methods a priority.[176] As early as 1927, Amani's director, William Nowell, included this subject among the topics his institute ought to investigate: "The most general native system of [soil] regeneration is based on shifting cultivation, usually with a weed or bush fallow. It is worth study to find exactly how its effects are produced, the degree of its efficiency, and its capabilities for improvement."[177]

East Africa's soil scientists met formally three times, in 1932 and 1934 under the auspices of the East African Governor's Conference and in 1935 at Oxford.[178] At their first gathering, they agreed that the system of shifting cultivation, which had been "frequently denounced," was "indeed possibly the only really permanent system available under certain circumstances." Acknowledging that knowledge of the "East African forms of shifting cultivation was inadequate," they endorsed the study of "local practice" and "traditions" surrounding these agricultural methods.[179] The task of coordinating their research efforts fell to Geoffrey Milne, the soil scientist at Amani.[180]

The discipline of soil science, or pedology, came into its own on the international stage during the 1920s.[181] This specialty underwent professionalization alongside tropical agriculture and tropical or "imperial" forestry. The key question for all three biosciences was what created the conditions for optimum fertility. Although the tropics had long been depicted as a region of abundant vegetation,[182] attempts to produce high-yield cash crops often met with biological and economic failure. William Nowell emphasized this point in an address on Amani's research efforts to the Royal Society of the Arts in 1934.[183] For more than thirty years, both the Germans and the British had sought to develop coffee plantations in East Africa. While some efforts met with success, attempts in the Usambaras and Rungwe districts in Tanganyika were "puzzling" and "expensive failures . . . under what seemed highly favourable conditions."[184] Amani's research officers assigned to explain this apparent anomaly discovered that the East African territories exhibited such "extreme variations in soil and climate" that these variations "localised both the problems and the value of experience in dealing with them."[185] Only when Geoffrey Milne undertook a chemical study of soil pH levels and F. J. Nutman examined the plants' physiology was it ascertained that the Tanganyikan failures had to do with "the high acidity of the soil," which caused unfavorable adaptations in coffee plants' root systems.[186] Ecological analysis demonstrated that the "micro" or "eco-climates" of Tanganyika influenced plant growth and yields through their direct effects on insects and parasites and indirect effects on soil conditions.[187]

Recognition of the low fertility of soils in much of tropical Africa was an important result of British pedologists' and agriculturalists' research in the 1920s and 1930s. Overly optimistic rhetoric about Africa's agricultural potential was noticeably dampened and often replaced by subtle critiques of European thought and practice. Upon his retirement in 1934, Alex Holm, formerly director of agriculture in Kenya, commented that the "acute" "agricultural problems of East Africa" were "related to soil fertility, plant nutrition, plant diseases, plant pests." Production for the "world's markets" actually exacerbated these difficulties, and the "elucidation of one problem often created another in its train."[188] Agricultural failures were openly acknowledged and their causes more systematically investigated. Researchers focused on the fallacy of tropical fertility:

> Land, even if the soil *per se* is very poor, in most cases brings good harvest at first due to the nutrient reserves accumulated over centuries and millennia. Catastrophe sets in usually during the third or fourth harvest. Numerous abandoned plantations in almost all colonial countries serve as vivid proof of what has been said. . . . More than 75% of all disasters suffered in tropical and sub-tropical agricultural enterprises must be related to improper selec-

tion of the soil. In the tropics and subtropics where the external appearance
is quite misleading and confusing, only a thorough acquaintance with soil
science can protect the farmer from grievous errors entailing great material
losses.[189]

As the studies in Northern Rhodesia had shown, many African cultivators
possessed useful knowledge of the soil. In a letter to the editor of *Nature*
in September 1936, Geoffrey Milne observed that "some six or seven of
the main types [of soils] are of sufficiently general occurrence [in Tangan-
yika] . . . to have given rise to a well-developed soil nomenclature in the
Sukuma language."[190] His colleague, B. J. Hartley, applied these insights to
the challenge of erosion, identifying what he called an "indigenous system
of soil protection." The "Erok method of sowing a listed crop between the
ridges carrying rotting crop residues is a development which would well repay investigation elsewhere under varying conditions of rainfall. Soil movement within the plot under this system is reduced to a minimum while at
the same time loss from the terrace is adequately controlled."[191]

Critiques of colonial agricultural policies and practices within British
Africa were consolidating along two fronts by the 1930s. The first was directed at Europeans' and white settlers' misconceptions of tropical soil fertility. Researchers in the technical services recognized the importance of agriculture for African economies, but they questioned the assumption that
financial rewards could readily be reaped. African soils were so varied that
methods that succeeded in one locale might be disastrous elsewhere. The
second critique was both sociological and environmental: since African agricultural practices were well adapted to differing natural conditions and
the philosophy of indirect rule prescribed a modicum of respect for indigenous practices, then the improvement of "native agriculture" should build
upon Africans' best practices. In a discussion of the causes of soil erosion
at the second conference of colonial directors of agriculture in 1938, C. J.
Lewin of Northern Rhodesia remarked that the papers presented showed
that, contrary to popular opinion, "erosion was almost invariably due primarily not to the native but to the European who had introduced tillage in
certain areas and had encouraged the production of economic crops."[192]

A consensus in support of Africans' agricultural practices and a nuanced
awareness of the relevance of the environmental sciences to "appropriate"
agricultural development emerged among scientific and technical researchers during the early 1930s. Most did not repudiate market-oriented production or challenge the overarching framework for colonial development,
but their fieldwork and reports undermined prevalent assumptions about

both. Research officers often became voices of moderation and restraint in their own territories as well as advocates of syncretic approaches that attempted to incorporate successful African practices into colonial agricultural strategies.[193] In the process, they had to take ethnography and the vernacular knowledge of African informants seriously.

When Uganda's agriculture department sponsored a series of nineteen surveys of soil and agricultural conditions in different districts, its research officers were advised to interview individual households to determine the baseline productivity of "native cultivation."[194] Cotton production had increased sixteen-fold between 1916 and 1936, and production of staple foods such as cassava had risen even faster.[195] Vested with the responsibility of overseeing agricultural development, the department thought it wise to ask how patterns of land use, social systems, and yields were correlated. They were surprised to discover that a handful of the more densely settled areas, which they thought might be "over-populated" and in danger of erosion, were sustaining soil fertility.[196] The director recommended incorporating "native custom that has stood the test of time" into regional policies, particularly in terms of crop rotations and soil protection measures.[197] Summarizing his experience as a researcher, T. R. Hayes drew attention to the breadth of the project: "the scope of the surveys was widened to include a larger field of subjects, such as diet, social organization and crafts. . . . The method of approach adopted . . . [was] that the people selected should be studied as a group of human beings, rather than as a collection of scientific specimens."[198] This attitude explains why so many of the reports included discussions of local residents' social life, "clan membership," educational levels, styles of food preparation, and kinship genealogies. The medical department joined forces with the agricultural survey to consider the relationship between "agriculture and health."[199] Harold Hosking, another officer in Uganda's agriculture department, observed: "The question of improving native food crops is beset with difficulties. With long established crops it is probably true to say that the native can teach us more than we can teach him."[200] Evidence of these ethnographic exchanges is visible in the department's lengthy summary of its research activities, which was published in 1940: more than half the book is devoted to "native agriculture" and subsistence crops, and another chapter concentrates on medicinal plants.[201] While the territory's research officers were more muted in their praise of Ugandans' agricultural techniques than those in Northern Rhodesia or Nigeria, they thought them significant enough to test and report upon in detail. As Grace Carswell has reported for the Kigezi district, the soil conservation measures the department pursued "were actually adaptations of methods already in use. This may explain why the policies were relatively accepted by farmers."[202]

Ecological Perspectives and Official Attitudes

The growing emphasis on "native agriculture" in investigations sponsored by imperial authorities and colonial states during the interwar period had a range of different rationales. In some places, the aim was to increase Africans' primary production both for domestic consumption and for export; in others, issues of land tenure and policies toward white settlement were involved. Yet attempts to institutionalize agroecological studies on an inter-territorial or pan-African scale were never a complete success, which meant that their recommendations were rarely adopted wholesale. At the root of this problem was not just lack of funding—an enduring concern when it came to British Africa—but also persistent behind-the-scenes debates over what kind of knowledge would be most useful for development. E. B. Worthington's work for the African Research Survey became so significant because he drew together the insights of researchers scattered throughout Africa with those of leading scientists in Britain and outlined an intellectual framework within which their various fields of expertise might cohere.

Worthington's first task for the African Survey, and by extension the Colonial Office, was to offer an overview of the relevant specialized disciplines and to recommend ways of basing imperial strategies more solidly on scientific foundations. He chose as advisors a substantial number of fellows of the British Ecological Society, as well as ecologically inclined leaders of research institutes both in Britain and in British Africa (table 3.7). He reviewed the research reports from the inter-territorial and pan-African coordinating conferences; the Amani Agricultural Research Station; the Ecological Survey of Northern Rhodesia, which he received directly from Colin Trapnell;[203] the Tsetse Research Department, which he called "essentially an ecological department" in his first drafts of *Science in Africa*;[204] the Ugandan agricultural surveys; and Faulkner and Mackie's book *West African Agriculture*, which both Julian Huxley and Joseph Oldham praised as of "first importance" for understanding recent agricultural trends.[205] His counterpart, Malcolm Hailey, took the trouble to read G. Howard Jones's book on "native farming" in Nigeria when it was published in 1937.

Officials working with the Colonial Office were sometimes unsympathetic to the ecological point of view. Francis Hemming, the secretary of Britain's Economic Advisory Council, objected to the position articulated in Worthington's draft chapters of *Science in Africa*, writing with some

Table 3.7 Advisors to the African Research Survey in 1938: Members of the British Ecological Society (*), Advocates of Ecological Methods (Φ), and Fellows of the Royal Society (§)

Name	Current position and previous location	Chapters advised
J. R. Ainslie Φ	Chief Conservator, Forestry Dept., Nigeria	4, 5, 6, 7, 8
R. Bourne Φ	Imperial Forestry Institute, Oxford	7
Prof F. T. Brooks * §	Botany School, Cambridge	5, 6, 11, 12, 13
Dr. J. Burtt Davy * §	Imperial Forestry Institute, Oxford; previously in South Africa	6, 7, 12
Prof P. A. Buxton *	Entomology, London School of Tropical Medicine; previously in Nigeria	10, 14
Prof G. D. Hale Carpenter *	Entomology, Oxford; previously in Uganda	8, 10
Sir David Chadwick Φ	Secretary, Imperial Agricultural Bureaux	entire volume
Dr. Fraser Darling *	former Deputy Director, Imperial Bureau of Animal Genetics	14
Dr. E. M. Delf *	Westfield College, University of London	17
Charles Elton *	Director, Bureau of Animal Population, Oxford	8, 14, 16
Clement Gillman Φ	Chief Engineer and Human Geographer, Tanganyika	2–14
Sir Richard Gregory §	Editor, Nature	entire volume
Sir Arthur Hill § *	Director, Royal Botanic Gardens, Kew	5, 6, 7, 11, 12, 13, 14
Julian Huxley § Φ	Secretary, Zoological Society of London	entire volume
C. J. Lewin Φ	Director, Agriculture Dept., Northern Rhodesia	11, 12, 13, 14
A. P. G. Michelmore *	Investigator, Committee on Locust Control; previously in Northern Rhodesia	10
Geoffrey Milne Φ	Soil Chemist, Amani Agricultural Research Institute, Tanganyika	5, 11, 12, 13, 14
William Nowell Φ	Director, Amani Agricultural Research Institute, Tanganyika	5, 6, 7, 11, 12, 13
Prof J. F. V. Phillips *	Botany, University of Witwatersrand; previously in Tanganyika	4, 5, 6, 7, 11, 12, 13, 14
Dr. J. Ramsbottom *	Keeper of Botany, British Museum	6
Dr. P. W. Richards *	Botany School, Cambridge	5, 6, 7
Dr. E. S. Russell *	Ministry of Agriculture and Fisheries	9
Prof E. J. Salisbury * §	University College, London	5, 6
H. C. Sampson Φ	Botanist, Royal Botanic Gardens, Kew	5, 6, 7, 11, 12, 13, 14
Prof. L. Dudley Stamp *	Geography, London School of Economics; previously in Burma	2, 3, 4
Prof. R. G. Stapledon * §	Director, Welsh Plant Breeding Station, Aberystwyth	6
C. F. M. Swynnerton *	Director, Tsetse Research Department, Tanganyika (d. June 1938)	4, 6, 8, 10, 13
Prof. A. G. Tansley * §	formerly at School of Botany, Oxford	6
Prof. D. Thoday *	Department of Botany, University College of North Wales; previously in South Africa	6, 12
Colin Trapnell *	Ecological Survey, Dept of Agriculture, Northern Rhodesia	5, 6, 11, 12, 13, 14
B. P. Uvarov *	Imperial Institute of Entomology	10

Source: "British Ecological Society, List of Members," *Journal of Animal Ecology* 7 (1938): 191–98.
Chapter topics: 2 Surveys and Maps; 3 Geology; 4 Meteorology; 5 Soil Science; 6 Botany; 7 Forestry; 8 Zoology; 9 Fisheries; 10 Entomology; 11 Agriculture-General; 12 Crop-Plants; 13 Plant Industry; 14 Animal Industry; 16 Human Diseases; 17 Health and Population.

urgency that it was "misleading to suggest that the time has come to concentrate almost exclusively on ecological investigations." Worthington had endorsed a recommendation made by John Phillips that Britain coordinate an ecological survey across its territories in South, Central, and East Africa on the grounds that it would help to "evolve methods of ecological investigation best suited to the conditions . . . in Africa" and place the challenges of agriculture, forestry, soils, and trypanosomiasis in their widest possible context. Phillips expressed concern that "the teaching of dynamic ecology" had been relatively neglected in British and overseas universities and stressed the need to train many more ecologists and provide them with skills "in such subjects as agriculture, forestry, veterinary or medical science, for the very good reason that without one or the other of these (or in special instances all) ecologists cannot be expected to realise and understand the great practical problems set them in the field, nor can they hope to work in the fullest and most intelligent co-operation with officers engaged in these particular professions. . . . The necessary basis is a dynamic ecological survey in which the co-operative team-work is one of the inspiring forces."[206] Hemming and his colleagues remained unconvinced, despite the fact that previous colonial secretaries and undersecretaries, including Leopold Amery and William Ormsby-Gore, had espoused similar arguments during the 1920s. Not only did Hemming regard embarking on such an "inflated and impracticable scheme" during the Depression "absurd," but also he thought that Phillips and Worthington "attach much too much importance to ecology and ecologists in Africa and much too little importance to systematics and systematists in Africa and elsewhere."[207]

Worthington responded, "With regard to your major criticism that I have overstressed ecology and underestimated the value of systematics, I don't know quite what to say."[208] The difficulty seemed to lie in the multiple definitions of ecology in use among practitioners and policy makers. During the 1930s, "ecology" encompassed at least four overlapping meanings, all of which Worthington used without much differentiation. For some, ecology was the relatively straightforward study of a species or group of species in its organic and inorganic environments (that is, biotic communities). For others, it was the study of interactions and changes among component parts of a wider "soil-vegetation unit," "ecosystem," or "ecological complex," which could include both human and nonhuman activity. For still others, it was a method by which the "interrelations" among particular "problems of applied biology" could be correlated and examined "as a whole." The field's most prominent British boosters, including Julian Huxley and Charles Elton, promoted ecology as a metascience that could fulfill the role of "super-coordination" among scientific disciplines and their respective objects of study.[209] John Ford pointed out the "confusion of thought" concerning the "ecological point of view" and posed the question, "In what way does ecology differ from other biological sciences?" He argued that ecologists' training to "reveal natural complexities" and "formerly hidden" patterns among "systems" and "integrated structures" did not always set them apart from other biological disciplines.[210] Worthington defended ecological methods, and in his next draft he wrote that ecological surveys were "the only way to produce . . . a good knowledge of the natural resources of Africa," on which economic development depended.[211] While he acknowledged that Phillips's "ambitious and comprehensive scheme of research . . . may need to be less elaborate," he endorsed an ecological approach that would consider not only the "soil-vegetation unit" but also "native agriculture . . . and animal husbandry," which were "among the most important natural factors in African ecology."[212]

This position found support among other government advisors. In his 1935 study, "Financial and Economic Position of Basutoland," Sir Alan Pim advocated an ecological survey as a precursor to a "programme of reclamation and conservation." "An experienced ecologist," with "the specialized type of training and technique" unique to this field, was "particularly suit[ed]" to a problem of such "great complexity." "He would be quicker, and surer, and better in every way in surveying and in making observations, and his advise as regards experiments . . . would be invaluable."[213] In a meeting of the colonial directors of agriculture in July 1938, the participants agreed that "ecological surveys should be a preliminary to the planning of land utilization."[214] The technical officers of the West Africa Commission of 1938–39, who included H. C. Sampson of Kew Gardens, one of Worthington's close advisors, referred to the "pioneering work undertaken by C. G. Trapnell and J. N. Clothier" in Northern Rhodesia and proposed a similar scheme of ecological surveys and the study of vernacular knowledge for West Africa. "We have in mind something much more far-reaching than the mere co-ordinating and filing of an indigestible mass of facts. We regard the central purpose of the proposed organisation as the ecological interpretation of the country and its mode of life. . . . This type of work has the great merit that it utilises the same kind of observations as the native uses instinctively in assessing the value of land. It is able to use local tradition and to employ African assistants to do work for which they are already well qualified."[215]

The reasons why these proposals for ecological studies were not fully implemented during the interwar period go beyond the financial constraints of the Depression. The discipline's comprehensive epistemology and the organization and priorities of the technical departments in British Africa meant that these plans were executed unevenly. The importance of the relationships between ecological scientists and technical officers and the officials of the colonial state is clear in the inter-territorial projects conducted by the Amani Agricultural Research Institute. In 1936 the acting director of Amani wrote to the secretary of the East African Governors Conference and to the Colonial Office asking them to support a "Conference of Ecological Workers" for East and Central Africa. The topics to be covered included the ecological study and control of disease-causing insects, especially tsetse flies and mosquitoes; the ecology of plant diseases; and the "eco-climatic factors under field conditions in the tropics" that had a bearing on both agriculture and disease.[216] Most of the replies received from the governments were positive, and many suggested additional topics to be considered. Yet several expressed a reservation most succinctly put by Kenya's government: "The proposed ecological conference would effect a regrouping of East African workers, whose fields of research are already covered by existing conferences . . . [in] Tsetse, Medical, and Veterinary Research. . . . If, however, it is considered that ecology as a science *sui generis* is sufficiently strong and well represented in East Africa to warrant periodical meetings of ecological workers, it is clear that, even though there will be overlapping, no institute the work of which impinges on this field can afford to remain outside such a conference."[217] Virtually every territorial researcher might have some reason to attend. Despite the endorsement of Frank Stockdale, who thought the conference would be "very useful," it was never held. Instead, ecological methods were pursued through the existing system of coordinating conferences and through teamwork in the field as well as by research institutes that, like Amani, had more comprehensive missions.

Science in Africa and the "Application of Ecological Principles"

Science in Africa became the major forum in which the view of ecology was articulated as a metascience that could synthesize the methods and findings of specialized disciplines and bring them to bear upon problems in the field. By demonstrating the fruitfulness of this synthetic approach to analyzing complex interactions, the book had a direct effect on postwar investigations in the East African scientific services as well as on the Scientific Council for Africa South of the Sahara. As one of Worthington's reviewers noted, "The ecological method in biology had not been developed in the days of the naturalist explorers, yet these men frequently noted relationships that have eluded deeper study until recent years. One of the chief purposes of this book, then, is to help to impart the ecological outlook to all future scientific investigations: and this now implies the fullest co-operation among specialists."[218]

An explicit imperative of Worthington's research for the African Survey, his advisors repeatedly emphasized, was to draw attention to the inter-connections among problems, policies, and scientific disciplines them-selves. After Professor E. J. Salisbury, a member of the Ecological Society, reviewed Worthington's draft chapters, he wrote: "The only general point I would make is that greater stress should I feel be given to the interrelated character of knowledge in the various branches. It cannot be too strongly urged that this has important practical application which unless appreci-ated as something more than a theoretical concept may well result in the hampering of progress in one field through the neglect of others."[219] Wor-thington adopted Salisbury's language almost verbatim in his introductory chapter to *Science in Africa*.[220]

The varied insights presented in *Science in Africa* and *An African Survey* was, in part, a result of their process of production. Successive drafts were circulated among hundreds of advisors who had several opportunities to comment on and criticize their content. The collaborative process discour-aged easy generalizations, sweeping solutions, and unfounded claims. It forced the authors to stress particularity, local specificity, and heterogene-ity in African conditions. The African Survey's contributors and advisors negotiated among multiple and competing strands of scientific research, public opinion, and official conviction to arrive at their conclusions. On almost every issue, they skirted between dominant and dissenting views, crafting unexpectedly moderate positions given their proximity to the cen-ters of colonial power. During the interwar period, the drive to control and improve colonial Africa's natural and human environments was tempered by sensitivity to the immediate and long-term consequences of imperial policy. The recognition of past failures was acute. Lord Hailey generalized the critique made by Faulkner and Mackie in Nigeria, saying that "exagger-ated beliefs" in the "natural riches" of the territories had led to "the loss of vast sums in misguided attempts to reap quickly without sowing."[221]

A sense of anxiety concerning Africa's changing environments pervades *Science in Africa*. The problems it drew to readers' attention remain depress-ingly familiar today: soil erosion, overstocking of grazing lands, deforesta-tion, new but problematic methods of cultivation, the general spread of disease, and changing epidemiological patterns. Worthington's definition of "environment" was very broad, encompassing the interactions between natural forces and human activities.[222] The book's recommendations for intervention usually stemmed equally from a desire to promote economic development and a determination to prevent potentially irreversible biosocial damage. Worthington described these changes in the context of "constructive and destructive factors" involved in the "balance of nature." In keeping with emergent thinking regarding the limits of homeostasis, he cautioned that "the balance of nature is not a simple balance, but a complex system of levers and links all balanced with each other, so that extra weight placed on any part of the system may cause the whole to change its equilibrium."[223]

The ecological system Worthington presented was dynamic and non-linear. The field was constituted by eighteen different subjects, organized in much the same way that ecologists represented food chains: the inorganic world of geology, meteorology, and soil science; the plant world of forestry, botany, and agriculture; the animal world of fisheries, zoology, entomology, and animal industry; and the human world of health, human diseases, population and anthropology (figure 3.4). Much of the necessary research could "only be carried out in Africa itself."[224] The importance of "local study" as well as of sensitivity to Africans' knowledge were recurring

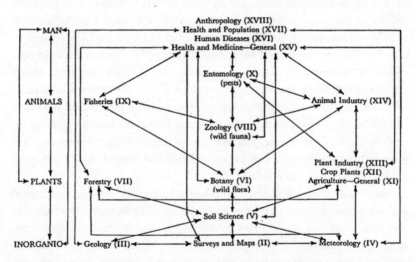

Figure 3.4. "Principal Relations between Subjects" considered in Worthington's *Science in Africa*. Figures in brackets denote chapter numbers. "The picture really presented by Africa is one of movement, all branches of physical, biological and human activity reacting on each other, to produce what biologists would refer to as an ecological complex" (Worthington, *Science in Africa*, 15). Source: E. B. Worthington, *Science in Africa* (1938).

themes. Worthington hoped that the hubris of the past was giving way to a modicum of humility. This critique of colonial practices was made most clearly with regard to agriculture.

> The soil-vegetation unit as a controlling factor in environment is only now being recognized by scientists, but it is significant that the *Africans themselves still know more about it than we*, for every cultivator bases his selection of plots for different crops on recognized associations between certain plants and certain types of soil. The deterioration and erosion of soil is perhaps the best example of any to show the rapidity of environmental change. This likewise *has been recognized by African farmers for centuries*, and on it have been built up the complicated systems of shifting cultivation which so admirably suit the soils and climate, provided the area of land is sufficiently large. There are many examples of individual tribes which have apparently evolved methods of avoiding soil erosion *which could hardly be bettered by science*.[225]

Worthington's synthesis of scientific specialties enabled him to form a set of seven guiding principles that are implicit throughout *Science in Africa*. First, knowledge of existing conditions is a necessary precursor to any intervention. Second, economic development strategies should build upon African methods already in place, especially in agriculture. Third, attempts to maximize economic yields should be balanced with efforts to sustain environmental fertility. Fourth, policies should be tailored specifically to different "localities." Fifth, "quick results" in research were "impossible"; "team-work," "interterritorial co-operation," and long-term efforts were required. Sixth, the management of particular problems is inherently complex. Finally, competing claims on Africa's land and natural resources should be adjudicated with caution. "Every branch of human activity, including cultivation, grazing, forestry, game preservation, mining and administration, involves the utilization of land, and the claims of various activities often come into conflict."[226]

These principles represented an attempt to mediate between ecological and sociological understandings of Africa's environments. Development was prescribed and constrained by human as well as natural conditions. This insight shaped the strategies the African Survey promoted. After describing the failure of efforts to raise imported European cattle in various regions, Worthington offered a general solution: "In the tropical parts of Africa, it appears desirable to breed local stocks selected for qualities appropriate for varying economic conditions, and for resistance to different diseases. . . . Breeding will have to be carried out for each different locality, on account of the prevalence of particular diseases in different areas."[227] Investigating the "naturally acquired immunity" of indigenous livestock and their adaptations to different environments was high on the research agenda.[228]

Science in Africa adopted similar positions with respect to fishing and African agriculture. "The chief consumer of African fish, . . . at any rate in the tropical regions, is the African native, and he is likely to obtain the best, most continuous, and cheapest supplies by the gradual improvement of the existing indigenous trade, rather than by the introduction of large-scale industries organized on modern lines."[229] Worthington and many of his advisors expressed a general mistrust of vast commercial enterprises because they often disregarded the welfare of indigenous people and their habitats. "There is a strong body of opinion, that [corporate agricultural] schemes of this nature are not entirely beneficial, and may even prove disastrous to native life in the regions concerned. The development of export crops at the expense of native food crops, which characterizes the company system, can easily lead to an unbalanced system of agriculture,"[230] especially "dependence on a single crop" that was vulnerable to "variations in market and climatic conditions."[231] "Native agricultural methods" typically involved the cultivation of multiple or "mixed" crops in confined areas. Africans were committed to horticultural diversity not only because it compensated for the vagaries of the weather but also because they knew "that many insect-borne diseases of plants, animals and man may be avoided or reduced by fairly close settlement."[232]

Worthington roundly criticized the misconception that in "tropical conditions" cultivators needed only to "remove indigenous vegetation and introduce crop plants. How far this idea is from the truth is shown from the mass of intricate research which has become necessary with almost every crop. As knowledge concerning tropical soils progresses, it becomes more and more evident that *to judge them from a European standpoint is entirely misleading.*"[233] European methods, characterized as capital-intensive, large-scale production, had adverse effects, especially soil deterioration. The European cultivator "may be even more destructive than the native since he works on a larger scale and aims at keeping a cleared area permanently under crops. . . . European cultivation also is usually cleaner from weeds than that of the natives and, therefore, is more liable to wash, as there are only the roots of the crop to hold the soil. It is certainly regrettable that some European-owned land in Africa is worked on a principle which is not worthy to be designated farming, but can only be termed soil exploitation."[234] Africans were already active in plantation agriculture, in particular in West

Africa where there were few white settlers, and Worthington acknowledged that drawing an arbitrary line between European and African methods of production was unjustifiable. "The real division probably lies between those methods which do and do not involve the investment of capital."[235] All development projects should begin on a small scale and be tested quite carefully.

Science in Africa captured a wide array of conversations among field scientists and technical officers that criticized previous colonial interventions. Worthington drew attention to an emerging consensus that many territories in British Africa had been poorly planned and inappropriately managed. "In a continent which has been developed almost wholly in the twentieth century, there might have been more room than elsewhere for [scientific] influence, but this has not been the case; economic development has taken the lead and often chooses the wrong turning. Science follows, but the pace is laboured, and falling behind she is neglected. . . . A development based on a real understanding of Africa's potentialities has hardly yet begun and will be impossible until the necessary scientific knowledge is recognized."[236]

It would be difficult to read this passage and not be reminded of the many ways the continent of Africa served as a "fruitful field . . . for experiment."[237] To deny or overlook the abundant evidence that many such experiments had adverse effects, however unintended, would be arrogant in the extreme. This is not the purpose. What this chapter has drawn attention to is how the production of scientific knowledge transformed colonial understandings of Africans and their environments. The auto-critique that emerged among research scientists and technical specialists at both the territorial and inter-territorial levels unsettled colonial certainties and opened up a space to take ecological specificities and subaltern knowledge more seriously. These studies, because of their detailed and empirical basis, had what were arguably liberalizing effects, relieving rather than concentrating inappropriate colonial interventions.

CHAPTER FOUR

A Medical Laboratory: Infectious Diseases, Ecological Methods, and Modernization

The science of medicine has to be built upon data which have been sought and found in the living bodies, habitations, and working places of men, and in the places frequented by parasites and insect vectors. . . . The last word must come from the records of happenings in the field.

—British Colonial Office, *Scientific Studies of Colonial Medical Service*, 1947

It cannot be emphasised too frequently that social, medical and educational services are the foundation on which the development of the African must be built. A great and wealthy Empire cannot afford to be mean about investment in the land and the people of Africa. If we are mean we shall leave to our children a legacy of chaos which history will justly attribute to the good intentions and the appalling stupidity of Western Civilization.

—Review of Annual Medical Report of Kenya Colony, 1935[1]

In late 1938, a wealthy British patron, Champion Russell, wrote to the Colonial Office (CO) offering a donation of £1,000 for an integrated health campaign in British Africa that would address chronic and infectious diseases as well as sanitary improvements. "I appeal for a Government experiment in a district selected by them—which would 'try out' the results of having a healthy population—as it were, on a laboratory scale. . . . It seems evident that the *prevention* of disease should be seriously tackled."[2] He had in mind a comprehensive approach to health that would deal not just with sleeping sickness, malaria, and hookworm, the three diseases he felt were the most threatening to indigenous populations, but also with water quality, agriculture, veterinary care, nutrition, and health education. Russell's primary exposure to British Africa had been in Uganda, where he witnessed firsthand the new educational facilities at the Mulago Hospital and the Makerere College Medical School, which had begun to train students from across East Africa for the medical and sanitary services.[3] He was also impressed with Uganda's "splendid work" in bringing sleeping sickness under control after the epidemics at the turn of the twentieth century. "The idea would be to have a model district, which might be eventually of use as a practice ground for young sanitary officers who have left Makerere and Mulago, before taking up a career." The results, he argued, would include "less incidence of disease . . . a lower death rate, improved physique—greater intelligence, more inclination to work, [and] more enterprise." Equally important, the data gathered would lead to new insights, "under present conditions in Africa," about disease distribution, susceptibility, and immunity. "Such an experiment in health is a scientific necessity."

The colonial secretary, Malcolm MacDonald, as well as several of the CO staff, liked the breadth of the proposal because it resonated with recommendations regarding health policy that they were reviewing in Lord Hailey's *An African Survey*. Indeed, a central conclusion of the chapter on "Health" was that "the task of the health services in Africa . . . extends beyond the application of modern scientific technique for the prevention and cure of disease; they share in the responsibility of other social services for the improvement of African conditions of life."[4] After discussing the matter with his subordinates, MacDonald decided to distribute Russell's proposal to the territories in East Africa because the directors of the six medical departments in the region would be meeting that summer to discuss new research priorities. In his cover letter, MacDonald drew attention to several "points of affinity" among Russell's suggestions and Lord Hailey's analysis. With respect to funding, however, he noted that one thousand pounds "would be insufficient to meet more than a small part of the cost of the experiment": lasting public health results could only be achieved, he told the medical directors, if the project were carried out for at least five to ten years and received substantial metropolitan support.[5]

When Russell's memorandum was circulated, several of the medical directors found themselves in the position of having to explain to their superiors that his ideas were by no means novel. "I am somewhat surprised," A. R. Paterson wrote from Kenya, "that [CO staff] should refer, or refer only, to Lord Hailey's book in this connection and I am profoundly disturbed that there can be even a suggestion that the Secretary of State may think that the institution of work on the lines indicated in Mr. Russell's memorandum would be something in the nature of an experiment."[6] Tanganyika's medical director, R. R. Scott, noted that he had long since understood that successful health interventions demanded a cross-section of specialists: "these include agricultural, educational, veterinary, medical, and administrative" officers.[7] All initiatives, agreed Scott and William Kauntze, the Ugandan director, required "the co-operation of the people themselves" and "the complete sympathy and understanding of their own leaders."[8] Kauntze, who had written the first draft of the health chapter for the African Research Survey, was adamant about the need to focus on preventive medicine.[9] All department directors had recently been given an opportunity to comment upon E. B. Worthington's three chapters on health in *Science in Africa*, stressing in their remarks "the close interdependence of all aspects of medicine . . . [including] (1) the causes of disease and disability[,] (2) the protection of individuals and the community from these causes, and (3) the provision of adequate living conditions whereby health may be maintained."[10]

Their greatest reservation about Russell's recommendation was the issue of scale. Russell had urged that a single district should serve as a laboratory for the health experiment, but medical administrators regarded this narrow scope as problematic. Nyasaland's director stressed "that all East African Medical Services are endeavouring, in cooperation with officers of all other Departments interested in the welfare of Africans, to achieve what Mr. Russell is aiming at, *over the whole of the territories under their control.*"[11] Paterson arrived independently at the same point: "Mr. Russell's memorandum is in short a very good summary of almost everything that ought to be going on in every district of every Crown Colony in Africa. In Kenya, at least, much of it is already going on and has been going on for a good many years, and I trust, no longer as a 'Government Experiment' but as a matter of settled Government policy." Additional funding for medical work was urgently needed, but it would be used for "an 'experiment' already in being, and not to inaugurate a new one. . . . What is now required in Kenya is not one 'model district' as an experiment, but intensification of work in each."[12]

The debates Russell's memorandum sparked provide an important glimpse into metropolitan and territorial thinking regarding health concerns in British Africa on the eve of the Second World War. The exchange reminds us that virtually every territory served as a space for biomedical experimentation on a massive scale. The substance and methods of these experiments took different forms in different places and were unevenly implemented, but no colonial state was exempt from the process. Biomedicine, to quote David Arnold, was indeed a "colonizing force in its own right . . . a potent source of political authority and social control."[13] Officials considered medical departments a crucial part of any imperial plan to ameliorate social conditions. Everyone accepted the premise that Africans were overburdened with disease and needed state intervention to overcome environmental and epidemiological obstacles that stood in the way of raising their standard of living. Medical efforts, they believed, would have the added bonus of creating more industrious workers.

Embracing broad definitions of health in principle if not always in practice, these experts argued that a modern approach to medicine meant collaborating across territorial departments, tackling cure and prevention simultaneously, and securing the trust and support of colonial subjects. These activities were considered so integral to state-building efforts in British Africa that Nyasaland's medical director urged that Russell's "experiment should not be considered purely a medical one." Improving health conditions "requires that the population should be effectively governed, educated, fed, clothed, housed, doctored, sanitated [sic], and further have means of finding sufficient money to provide for all other essential needs."[14] Precisely how their states would achieve these goals remained an open question, but medical directors' belief in the need to achieve them was unequivocal.

Finally, the correspondence highlights that there was a consensus that integrated efforts at "applied hygiene" took time, up to ten years according to several of the respondents, and even then there was no guarantee of success. As Scott from Tanganyika recalled, "Experiments with a similar object have been launched in the past, but have generally failed through some unforeseen cause, or through lack of sufficient funds."[15] Like Paterson, he implied that the British government had a duty to provide the funds necessary to augment medical projects. At the conclusion of their discussions, the colonial secretary agreed and recommended a dramatic increase in grants toward public health and medical research through what became the Colonial Development and Welfare Act of 1940.[16]

The exchange over Russell's proposal raises a number of questions that this chapter will address in depth. How did colonial states and imperial authorities develop their outlook on health and disease in British Africa? What kinds of medical research took precedence in the decades leading up to the Second World War, and in what ways were they connected to imperial aims regarding colonial development? Did biomedical specialists share with agricultural investigators an interest in ecology and ethnography? If so, what kinds of therapeutic and epistemological entanglements were produced as a result? Is it possible to identify parallel patterns of vernacular exchange and appropriation in biomedical undertakings as there were in agricultural ones? Finally, how did the African Survey play a role in mediating these interests across metropolitan and colonial sites?

This analysis takes into account interactions among territorial, imperial, and international policy making and practice simultaneously. It considers, in particular, the predominance of biomedical *fieldwork* in British Africa, a primacy the Colonial Office emphasized when it reviewed the results of its research officers between 1930 and 1947:

> Advances in the understanding of natural phenomena depend upon the availability of measurable facts about them, and the chief source of such information is the field worker. . . . This is often over-looked. Laboratory experiment has its renowned place, but the essential part played by the field worker should not be forgotten. His data provide important material for the scientific imagination, reflecting as they do the inherent variability of biological events.[17]

By the interwar period, tropical Africa had become one vast region for biomedical research; it could accurately be called a "field laboratory." Yet, the intensity and depth of this work was often far less comprehensive than biomedical efforts in industrial countries.

Imperial and International Research Priorities and Territorial Divisions of Labor

In spite of the medical directors' resistance to Russell's suggestion, an important division of labor with respect to biomedical research already existed in British Africa. Beginning in the mid-1920s, metropolitan authorities singled out various dependencies for public health and medical grants, either through the Empire Marketing Board or the Colonial Development Fund (CDF). These projects turned colonial states into experimental sites for long-term field studies. Kenya and Nigeria were chosen for focused investigations into the nutritional status of nomadic versus sedentary populations: Kikuyu and Maasai in Kenya, and Hausa and Fulani in Nigeria.[18] Justifying their selection of Kenya, the Civil Research Committee wrote that it "possesses conditions peculiarly favourable as a site for an intensive dietetic experiment." According to the territory's medical director, the study would produce results "for the benefit of the Empire at large."[19] Uganda served as a key venue for the League of Nations International Commission on Sleeping Sickness (1925–27) and then for an inter-territorial laboratory for the study of human trypanosomiasis (1927–36). Explaining the choice of location, H. H. Scott, of the Bureau of Tropical Diseases, observed that "it was thought that the clinical conditions and therapeutic measures could be well studied in the French and Belgian Congo, the behaviour of the parasite in the insect vector in Uganda; in other districts measures of control and eradication."[20] The West African territories received substantial grants from the Rockefeller Foundation for research projects on yellow fever, while the East African territories were awarded a grant by the CDF to establish a new Medical Research Unit in the region.[21] As early as 1921, the Colonial Office made it clear to Rockefeller staff that British officials would be willing to offer them access to "fields of research in which large scale experiments may be conducted," as long as this cooperation led to the "advancement of human well being in those parts of the World that we have the power to influence."[22] By the time they were discussing Russell's proposal, Nyasaland had been selected for a multidisciplinary "nutritional survey," with several village sites identified for intensive agricultural, biochemical, economic, and sociological analysis.[23]

None of the territories in British Africa, however, received anything approaching the sums of money allocated by the CDF in the interwar decades to Tanganyika, a mandate of the League of Nations. Between 1925 and 1940, Tanganyika became home to research institutes investigating tuberculosis, malaria, and sleeping sickness and was awarded more than 45 percent of all biomedical grants made to Britain's colonial dependencies in Africa (table 4.1).[24] Of that amount, 75 percent went toward studies of African trypanosomiasis, with the largest share of the awards (£286,037) granted to the Tsetse Research Department, which had been founded in 1927 (table 4.2). No other public health or biomedical institute in British Africa received such substantial metropolitan support during this period.[25] Examining its epistemic frameworks and intersections with development reveals a good deal about imperial research priorities. Indeed, much like Northern Rhodesia's Ecological Survey, the Tsetse Research Department played a crucial role in interpreting local conditions and translating local knowledge for metropolitan decision makers.

The turn-of-the-century epidemics of trypanosomiasis prompted unprecedented levels of biomedical fieldwork within tropical Africa. Between 1901 and 1914, European powers launched seventeen different sleeping sickness commissions, which were dominated by British sponsors (table 4.3). Imperial officials were alarmed not only about human survival but also about the habitability of large swaths of land; many believed that

Table 4.1 Colonial Development Act, public health, and medical grants, 1929–39

Territory	Grant total (£s)	% of African grants	% of total grants (£1,077,242)
1. Tanganyika	404,109	45.7	37.5
2. Nigeria	161,250	18.2	15.0
3. Nyasaland	124,598	14.1	11.6
4. Bechuanaland	37,665	4.3	3.5
5. Sierra Leone	37,500	4.2	3.5
6. Kenya	33,080	3.7	3.1
7. Northern Rhodesia	23,725	2.7	2.2
8. Zanzibar	22,300	2.5	2.1
9. East African Med. Res.	18,750	2.1	1.7
10. Swaziland	13,241	1.5	1.2
11. Somaliland	4,200	0.5	0.3
12. Uganda	3,332	0.4	0.3
Totals	883,750	100.0	82.0

Sources: These figures are compiled from the ten interim reports of the Colonial Development Fund, 1929–39; Command Papers 3540, 3876, 4079, 4316, 4634, 4916, 5202, 5537, 5789, 6062.

Table 4.2 Tanganyika Territory public health and medical grants, Colonial Development Act, 1929–39

Grant area	Totals (£)	% of total
1. Tsetse Fly and Sleeping Sickness Research	302,763	75.0
2. Malaria Research	34,900	8.6
3. Malaria Control	27,533	6.8
4. Medical Training School for African Dispensers and Sanitary Inspectors	20,000	5.0
5. Tuberculosis Research	13,913	3.4
6. Tsetse Fly Control	5,000	1.2
Total	404,109	100

Sources: Command Papers 3540, 3876, 4079, 4316, 4634, 4916, 5202, 5537, 5789, 6062.

economic development depended on stopping the epidemics. Sleeping sickness was also a catalyst for the first attempts in colonial Africa at transnational coordination in biomedicine, with the 1907 International Sleeping Sickness Conference in London and the founding of the Sleeping Sickness Bureau the following year.[26] Although it is difficult to assess the costs of these undertakings, it seems likely that before the First World War far more metropolitan funds were spent on trypanosomiasis research than on any other infectious disease in the newly acquired African territories. Individual colonies had other health concerns of course; in the early decades of

Table 4.3 African field research: Scientific and government commissions on sleeping sickness

Date	Location—sponsoring organization	Sponsoring country
1901–2	Angola—Ministry of the Navy and the Colonies	Portugal
1902–3	Senegambia—Liverpool School of Tropical Medicine	UK
1902–3	Uganda—First Sleeping Sickness Commission of the Royal Society	UK
1903–6	Uganda—Second Sleeping Sickness Commission of the Royal Society	UK
1903–5	Congo (Belgian)—Liverpool School of Tropical Medicine	UK
1906–8	Congo-AEF—Service de Santé des Troupes Coloniales Françaises	France
1906–7	Tanzania—German East African Sleeping Sickness Commission	Germany
1907–8	Sudan—Foreign Office Sudan Sleeping Sickness Commission	UK
1907–8	Principe—Colonial Health Department (Angola)	Portugal
1907	Rhodesia/Zambezi—Liverpool School of Tropical Medicine	UK
1908–12	Uganda/East Africa—Third Royal Society Sleeping Sickness Commission	UK
1908–10	Togo—German Togoland Sleeping Sickness Commission	Germany
1909	Equatorial Guinea—Spanish Commission	Spain
1910–12	Congo (Belgian)—Scientific Mission to Katanga	Belgium
1911	Gambia—Liverpool School of Tropical Medicine	UK
1911–13	N. & S. Rhodesia (Zimbabwe & Zambia)—British South Africa Company	UK
1912–14	Nyasaland—Royal Society Sleeping Sickness Commission	UK

Sources: Pieter de Raadt, "The History of Sleeping Sickness," in *Protozoal Diseases*, ed. H. M. Gilles (London: 1999), 253; H. Harold Scott, *A History of Tropical Medicine*, vol. 1, 511–17; C. A. Thimm, ed., *Bibliography of Trypanosomiasis* (London: Sleeping Sickness Bureau, 1909); and *Reports of the Sleeping Sickness Commission of the Royal Society* nos. 1–16 (1903–15).

state building, yellow fever, malaria, smallpox, plague, yaws, venereal disease, and even mental illness all attracted official attention. When it came to pan-African priorities, however, trypanosomiasis was at the top of the list. Sleeping sickness investigations gradually laid the groundwork for new biomedical infrastructures that linked Europe and Africa.

When the League of Nations Health Organization (LNHO) was founded in 1921, sleeping sickness quickly became a cornerstone of its work in tropical Africa as well. The League published four special reports on the subject between 1922 and 1928, hosted two international conferences in London and Paris, and sponsored the International Commission in Uganda.[27] Some administrators, including Assistant Secretary of State William Ormsby-Gore, saw the LNHO as a way to sidestep parochial and shortsighted approaches to infectious diseases.[28] Indeed, the League's work on sleeping sickness was part of a broader movement to encourage ecologi-

Table 4.4 African trypanosomiasis research and coordination: Inter-territorial and inter-imperial efforts, 1907–36

1907	International Sleeping Sickness Conference (London)
1908	Sleeping Sickness Bureau founded (became Tropical Diseases Bureau in 1912)
1913	British Parliament Inter-Departmental Committee on Sleeping Sickness
1922	League of Nations Appoints Expert Committee-Sleeping Sickness in Tropical Africa
1925	League of Nations International Sleeping Sickness Conference (London, May) Imperial Entomological Conference (London, June) Civil Research Committee founds Tsetse Fly Sub-Committee (June)
1926–27	League of Nations International Sleeping Sickness Commission, Uganda
1927	Tsetse Research Department (for East Africa) Founded, Tanganyika
1928	League of Nations International Sleeping Sickness Conference (Paris)
1927–36	East African Institute of Human Trypanosomiasis Research, Entebbe, Uganda
1933	East African Coordination Conference of Tsetse and Tryps Research at Entebbe
1936	East African Coordination Conference of Tsetse and Tryps Research at Entebbe

cal analyses of the disease. Ormsby-Gore told the audience in his opening remarks to the 1928 International Sleeping Sickness Conference:

> Let me at once emphasise . . . that the problem of trypanosomiasis is very varied, but must be regarded as a whole. The fact that the human and the animal diseases are conveyed by the same family of blood-sucking flies— the *glossinae*—and that the protozoan organisms which the flies carry are so similar, makes it absolutely essential that both the medical and veterinary problems should be discussed in their widest ecological bearings.

Pointing to the studies conducted by the Tsetse Research Department in Tanganyika under the leadership of Charles Swynnerton, Ormsby-Gore asserted: "One thing is clear from the work that he has carried on for the last five years, namely, that no one form of attack upon the tsetse fly is universally practicable. . . . No one method can be universally applied, but each administration faced with the problem has much to learn from the experiments carried out by its neighbors."[29]

The League's coordinating efforts on sleeping sickness were a direct stimulus for the first pan-African health conference held in South Africa in 1932 (tables 4.4 and 3.3). Medical administrators associated with the League began to stress in the late 1920s that it was not enough to focus on a single disease; according to the Health Committee, those interested in controlling trypanosomiasis "might with advantage devote attention to the general health conditions of the native population in Africa."[30] Attended almost solely by the medical directors from British Africa, the 1932 gathering provided these experts with a forum to present a collective statement on links among economic conditions, infectious diseases, and rural hygiene. The 1924 East Africa Commission, on which Ormsby-Gore served, had made a similar point, but this was the first time medical directors had an opportunity to express their own opinions on these questions and begin to set pan-African priorities. "No community can be healthy," they emphasized, "unless its economic status is sufficiently high to provide at least reasonably effective housing and particularly a balanced and sufficient [food] ration throughout the year."

> It may therefore be said without fear of serious contradiction that the first task before the administrations of predominantly native territories is the raising of the economic status of the population. In an under-nourished population, especially if it is subjected to periods of famine or semi-famine, the mere treatment of disease, no matter how effectively and widely carried out, will achieve but negligible results.[31]

The high priority placed on trypanosomiasis research is evident also in British funding patterns: nearly half of all CDF grants to British Africa between 1929 and 1939 were for sleeping sickness (table 4.5). Following the war, funding ratios decreased somewhat, but overall spending increased. Between 1940 and 1960, Great Britain still allocated more than one-fourth of its colonial development funds earmarked for biomedicine to trypanosomiasis, which amounted to three times more than was spent on malaria and eleven times more than was spent on yellow fever, typhoid, leprosy, and tuberculosis combined.[32] As CO officials acknowledged when explaining these priorities, they saw the control of trypanosomiasis as being directly linked to colonial development and to "preventive and social medicine" in rural areas.[33] The Anchau settlement scheme in northern Nigeria, launched in 1937, was a "model" effort to combine curative, preventive, investigative, and development activities. Its central coordinator, Thomas Nash, had previously been employed in the Tsetse Research Department in Tanganyika.[34] As technical officers moved from one colonial site to another, they often took formative ideas and techniques with them. Ecological methods, which specialists viewed as highly adaptable to local conditions, were among the key tools of their trade.

Historians of medicine concur that the early twentieth century marked a moment when biomedical specialists became optimistic that infectious diseases could be conquered, that all it would take was finding the right "magic bullet."[35] Those who have focused on colonial Africa have suggested that the germ theory of disease ensured that health interventions across the

Table 4.5 Tsetse fly and sleeping sickness research and control grants, Colonial Development Act, 1929–39

Grant area	Totals (£)	% of total
1. Tanganyika	307,763	70.7
2. Nigeria	102,500	23.5
3. Kenya	11,860	2.7
4. Bechuanaland	7,000	1.6
5. Uganda	3,332	0.8
6. Northern Rhodesia	3,000	0.7
Totals African Tryps	435,455	100.0
Tryps grants as % of CDF grants to British Africa	883,750	49.3

Sources: Command Papers 3540, 3876, 4079, 4316, 4634, 4916, 5202, 5537, 5789, 6062.

continent were narrowly conceived. Maryinez Lyons has argued that "medical services tended to be more vertical than horizontal," with "medical campaigns launched against single diseases [rather than] health programs address[ed] to a broad spectrum of public health issues." Approaches to disease neglected "relationships between humans and their environments" in favor of "powerful theories of the 'biological determinism of disease.'"[36] The ecologist John Ford explicitly reinforced this perception in his discussion of the colonial history of trypanosomiasis control when he criticized "Western science" for its "piecemeal approach to disease."[37] Randall Packard has likewise asserted that "despite a great deal of rhetoric around the health of the empire, little change occurred in the direction or definition of health in the tropics between the [world] wars. Health remained defined as the absence of disease and the control of disease continued to be viewed in narrow technical terms."[38]

Closer attention to the myriad efforts at research and disease control in colonial Africa tempers these assumptions and highlights colonial states' emphasis on interdisciplinary approaches to disease and public health. Lyons and Packard are not wrong to stress the vertical nature of health care and the gap between rhetoric and reality, but overemphasizing these patterns prevents us from seeing that scientists and administrators not only developed critiques of their own but also actively worked to redefine health in dynamic and highly localized terms. Transdisciplinary approaches to health were advocated in Britain's colonial dependencies, in some cases years before they became normative in metropolitan institutions concerned with Europeans' well-being. Medical departments and scientific institutes within tropical Africa were just one site for these efforts. Interterritorial conferences, advisory committees and clearinghouses, as well as inter-governmental organizations also influenced these developments. In fact, to separate territorial from imperial or international efforts obscures the necessary part each institutional level played in producing these perspectives. So, to Shula Marks's question, "what was colonial about colonial medicine?" we ought to add, what was *imperial* about international medicine, and how do we disentangle these relationships?[39]

In her excellent study of medicine in the Anglo-Egyptian Sudan, Heather Bell has suggested that "international medicine, in its eagerness to cross international borders, was fundamentally at odds with . . . colonial medicine. For colonial medicine was ever preoccupied with creating a country, protecting a profession, and controlling disease by erecting and reinforcing boundaries, literal and figurative."[40] Bell is right to argue that biomedical work was constrained by the structures of colonial states. Their budgets, facilities, staffing, and borders shaped the kind of work that was done in each territory. At the institutional and epistemic levels, however, Bell overlooks central connections between colonial and international trends. A particular kind of "international medicine" emerged directly out of transnational endeavors in colonial Africa and rested largely on an imperial infrastructure. In many respects imperial concerns constituted and created international medicine: for much of the colonial period they were virtually one and the same thing. This process was dominated in the interwar decades by the British territories and was facilitated by the Health Organisation of the League, which functioned to surmount the epidemiological "difficulties arising at international boundaries."[41] Without representation of the United States in the League, the European powers served as the international gatekeepers to much of sub-Saharan Africa, with Britain at the forefront of these efforts. While the Rockefeller Foundation's International Health Division entered the fray unofficially through financial contributions and regional programs, the accumulated experience, knowledge, and most important, *control* of these undertakings remained largely in British hands.[42]

Africa's almost total legal subordination encouraged specialists and politicians to think of the continent in ways that transcended the colonial state and framed interventions in terms of inter-territorial and international collaborations. During one of the many exchanges about sleeping sickness in the 1920s a member of the League's Secretariat argued:

Surely the need for international action in Africa in dealing with problems of common interest is, if anything, even greater and more urgent than in Europe, for fortunately in Africa, which has not yet become an organised continent, political frontiers do not yet have all the disadvantages which they present in Europe. . . . I should think that such a combined effort [of sleeping sickness control] on a big scale . . . would be a practical constructive demonstration of the larger international view of European trusteeship for Africa.[43]

The aim to achieve interdisciplinary and transnational approaches to health care was at times far more prevalent in colonial Africa than it was in several European states.

Colonial Modernity, African Therapeutics, and the Status of Medical Pluralism

Scientific studies designed to address agricultural development in British Africa often had the unintended effect of placing Africans' agricultural practices in the spotlight. As technical officers combined ecological and ethnographic interests in the field, they began to codify and valorize "indigenous knowledge," although usually in a way that continued to sanction bioscientific standards of efficacy and accuracy. Vernacular science, in turn, drew attention to the dangers of capital-intensive cash crop production and paved the way in at least some locations for bottom-up strategies that took heed of existing cultivation techniques.

By contrast, imperial administrators' interest in African therapeutic traditions was far less direct and far more intolerant, at least on the surface. Medical departments tended to grow at the expense of colonial subjects' collective knowledge of healing and disease control. In most instances, officials considered a "modernizing" agenda incompatible with the coexistence of what is now often referred to as popular or traditional medicine.[44] Yet, not all medical experts held pejorative opinions of indigenous therapeutics. When E. B. Worthington referred to African healers as "quacks" in the draft of one of his chapters, Tanganyika's medical director, R. R. Scott, and its senior pathologist, Burke Gaffney, replied that "native practitioners are not 'quacks' to their own people: and western medicine still has much to learn about the treatment of [disease] to arrogate to ourselves an exalted position in connection with its cure."[45] (Worthington removed the comment from his final version.) Kauntze included a similar remark in his draft "Health" chapter for An African Survey, which appeared in the final volume: "Not all those who practise native medicine in Africa can be dismissed as witchdoctors; many are much respected and it is indeed possible that a study of the herbs used by some of them might add to the list of remedies, such as quinine, which the pharmacopoeia owes to primitive medicinal practices."[46] Yet, the chapter continued, "the sanitary services must now undertake an education in health matters which will *replace* the advice given by native practitioners and also many of the prescriptions of African custom, as conveyed in initiation ceremonies and the like."[47] In the implicit "contest" between colonial experts and popular healers, the African Survey, in alliance with medical departments and the Colonial Office, wanted to declare biomedicine the winner.

For a host of reasons, however, a singular victory for biomedical frameworks was far easier to assert than to achieve. Medical anthropologists and social historians have shown that in their "quest for therapy" individuals and kinship networks in Africa retained considerable agency in deciding which kinds of therapeutics they would pursue.[48] As Steven Feierman has pointed out, "No colonial power and no independent African state has ever intervened decisively to destroy popular healing."[49] Therapeutic practices across colonial Africa were flexible and heterogeneous. Colonial states and the experts who served them often dealt with medical pluralism in their territories in inconsistent and contradictory ways. In British East Africa, for instance, laws that banned "witchcraft" and "fetish belief" existed uneasily alongside medical licensing laws, allowing administrators to undermine healers' powers but leaving room for indigenous practitioners to continue to offer their own "systems of therapeutics" as long as they limited their efforts to "the community to which [they] belong."[50] Conflicting laws not only presented officials with choices between prohibiting and accommodating healers but inadvertently also raised questions about cross-cultural epistemologies.

Europeans in colonial Africa often found it impossible to avoid acknowledging competing worldviews: magistrates had to deal with these questions on a regular basis in their courts, physicians and fieldworkers had to confront them whenever they tried to gain the trust of local populations in health campaigns, and administrators and missionaries were exposed to them in their daily work among the people they were seeking to control or convert. In 1930, Uganda's acting governor called the Colonial Office's attention to the weakness of British states when it came to legislation related to Africans' healing "customs."

> Direct prohibition is valueless unless it can be enforced; and enforcement . . .
> depends upon the co-operation of the people themselves. As the native is
> not represented in the central machinery of Government, legislation vitally
> affecting that population can be passed without opposition; but when it is
> passed it will remain a dead letter unless native public opinion accepts it, or

unless the protective power devotes to its enforcement an effort which would almost always be beyond the resources of the local Government. The dead weight of passive resistance, which the African is capable of bringing to bear, can only be appreciated by those who have experienced it.[51]

Even when colonial laws staked out unequivocal positions, state leaders often had to admit that they were unable to enforce them systematically.

A final factor in biomedicine's mixed success was the ascendance of anthropology, whose adherents took a growing interest in studying Africans' medical and natural knowledge within wider contexts of culture and cosmology. Most of these professionals interpreted Africans' explanations of illness, misfortune, and health in light of existing bioscientific frameworks, yet their publications often drew attention to the coherence and logic of different therapeutic traditions. By taking Africans' worldviews seriously, anthropologists unsettled popular depictions of these ideas as irrational, superstitious, and unscientific. Similarly, technical and administrative officers in virtually every British territory undertook sporadic ethnographic research into "native" *materia medica*, diet, and disease control. These projects, despite their ad hoc approach, could inadvertently validate the usefulness of vernacular knowledge. In 1909, for example, Sierra Leone's small medical department gathered information on various beliefs about land use and health practices and noticed that the state laws against "native fetish" failed to distinguish helpful from harmful "local customs." They issued a directive to all medical and political personnel, which they then forwarded to Lord Crewe, the colonial secretary. Crewe reviewed the findings and thought they were significant enough to circulate to all the governors of British African territories.

> It will be found that where a town is governed by a good chief, the chief principles of sanitation are observed. Moreover, close questioning of the older men will show that these principles are strictly enjoined by native law. It is important, therefore, to remember that an unhealthy town is one wherein native law and customs are departed from: the chief should be supported to the utmost by the Political and Medical Officers in enforcing the native laws with respect to sanitation. . . . Many customs and many apparent superstitions are connected with the preservation of health.[52]

These kinds of vernacular translations between biomedical and indigenous worldviews usually had highly instrumental aims: to make the state and its subsidiary departments work better. If officials could achieve these goals by paying closer attention to colonial subjects' ideas and practices then, on occasion, they would. Rarely, however, did their efforts result in policies that explicitly sanctioned medical pluralism: they were willing to tolerate *cultural* relativity for pragmatic reasons, as they did in adopting indirect rule, but *epistemic* relativity posed serious problems. Yet, much like African healers who gradually made use of the tools and ideas of colonizers, a process Julie Livingston aptly calls "entangled therapeutics," adherents of biomedical logic were also able to incorporate African ways of knowing within their own disciplinary frameworks. These dynamics tended to produce dual effects. First, the act of appropriation was often done in such a way that it could be erased over time, making whatever debts field officers owed to their ethnographic subjects and research assistants invisible. Second, it created lasting, low-level tensions over the boundaries between science and nonscience and helped to establish new fields of research that were designed to straddle these boundaries—ethnomedicine and medical anthropology. The first effect is explored in more detail in the remaining sections of this chapter, which include an analysis of ecological fieldwork and the shift toward rural health care, while the second, given its close ties to anthropological research, is considered briefly in chapter six.

Ecology, Epidemiology, and Tropical Medicine

"Where did the modern, ecological understanding of epidemic infectious disease come from?" asks Andrew Mendelsohn in his history of German and English epidemiological theories during the interwar period. While conceding that the "obvious answer would seem to be that it came from ecology," he rejects this explanation: "How indeed is one to imagine that the fledgling ideas and methods of upstart population ecology, or the premises of parasitology, which were of uncertain relevance to bacterial and viral disease . . . could have conquered bacteriology?" He argues that in fact scientific theories of epidemics "became complex," and epidemiology more syncretic, through intellectual currents within the discipline of bacteriology.[53] He locates this development in the decade following the First World War, a time when the influenza pandemic called into question existing explanations of the "causes and nature of [human] epidemics."[54]

The full story behind the question Mendelsohn posed, however, remains only partially told.[55] As he acknowledges, during the interwar period many sciences were affected by such concepts as "holism, complexity, equilibrium, web of causes, and system," suggesting that there had been "a broad intellectual transformation" among scientific disciplines and their adherents.[56] The evidence from African and colonial contexts supports this interpretation but locates the origins of the shift earlier than he does. Ecological ideas, principles, and research, fragmented though they may have been, clearly had an effect on medical researchers, including epidemiologists, bacteriologists, and public health officials.[57] A handful of studies of African trypanosomiasis at the turn of the twentieth century, for instance, explicitly mentioned "ecological conditions" as part of their analyses. Yet, the converse of this pattern was also true: adherents of ecology, in particular those concerned with questions of disease and health, drew upon the work of medically trained researchers in part because their arguments were bolstered by this evidence. A strong case could be made that the "new epidemiology" being promoted in Britain, which supported the idea that epidemics should be examined from the "bird's eye view of all from an aeroplane," was in many respects an organic bedfellow to that other synthesizing and "aerial" science, ecology.[58]

Tropical medicine was central to this development. Historians have already made the point that this young discipline emerged equally from the biological and the medical sciences. Tropical medicine, Michael Worboys observed, was "structured around the life-cycle of parasites" and "required detailed knowledge of the taxonomy of vector species and ecological management, which found application in the tropical environment."[59] However much the different European powers might have seen one another as competitors, their experience of sleeping sickness epidemics was a shared one. All classified the disease as endemic to "that portion of the African continent lying between the Tropics of Cancer and Capricorn" and devoted considerable resources to its study in their institutes of tropical medicine.[60] By the second decade of the twentieth century, tropical medicine was shifting gradually away from a linear understanding of disease causation, in which microbes alone were targeted as the culprit, toward a more integrated and comprehensive approach. The doyens of the discipline, Ronald Ross and Patrick Manson, heralded these shifts in their public speeches and published works.[61] By 1925, Britain's Medical Research Council offered its own endorsement of the field's breadth: "tropical medicine is no separate branch of medicine, but touches all the fields of medicine and needs the services of all the medical sciences."[62]

A unifying preoccupation of ecology, the "new epidemiology," and tropical medicine in this period was to discern how human, animal, plant, and parasitic organisms interacted. Julian Huxley, one of the key architects of this synthesis, following his research trip to East Africa observed that "organisms . . . have no biological meaning apart from their environment."[63]

In both the field and the laboratory, practitioners began to emphasize "the inter relations of different parts": species diversity to climate, epidemic patterns to population levels, and disease distribution to geography.[64] Scientists in all three disciplines shared a fascination with the question of whether it would be possible to "eradicate" diseases permanently, or only to control them. During the interwar period, approaches in all three specialties began to converge around the analytic categories of "environment" and "disease."

The idea of the elimination of diseases can be traced back at least to 1801, when Edward Jenner pronounced that "vaccine inoculation" would result in "the annihilation of the Small Pox." Not until the late nineteenth century, however, was the term *eradication*—meaning to pull out by the roots—applied in a systematic manner to infectious diseases. Some of the earliest attempts at eradication were directed toward diseases of domestic animals, such as bovine pleuropneumonia and rabies. These tended to be regional efforts, undertaken in such places as the United States, Britain, and South Africa, and were often effective on a limited geographical scale. Only in the early twentieth century were transnational disease campaigns first conceptualized under the rubric of eradication. The best known was the Rockefeller Foundation's Commission for the Eradication of Hookworm Disease, founded in 1909 and phased out after 1922, which sponsored projects in over fifty countries that had limited success. The 1915 Rockefeller Commission for the Eradication of Yellow Fever was frustrated both by a lack of knowledge of the "jungle" variety of yellow fever and by measures that proved inadequate to reduce the mosquito vector (*Aedes aegypti*).[65] How did these early twentieth-century conceptual shifts affect imperial and international confidence in eradication as a control strategy? At that time, experts in epidemiology and tropical medicine began to focus more on the site specificity and unique attributes of diseases than on their universal features. To understand the connections, we must explore the research interests and activities of field officers in British colonial Africa.

Vernacular and Ecological Approaches to Disease Control

Just after the First World War, Charles Swynnerton, game warden of Tanganyika, undertook a study of the "tsetse fly problem" in North Mossurise at the invitation of the Portuguese administration, the Mozambique Company.[66] Swynnerton was a fellow of the Linnaean and Entomological Societies in Britain and had prepared a series of articles on the history of forestry in East Africa, including a study of the "medicinal uses of plants."[67] Born in

India in 1878 and educated in Britain, he had moved to Southern Rhodesia when he was only nineteen, arriving at the height of the rinderpest cattle epidemic in 1897.[68] Beginning as a farmer, Swynnerton was a naturalist by inclination and quickly took up the study of Southern Rhodesia's environments. An advisor to the agriculture department later described his large farm as "nothing less than a private research station, and apart from rubber, he experimented with many crops and plants, such as coffee, fruits, forest trees, fibre, oil and forage plants."[69] Swynnerton's research on forestry had sensitized him to the need to cultivate elderly informants, including African healers, since they could help him develop a picture of environmental changes over time. He often juxtaposed their recollections to the published accounts of earlier explorers and naturalists who had traveled through the region. Although at first he worried that some informants might not be reliable witnesses, he relied on their accounts as he developed explanations for how and why species distributions in the region were changing.[70]

Swynnerton conducted oral interviews during his research trip to Mozambique as well. Not only did he travel with a retinue of assistants from his own farm, but also he worked with a number of "native informants" who interviewed elders when he himself did not speak the local language. What he learned during his three months in Mossurise had a powerful influence on his future investigations. It quickly became apparent to Swynnerton that Zulu residents of the region had once developed a system of managing the tsetse fly on their own. "The area west of the Sitatonga hills was, under the Zulu domination, the scene of a particularly fine experiment in the banishment of tsetse [flies]. It was not difficult to obtain the details of this experiment, as most of the older natives had . . . taken part in it."[71] Through carefully timed annual grass fires, which served as "a grand general cleanser and disinfector of the country," residents had been able to rid the region both of the habitat in which tsetse species bred and of the "noxious insects," including ticks, that posed a threat to human and animal health.[72] According to his informants, "under the Zulu, burning was the subject of regulation . . . a late, thorough burn aimed at and usually achieved. Under the white man every one burns when he pleases."[73] Umzila, the Zulu leader, also encouraged dense settlements and the careful culling and regulation of game in the region. "Every one of my informants described most graphically the result of this concentration. The bush disappeared and the country became bare, except for the numberless native villages and a continuity of native gardens."[74] Without tsetse flies or game near their settlements, they were able to rear their cattle without fear that they would be lost to *nagana*, the animal variant of trypanosomiasis. According to Swynnerton, "The Zulus . . . knew the fly well and the disease caused by it, and they regarded the proximity of game as dangerous for cattle."[75] The bottom line for him was clear: "Umzila's principle . . . would be well worth consideration and investigation." Encouraging similar interventions in Tanganyika, he believed, might even promote greater affinity with "the natives as late burning represents their own old custom and . . . they still speak of it as the correct method."[76]

In 1921, Swynnerton prepared a memorandum on "tsetse control" for Tanganyika in which he advocated testing the strategy of employing fires late in the growing season and encouraging agricultural settlements in their train. "This burning," he told an audience at the 1925 Imperial Entomological Conference in London, "had been carried out . . . for at least one hundred years, but not systematically."[77] His experiments were designed to examine whether these methods could be more widely and effectively applied. The first site he selected was in the Shinyanga District of Tanganyika, where the Sukuma people were known to take an active interest in imperial administration. In this region, people, cattle, tsetse flies, and game including antelopes, zebras, wildebeests, gazelles, and giraffes occupied close quarters. According to his interviews with residents in the region, Shinyanga was also thought to have "the best soils and the best pasture," which they were keen to use for their cattle and agriculture if at all possible.[78] Because the Colonial Office was considering plans to construct a railway through Shinyanga, it made sense to Swynnerton to see if there were ways to protect the region from "depopulation" as a result of disease (see plate 8).[79]

Swynnerton's Mozambique trip also prompted him to rethink scientific disciplines. He came to realize that "the practical study of tsetse is a matter for the botanist and oecologist rather than the unaided entomologist." Some years later, he argued that "only by an investigation on ecological lines, never applied to [trypanosomiasis research] previously, were these problems likely to be solved."[80] Given Swynnerton's participation in British and South African scientific networks, it seems likely that he developed his appreciation of ecology from botanists.[81] His first detailed report for Tanganyika sketched the "oecological divisions" of the territory and paid close attention to the relationship between fauna and flora.[82] His responsibilities for game, combined with the growing awareness among disease specialists that wild animals could harbor parasites threatening to humans, pushed Swynnerton to consider connections between disease and "oecology" more carefully. "We should begin [our control schemes]," he recommended, "with a close botanical and oecological study, and we could produce Umzila's results with a half, a quarter or an eighth of the [labor] force used by him."[83] As Swynnerton noted, this approach was novel both in field research and in understandings of indigenous practices.

Colonial officials were beginning to appreciate the intractability of the problems involved in combining game preservation with disease control. In a 1911 review in *Nature*, Harry Johnston observed that "about three years ago,"

> It was suspected that in many parts of Africa the existence of big game was actually prejudicial, and even dangerous, to the coexistence of the human race, black, white, or yellow. It seemed as though other creatures than man and monkeys must act as reservoirs of micro-organisms, especially trypanosomes, provocative of disease. Consequently, so long as they coexisted with man, the various species of tsetse-fly, of tick, and flea, would, even if infected human beings were isolated, have always the means of renewing their supplies of disease germs.[84]

In 1913, Parliament appointed a joint committee to undertake a systematic review of the evidence for possible disease reservoirs, calling upon several members of the Royal Society's sleeping sickness commissions, including David Bruce, Geoffrey Carpenter, and Lyndhurst Duke, as well as Johnston, John Kirk, and other officials and scientists. Although many experts believed it would be extremely difficult to attempt to control trypanosomiasis by "eradicating" fauna, this line of attack was adopted by the leaders in Southern Rhodesia.[85]

Swynnerton wished to avoid such drastic measures and sought to find other methods that could balance the different needs of colonial states and their subjects in the region. He explained the rationale behind his approach to his colleagues at the Imperial Entomological Conference: "There were twenty species of tsetse flies [in East Africa], each having its own requirements and existing under a variety of conditions. Thus there might be fifty different tsetse problems." It made little sense to address every problem identically since the natural and social conditions of the affected regions were so different. After his fieldwork in Mozambique, he had concluded that "no investigator can at present lay down the law for another area than his own."[86] Taking this line of reasoning further, he argued that cooperation both among scientific disciplines and across political boundaries was essential. "Each piece of work required careful preliminary study from the entomological, botanical and zoological points of view, so that co-operation between the various branches of science had to be arranged for. . . . International co-operation was necessary, as political boundaries were often not natural ones. Inter-colonial co-operation [had] also [to] be arranged."[87] In Swynnerton's framework, disease control would involve working closely with local populations to discern their own understanding of their history and environments. He employed a number of African naturalists in his research institute who were, by the early 1930s, initiating and controlling experimental schemes of their own. Swynnerton's penchant for taking part in the social festivities associated with his "tsetse experiments" earned him the nickname "Mayoka" (snake) since he was an avid participant in the Sukuma Snake Society dances.[88]

The Bionomics of Tsetse Flies and Early Bioscientific Fieldwork

By the second decade of the twentieth century, every territory in British Africa had had some experience with trypanosomiasis research, many through both the Royal Society commissions and state-building endeavors. Since the 1901 epidemic was first noticed in Busoga along the shores of Lake Victoria, Uganda had become a headquarters for inter-territorial research on sleeping sickness.[89] Uganda's experience with human epidemics, including venereal diseases and yaws as well as trypanosomiasis, explains, in part, why it became the second territory in British Africa, after the Anglo-Egyptian Sudan, to pursue systematic training of Africans as part of its medical services. As Diane Zeller has shown in her case study of biomedical work in the Buganda districts, colonial officials pursued close collaborations with Ugandan leaders and relied heavily on them to negotiate resettlement schemes with local populations. Had Bugandan leaders not issued their own regulations on sleeping sickness alongside those of the colonial state, many of the state's measures would not have succeeded to the extent they did in separating island and lakeshore populations from infective flies.[90] During the First World War, colonial officials turned to Bagandan leaders in order to recruit more than one thousand individuals to make up a "native medical corps." In 1923, these foundations led to the establishment of Makerere College for medical training.[91]

Organized studies of the "bionomics," or environmental conditions, of the different tsetse fly species got underway in association with the third Sleeping Sickness Commission of the Royal Society. Their primary author was Geoffrey Carpenter, a junior colleague of Swynnerton, who had received his MD in tropical medicine from the London School in 1910 and joined Uganda's Medical Service following the war. He remained in the territory for twenty years, leaving in 1930 to take a post at Oxford University where he helped to supervise the work of John Ford (whose first post in British Africa was with the Tsetse Research Department in Tanganyika). In addition to his scientific articles,[92] Carpenter published a popular narrative of his endeavors, *A Naturalist on Lake Victoria with an Account of Sleeping Sickness and the Tse-Tse Fly* (1920).

Carpenter spent much of his time between 1910 and 1914 making observations on the recently depopulated islands of Lake Victoria, in the hope that "the presence or absence of *Glossina* [tsetse flies] might be found to be correlated with definite factors."[93] Members of the Royal Society's commission had been alarmed to learn that the tsetse species along the lakeshore were still infective nearly two years after the islands had been forcibly abandoned by local residents.[94] Seeking to address this puzzle, Carpenter was concerned not just with the life history of the "tse-tse" and its "natural enemies" but also with "the relations between Tse-tse, Trypanosome, and the 'alternative hosts' of the latter from which the fly acquires it."[95] To test various hypotheses, he explored tsetse breeding sites and feeding patterns; by examining the blood of the islands' fauna, he discovered that antelopes were their favorite. He correlated climate with fly population levels, finding that flies were less numerous during the hotter and humid months. Tracking the distances flies could travel, he found that they used fauna, including humans, for long-distance travel.

Carpenter concluded that to "exterminate Sleeping Sickness two animals must be kept from each other—the Situtunga antelope, from which the fly obtains the Trypanosome, and the fly, which inoculates the Situtunga with the Trypanosome. Each without the other is harmless."[96] Like Swynnerton, Carpenter believed that game extermination experiments would prove ineffectual. He explained to the 1913 parliamentary committee that not only would it be virtually impossible to destroy all fauna on which flies fed, it would also be hard to keep infected flies that had come from other areas out of experimental sites.[97] For field researchers, geographical scale was a key challenge. The 1912–14 Royal Society Commission to Nyasaland came to the same conclusions: "The problem cannot be attacked with any chance of success from the side of the fly alone." There were nearly 5,000 square miles of "fly-country" in the area that they had surveyed, and "tsetse flies are numerous and deposit their larvae all over the country."[98] Laboratory scientists underscored these points, acknowledging that while they might be able to determine lines of infection with some certainty in the laboratory, "whether it is [this way] in nature or not is quite another question."[99]

There had been some success with tsetse eradication in tropical Africa before the First World War. The Portuguese on the island of Principe, off the

West Coast of Africa, had made use of various traps, including tar-painted strips on the backs of laborers, which had managed to rid the island of the tsetse species that caused disease.[100] When the 1924 Parliamentary East Africa Commission reviewed this experiment they too emphasized the issue of scale: "it is one thing to deal with an island and altogether another proposition to deal with a continent."[101] They could have added an observation that the Parliamentary Joint Committee made in 1913: Principe's tsetse flies were imported from the mainland and were not originally endemic to the island, which meant that their "eradication" consisted of destroying a finite number of flies that had a limited sphere of entrenched breeding sites. It also meant that there were no faunal populations, besides the imported cattle, that might be infected with trypanosomes.[102] When working on the scale of a continent these distinctions mattered a great deal.

During their tour of the East African territories, William Ormsby-Gore and his colleagues from Parliament had met Swynnerton, Carpenter, and several other scientists, including Tanganyika's veterinary pathologist, H. E. Hornby, Uganda's bacteriologist, Lyndhurst Duke, and Nyasaland's medical entomologist, W. A. Lamborn. Stemming the disease seemed to them an "urgent necessity" since it obstructed economic development. The commissioners were concerned by reports that tsetse fly populations were extending their territory. They recommended not only further bionomical studies and fly surveys but also field experiments on fly extermination.[103] The commission's recommendations were considered on several fronts. In 1925, when the British Parliament approved the creation of the Committee on Civil Research, one of the first subcommittees they established was on the tsetse fly. Ormsby-Gore described it as "the trypanosomiasis committee for the whole of Africa, South, West, and East."[104]

Deepening Ecological and Vernacular Investigations in the Tsetse Research Department

The years between 1925 and 1935 marked a central shift in tsetse and trypanosomiasis research in British Africa. During this period, metropolitan funding bodies began to provide far greater sums to support extensive field experiments. At the same time, ecology replaced bionomics as the central organizing principle for epidemiological fieldwork. While bionomics was by definition the study of organisms, usually insects, in their environments, in practice it was often much broader, as Carpenter's far-reaching investigations indicate.[105] In 1935, a Colonial Office administrator remarked, after looking up the word *ecology* in the *Encyclopedia Britannica*:

"It appears to mean all . . . those factors which go to maintain or destroy animal and/or plant life and is apparently the modern jargon for our old pal 'bionomics'!"[106] The shift was not merely semantic, however; it was accompanied by methodological, institutional, and financial changes, which flowed in part from institutional reconfigurations in the metropole.

Shortly after the Tsetse Sub-Committee was created in Britain, it designated three territories, Tanganyika, Uganda, and Nigeria, as the coordinating centers for all "long-range fundamental research" on trypanosomiasis.[107] Individual territories' medical and agricultural research facilities would continue to conduct projects tailored to meet each state's needs.[108] According to Walter Fletcher of the Medical Research Council, work done in institutes sponsored by metropolitan funds would address "wider scientific problems that affected Africa as a whole." In the case of trypanosomiasis, Fletcher noted, it was important "that the territorial boundaries be disregarded because they have no existence in [the] face of common scientific problems."[109] While Uganda and Nigeria played important parts in sleeping sickness and trypanosomiasis investigations before the Second World War, Tanganyika ultimately assumed the most prominent place in inter-territorial research activities.[110]

In its first year, the Tsetse Sub-Committee recommended that Tanganyika be allocated a five-year grant of £70,000 from the East African Loan Scheme and that Charles Swynnerton be asked to direct efforts on tsetse fly studies. By 1927, the Tsetse Research Department was founded in Shinyanga, Tanganyika, the site of Swynnerton's first experiments. When the East African grant expired in 1931, the department's research results were considered significant enough to warrant continued funding, and another £75,250 was awarded over the next seven years through the Empire Marketing Board and the Colonial Development Fund (CDF). In 1938, the CDF made a further allocation of nearly £208,000, which was intended to help the department scale up its field experiments for another seven years. Tanganyika's efforts to study trypanosomiasis even caught the attention of the British filmmaker Thorold Dickinson who selected the territory as the site for his movie *Men of Two Worlds*, whose plot revolved around the state's development scheme to combat a sleeping sickness epidemic. Before he made the film, Dickinson directed a short documentary on Swynnerton himself.[111]

When the Colonial Development Fund decided to support Northern Rhodesia's Ecological Survey, its advisors drew attention to the intersections its results might have with the "tsetse research" in Tanganyika. Collaborations between the two projects began early in the 1930s and were concentrated initially in the Abercorn District of Northeastern Rhodesia, where Unwin Moffat still worked.[112] The person charged with conducting the field surveys was Swedi Abdallah, one of the Tsetse Department's "native staff," who traveled to Abercorn in September of 1935 "with two assistants" to carry out "necessary preliminary work on the distribution and density of the flies."[113] He returned in 1937 to conduct a second "full survey."[114] Swynnerton and one of his field botanists, B. D. Burtt, followed suit. Tanganyika also loaned Amani's botanist, Peter Greenway, and its soil scientist, Geoffrey Milne, to the Ecological Survey, rounding out the interterritorial collaborations.

Swynnerton had been conducting small-scale experiments in tsetse control across Tanganyika since the early 1920s. After a 1922 epidemic of sleeping sickness in the Mwanza District, just north of Shinyanga, he and Lyndhurst Duke, Uganda's trypanosomiasis specialist, advised the government on its possible origins and the best methods of its control.[115] This project drew on oral interviews, which suggested that the disease in humans was a relatively recent occurrence in the region. With the creation of a formal Tsetse Research Department, Swynnerton expanded his staff and hired a "trained ecologist." John Phillips, a young South African who was a forest research officer in Knysna, had received his doctorate from Edinburgh in forestry and botany with an emphasis on ecological methods. From 1927 until 1932, he served as deputy director and ecologist.[116] The driving vision behind the department was that ecological science could help untangle the challenges of African trypanosomiasis, a problem they considered "as bewildering in its complexity, as comprehensive in its interrelations" as any problem then being confronted in Africa.[117]

As a result of the colonial development loans and grants, the Tsetse Research Department employed twenty-one European staff by 1930 and rivaled in size many other technical departments across the African territories.[118] After a decline in numbers during the Depression, the European staff rose to twenty-six by 1938; John Ford was the most notable new recruit. The pyramid structure of its staffing was even more important. In the mid-1930s, the department employed 122 "African Assistants," and by 1938 that number had increased to 250.[119] These individuals were paid to clear brush, catch flies, collect pupae, assist in laboratory experiments, set traps, oversee large-scale fieldwork, and keep daily records of game and tsetse movements. All members of the senior staff adopted ecological methods in their research; publications by Swynnerton, Phillips, Thomas Nash, B. D. Burtt, Charles Jackson, H. Harrison, H. Hornby, and John Ford were particularly significant. Nash, Jackson, and Ford had all been mentored by members of Britain's Ecological Society before they went to Tanganyika. To this list we should add Abdallah, as well as two of his counterparts in field experiments: Milambo Kazila, who directed tsetse research on Maboko Island on Lake Victoria near Kisumu, Kenya; and Makashasha bin Sapila, who supervised large-scale clearing experiments near the Shinyanga headquarters.[120] While none published in scientific journals, each took the lead on projects central to the department's mission. In 1935, Swynnerton and Kazila were joint recipients of Tanganyika's Jubilee Medal, while the governor awarded Abdallah the King's Certificate of Honour and Badge for his work in the department.[121]

Thanks to Swynnerton's own interests, the department tried to combine ecological research and ethnographic understanding. Oral interviews helped Swynnerton develop a history of the effects of disease in each region he visited. During his travels in East Africa, he learned of the disparate consequences of the rinderpest epidemics at the end of the nineteenth century. Some populations had been hit hard and shared vivid memories of cattle loss and ensuing famines, while other groups explained that they had managed to avoid serious devastation. The unevenness of the epidemics influenced population and cattle distributions, patterns Swynnerton was then attempting to reconstruct. He also found it helpful to know the genealogies of different leaders and the nature of their interactions with other groups. Most of the articles he wrote that dealt with trypanosomiasis included a historical account of the peoples in the region, which he brought to bear on the kind of control schemes he recommended. Swynnerton found that ethnographic information could teach him new ways to think about sleeping sickness and *nagana* control. Learning about the Zulu late-fires strategy shaped the way he conceptualized involving Africans in "reclamation projects." In 1923, during his first campaign, he held two preliminary meetings with "Sultans and natives" to hear their concerns about tsetse fly encroachment and cattle rearing. Once the leaders had affirmed that it was the right time to act, Swynnerton scheduled a series of clearing campaigns designed to test their late-burn hypothesis. "The people came out well," Swynnerton reported. Leaders brought men to clear the margins of their own localities; "in all, more than ten thousand men engaged in these operations."[122]

It is not entirely clear just how much control Swynnerton actually exercised over these measures. As the campaign tapped into techniques that people in the region thought were correct, elites might have seen these activities as serving their own needs as much as those of the colonial state. Had they not, it is doubtful that regional leaders would have continued to mobilize such large numbers each year. This point was reinforced during the testimony of Chief Makwaia before the Parliamentary Joint Commission on Closer Union in East Africa in 1931. While Makwaia objected to forced labor because he thought it an inappropriate assertion of state authority, he saw the annual tsetse fly endeavors differently. With Philip Mitchell, Tanganyika's native secretary, serving as his translator, Chief Makwaia emphasized that with regard to "the tribal activities for clearing tsetse bush,"

> there is no payment for that; that the tribe turn out to clear the bush for themselves because it is their own grazing lands. . . . The people who go to work in this way are fed at the expense of the Native Treasury, . . . [through] which they make provision . . . for cattle to be bought, and other rations for the people who are doing the work. The whole tribe turn out together, the Chief and everybody. He says that in carrying out this work, they receive assistance from the Tsetse Research Officer and from the Officer of the Game Department, who help them in this reclamation work.[123]

Clearing activities could bolster leaders' authority and power, but they also enabled people to use land that might otherwise be unavailable. Ethnographic sensitivities and vernacular science mediated these colonial arrangements.

Swynnerton acknowledged that he learned a great deal about natural conditions from the people who had lived in these environments for a long time. He drew on the expertise of colonial officials with naturalist inclinations as well as on that of Africans.[124] An explicit part of the department's methodology was to draw upon informants' knowledge and, when possible, to triangulate it with other kinds of evidence. This approach helped to ensure that his epidemiological and ecological reconstructions were as accurate as possible.

Swynnerton quickly learned that relying on African research assistants had both epistemic and sociological rewards. His earliest recruit to tsetse research was Swedi bin Abdallah, whose personal history remains obscure although his name suggests a Muslim background. He may have begun his career as a naturalist in the German administration; his first appointment in 1920 was as Swynnerton's "head fly boy," a designation that indicates he was not a beginner.[125] Abdallah had several claims to fame during his time as a research assistant. First, he helped the sleeping sickness service identify actively ill individuals, which was difficult and sometimes dangerous because communities might go to great lengths to protect their ill from detection. According to Swynnerton, Abdallah was persistent in tracking down sick people: "Following up a difficult case in which the relatives of a sick man kept moving the latter from one part of the district to another in order to avoid the sleuth, Swedi finally overtook him just after death and still, despite threats, took his blood-smears." (The intrusiveness of biomedical research was not mitigated by its being carried out by Africans.) Abdallah was also responsible for identifying a new species of tsetse fly. Swynnerton was full of praise, not quite realizing the irony that he became the beneficiary of Abdallah's insights: "It was to Swedi that the identity of what was then the undescribed fly, which was subsequently described as *G[lossina] Swynnertoni*, was first referred . . . and who, unlike the European investigators before him, declared it to be totally distinct from *G. morsitans*." In a 1925 field investigation, "Swedi bin Abdallah . . . showed, and Mr. W. H. Potts, Mr. B. D. Burtt, and Dr. Wallace confirmed, that low continuous thicket in adequate width is not favoured by *G. morsitans*." Abdallah's insight, combined with Swynnerton's previous studies on regions that went unburned, "suggested [new] possibilities for control."[126]

The department then compared late burning "experimental plots" to plots that had not been burned at all. Scale played an important role: the unburned control plots ranged in size from 2 to 150 square miles (figure 4.1).[127] One of the longest standing sites on the Samuye Hills, Blocks 7c and 7d, which took up approximately 35 square miles of land located near the Shinyanga Headquarters, was under the supervision of Makasha-sha bin Sapila. Sapila had begun his employment in the department working on the annual clearing schemes, recruiting and supervising up to five hundred volunteers at a time.[128] His diligent efforts were rewarded when he was selected to coordinate one of the fire-exclusion sites where, according to senior entomologist W. H. Potts, he was "singularly successful in the control of fires in his areas, which [had] been under him since the beginning of the experiment." In Sapila's interactions with the neighboring villages "he has been well supported by the local native authorities, who, having been shown the effects of this method in the early experimental areas, are now enthusiastic in the support of its application to their own country."[129] Indeed, large numbers of people in the region took part in these experimental schemes.

At the time of Swynnerton's death in mid-1938 (in a plane crash in which both he and Burtt were killed), Swynnerton believed that a combination of late burning and no burning would prove most effective in controlling fly populations and managing human-fly and cattle-fly contact. These insights were gained after more than a decade of field research, oral

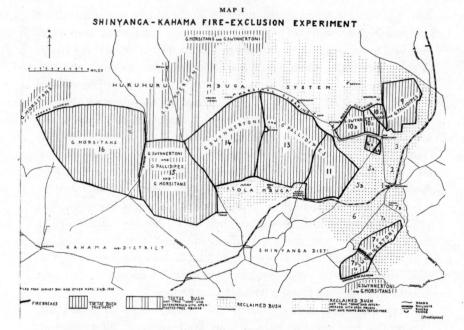

Figure 4.1. "Shinyanga-Kahama Fire-Exclusion Experiment," 1929–38 (total experimental area: approximately 500 square miles). Source: Tanganyika Territory, *Tsetse Research Report, 1935–1938* (Dar es Salaam: Government Printer, 1939).

interviews, systematic experimentation, and productive interactions with leaders in the area. John Phillips characterized the Tsetse Research Department in this period as being "intimately interlinked, indeed almost in symbiosis, with the chiefs and tribesmen of some of the lesser known regions of the country."[130] Vernacular approaches to disease control grew out of these interactions. The relative weakness of colonial states created the conditions for new kinds of epistemological entanglements.

Transnational Networks and Inter-Imperial Circulations

The significance of the Tsetse Research Department's multidimensional approach to trypanosomiasis was not lost on other medical and ecological observers. Without the discussions of Swynnerton's work at the Imperial Entomological and League of Nations conferences as well as in the Civil Research Committee's Tsetse Sub-Committee, his department would not have been selected to receive such large sums of money from the Colonial Development Fund. Swynnerton's first full-length report of his department's research results, prepared for the Tanganyika government and then sent to the Colonial Office, was selected for publication as a special volume of the Royal Entomological Society's journal. One of Amani's research officers, T. W. Kirkpatrick, whose analysis of "eco-climates" was brought to bear on the tsetse investigations, reviewed the report approvingly and noted the overlap of interests between Amani and the Tsetse Department.[131] Kirkpatrick praised Swynnerton and his research team for tackling the problem "principally from the viewpoint of practical ecology."[132] An ecological entomologist at the London School of Hygiene and Tropical Medicine (who had supervised Thomas Nash's PhD research), Patrick Buxton was impressed with how many factors, including ethnographic, were taken into account. "The point of view of this group was 'ecological,' that is to say, they sought for the solution to their problem in the field rather than the laboratory, and endeavored to see the fly's place in nature in front of a changing background of climate, plants and trees, migrations of game and shifting agriculture. . . . The men in the field give consideration to problems of anthropology and village economics and do not forget the danger of increasing the erosion of valuable soil."[133]

West African territories were quick to seize upon the Tsetse Research Department's insights and methods as well. As early as 1928, the newly appointed director of the medical services in Nigeria, Walter Johnson, decided to undertake a tour of African colonial dependencies in order "to observe general medical organisation and methods of trypanosomiasis control."[134] Of all the research efforts he observed, those in Tanganyika received Johnson's highest praise. In his forty-page report covering the Belgian Congo, the Sudan, Uganda, Tanganyika, and French Equatorial Africa, he devoted one-quarter to describing the Tsetse Research Team's "objective, dynamic, comprehensive, [and] balanced" program. *Any attempt to lessen the conceptional scope of the investigation, would result in a general weakening of the possibility of the team's ultimate success. . . .* A concept embracing [a] purely pathological, histological, veterinary, medical, or entomological basis of research, it is believed, would fail to achieve more than a fraction of that knowledge which the biological-ecological concept outlined in this communication should lay before us."[135] Johnson later recruited one of Tanganyika's young ecological entomologists, Thomas Nash, to the Nigerian Medical Department. Nash conducted research on tsetse flies and trypanosomiasis there for the next twenty-six years, ultimately as the director of the West African Institute of Trypanosomiasis Research.[136] One of his significant undertakings in the mid-1930s was to conceptualize and launch the public health project at Anchau in northern Nigeria to address the high prevalence of sleeping sickness in Zaria Province. This project was known officially as the Anchau Settlement Scheme, but was dubbed Takalafiya, "to walk in health" in Hausa. Nash characterized this project as both an "ecological study" and an "experiment in rural development."[137] Although not as ethnographically inclined as Swynnerton, Nash sought to build upon Hausa social preferences and hoped to encourage their "local industries."

In a nuanced analysis of the history of development in northern Nigeria, Raufu Mustafa and Kate Meager have offered a critique of the Anchau scheme in which they argued that Swynnerton's strategies, and by extension Nash's, were "either counter-productive . . . or of doubtful effectiveness." They are also critical of the scheme's expense and its "sense of power and social engineering."[138] However, they acknowledge that sleeping sickness incidence in the area declined radically.[139] Anchau was one among a handful of projects that had the sanction of both the colonial state and inter-imperial coordinating bodies. Its attempts to bridge social and ecological sensitivities have yet to be fully explored.[140] These perspectives were largely a by-product of imperial debates about the need for multidimensional research. After the Second World War, the Colonial Office advisory committee explained that the "Anchau Settlement in Northern Nigeria has demonstrated convincingly the ability of rural Africans to maintain high standards of communal hygiene in well-planned surroundings. Originally designed as a means of concentrating the rural population for tsetse-exclusion purposes, it is now recognised as a model creation in the field of social medicine."[141] The paternalistic attitude evident in this assessment should not blind us to another insight: the move in British colonial Africa toward social medicine was rooted both in infectious disease control and in ethnographic translations of Africans' ideas and practices. (The term *social medicine* is used here to convey sensitivity to human welfare and preventive medicine, which, as Paul Weindling has pointed out, can also be "technocratic and elitist."[142])

Focusing on these entanglements and circulations should not be taken to suggest that colonial interventions were benign. Rather, it helps us appreciate that the exigencies of colonial administration, combined with a relatively novel configuration of disciplines, prompted field researchers to think about disease and well-being in flexible and ethnographically aware ways. These conceptual frameworks circulated inter-territorially through conferences, clearinghouses, advisory committees, and specialist reports. Once methods were embedded in new places, however, their intellectual debts to other territories and peoples were not always obvious.

Questioning Eradication and Localizing Knowledge

A recurring motif of trypanosomiasis research was the sheer intricacy of the disease and its "curiously localized incidence."[143] Swynnerton and other researchers repeatedly reiterated "the magnitude and complexity of the problems that a tsetse investigation has to solve."[144] Writing privately to the Colonial Office in 1936, Swynnerton elaborated on the multiplicity of variables that had to be taken into account—the variety of tsetse species, the different strains of trypanosomes, the diverse flora and fauna, changing settlement patterns—and their interactions in specific localities. "In these two respects—diversity and extent, no other . . . problem approaches it [trypanosomiasis]. And it brings in for necessary study an astounding variety of subjects":

the fly itself—each species separately, its distribution, habits, movements, senses, requirements, which differ with the season—reactions and economic status; protozoological problems; man, his habits in relation to the flies and theirs in relation to him; the domestic animals in their relation to the fly and to trypanosomiasis; nearly the whole vertebrate population . . . ; the various vegetational types, each being friendly or inimical to one or the other of the numerous species of tsetse . . . ; geology and elevation . . . ; the soil in relation to the latter . . . ; meteorological phenomena generally in their effect on the harbouring vegetation and soil . . . ; the study of eco-climates . . . ; administration—for breaking the contact of the natives with the infection . . . ; medicine, human and veterinary, for curing or protecting them and their cattle; agriculture and development generally—as one means of expelling the flies and consolidating gains; [and] engineering . . . chemistry . . . forestry . . . [and] game-control . . . [as well as] laboratory work and large scale experimental attack in the field.[145]

The complexity of the problem, Swynnerton suggested, tempered optimism about the ease with which African trypanosomiasis might be controlled or eradicated.

As Ormsby-Gore cautioned specialists gathered for the 1928 League of Nations conference that "no one method [of control] can be universally applied," Geoffrey Carpenter was busy helping to draft a revised sleeping sickness ordinance for Uganda. The introduction of new drug treatments, such as tryparsamide, which had been developed at the Rockefeller Institute in the United States and field tested by Louise Pearce and others in the Belgian Congo, had given physicians hope that they might be able to pursue preventive and curative strategies simultaneously.[146] Yet it was also becoming apparent that, even with new treatments, this disease complex would remain particularly difficult to control. In his report for the medical department in 1929, Carpenter explained that

> the ideal of extirpation of the human trypansome cannot be realised in Uganda, since it requires a degree of interference with the normal commercial development which would not be tolerated. . . . Our policy must be, by force of circumstances, to recognise sleeping sickness as an endemic disease to be suppressed as much as possible by seeking out and treating cases where they occur, but with a full appreciation of the fact that every endemic focus is a possible starting point for a spreading epidemic unless the degree of contact between fly and man is rendered smaller than is required for an epidemic.[147]

Uganda's governor quickly endorsed these principles and sent them on to the Colonial Office where they received official approval.

Carpenter's views are similar to John Ford's argument that control rather than eradication was the best approach to trypanosomiasis. Although Ford's perspective may have been a minority view when he wrote his magnum opus in the early 1970s, it was actually the norm at the height of inter-imperial and transnational coordination during the interwar period.[148] Ford was introduced to this perspective long before he ever set foot on the African continent through his work with Charles Elton and Geoffrey Carpenter in Oxford during the 1930s. In fact, parallel and mutually reinforcing changes in attitudes to disease eradication were underway in British Africa and in metropolitan centers of learning.

When Ford moved to Oxford in 1929, his supervisor, Charles Elton, had taken up an extended discussion with Julian Huxley on the prospects of founding a Bureau of African Ecology there. Huxley and Elton were pivotal players in the interwar effort to establish the field of ecology on a firm footing: Huxley as a mentor, popularizer, and booster for the field; Elton as a researcher, editor (*Journal of Animal Ecology*), and founder of institutions (Bureau of Animal Populations, Oxford). Their social and intellectual networks were extensive. A number of other individuals provided important stimulus. The demographer Alexander Carr-Saunders, author of the widely influential book, *The Population Problem: A Study in Human Evolution* (1922), was one of Elton's supervisors at Oxford and helped coordinate the scientific side of the first expedition to Spitsbergen Island in 1921 on which Elton served.[149] The statistician and eugenicist Ronald Fisher, author of *Statistical Methods for Research Workers* (1925) and *The Genetical Theory of Evolution* (1930), helped train or mentor researchers bound for the Colonial Service, including Charles Jackson and Thomas Nash who went to the Tsetse Research Department in Tanganyika.[150] What all these men shared was an interest in relating population patterns—densities, growth, and fluctuations—to wider environmental changes.

Huxley's interest in relationships between ecology and disease emerged in a short review of a utopian novel by H. G. Wells, *Men Like Gods*. Wells was not only the preeminent science fiction writer of his generation but also a popularizer of science, and he was profoundly convinced of the liberating possibilities of scientific knowledge.[151] Wells's views of the natural world had been deeply influenced by his friendship with Ray Lankester, whose writings on sleeping sickness and evolution had found a wide audience in Britain in the first decade of the twentieth century.[152] Seizing upon Wells's optimism, Huxley wrote that "the triumphs of parasitology and the rise of ecology have set him thinking; and he believes that, given a real knowledge of the life-histories and inter-relations of organisms, man could successfully proceed to wholesale elimination of a multitude of noxious bacteria, parasitic worms, insects, and carnivores."[153] The hope that science would enable humans to predict and control the natural world to such an extent was tempered somewhat by the new conceptual tools of ecology. Understanding species interrelations was no simple affair; already scientists and land "managers" were discovering that the precise kinds of controls being described could sometimes have unpredictable consequences. However, wrote Huxley, "Mr. Wells does not need to be reminded [of this] . . . his Utopians proceed with exemplary precautions."[154]

In 1927, on the basis of their shared interest in biological relationships, Julian Huxley, H. G. Wells, and Wells's son George teamed up to write what became a three-volume study titled *The Science of Life*. In the chapters on "The Science of Ecology," and "Life under Control," they elaborated on the challenges involved in understanding as well as predicting dynamics in the natural world and drew attention to important distinctions between temperate and tropical environments.[155] "To work out this web of inter-relations in detail for a whole community is all but impossible even in our temperate regions—let alone in the richer tropics." Because biological interrelations were so unstable, they argued, to tamper with the "web's weaving, whereby a twitch on one life-thread alters the whole fabric, . . . it behooves the would-be benefactor of humanity to proceed with caution; if he is not careful, he will do infinitely more harm than good."[156]

The Science of Life had far-reaching effects on lay and science audiences. Macfarlane Burnet, who later received the Nobel Prize for his work on immunological tolerance, claimed that it "had a major impact on my thinking. It made me particularly interested in the ecological aspects of microbiology and epidemic disease, the field I was then just entering as a laboratory worker."[157] In an article written for the *British Medical Journal* in 1936, Burnet wrote that since the First World War, "the most characteristic development of epidemiology has been the adoption of what may be called a more oecological point of view, in which the activities of the two organisms concerned—man and the pathogenic micro-organism—are both considered from the point of view of survival of the species."[158] That same year, Burnet began his book *Biological Aspects of Infectious Disease* with a section on "The Ecological Viewpoint," which owed its framework and many of its examples to *The Science of Life*.[159]

Burnet also acknowledged the formative influence of Charles Elton. In 1927, at Huxley's urging, Elton wrote one of the first full-length textbooks on animal ecology, in which he devoted an entire chapter to a discussion of the ecology of parasites and their relationships to "complex food chains." In Elton's schema, parasites were treated as dynamic organisms that shaped and were shaped by their relations with other organisms and their environments. In this context, epidemics were defined as a consequence of parasites' "power of multiplication" creating a situation in which they were able to overwhelm their hosts. Elton took issue with the idea that any simple "balance of nature" existed with respect to the regulation of animal and parasite populations. His objection stemmed from field studies in which he realized that it was extremely difficult "to predict the effects of variation in numbers of one species upon those of other ones in the same community."[160]

Behind the discussions of prediction in both *The Science of Life* and *Animal Ecology* lay a deeper issue that was central to the debate about control: the way in which disease was to be defined. Considering disease as the result of population dynamics among organisms and their environments, which were fluctuating and unpredictable, called into question two key assumptions: first, the idea that simple relationships between species and their environments or among species were the norm; and second, that it was realistic, in the case of diseases that demonstrated nonlinear and complex dynamics, to believe that they could be "eradicated." By challenging these assumptions, Elton argued that epidemic disease, whether among humans or animals, should not be considered an aberration, but rather "a normal and frequent phenomenon in nature."[161] As a result of this redefinition, he believed that "our general attitude towards health in the human population . . . [and] existing theories as to the origin of human disease in history may have to be reconsidered."[162] Since humans are themselves organisms and interact dynamically with other species in their environments, the phenomena of disease in human populations should be viewed in terms of interrelations among organisms rather than as a simple chain of cause and effect.[163] In the editor's introduction to *Animal Ecology*, Huxley drew attention to the implications of these findings for medicine and disease control:

> Under the magic of the germ-theory and its spectacular triumphs, medical research on disease was largely concentrated upon the discovery of specific "germs" and their eradication. But as work progressed, the limitations of [this] mode of attack were seen. *Disease was envisaged more and more as a phenomenon of general biology, into whose causation the constitution and physiology of the patient and the effects of the environment entered as importantly as did the specific parasites.* . . . In other words, a particular pest may be a symptom rather than a cause; and consequently over-specialisation in special branches of applied biology may give a false optimism, and lead to waste of time and money through directing attention to the wrong point of attack.[164]

It was in this context that Elton and Huxley began to discuss the creation of a Bureau of African Ecology and the importance of research on "medical ecology in relation to disease."[165] Huxley was the prime mover behind this shift in attention, which was the direct result of his 1929–30 tour of East Africa on behalf of the Colonial Office Advisory Committee on Native Education. Like the East Africa Commission before him, Huxley had been captivated by the issue of sleeping sickness and African trypanosomiasis.[166] While in the field, he also had a chance to meet with staff at the Tsetse Research Department in Tanganyika and with Geoffrey Carpenter in Uganda, whose sleeping sickness ordinance had just been approved.

Huxley had already observed that "especially in the tropics . . . climate gives such an initial advantage to [humans'] cold blooded rivals, the plant pest and, most of all, the insect."[167] Since eradication was increasingly seen as a problematic concept, the alternative would be to research "the ecological status" of vector species and parasites in relation to their hosts and environments. This program would form the basis of the Bureau of African Ecology that he envisaged: "The tsetse problem is broadly speaking an ecological one. The abundance and movements of game, tsetse-fly habits, the ecological peculiarities of different kinds of bush, the effects of burning, grazing and planting, soil-erosion, and afforestation and even bee keeping all enter the ecological nexus. . . . [Soil erosion] is immediately seen to be connected with tsetse research on one side, forestry on another, and native customs ("human ecology") on a third."[168] Huxley's conceptualization of ecological and disease patterns was far reaching. Positioning humans within nature, he saw human, medical, and animal ecology as intimately related.

Taking a similar view, Elton was concerned with breaking down the disciplinary and institutional boundaries that inhibited insight. As the two men corresponded about the Bureau of African Ecology, Elton had begun to examine the research on trypanosomiasis in the Lake Victoria region of Kenya, Uganda, and Tanganyika. He remarked to Huxley that although the work was "in itself first-rate," "I have been astonished at the lack of coordination of the information already in existence." Elton's primary interest was to correlate tsetse fly populations with climatic cycles and epidemic outbreaks of the disease. The evidence he reviewed confirmed his concerns about plans to develop the region economically: "It follows that there is a [cyclical] fluctuation in potential sleeping sickness conditions and that any plans . . . to control the lake waters [for development] will have to reckon with this fact. We therefore have a) medical, b) meteorological c) engineering people, all more or less aware of each other's work, but unconscious of its significance. This seems to be the role of ecology in the next fifty years: super-coordination."[169]

The ecological perspective challenged the idea that a vector-borne disease could be eradicated. The crux of the issue rested on distinctions between eradication and control. Imagining all of nature as a dynamic and interrelated "web of life" did not preclude the idea that humans could understand and control that web, but it made matters of control infinitely more complicated.[170] The challenges posed in any effort to eradicate a species seemed overwhelming, particularly in the face of admitted ignorance regarding interrelationships among parasite strains, vectors, hosts, and their environments. Even when sufficient information was thought to exist, financial, institutional, and social factors proved to be significant obstacles.

Two additional challenges to the concept of eradication stemmed from developments in bacteriology and the emerging science of immunology. The first involved the insight that vulnerable organisms, including humans, underwent internal adaptations that could cause them to be partially or completely protected from infections (acquired immunity); the second arose from the realization that strains of disease could change, sometimes quite rapidly, which gave them unstable characteristics. Both features were dealt with at some length in *The Science of Life*.[171] These developments had a direct bearing on the activities of researchers in the British African medical services in the interwar period, many of whom adopted an extremely cautious approach to the idea of eradication. Aware of the extremely complex vector-parasite-host relationships in malaria and trypanosomiasis, investigators advocated careful experimentation to find effective methods of control, rather than aspiring to eradicate the disease or its vectors.[172] In this respect, the sciences of ecology and epidemiology came together in the African field.

Official skepticism about eradication was demonstrated in two reports of Britain's Tsetse Sub-Committee. In 1935, the committee observed: "It is idle to suppose whatever progress is made by research, that a point will shortly be reached where by the application of simple rules of thumb the tsetse fly can be rendered harmless or can be eradicated."[173] Three years later, when endorsing substantial increases to the grants for Tanganyika's Tsetse Research Department, the committee explained: "The different aspects of the matter are so closely interrelated that only limited progress can be made by the investigation of any one aspect of it in isolation. . . . The problem is in the full sense a scientific one, not insoluble, but requiring for its solution a considerable period of investigation."[174]

Indeed, it was often field officers who first expressed these views. R. A. Barrett, who had worked in Uganda since 1928, was responsible for studying epidemiological patterns in the West Nile district of the territory bordering the Belgian Congo and the Sudan, where sleeping sickness epidemics had the most serious effects.[175] One of his initial tasks was to establish a relationship with the "native authority" in order to devise preventive and treatment measures that would stem epidemics. The best way to keep the region's 50,000 people out of direct contact with tsetse flies was to establish selective clearings near the Nile river system where the flies typically bred. In 1934, echoing Carpenter's analysis from 1929, Barrett reported that "for the effective transmission of sleeping sickness in a community, a minimum breadth of contact between fly and man is necessary." He stressed the contingent elements in play that made the disease so difficult to manage.

> Breadth of contact is essentially dependent on numerous biological factors which vary not only in different localities, but at different times in the same locality. The most essential of these factors are human and fly density, the domestic and economic activities of the human population, and fly activity in relation to its human host. . . . These factors are closely interwoven with each other, and are to a certain extent, mutually interdependent. . . . Their sum total in a given area results in what may be termed an epidemic potential the degree of which determines the nature and extent of an outbreak of

sleeping sickness should a strain of trypanosome of a certain inherent trans-
missibility be introduced.

Barrett concluded that "*a knowledge of local conditions,* is a most important
point in the epidemiology of the disease, and one to which every consideration should be given in estimating the probability and extent of an outbreak occurring in any given time."[176] Independently of Huxley and Elton,
Barrett had arrived at a synthesis of human and animal ecology that he
brought to bear on his recommendations for control.

Biomedical investigators increasingly included endemicity, social organization, population densities, habitat changes, and individual suscepti-
bilities within their analytic frameworks.[177] Trypanosomiasis *control* efforts,
however, did not always reflect this level of integration; not until the 1940s
did experts begin "to consider where research and experiment end and
where actual operational control and reclamation measures begin."[178] Yet a
more comprehensive analysis emerged through research coordinating conferences and concomitant training programs that stressed interdisciplinary
teamwork and local specificity.

The analysis presented here is not a complete history of research on African trypanosomiasis.[179] It says little about disputes among scientists, or
about their long-term negotiations with the people and administrations
in the territories in which they worked. Nor does it explain how ecological methods and ethnographic sensitivities could become ubiquitous in
research projects and yet have a seemingly negligible impact on medical
treatment programs. What it demonstrates is that investigations of epidemiological patterns that were unique to tropical Africa helped to transform
biomedical understandings of disease causation. Moreover, technical officers who studied these patterns selectively incorporated the knowledge and
expertise of local populations into their analyses. The reason scholars have
often overlooked or even denied the existence of these vernacular translations is that their traces are most visible only when one examines the "machinery of knowledge" that connected territories to the metropole. Once
vernacular knowledge began to circulate, technical officers, scientists, and
administrators often forgot its African origins.

Synthesizing Metropolitan and Territorial Perspectives
through the African Research Survey

The medical services in tropical Africa occupied an unusual position within
the imperial scientific infrastructure: they far surpassed other technical de-

partments in size. By the close of the 1920s, medical staff made up over 50 percent of technical personnel (see Appendix, 1929). They also had a less direct relationship to initiatives promoting economic development than agriculture, forestry, and geology departments did. Indeed, there was a recognized tension between concern for Africans' well-being and concern for their economic productivity. When reviewing Kenya's Medical Department Report for 1935, one observer commented, "It seems to me important to remember that behind the mass of statistics—for instance, 999,796 out-dispensary attendants—there lie countless stories of human suffering and misery."[180] More was at stake in medical officers' work than the instrumental aim of improving their subjects' ability to labor.

Britain's Colonial Office had struggled with these tensions since the turn of the twentieth century. Advisory committees and semiofficial bodies such as the Sleeping Sickness Bureau (later Bureau of Tropical Diseases) were often forced to acknowledge privately that conditions on the ground in British Africa were grave. Medical departments lacked the resources and staff that would have been required to meet the needs of the people they were meant to serve. At the same time, theories of disease causation and control were in a constant state of flux, which posed real challenges for those administrators charged with devising state policies. Opening the first coordinating conference on tsetse and trypanosomiasis research in 1933, the governor of Uganda emphasized the difficulties of translation that existed between administrators and scientists. "You tell us of diseases which are to us mere names; you use long words in unintelligible foreign languages; and when in addition you speak to us with the different voices of [multiple] territories it is, perhaps, not unnatural that we become a little bewildered."[181]

Imperial administrators also faced the question of how to link research on single diseases to broader rural health care efforts. The deputy director of Kenya's sanitary service told a meeting of the Royal Society of Tropical Medicine in 1928, "The amount of work so far undertaken is but small in comparison with what remains to be done; and in rural Africa especially we are working in a field where public health problems now present themselves in circumstances of which we have had no previous experience."[182] The push for medical departments to undertake "Research on Problems of African Welfare" came from directors themselves, from specialists advising the League of Nations, and from metropolitan scholars and critics who viewed health conditions as a litmus test for how well or badly specific colonies were being managed.[183] Studies of nutrition and of malaria, as well as African trypanosomiasis, played a direct role in the shift toward social medicine. Participants at the 1933 East African medical research conference summarily stated, "Ill-health in Africa is to no small extent a matter of poverty, poor nutrition, and fly-borne disease."[184] Three years later, during the second coordinating conference, A. R. Paterson detailed for his audience the significant transition he felt they were helping to inaugurate. In the early decades of colonial rule, Paterson observed, the imperial powers had concentrated to a great extent on "the elucidation of the natural history of the causal organisms of disease and of the vectors of these organisms and to their control by means of alterations of their environment." More recently, specialists had broadened their outlook and begun to focus on "the nature of the patient himself, that is the African, and to his reactions to infection and treatment and to his own environment and to the strains and stresses carried by the great changes in that environment which are now taking place as a result of development in East Africa."[185] "Medical research in the tropics" had, unfortunately, been preoccupied with research on the biological sources of epidemics, but the East African and pan-African conferences were enabling scientists "to break away from that older tradition."[186]

This shift in perspective is evident in medical experts' discussions and recommendations concerning malaria during the second Pan-African Health Conference in 1935. Specialists' approach to malaria mirrors attitudes toward trypanosomiasis and underscores the emphasis placed on poverty and development. Malaria, they noted, "occupies one of the foremost places—if, indeed, not the foremost—among the infective diseases of Africa as a cause of mortality and morbidity in the indigenous populations."[187] As knowledge of the different strains of the disease increased and studies were undertaken to unravel its lines of transmission, researchers around the world began to note important local variations. Studies in Europe and Asia were particularly significant to this process.[188] In Africa, both the eastern and southern territories played important roles: in 1928, Colonel Sydney James visited Kenya and Uganda in the wake of epidemics to research antimalarial measures on behalf of the Colonial Office; between 1929 and 1932, an epidemic in Natal had taken the lives of over 13,000 people.[189]

One of the principal debates concerning malaria in the 1930s was whether it should be controlled directly via short-term campaigns to eradicate specific mosquito species, or indirectly through long-term social and sanitary measures. Conference participants acknowledged that in Africa the "prevention and control of malaria" had really "been attempted only in the case of populations inhabiting comparatively circumscribed areas—in towns and at ports, on plantations, in connection with large engineering or industrial projects, or in some areas of European farming settlement." The "outstanding" priority for the future, they concluded, was "to devise methods and means of lessening the incidence, or the effects, of the disease among the great indigenous rural populations of the continent."[190] This "vast undertaking" would "require financial provision which few African territories can at present afford from their own revenues." A successful endeavor would have to couple long-term "economic advance" of the "African peasant" with carefully researched and selective control schemes, "for the problems of malaria and the methods of approach to these problems must vary from place to place and from time to time."[191]

Particularly important to conference participants was the fact that malaria "under African conditions" was so poorly understood. Questions of "tolerance" and "immunity" presented real difficulties: was there a "danger" to "indigenous populations" of interrupting immunity by the application of "occasional drug treatment" or by "species eradication"? G. A. Park Ross reported, "We have felt diffident about applying insecticidal control on the grand scale to native areas infected with both A[nopheles] funestus and A. gambiae, lest any degree of immunity acquired by the native population be impaired."[192] Research into these and other questions—such as malaria's contribution to child mortality, its effect on children's "mental development," and the impact of "various therapeutic substances on various [parasite] strains"—would help shed light on the nature of the disease in Africa. With respect to control, the conference advocated continued research and controlled experiments with "insecticidal sprays," which had been used with some success in the Natal epidemic.[193] Their concluding recommendations returned to the importance of development: "While research along the lines suggested in the foregoing resolutions is an urgent matter, it must not be forgotten that, without raising the economic status of the vast bulk of the population of Africa as a whole, there can be no hope of applying successfully on a continental scale the results of research or of markedly improving the position of great populations with regard to malaria as a disease."[194] Far from offering a purely technocratic solution, medical directors wished to use the opportunity of their pan-African gathering to highlight economic and political factors that they believed were equally critical to public health efforts.

"Science in Africa" and Medical Syntheses in the "African Survey"

Nearly one-fifth of Science in Africa was devoted to the issues of public health and infectious diseases. Of the fifty-four advisors who read and commented on these three chapters, twenty were current or former directors and senior officials in African medical departments, including several who had attended the East African coordinating conferences and the two Pan-African Health conferences. The remainder was comprised of senior researchers and administrators in the fields of tropical medicine, nutrition, public health, and hygiene. The influence of medical officers is particularly apparent in the two framing chapters on "health and medicine" and "health and population." It was likely they who prompted Worthington to begin these chapters with a critique: "In Africa, as elsewhere, much of medical activity in the past has been devoted to the treatment of disease rather than the promotion of health, and there has been a tendency to separate work into categories such as curative medicine, preventive medicine, and other social services. To-day, however, expert opinion appears to be in agreement as to the close interdependence of all aspects of medicine."[195] In summarizing the medical research concerning these issues, he concluded that "it cannot be over-emphasized how great is the importance of a thorough knowledge of tribal custom and native attitudes." Ill-informed interventions might meet with "passive resistance," which weak states could hardly afford, and "any general plan for social amelioration . . . must be left to those with intimate local knowledge."[196]

One of the noteworthy features of Worthington's conceptual framework was his emphasis on "the internal environment of the human organism," which he considered integral to any discussion of Africa's "changing environment." Both in terms of nutrition and immunities, the human body had to be recognized as "capable of adaptation" and change.[197] Disease could then be defined as a function of interactions within the human body and among populations and their surroundings. Conceived in this way, health involved the wider socioeconomic context. Worthington endorsed the philosophy that "the mere treatment of disease is insufficient" in Africa; "the economic status of the community" was crucial, and any effort to improve health required addressing "insanitary conditions" and raising Africans' "standard of living."[198] While confident that many diseases were likely to "disappear with the introduction of improved social conditions," he was careful to acknowledge that complete eradication of endemic diseases such as trypanosomiasis and malaria was an unrealistic goal. Worthington's summary of scholars' understanding of malaria shows that reservations about the possibility of eradication were widespread and closely connected to the limits of biomedical power. While both trypanosomiasis and malaria were vector-borne diseases that varied considerably across sites, humans appeared to develop certain kinds of immunity to the strain of malaria in their locale, which did not seem to happen with trypanoso-miasis. "With regard to Africa as a whole, the upholders of treatment as a chief method of attack [against malaria] have stressed that since it is impos-sible to eradicate *Anopheles* [mosquitoes] everywhere, its eradication from limited areas might do more harm than good."[199] The problem revolved around the question of acquired immunity. To eradicate the vector, and by extension, the disease threat, in one place might create the possibility that "children grow up without attaining partial immunity, and subsequently move into an infected area [where] they are liable to contract the disease in a more serious form. According to this view, if it is conceded that the dis-ease cannot be stamped out completely, the great aim must be to develop immunity to all forms of the disease by a better standard of living, together with improved therapeutic methods."[200] Therapies, "like every other aspect of the disease," were found to be a "much more local and individual prob-lem than has hitherto been thought."[201]

Worthington encouraged medical departments to continue health and agricultural surveys to ascertain the kinds of diseases present, standards of nutrition, and variations of "local foodstuffs," as well as particular meth-ods of medicinal treatment. Gesturing to his own research on nutrition, he underscored the "importance of [having a] full understanding of tra-ditional native attitudes towards different articles of native diet and their preparation," before colonial states attempted to improve them.[202] Recent nutritional research had shown that Africans consumed certain foods, in-cluding earths, which contained quantities of minerals and other nutrients that would otherwise be insufficient in their diet. In Nigeria, for example, vitamin-rich leaves from the baobab tree were crushed and prepared in a soup. This technique "provides a striking illustration of the dietetic value of a customary native practice. Precautions are always taken to avoid direct sun on the leaves during the drying process, a practice for which a very sound reason has been elucidated by laboratory experiment: it has been shown that sun-drying as opposed to shade-drying destroys the vitamin content of the leaves."[203] Studies undertaken by botanists and foresters also demonstrated that "many trees and shrubs are used by natives for medici-nal purposes."[204] Following the principle that it was best to build upon what already existed, Worthington reiterated the importance of teamwork among technical departments, including social anthropologists. Careful re-search might unearth evidence that would help to explain how "in spite of the prevalence of diseases in Africa, for which malnutrition is partly responsible," many Africans were still able to attain "a high level of phy-sique and health." "In this respect," he concluded, "Europeans may have something to learn, as well as to teach."[205] Sir Robert McCarrison, who had served as the nutrition research director with the Indian Research Fund, suggested that "this section of the Report has an educational value by no means confined to Africa. It has much to teach this country and others."[206]

In 1937, A. R. Paterson, the director of Kenya's medical department, was asked to comment for the East Africa Commission on Higher Education on what kind of training Africans should receive for future work in medical departments. His final report minced no words: East Africa urgently needed fully qualified African doctors. Deploring "Western Civilization's appalling stupidity" in this respect, he admitted that "little more than a beginning has been made . . . towards the relief of suffering . . . [or] towards the increase of the standard of health." "A tenfold multiplication of anything that has hitherto been done . . . might be of such effect that we might be able to say, or think, that we were at last fairly on the road towards beginning to grapple with the problem with which we are confronted."[207] Thanks in large part to the collective contributions of medical personnel, the African Survey concluded that Africans themselves ought to be given increased access to medical education in order that they might meet the need for "more highly trained auxiliary staff capable of efficient diagnosis and treatment."[208] Rather than limit the field to the current level of training that prevailed in most territories—wound dressers, nurses, midwives, and drug dispensers—they called for "the training of a large number of Africans up to the same standard as the Assistant Surgeon in India."[209] Since the continent lacked facilities "designed to afford Africans an opportunity of gaining full medical qualifications in the sense in which that is understood in Great Britain," the Survey recommended expanding existing training facilities at Yaba in Nigeria, Mulago in Uganda, and the Kitchener School of Medicine in Khartoum.[210] The following year, a joint meeting of Kenya's, Uganda's, and Tanganyika's medical services urged that a complete training program be established.[211] Only bolstering educational institutions *within* Africa would enable medical departments to recruit candidates for this "intermediate class" of staff in sufficient numbers. The Survey's authors identified the lack of existing opportunities and insufficient funds as the main obstacles to enabling "Africans to take a full medical degree."[212] The report in essence endorsed a move that was already afoot in many British territories to "Africanize" the medical departments.

The African Survey also drew attention to epistemic shifts that were taking place within biomedical research across British Africa. Long-standing concerns with hygiene and public health were gradually being coupled to significant new insights within ecological science, immunology, nutrition, and epidemiology. Arguably, ecological critiques of simplistic assumptions of disease and its causes that were popularized during the 1960s and 1970s owed their origins, at least in part, to experiences in the mainstream of *colonial* societies.[213] Ironically, critiques of the political economy of health also had roots in African experiences. Some scholars might interpret this viewpoint cynically and argue that medical departments' shift toward social medicine served as yet another means for the colonial state to intrude in Africans' lives. Others might argue that attention to local knowledge and needs, in light of the simultaneous exclusion of "popular healers" from the ranks of the medical profession, was an exercise in appropriation rather than sensitivity. There is a degree of truth in both those positions. Yet we must also be aware that medical departments were among the first to develop a more comprehensive analysis of environmental, economic, and biological factors affecting "the causes of disease and disability." When the delegates to the 1935 pan-African health conference spoke of the "deplorable neglect of African problems" they were actively challenging the status quo.[214]

Compare, for instance, Lesley Doyal's critical assessment of medicine in East Africa during the colonial period with that of the African Survey. In 1979, Doyal wrote: "It is important to appreciate that in the absence of the fundamental social and environmental constituents of health, a predominantly curative [medical] policy has serious limitations."[215] In *An African Survey*, under the heading "the social aspects of health," the authors stated: "It is clear that poor physical conditions, due to poverty and environment, play an important part in the story of disease and sickness in Africa. The gradual realization in other continents of the relationships between environment and health is reflected in the growth of a similar outlook in Africa."[216] When the heads of the medical services stated that the "mere treatment of disease was insufficient" and that the economic context of health problems needed to be taken into account, they were anticipating several important strands of the "political economy of health" critique. Indeed, they would probably have been surprised to learn that so many scholars and social critics would later accuse them of being oblivious to these patterns, when they constantly confronted them.

Perhaps more interesting, in this context, is the attention researchers were being encouraged to pay to the substance of African medicinal and dietetic practices. Of course, this was uneven and sporadic attention, but the guiding premise behind it was that biomedical healing and nutritional improvements should be grounded in a thorough knowledge of "existing conditions" in Africa. So while popular practitioners might be excluded,

Africans' collective knowledge was not. In this sense, medical services were explicitly bridging the gap between "technoscience" and "ethnoscience." Their objective was neither profit nor patents, but patients. Was appropriation part of the pattern? Yes. Were there asymmetrical relationships between informants and researchers? Yes. Was it too uncoordinated to make much of a difference to the lives of most Africans? Probably. Nonetheless, the model of health care provision and biomedical research articulated by the medical services and reiterated in the pages of *Science in Africa*, and to a lesser extent in *An African Survey*, was a practical effort to address the myriad health problems affecting Africans' lives.

This said, one should avoid interpreting these viewpoints as the sum total of medical interventions. The point is not to deny the extreme epidemiological and sociological upheavals that colonialism caused, nor is it to ignore the complicity of medical departments in colonial aims when they collaborated with administrators to produce, for instance, healthier laborers. But this understanding does not seem incompatible with another insight: knowledge produced within colonial processes could be simultaneously critical and original. The strong empirical thrust of interwar research, combined with relatively new synthesizing methodologies, rendered old practices and assumptions, if not obsolete, then at least open to question. As biosciences were brought to the African colonial field, their adherents developed new insights—about the intricate environmental and social relationships that gave rise to vector-bone diseases—which they placed in the foreground of their analysis. In fact the history of British attempts to understand African trypanosomiasis, among other diseases, reveals a central but hitherto unappreciated epistemic framework within which the very rhetoric regarding the "ecology of disease" first found expression. Because these intellectual shifts were rarely presented as radical critiques, their underlying challenge to the colonial system can sometimes be overlooked.

CHAPTER FIVE

A Racial Laboratory: Imperial Politics, Race Prejudice, and Mental Capacity

In these days . . . of strong racial passions, of organised racial aggression, and of corresponding legitimate resentment and profound anxiety on the part of the non-Caucasian races, the use of the prestige of science for obviously partisan aims is dangerous politically and socially. . . . It seems doubtful to me . . . whether it is really a benevolent gesture to "extend the hand of science" to a fellow-creature only to deal him a knock-out blow: to brand him with intrinsic racial inferiority; to insinuate that education leads him into dementia precox. . . . All these implications do as much harm to the reputation of Anthropology or Psychology, as they offend the sense of fairness of a European. They will be resented by all intelligent Africans.

—Bronislaw Malinowski, *letter to the Times*, 1934[1]

In the summer of 1934, the British Colonial Office received a funding proposal from the Kenya Colony for an extended examination of African "mental capacity" and "backwardness."[2] The research would concentrate principally on physical and environmental factors that were thought to affect "African cerebral development, quality, and reaction" with the intention of shedding light on Africans' "capacity for mental development" and "educability."[3] Through an unusual twist of events, the prime minister, Ramsay MacDonald, was informed of the proposal and decided to ask the leadership of the African Research Survey, rather than any government agency, to determine the "importance and urgency" of the colony's plans.[4] MacDonald's decision was not as arbitrary as it might appear, for the African Survey's mission to "secure an improvement in the system and methods by which research [could] be undertaken, and knowledge acquired and distributed" was increasingly well known.[5]

By tracing the origins and reception of the East African proposal and its entanglements with the African Survey, this chapter places the history of racial science in British Africa in the wider context of imperial and transnational research priorities. In the process it explores various fault lines between racial sciences, racial ideologies, and racial states. What makes the Kenyan case interesting is not only the attention it generated but also the fact that it ultimately failed to receive official support. In colonial contexts, failure can be as revealing as success. In this instance, explaining why the mental capacity proposal was rejected helps to highlight scientific practitioners' and colonial administrators' shifting ideological and epistemic commitments. It also underscores three interrelated developments of the period that might otherwise go unnoticed: first, the growing preoccupation among a circle of researchers and administrators in Britain and Africa with the issue of "racial prejudice" and "race discrimination"; second, scholars' simultaneous challenges to the óntological validity of the concept of "race," that is, the question of whether it described anything "real" or scientifically meaningful; and, finally, the eclipse of a particular kind of racial research in British colonial Africa in favor of other kinds of biological and sociological interpretations of African peoples. Although racial preoccupations in science did not disappear from colonial studies of British Africa when this proposal was rejected, they were certainly dramatically transformed by it.

In the literature on colonial Africa, surprisingly little has been written about the history of racial *research*, while almost every major study of colonial conquest and state building tacitly assumes that racial *ideologies* played a significant role in justifying European overrule.[6] There is much truth in this assumption, but it leaves unexamined how changes in racial theories and racial science affected the character of colonial rule. If indeed the concept of race helped shore up a rationale for European imperialism, must we assume that challenges to that concept came only from outside these power structures? Should we also expect that only those adversely affected by racial science and racial ideologies would mount the most serious critiques? Although it is tempting to answer "yes" to both questions, this chapter demonstrates that, in fact, a more accurate answer would address the way interactions between imperial administration and scientific research helped to produce critiques not only of racial science but also of "racial domination."

While the category of "Africans," or "Negroes," had been pivotal to European racial classification systems since the eighteenth century, it was not until the late nineteenth century that scholars actually began to conduct their work across several sites *within* Africa to address these questions. Prior to this point, most of the discussions of "Negroes" typically drew more evidence, both anecdotal and substantive, from experiences in the United States, the Caribbean, and Latin America, emphasizing African *diasporas*.[7] Only during the first half of the twentieth century did racial research, as a discrete and deliberate scientific activity, come to be pursued systematically in different parts of the world. Yet even then, the sheer volume of studies done in the United States still dwarfed the number of studies done across tropical Africa.[8] Specialized research of this nature was more easily undertaken in areas with well-funded and autonomous scientific and educational facilities. Within British Africa, this pattern bears out in the disproportionate volume of studies on race conducted in the Union of South Africa compared to the dependent territories overseen by the Colonial Office. Clarifying just how central or marginal racial questions were to different kinds of scientific research makes it possible to address the extent to which and the ways in which tropical Africa did indeed serve as a "racial laboratory."

Racial States versus Racial Science: Imperial Politics and the Origins of a Discourse on Race Prejudice

Colonial states in tropical Africa were racial states from the outset. David Goldberg, from whom I am borrowing the concept, has defined racial states as "historically . . . engaged in the constitution, maintenance, and management of whiteness. . . . Racial states, in short, are states . . . in which white rule prevails and is shored up by a world racial order of white dominance." Goldberg explains that white rule is inherently expansionist, taking shape in "colonialism, imperialism, the Third Reich, anti-communism, globalization."[9] For most regions of tropical Africa, the creation of racial states was a departure from more fluid political dynamics of the nineteenth century.[10] Through new laws, bureaucratic structures, and administrative practices colonial states gradually routinized racial categories and applied them to social and economic relations as well as physical reproduction. Scholars and critics recognize these phenomena in census categories, patterns of employment and land appropriation, distinctions between citizens and subjects, pass laws and color bars, differential pay scales, segregated medical and educational services, regulations restricting access to certain jobs and markets, marriage and sexual union legislation, the creation of "native reserves," disproportionate and unequal punishments, and even definitions of development itself, which often rested on assumptions of Africans' "backwardness."

Just how officials and scientists went about achieving white rule and how their strategies and anxieties differed across the continent is what makes the study of racial states in colonial Africa so revealing. Indeed, while state building and scientific research were conducted in tandem, they did not always address the same kinds of racial ideas or reinforce identical policies. Racial thinking could take stronger and weaker forms, in much the same way that colonial states could employ multiple and contradictory definitions of race, tribe, and ethnicity.[11] Chapter one has shown that climatic and epidemiological anxieties, which were often defined in racial terms, shaped European settlement patterns and land selection in various territories. However, attitudes toward segregation and miscegenation varied widely and were often intensely debated in each territory as well as in metropolitan centers.[12]

Different disciplines tended to approach racial questions from different starting points. Physical anthropologists and evolutionary theorists were typically preoccupied with the question of human origins and biological differentiation. By the last third of the nineteenth century in Britain, the idea that there had been multiple sites of human ancestry, known as polygenesis, had given way to the idea that humans emerged in one region alone, monogenesis. Just where this point of origin was located remained in doubt, but for those who speculated on the question Eurasian sites were still the most popular choice. Charles Darwin was one of the few theorists to posit that in the future evidence might point to Africa, but he did not push this idea.[13] Even among specialists who embraced monogenesis, the exact number and definition of races or varieties remained an open-ended question. Did they correspond to continents, language groups, skin color, or nations? Should they be defined by their ability to interbreed? Just how fixed or fluid were their qualities? That during the height of the colonial period in tropical Africa Eurasian explanations came to be discredited and the idea of African origins became more mainstream is one of the many ironies surrounding the intersections between racial science and racial state building in African imperial history.[14]

Another set of disciplinary questions at the root of racial classification systems had to do with appearance and physiology. How were differences in skin color, hair type, facial features, stature, and even skull shape to be explained? Why did some populations survive diseases that seemed to be lethal to other groups? Did different races' bodies behave in the same way under identical conditions? What were the patterns of inheritance and heredity when different groups intermarried and reproduced? How should changes in populations be explained if and when they appeared to "degenerate" physically? Were there distinct "tropical" and "temperate" races, and what role did climate play in affecting survival rates in different regions of the world? Did Africans and Europeans share the same psychology? If not, were biological or sociocultural factors responsible for the differences? These questions were addressed not just by physicians or physiologists but also by specialists interested in medical geography, climatology, natural history, anthropology, statistics, immunology, tropical medicine, psychiatry, neurology, psychoanalysis, and philosophy and political theory.

The varied evidence and arguments available about race meant that people could, and usually did, find support from legitimate scientific sources for whatever position they wished to uphold. In 1900, Karl Pearson, arguably the father of biometrics, expressed a deep and abiding faith in racial hierarchies in his book *The Grammar of Science* and connected this faith to imperatives for research and development:

> It is not a matter of indifference to other nations that the intellect of any people should lie fallow, or that any folk should not take its part in the labour of research. It cannot be indifferent to mankind as a whole whether the occupants of a country leave its fields untilled and its natural resources undeveloped. It is a false view of human solidarity, a weak humanitarianism, not a true humanism, which regrets that a capable and stalwart race of white men should replace a dark-skinned tribe which can neither utilise its land for the full benefit of mankind, nor contribute its quota to the common stock of human knowledge.[15]

Just two years later, Robert Cust, the linguist and geographer, optimistically declared that "science has told us that the five Races, white, black, brown, yellow, red, differ only in the colour of their skin, but are the same in *body*, *mind*, and *soul*, capable of the same crimes, amenable to the same reasonings, susceptible of the same spiritual inspiration." On the issue of inferiority Cust answered with an emphatic protest: the idea that "whiteness" and superiority were connected ought to be discarded.[16] Significantly, Cust had decades of experience in India and a long-standing interest in Africa, while Pearson's primary exposure to the world was through Europe, especially Germany. Racial science was as deeply embroiled in political struggles as many other fields of expert knowledge.

At the beginning of the twentieth century, the very precepts that served to support racial hierarchies were being challenged more actively from many quarters of the world. This questioning arose in part from geopolitical recalibrations, including Japan's victory over Russia in 1905 and China's declaration of a republic in 1912. Yet it also stemmed from more systematic critiques of "racial prejudice" and "race hatred" in the context of national and colonial experiences in Asia, the Americas, and Africa.[17] During a speech before Britain's African Society in 1903, the Caribbean and West African intellectual Edward Blyden encouraged his listeners to scrutinize conventional wisdom with respect to race. "The unthinking European partly from superficial knowledge and partly from a profound belief not only in an absolute racial difference, but in his own absolute racial superiority, rushes to the conclusion that this difference of external appearance implies not only a physical difference, but an inferior mental or psychological constitution."[18] Blyden considered scientific research the remedy to these biases and advocated, as early as 1878, that Europeans undertake "studies of African psychology" rather than racial difference.[19] Even someone as pro-imperial and pro-"white" as Alfred Milner, high commissioner for South Africa, felt it necessary to draw attention to the dangers of race prejudice during a 1903 discussion of the Chinese Labor Ordinance in the Transvaal.[20] A group that called itself the "White League" objected to importing workers from China and India to the Transvaal on the grounds that they would corrupt South African society and usurp jobs from whites and Africans alike. While Milner told them he was sympathetic to their concerns, he also worried that their views would lead to "a general crusade against Asiatics."

> It is utterly contrary to justice and common sense, it is repugnant to the most honourable instincts and traditions of the British race to treat, or to speak of, Asiatics as a body, including the best of our Indian fellow subjects, as if they were so many raw barbarians. The attitude is unjustifiable, and it is foolish. . . . Every community has this right of self-preservation. We only obscure the issue and weaken the case by importing race-prejudice and colour-prejudice into the discussion.[21]

Milner could have advocated the legislation without any recourse to the idea of prejudice, which means we must explain why he did so. During the early years of the twentieth century, changing power relations between residents within South Africa and the vocal criticisms of British and South Asian politicians and scholars forced officials to take greater notice of the competing claims of their various imperial subjects.

Ironically, members of the "White League" were making the same arguments that African intellectuals were then championing in the Gold Coast: Chinese workers were being imported, *both* groups argued, not because Africans would not work but because they were not being offered a fair wage or decent working conditions.[22] Both groups stressed the chaos and friction that they believed would ensue if Chinese populations were allowed to increase dramatically in either the Transvaal or the Gold Coast.[23] While these critics were largely correct about the class basis of colonial interest in Asian labor, Milner's comments illustrate just how a discourse linking race and prejudice was becoming part and parcel of any discussion of imperial and world politics by the early twentieth century. Liberal and socialist theorists such as John Hobson and Sydney Olivier felt it necessary to address the question of "white superiority" and "race and colour prejudice" in their respective books on *Imperialism* (1902) and *White Capital and Coloured Labour* (1906).[24] Both men submitted essays critical of existing conditions to the 1911 Universal Races Congress in London and expressed their concerns about new "inter-racial" problems during the Congress's debates. Even the novelist and scientific popularizer H. G. Wells devoted a chapter of his 1905 book *A Modern Utopia* to the question of "Race in Utopia," expressing disapproval of "race prejudices" and "superstitions."[25] These trends existed outside of Britain and its imperial sphere as well. Jean Finot's 1905 book *Les préjugé des races* (Race Prejudice) was part of a wider Francophone conversation on the same subject.[26] Josiah Royce of Harvard published an essay, "Race Questions and Prejudices," that took up various nations' "imperial ambitions" as well as his own country's patterns of discrimination.[27]

Whether intentionally or unintentionally, condemnations of race prejudice and calls for fair treatment began to destabilize the supposition that colonial rule was but a logical extension of the laws of nature. Hobson, one of the most strident critics of "crude biological sociology," argued that there was no "scientific validity" to the idea that empire was either inevitable or necessary for social progress.[28] "White rulers" in imperial states, he argued, were "parasites": "they live upon these natives, their chief work being that of organising native labour for their support. . . . This holds of all white government in the tropics."[29] Hobson, on the one hand, was concerned more with empire than race, but by analyzing them together he produced a critique of "white domination." Jean Finot, on the other hand, was more interested in race than empire, but his challenges to "the gospel of inequality" led him to dispel an imperial mystique.

> The conception of races . . . has cast as it were a veil of tragedy over the surface of the earth. . . . We see the immense amount of nonsense connected with the racial theories of peoples. . . . On the ruins, therefore, of the belief in superior and inferior races, the possible development and amelioration of all human beings arise. . . . The principle of human equality takes away the

right of killing so-called inferior people, just as it destroys the right claimed by some of dominating others.[30]

Royce was equally dismissive of the ostensible accuracy of scientific research: "when men marshal all the resources of their science to prove that their own race-prejudices are infallible, I can feel no confidence in what they imagine to be the result of science. . . . Our so-called race-problems are merely the problems caused by our antipathies."[31] Echoing these sentiments, Wells distinguished between what he called "bastard science" and legitimate investigations. "'Science' is supposed to lend its sanction to race mania, but it is only science as it is understood by very illiterate people." Pointing to several recent studies by physical anthropologists and geographers, he concluded, "there is probably no pure race in the whole world. The great continental populations are all complex mixtures of numerous and fluctuating types."[32]

Field officers and governors in British Africa were becoming aware of the "race question" as well and often associated it explicitly with their fragile balance of power.[33] This concern could lead them to challenge the wisdom of scientific recommendations even when these lent support to white rule and social domination. Nowhere was this more evident than in debates over racial segregation and biomedical justifications for spending state money to insure that European populations had separate commercial and residential quarters. When the sanitary officer in northern Nigeria recommended segregation of the "races" in 1911 as a strategy to manage infectious diseases, he quickly added that "the principle of the separation of native from European communities is based not on caste or colour prejudice, but on physiological idiosyncrasy." "We have reason to believe that the native African . . . may, without any apparent harm to himself, act as a continual source of malarial infection to his European neighbour."[34] In spite of the fact that these claims had the backing of the Colonial Office's Advisory Medical and Sanitary Committee for Tropical Africa, the governors of Nigeria and the Gold Coast were unmoved.[35] Gold Coast governor Hugh Clifford protested:

> The native inhabitants of the Colony are the principal tax-payers. It is their money which it is proposed thus lavishly to expend for the purpose of protecting the health of Europeans . . . resident in their midst. . . . Reasonable expenditure on sanitary precautions designed to improve the public health is obviously money well laid out. . . . But to expend money far in excess of actual requirements whenever land is acquired for European occupation, and

to spend on this service, moreover, large sums which are so urgently needed for the development of the Colony in other directions, appears to me to be so selfish a policy that, viewed from the standpoint of the native tax-payers, it will not bear the test of examination.[36]

As Clifford knew only too well, colonial states were almost always cash strapped and far weaker than their metropolitan counterparts. Even Frederick Lugard, who was less vehemently opposed to segregation than his successor in northern Nigeria, Hesketh Bell, had to agree that unpopular interventions made governing far more difficult. As Bell put it, "it is not practicable to lay down the hard and fast rules suggested by [the senior sanitary officer]. Such restrictions could only be enforced by legislations, and would, I believe, arouse strong opposition in many quarters."[37]

A discourse focusing on the foundations of race prejudice and ideas of white supremacy was insufficient to undermine the social hierarchies of colonial states, but it did enable individuals to draw attention to questions of social equality. Colonial subjects themselves were important voices in this conversation. When a group of West African physicians, who had trained in the United Kingdom, submitted a memorandum in 1909 to Britain's Colonial Office objecting to being excluded from the West African Medical Service, they argued that the European officers who had helped to craft this policy suffered "from racial bias" and the "contaminating influence of others who entertain bitter hostility and race hatred against the native races." Europeans' views were "the outcome of racial prejudice and professional jealousy and a tissue of baseless assertions." What the petitioners found most disturbing was the suggestion that while they might be deemed competent to treat other Africans, their skills were considered deficient when it came to treating Europeans. "Are the lives of the native subjects of His Majesty the King not of any value because they are black?" Having been taught themselves by "white men" and having trained using "European human subjects" these physicians felt comfortable concluding that "the human system or constitution is the same in white and black, and that in medical science the difference of race or colour is an immaterial point." (In this sense, they were endorsing the views of Louis Sambon, a bacteriologist who had declared in 1903 that, as a result of recent research on sleeping sickness, which affected both Europeans and Africans alike, "our ideas concerning the relation between race and disease have been totally changed. We know now that there are no purely ethnic diseases."[38]) The *Sierra Leone Weekly News* summarized the controversy in an editorial: "The foundation of this nefarious intention is race-prejudice, which is against all human reason; and when once this madness takes possession of any people, they do not hesitate to do the most contemptible and the meanest things for the purpose of creating some semblance of evidence to justify this prejudice."[39]

Alex Fiddian, the Colonial Office representative assigned to present an overview of the problem, acknowledged that the petition was "a fair statement of the case of the native medical officers." Still, he was hard pressed to find an easy solution because in his view there were too few fully trained African physicians to make much of a difference in the West African Medical Service and no system to train a subordinate tier of medical officers. Like many officials, Fiddian recommended against sending Africans to the United Kingdom for medical training. He would prefer to see "a much larger contingent, derived from parts of West Africa, and educated and trained in the country i.e. without being spoilt by the attentions usually lavished on black men by a large proportion of the population of the United Kingdom, [and who therefore] might be less ambitious and less conceited, and at the same time, useful popularizers of medical and sanitary reform."[40]

As Goldberg reminds us, racial states were "engaged in definition, regulation, governance, management, and mediation of racial matters they at once help to fashion and facilitate."[41] While British colonial states had an urgent need for more medical and scientific staff and most official reports encouraged employing Africans, administrators considered the effects these changes would have on the social order. By regulating Africans' contact with European institutions and limiting their exposure to situations that would convey a sense of social parity, officials hoped they could maintain the balance of power. In advising against educating Africans in botany in England, the director of Kew Gardens cautioned, "As soon as these young men find themselves treated on a footing of equality with young Englishmen, their prospects of future usefulness [to the colonial state] is destroyed."[42] Fear of social instability, rather than doubts about Africans' inherent ability, prevented training programs outside or within British tropical Africa from developing more rapidly.[43]

Some colonial officials dissented. Hugh Clifford complained to the Colonial Office shortly after assuming his position as governor of the Gold Coast that "the refusal to employ qualified native practitioners under Government is something resembling a real grievance, as it practically excludes natives of this Colony from one of the two principal learned professions."[44] When British medical officers questioned Africans' skills, Clifford retorted, "Government in this matter has not in the past offered any inducement to the better type of native in the Gold Coast to devote himself to the study of medicine, since at the present time no prospect of employment under

Government is afforded to native medical practitioners." He reminded the Colonial Office that "ere long, I think, Government will have seriously to consider the obligation under which it is to provide skilled medical assistance for the people of this Colony in localities other than the principal European centres." The only way to do this affordably would be to employ African physicians and create a proper "Native Medical Staff."[45] The imperatives of empire and development could occasionally supersede residual concerns about social hierarchies.

By the start of the 1920s, the international and imperial landscapes looked very different from 1903 when Edward Blyden had first addressed Britain's African Society on racial questions. Yet, concerns about race prejudice had only been sharpened in both spheres. Japan had pushed hard for an amendment on racial equality in the League of Nations' covenant. Its proposal was ultimately rejected by the Australian, British, and U.S. negotiators, but not before it caused considerable behind-the-scenes discussion about "race discrimination" and the appropriateness of "white races" acting as "lords of the earth."[46] India, which had never been a full member of Britain's prewar Colonial Conferences, was now an official signatory to the League of Nations. This political change forced imperial administrators in places like Kenya and South Africa to consider much more seriously the rights of Indian nationals.[47] Would South Asians be permitted to vote, own land in the "white highlands," run businesses, serve on legislative councils, and live where they wished? Or would they be subjected, as the Indian government feared, to "racial stigma, prejudice, and antagonism?"[48]

In an ironic twist in the Indian controversy, Britain's colonial secretary was now Alfred Milner, who was more interested in supporting white settlers in Kenya than in advocating on behalf of Asian interests. Nonetheless, he felt it his duty to avow publicly that in "East Africa . . . it is the definite intention, of the British authorities to mete out even-handed justice between the different races inhabiting those territories." The Indian population in Kenya already surpassed Europeans by at least two to one.[49] If these two groups were treated on equal footing politically, the white settlers were bound to lose a range of rights and privileges that they had come to take for granted in their increasingly racialized state. Kenya's acting governor, Charles Bowring, only reinforced Asians' fears when his chief secretary wrote to the Indian Association in Nairobi in mid-1919 that "His Excellency [the Governor] believes that, though Indian interests should not be lost sight of, *European interests must be paramount* throughout the Protectorate."[50]

In 1921, at the first Imperial Conference convened after the war, these concerns were important enough for the representatives of the Dominions, with the significant exception of South Africa, to pass a resolution "recognis[ing] that there is an incongruity between the position of India as an equal member of the British Empire and the existence of disabilities upon British Indians lawfully domiciled in some other parts of the Empire."[51] Not only were the Indian and Colonial Offices coming into conflict on this question, but metropolitan campaigners—including Joseph Oldham, Gilbert Murray, John Harris, Norman Leys, William McGregor Ross, and Sydney Olivier—who had been trying to ensure that "native rights" were handled justly, also found it necessary to ratchet up their efforts. By the time the new colonial secretary, Lord Devonshire, was able to make a public pronouncement on the matter in July 1923, the default principle of European paramountcy had been pushed aside in favor of an official position that resonated with the League of Nations' endorsement of trusteeship. "Kenya is an African territory," the White Paper declared, "and His Majesty's Government think it necessary definitely to record their considered opinion that the interests of the African natives must be paramount, and that if, and when, those interests and the interests of the immigrant races should conflict, the former should prevail."[52]

Focusing on African interests, Robert Maxon has argued, helped neutralize the competing claims of Asians and Europeans while simultaneously satisfying public pressure to avoid the "exploitation of the natives." Even though concerns for social justice were not the driving motive behind the 1923 policy, it nonetheless came as a severe blow to settlers' sense of entitlement and security.[53] It also brought to the fore the issue of racial hierarchies across the empire. This was not lost on the permanent undersecretary of state in the Colonial Office who recorded, in the midst of the controversy, that they were witnessing "the struggle going on at present between two rival ideals—the one tending towards equalities of races and communities, the other insisting on the maintenance of white supremacy. The latter ideal till recently undisputed, even now in practice dominant . . . The former ideal, young and growing, prevails only occasionally."[54]

In the fallout from Devonshire's Declaration, racial science received a blow as well. The Indian government and Asian residents in Kenya objected to a 1913 proposal to segregate urban centers into quarters for "Europeans, Asiatics and Africans" to prevent "the extension of disease from one race to another."[55] In a 1920 dispatch, Milner endorsed this view on social grounds: "My own conviction is that in the interests of social comfort, social convenience and social peace, the residence of different races in different areas . . . is desirable, and so far from stimulating it is calculated to mitigate hostility and ill-feeling."[56]

The Indian government, after remarking that these proposals were "bitterly resented not only by Indians in East Africa, but by educated opinion throughout India," argued that the entire scheme of commercial and residential segregation was "irrational, inconvenient, and impracticable." Although distinct cultural groups might wish to reside in different areas, "compulsory segregation implies a racial stigma." Pointing to Uganda's 1920 Development Commission, which decided against racial segregation as a strategy of town planning, they observed that sanitary segregation on racial grounds really seemed another means to benefit Europeans: "the best sites will be allotted to the race which is politically most powerful."[57] In response, the British government conceded that there would be "no segregation, either commercial or residential, on racial lines." The Kenya government and its municipal authorities retained the "power to impose at their discretion sanitary, police and building regulations, subject to these regulations *containing no racial discrimination as such*."[58] Kenyan cities did not avoid all forms of segregation, and biomedical interventions were still based on racial assumptions. But the efforts of these pressure groups made imperial administrators less comfortable with giving these measures official sanction. The issue of "racial discrimination" had become an explicit factor in inter-imperial decision making.

There is, however, a paradoxical twist to this story. The various critiques of racial prejudice in circulation during the first decades of the twentieth century often inadvertently reified the idea of race and sidestepped questions of *difference* in favor of critiques of *power*. Challenging the foundations of "white supremacy" did not necessarily undermine theories or research priorities designed to explicate human differences. "Races" remained a reality even among those who wished to attack their legitimacy or contest their scientific foundations. Meanwhile, scientists in different disciplines were continuing to redefine race, so these debates rested on shifting sands. The resurgence of research on heredity and its connections to biological studies of genetics (or germplasm) in the interwar period was an unanticipated development. These hereditary concerns drove the Kenyan research proposal into Africans' mental capacity.

Ambiguities of Racial Science: A Genealogy of the "Mental Capacity" Proposal and the Mixed Fortunes of Intelligence Tests

Thus far this chapter has focused more attention on racial *politics* than racial *science*, partly to highlight the different ways racial ideas intersected in international and imperial realms. In both contexts, race was a malleable concept that could refer loosely to skin color, ancestry, continent of origin, civilizational status (i.e., advanced and primitive races), and sometimes all of these dimensions. Those who invoked the idea of race prejudice usually had in mind biases against certain phenotypic traits, but they could just as easily link these to cultural and national qualities. In fact, the point of their critiques was that prejudices were objectionable because they were imprecise and inaccurate. There was also often slippage between singular and plural forms of the concept, that is, the "yellow" race and the races of Asia, the "Negro" race and the races of Africa, and the "Caucasian" race and the races of Europe. In some accounts the number of races around the globe was fewer than ten, while in others the number seemed almost limitless because the term designated different population groups within circumscribed geographical boundaries. Similarly, ethnological maps could depict racial "stocks," which usually encompassed four or five categories, such as Bantu, Negro, Hamite, and European, or tribal and linguistic groups, which involved dozens if not hundreds of categories. In African settings colonial administrators and their critics used the terms tribes, races, peoples, primitives, and natives loosely and relatively interchangeably.

Discussions about equality, prejudice, and inferiority affected scientific research as well as political debates. The most significant controversy over racial research in British Africa emerged in the early 1930s. The previous decade had witnessed a large influx of physicians and biologists to the colonial territories (see Appendix). Few were hired to conduct research with explicit racial boundaries, but their work was infused with racial assumptions. Colonial states tended to naturalize the norms of "whiteness" and of European cultural standards; scientists and technical officers tended to do the same in their research and day-to-day professional activities. The question we must answer is why some kinds of racial classifications and racial research were considered uncontroversial, while other kinds touched off vitriolic debate. This matter teaches us not only about the politics of knowledge but also about the norms that underpinned colonial development and scientific research.

In large part as a result of the political and economic attention Kenya attracted in the early 1920s, East Africa came to be conceived in the minds of humanitarian and imperial advocates alike as an ideal arena for social and scientific experiment. Kenya's problems, although atypical in several respects, were viewed as a stand-in for those of Britain's African empire as a whole. The pressing need, advocates argued, was to define and investigate these problems systematically. The 1924 Parliamentary Commission on East Africa gave a special impetus to this end. During their tour of

East Africa, the commissioners had been struck by the tensions between those who were interested in white settlers' rights and those who believed strongly in Africans' rights. While their final report unequivocally favored white rule, which they referred to as another form of "aristocracy," they also felt obliged to acknowledge that "the white man's leadership in this co-operation [toward development] is not due to any inherent right on account of the colour of his skin."[59] For those sensitive to fault lines within the empire, this declaration was a nod to key differences that were emerging between policies pursued in South Africa and those taking shape in East Africa.

Shortly after the commission's report was published, a number of individuals began to work on a proposal to establish a government-sponsored Institute of Research in Kenya, including Joseph Oldham, Lionel Curtis, Frederick Lugard, all of whom later served on the general committee of the African Survey, and the recently appointed governor of Kenya, Sir Edward Grigg. The central purpose of the institute was to support efforts to develop the region economically. The most obvious source of funding was the East African loan, which had been one of the commission's recommendations. The sums under discussion were substantial: £9.5 million for economic development and £500,000 for research, "to ensure that the nine and a half millions [were] wisely spent."[60] The research agenda remained open for debate. Oldham was particularly keen to consider branches of knowledge relevant to Africans themselves. In a memorandum on the institute, Oldham expanded on his concerns in a subsection of the proposal titled "Study of the Native as a Human Being":

> If it is necessary for the successful development of East Africa to undertake the scientific study of the agricultural possibilities of the soil, and of the diseases which menace human, animal and plant life, it is no less essential to study the human beings whose ideas, feelings and desires must be a dominant factor in determining the future of East Africa. Practical recognition has already been given to the former necessity; no separate and distinct provision has yet been made for the latter.[61]

Perhaps because he was navigating national, imperial, and international networks at the time, Oldham's proposal for East African research was comprehensive and far reaching. In a lengthy memorandum to William Ormsby-Gore at the Colonial Office, he stated that the Kenyan and British governments had an opportunity to "make a more vigorous and systematic effort than has yet been attempted, to understand the real nature of the forces at work in Africa." The problems he identified as most in need of investigation included: land and "native ideas regarding land tenure"; health and the possibility of "a declining and physically inefficient population"; agriculture and "native methods of production"; labour and "native conditions as a whole"; and finally, "native beliefs and customs . . . [and] knowledge of the native mind."

> An experiment of this kind made in any part of Africa would yield valuable results, but there are strong grounds for thinking that it could be made with greater advantage in Kenya than anywhere else. . . . If the experiment were well-conducted and successful, the eyes of the outside world would be turned towards Kenya not, as at present, in criticism and censure . . . but with interest, expectation and gratitude for a contribution to the improvement of the relations between white and black, which is a matter of world interest.[62]

Oldham had begun to think seriously about the connections between racial science and racial politics in 1921, when he started work on his book *Christianity and the Race Problem*. He corresponded actively with Norman Leys, a physician who had worked in Kenya for nine years and was beginning a public relations campaign against colonial policies in Kenya, which he believed were exceedingly unjust. Leys's views on racial questions already leaned heavily against settlers' influence in the territory; they were pushed even more strongly in that direction after he met W. E. B. Du Bois at the Second Pan-African Congress in London that same year. Writing to Oldham afterward, Leys remarked that he had been "more deeply moved than ever in my life since childhood."[63] While Leys believed he was politically more radical than Oldham, both men increasingly saw conflicts between races as the result, not of inherent differences, but of economic and political competition. They had much in common with Gilbert Murray, who began to condemn race domination in the early 1920s.[64]

Shortly before his book appeared, Oldham explained his reasons for writing it in a letter to one of his missionary colleagues based in Nyasaland. He had become increasingly worried because "doctrines utterly hostile to all that . . . I believe in are being propagated." "From American universities, of all places in the world, there is issuing quite a stream of books repudiating in the name of modern biological and psychological science Christian ethical ideals and openly defending a policy of exploitation. This kind of literature, especially when it appears in the guise of scientific doctrine, is capable of poisoning . . . the public mind."[65] Oldham was not opposed to exploring the "biological aspects of race," but he objected to what he characterized as the "pseudo-science" that underpinned some of these new studies.[66] Only "impartial and objective" research could determine whether there were differences in the "inborn racial capacity of different [human] stocks." "Firstly, there must be some means of measuring accurately native mental qualities. Secondly, there must be agreement regarding the standard to be applied; it is necessary, that is to say, to reach agreement as to what constitutes superiority."[67] Oldham remained uncertain that it was necessary to determine the existence or absence of innate difference, however; other considerations, both ethical and scientific, might be deemed more important in the long run.

In the autumn of 1925, Oldham helped to organize a panel at the annual Church Congress on racial questions, which included social critics as well as colonial administrators. During his own presentation, he reiterated the conclusions of his book and emphasized economic strife as a primary factor in racial conflict. "If [the panelists] were asked how the racial problem presented itself to the mind of the non-white people," he told the audience, "it would be found that it resolved itself very largely into a claim for equality." One of his co-presenters, Lord Willingdon, who had served as governor of Bombay and Madras, went a step further: "the white races must realize the necessity of treating all coloured men in a spirit of absolute equality, and relinquishing the attitude of colour superiority."[68]

When Oldham presented his proposals to the East African governments in Nairobi in 1926 he emphasized that comprehensive research into the human dimension of colonial problems might help them do their work more effectively. Considering Oldham's suggestions in relation to research priorities for all the territories, the governors decided that they could not afford to spend "money to appoint an organization to study the effect of European civilization on the native mind." Instead, they would appoint a statistician who would begin to assemble demographic data about the inhabitants of the territories.[69] They also resolved to ask their respective technical services in medicine, veterinary science, and agriculture to begin to coordinate their research programs.

Several cooperative efforts were already underway: the Amani Agricultural Research Station in Tanganyika; in Uganda, an internationally sponsored sleeping sickness study and a proposed Human Trypanosomiasis Research Institute for East Africa;[70] in Kenya, a multiyear study of the nutrition of indigenous inhabitants.[71] None of these efforts included psychological, sociological, or anthropological research, the subjects that might shed light on the human side of economic development. This lack concerned Oldham and many of his colleagues in Britain, especially those affiliated with the Colonial Office Advisory Committee on Education in Tropical Africa and with the recently founded International Institute of African Languages and Cultures (IIALC), both of which were preoccupied with research in the human sciences. It also worried a physician and amateur psychologist just beginning to make his presence felt in Nairobi: Henry Laing Gordon.

Gordon's first intervention on these questions came while he was still squarely among the settler population, having moved to Kenya in the mid-1920s as a private physician. As the East African governments were beginning their efforts at coordinating research, Gordon gave an address on "The Mind of the Native" to the Kenya and Uganda Natural History Society in Nairobi.[72] Along with numerous officials and scientists at the time, including the director of Kenya's medical department who was preparing a memorandum for an Imperial Research Service, he was convinced that a good imperial trustee requires knowledge of "the facts."[73] Although at this point his rhetoric was still close to those advocating ethnographic studies of African "mentalities," his emphasis was not so much on psychology as on physiological studies of the brain as "the servant of mind."[74] A sustained inquiry of Africans in Kenya, Gordon argued, would help reveal "possible sources of persistent racial differences produced in the long ages of human life before the brief period we call history began." This knowledge, he believed, was essential for effective "native policy."

Although the East African research institute never became a top priority for either the Civil Research Committee, which was charged with evaluating it, or the East African governments, which viewed agricultural, nutritional, and epidemiological research as far more important in the short term, it helped to catalyze a set of projects that eventually prompted renewed conflict over the question of racial science. Oldham continued to press for research into the human consequences of African colonization, developing a strong relationship with the social anthropologist Bronislaw Malinowski, whom he recruited to play a more active role in the IIALC.[75] Through his efforts with the Colonial Office advisory committee on education, Oldham also helped to usher in a trial project in Kenya, sponsored by the Carnegie Corporation, to experiment with "the wholly new idea" of mental testing as an aid in selecting individuals for higher education and government service in Africa.[76]

Meanwhile, Gordon continued to pursue his interests in psychology and neurology. In 1928, he became the visiting physician at the Mathari Mental Hospital in Nairobi where he encountered both African and European patients. During this period, he deepened his interest in the neurological dimensions of mental illness in Africans and developed a close working relationship with Kenya's medical pathologist, F. W. Vint.[77] Summarizing his outlook on the possibilities for research in 1929, Gordon stressed their explicit racial dimension: "the question of racial differences of the nervous system—physiological, histological, sensory-motor, psychological—opens a virgin field for investigations of world-wide interest." His proposals would, he hoped, "attract and keep busy the new spirit of Scientific Colonisation."[78] Gordon became increasingly focused on eugenics, opening one of his articles with the precept, "Control of human breedings must come before betterment of human brains. Lacking good endowment education fails."[79]

At the beginning of the 1930s, in spite of their differing emphases, there was little obvious antagonism between the research plans pursued by Gordon and Vint on the one hand and those of Oldham and Malinowski on the other. By the middle of the decade, they were on opposite sides in a public and heated controversy over research into Africans' mental capacity. Proponents on each side saw the debate as a test of imperial research priorities as a whole. The institution brought in to adjudicate the debate, the African Research Survey, was asked, in effect, to navigate the increasingly volatile terrain of the human and racial sciences vis-à-vis Africa. Its final verdict on these questions helped lay the foundations for research in and on British Africa well into the 1940s and beyond.

Race, Eugenics, and Social Anthropology: Reactions to the Kenyan Research Proposals

At the crux of the mental capacity debate was a range of questions relating to equality and difference: Were Africans to be treated the same as Europeans? If not, should it be assumed that they were inherently different, and was there evidence to substantiate that assumption? These questions came to the fore when H. L. Gordon visited Britain in the autumn of 1933 and first made public his and F. W. Vint's efforts in Kenya.[80] Gordon arrived in England in October on special leave from his post in Nairobi to present the results of his research and secure further financial support.[81] With the exception of a very brief notice in the *British Medical Journal* in 1932 on the "Brain and Mind in East Africa," little was known of his investigations outside Kenya, particularly since he had pursued them independently of any official research program.[82] As he told an audience at the British Eugenics Society, the main question that had prompted his inquiries was "whether or not natives of the [Kenya] Colony possessed the same mental capacities as white men."[83] Puzzling over different explanations of Africans' "backwardness," he elected to answer his question through a preliminary examination of Kenyan cranial capacities. After giving several talks to national scientific and political societies during his stay, just before his departure Gordon submitted an article to the *Times* titled "The Native Brain: Observations in Kenya, a Comparison with Europeans," in which he amalgamated and summarized the substance of his and Vint's research.[84]

An examination of approximately 3,500 skulls led him to conclude that the average adult male "native" had a cranial capacity at least 150 cubic centimeters smaller than that of an average European. Organizing the data by age, he determined that at puberty the African skulls he had examined ceased to develop, whereas European brain capacity rose "steeply." Moving into the medical laboratory, where Vint conducted his studies, Gordon reported that Vint's postmortem examinations of more than three hundred brains indicated that the "cortex" of native brains showed a "quantitative deficiency" at the cellular level when compared with those of Europeans. Gordon's final conclusion was also the most controversial: only those Africans who had "received some kind of European education" exhibited the "mental affliction known as *dementia praecox*," referred to in other contexts as schizophrenia.[85] These findings, he argued, were immediately relevant to debates concerning colonial development and trusteeship: only if administrators had an empirical grasp of the biological differences between Africans and Europeans could they develop effective educational and social programs suitable for Africans.

In certain respects Gordon can be described as an old-school eugenicist who clung to assumptions that were increasingly discredited or under assault both within and outside eugenics societies in Britain and North America.[86] In Nairobi, just a few months before his departure for England, he elaborated on these tenets in an address on "Eugenics and the Truth about Ourselves in Kenya."[87] His central argument was that colonial trustees overlooked human heredity when they developed their medical, educational, and legal policies: "nurture" took precedence in the absence of any knowledge of "nature." Drawing parallels between animal and human "stock," he lamented that "the worst of our [white] race" was allowed to reproduce far too rapidly while the "best" declined in numbers. "Subcultural groups" of all races—"white, Asiatic, and African"—were proliferating and "we do nothing about it"; "if this is allowed to go on we cannot avoid sinking to a level beyond rescue."[88] Fundamental to Gordon's research was the belief that "racial backwardness," which he used as a synonym for inferiority, was the norm among Africans, "for races of men differ in body, brain, and natural ability just as individuals differ."[89] Although he conceded during his talk to the Eugenics Society in London that further investigation would be required to show whether environmental factors influenced mental development, he believed it was evident that "cerebral deficiencies" in Africans had a "genetic basis"; "even the best of the natives are on a lower biological level than the average European," he asserted.[90]

Concerns over racial degeneration and "social decay" were widespread in both Europe and Africa in the late nineteenth and early twentieth centuries and laid the foundations for many eugenics programs.[91] Francis Galton's experiences in Africa arguably helped produce his interest in innate biological difference in his home country.[92] While in Europe these preoccupations focused on differential birthrates between the wealthy and the poor, as well as the "degenerate" stock of criminals and the feebleminded, in tropical Africa commentators had grown equally if not more concerned with the question of whether certain African populations were "dying out."[93] Speculations over the causes of demographic decline ran a gamut of explanations: infertility, psychic trauma, epidemics and malnutrition, cultural practices, and even colonization itself.[94]

In African contexts, lurking behind these preoccupations was an equally intense anxiety about the place of European populations in the tropical territories and their lack of resilience in the face of climatic and epidemiological threats.[95] In Kenya, several settler physicians had begun to explore the biological consequences of "climatic conditions" on European schoolchildren in the 1920s.[96] At a meeting on white settlement in the highlands in 1929, participants discussed the effects of the Kenyan climate on residents, the lead speaker remarking that "the Highlands are in no sense a Colony medically speaking . . . neither a white man, and far less women and children can live here healthily without [returning] home."[97] The question many posed, including Gordon, was the extent to which these were permanent and irreversible changes. The very presence of settlers in East Africa's "highlands" was referred to as an "experiment" in the literature on imperial health throughout the first several decades of the twentieth century.[98] Jan Smuts called the region a "laboratory" during his speeches at Oxford University that same year: "It is even possible that just as in the biological world new types are evolved in a new environment, so a new human type may in time arise under the unusual climatic conditions of Eastern Africa. . . . The human laboratory of Africa might yet produce strange results, and time alone could show whether or not the experiment was worth while in the interests of humanity."[99]

Taking the lead from Smuts, several of the speakers at the Oxford conferences on African problems repeatedly turned the discussion to the rapid social changes occurring across the continent and underscored the need to investigate them more fully. Leopold Amery, the former colonial secretary, pointed out that Europeans in Africa functioned as a "ruling race" and mused "whether in the last resort the best could be got from the African without admixing his blood with other stocks. This in turn led him to reflect on the degeneration noticed in, for example, the Portuguese colonies." Dietrich Westermann, the German linguist and collaborator with Oldham on the International Institute for African Languages and Cultures, concurred, mentioning the "problem" of "detribalization" and the apparent depopulation in Sierra Leone and Liberia. Africans, he claimed, were "losing the basis of their existence (the tribe) . . . [and] had lost 'lebenslust, lebensinhalt, lebenszweck' [love of life, content of life, and purpose in life]. When a race no longer knew what it was living for, it might well be in danger of decay," he asserted.[100] While Amery and Westermann were troubled more by "detribalization" and the labor implications of "depopulation," their wider conversation, concentrating on how one "race" should govern another and the potential for improvement among the "natives," resonated with the concerns Gordon himself expressed.

A number of natural and human scientists in Britain reacted swiftly to Gordon's publications, expressing several cautions and, frequently, sharp criticisms. Among this group were Louis Leakey, the Kenyan-born archaeologist; Julian Huxley, who was working to reform the British Eugenics Society; Cyril Burt, also a member of the Eugenics Society and a psychologist at the University of London; Lancelet Hogben and J. B. S. Haldane, two London-based geneticists who had already done much to discredit tautologies in eugenics arguments; and Bronislaw Malinowski, Joseph Oldham, and one of Malinowski's students, Meyer Fortes, who was then at work on the applicability of intelligence tests across "racial" groups.[101] The criticisms coming from members of the Eugenics Society serve to reinforce several historians' arguments that in the interwar period eugenics came under increasing scientific attack even by its adherents and sympathizers.[102]

The first and most frequent criticism focused on Gordon's methods. Leakey not only challenged Gordon's ability to get accurate skull measurements from living subjects but also questioned whether the postmortem subjects were in any sense normal if these skulls were acquired from hospitals, prisons, and mental asylums.[103] Leakey regarded skulls as useful for explaining human evolution, not intelligence.[104] He favored instead research in social anthropology. In a 1930 proposal for an East African research institute, he wrote that the "most pressing and important problem" was the "proper study of the social customs and organisation of the tribes of East Africa."[105] His views were influenced by his parents' position and perspectives as well as by his own scientific interests. His father, Reverend Harry Leakey, served as a "representative for native interests" on Kenya's Legislative Council and had recently submitted a memorandum to the Parliamentary Joint Committee on Closer Union voicing objections to the "status quo" in Kenyan governance. Acknowledging the competing claims of settlers and Africans, the elder Leakey tried to strike a balance between a "pro-native" and "pro-settler" stance, but made clear that he was opposed to everything "which is . . . selfish and tends to increase racial hatred and prejudice, instead of breeding good-will and contentment. . . . So long as the black man thinks that the white man is trying to 'do him down', and wants to steal his land, or keep him a hewer of wood and drawer of water for ever, of course there will be trouble."[106] Like his father, Louis Leakey was skeptical that education was a forerunner to "detribalisation." Just because people received education did not necessarily mean they would eschew their cultural or familial ties.[107] His own experience among his Kikuyu age group was testimony to that.

Julian Huxley shared Leakey's concerns about Gordon's research methods and drew attention to the difficulties of comparing Africans with Europeans, whose body mass was often substantially higher. If this difference were taken into account, Huxley pointed out, it might be found that Africans and Europeans had identical brain sizes relative to their body weight. He worried that Gordon drew his evidence predominantly from one population group: "the variety of racial types in East Africa is so great that a good deal of comparative work will have to be done before any conclusions can be drawn."[108] Using a monolithic category of "*the* African" would not address the possible differences that existed between heterogeneous groups in the region. Huxley's sense of "races" encompassed far more than the four or five usually employed in classification systems.

The second criticism, expressed by almost all the writers, addressed Gordon's arguments and was much more damning. Burt wrote in his rejoinder, "Few scientists will accept the conclusions he has drawn . . . [which] are so contradictory to current theories and current practice that both the data and the logic on which they rest demand the closest scrutiny."[109] The principal problem was that correlations between skull size and "inborn mental capacity" had already been rejected as erroneous. Meyer Fortes, a psychologist turned social anthropologist, called Gordon's work "irresponsible . . . because it exploits long abandoned scientific fallacies in the interests of his own prejudices."[110] Lancelot Hogben and J. B. S. Haldane were equally adamant about this matter; in their recently published research on human heredity, they emphasized the complexity of genetic interactions with the environment.[111] New studies had demonstrated that genes were not fixed properties and could be altered depending on their external conditions. Gordon's generalizations, they felt, were based on faulty methodology and assumptions, as well as untenable chauvinism.[112] Hogben wrote sarcastically in the Royal African Society's journal, "There is an obvious danger in assuming that a grandfather clock is necessarily a more sensitive instrument than a wrist-watch, or imagining that empty heads are any less empty because they are large."[113]

Hogben, who had spent 1927–30 in Cape Town as professor of zoology, was a veteran of controversies regarding how Africans were presented to the British public. Shortly after his return to Britain in 1930, he co-wrote a letter to the BBC sharply criticizing a pamphlet it had distributed titled "Africa—the Dark Continent." He and his co-signers, who included Sydney Olivier and Norman Leys, were concerned that readers were presented with a "misleading" picture of modern life in Africa. Most troubling was the pamphlet's "central thesis" that "'there is a fundamental difference of mentality' between Africans and Europeans." "It is our experience that this last statement is quite untrue. All the evidence that we are aware of supports the view that Africans and Europeans have the same natures and are of the same average intelligence."[114] Scientifically as well as morally, Hogben was committed to equality in mental capacity and mentality.

Julian Huxley, who was by then a member of both the African Survey's general and executive committees, offered a more ambiguous response to Gordon. His reply in the *Times*, while not explicitly antagonistic, undermined much of Gordon's evidence. At the same time, he and several prominent members of the Eugenics Society wrote privately to the undersecretary of state for the colonies supporting "further and fuller investigations into the matter . . . by a team of scientific experts." This position requires explanation. From Huxley's 1929 tour of East Africa, he had become uncomfortably aware of what he referred to as Europeans' "first and greatest assumption, of black's considerable and inherent inferiority to white." He condemned this view as based on "race and class prejudice, and the all but universal human tendency either to exploit or to patronize those who have less power and less knowledge than ourselves."[115] According to Huxley, the evidence simply did not exist to establish "inherent inferiority," and the assumption was counterproductive to effective colonial administration.

Yet Huxley remained committed to scientific investigations of difference between Europeans' and Africans' physiology and intellect. He and his colleagues in the Eugenics Society thought that Gordon's and Vint's findings seemed to "show the existence of a definite degree of inferiority" between the average African and the average European. "It is, however, quite uncertain how much of the difference is truly genetic, and how much due to environmental conditions, notably nutrition and disease."[116] Averages were scientifically suspect since they cannot capture the broad range of abilities in any particular population. In a frequently cited passage, Huxley declared that even if it were someday shown that differences in intelligence did exist, "the great majority of the two populations [Africans and Europeans] will overlap as regards their innate intellectual capacities."[117] They suggested rigorous research into the relationships among diet, disease, and ability. In his publications, Huxley made clear that he believed "improper and inadequate diet," as well as vulnerabilities to disease, rather than any inherent intellectual inequality, were at the root of "backwardness among the native inhabitants of Africa."[118] Although he held onto the idea of backwardness, he never embraced a priori arguments of innate inferiority.

While many of the respondents agreed that "the educational policy of the Empire" would benefit from a deeper understanding of the mental attributes of Africans, several doubted whether anything but "racial prejudice" was at the root of Gordon's research.[119] Bronislaw Malinowski was particularly uneasy with the idea that anyone would argue that human differences were the result of permanently fixed, biological properties.

No correlation between the weight of the human brain and the cultural efficiency of an individual or of a race can in my opinion be established. . . . It seems doubtful to me . . . whether it is really a benevolent gesture to "extend the hand of science" [to quote Gordon] to a fellow-creature only to deal him a knock-out blow: to brand him with intrinsic racial inferiority; to insinuate that education leads him into *dementia precox*; . . . All these implications do as much harm to the reputation of Anthropology or Psychology, as they offend the sense of fairness of a European. They will be resented by all intelligent Africans.[120]

Malinowski had spent much of the previous decade promoting a new kind of research in anthropology that would be more directly relevant to colonial affairs. As he described the situation to Oldham in 1927, "The conflict of race and culture is rapidly becoming the burning problem of world politics. Anthropology, which purports to study races and cultures, should obviously provide some directing principles or at least a dispassionate and scientific attitude towards the issues."[121] Particularly during the late 1920s, Malinowski's description of anthropological science was neither strictly social nor purely biological, but instead drew attention to areas of overlap, even fusion. The discipline's subject matter "touches both biology and the science of society"; it "unites the knowledge of the human organism and race with that of culture."[122] Yet his explicit research program, with respect to Africa, increasingly stressed society and culture and evinced critiques of biological and racial preoccupations. Unlike geneticists and physical anthropologists, who were more willing to offer biological explanations of human differences, social anthropologists like Malinowski preferred to see these as the result of culture. "The better we know people of an alien race, the more we realise that their minds work even as ours do, though within a different cultural setting."[123]

Their reactions to Gordon make clear that Huxley's and Malinowski's priorities for research did not coincide: Huxley was far more willing than Malinowski to support investigations of Africans' "mental capacity" using the tools of biology and medicine. Biologists like Huxley liked to look inside organisms and see what made them tick; anthropologists of Malinowski's ilk preferred to consider how humans interacted socially and culturally.[124] As early as 1917, Malinowski had written to James Frazer, "I am not making any physical measurements of the natives, because I never could see any philosophical value of these studies, moreover they can be done a long time after the customs, original ideas and social organization of a people have vanished for good."[125] By the late 1930s he was working actively to distance anthropology from "the effervescence of racial doctrines" and call into question "the concept of 'stability of race', as well as its genuinely genetic character."[126] Given their prominent roles in the African Survey, we must wonder: if these two men had been asked to decide the future direction of African research, whose point of view would have triumphed?

Yet there was another significant body of thought on these issues: that of colonial officials themselves. The only nonscientist to respond immediately to Gordon's article in the *Times* was the former parliamentary undersecretary of state for the colonies, Thomas Drummond Shiels, who expressed concern that "rash generalizations" in scientific research about Africans' alleged "mental inferiority" might divert attention from the real task at hand: "to give opportunity for the African to rise to his fullest possible stature. . . . The proof of the brain is not in its anatomy or histology, but in its product."[127] Shiels had been invited to take part in a discussion two years earlier concerning the possible areas for research in the human sciences that the African Survey might pursue. During the conversation he cautioned those present, including Malinowski, Huxley, and Lord Lugard, that "we had to be careful that all this study and research on the African in Africa did not merely contribute to retention of the native in an inferior position."[128] This combination of administrative concerns and shifting scientific understandings of race in the 1930s effectively tipped the imperial scales away from biological or even psychological research and toward sociological interpretations of African peoples. Because most inter-imperial and state-sanctioned research had to be vetted by the Colonial Office and the advisory bodies established to oversee the dispersal of colonial development research grants, this shift was of great importance in the long run. While research continued to be inflected with racial concerns, studies explicitly designed to investigate racial differences were but a small fraction of the total research in the human sciences.

Brain Physiology and the Design of Intelligence Tests: Developments in Kenya

Despite these metropolitan objections, Gordon returned to Kenya convinced that his presentations and lobbying had had a great effect. During his return voyage, on which he embarked before the majority of critics' views were published, he wrote to the director of Kenya's medical department: "Without exception, these eminent scientists expressed high opinions of the work by Dr. Vint and myself, agreed as to the great importance of our results, and gave their support without hesitation to my plea for a big team research" effort. Gordon was evidently unaware that scientists' support for further research did not necessarily mean they endorsed his methods, assumptions, or conclusions. He did note that the senior officials he met with at the Colonial Office "gave no encouragement" and "ridiculed the letter" from the Eugenics Society, reactions that had far more significance than he realized.[129] Still, he felt the next task was to develop an official research proposal from the Kenya government to be presented to Britain's Economic Advisory Council.

In Kenya, reception of Gordon and Vint's work was more positive, but it did encounter criticism. In addition to Leakey's vocal refutations, a group of educational experts, including the director of the Jeanes School, James Dougall, the school's temporary psychologist, Richard Oliver, and the colony's director of education, H. S. Scott, registered their reservations. Even a member of the medical service, H. C. Trowell, who was then developing plans for a training program for Africans, weighed in on the matter. The controversy had begun in March 1932 following a talk delivered by Vint to the Kenya branch of the British Medical Association on deficiencies in the "Cell Content of the Prefrontal Cortex of the East African Native." Considerable debate followed concerning "the native brain-capital available for native progress."[130] Although Vint was careful not to draw elaborate conclusions from his research, those who responded to his presentation seemed prepared to accept the idea that African brains were qualitatively inferior to Europeans' and, therefore, that their intellectual abilities were also inferior.[131]

H. S. Scott, who later helped to draft the chapter on "Education" for the African Survey, was unsettled by these hasty conclusions.[132] Scott considered attempts to correlate intelligence and brain weight a matter of "controversy" and raised doubts about Vint's methodology, including differences that might exist among different "tribes" in Kenya, variations between body and brain size that Vint did not discuss, and the effects of the "physical and social environment" that might influence brain development. Calling the implications of Vint's results "profoundly disquieting," Scott nonetheless decided to "take them at face value" in order to highlight other factors that he believed influenced intellectual development:

> [Let us suppose] the native arrives to within 16% of the mental capacity of the average European. Is that really astounding? If you took an average European family or group of families, and bound them down for generations with a burden of malaria, and riddled them with worms—to mention only two of the environmental disabilities under which the African suffers— would you be astonished to find that such a group of Europeans failed to reach within 16% of their more fortunate fellow Europeans? Is it not evident that so far from discouraging us from further effort, the facts disclosed must inevitably spur us to giving the East African at least not less opportunities than he now enjoys of mental development both before he is born and in his childhood?[133]

At issue for Scott was how Africans were to gain access to higher education in Kenya, which he had been pursuing since 1929 when he developed a detailed proposal concerning the "Education of Africans."[134] His fear was that policy makers would use scientific evidence that suggested African inferiority to justify reducing support for further education. Calling himself a "convinced environmentalist," he sought to improve standards of life for Africans, particularly in health care and nutrition, so they would benefit socially and intellectually. "The evidence as far as it goes does seem to indicate the existence of a large number of Africans who, given good environmental conditions and fair training, are of relatively high educable capacity."[135] As part of a seeming compromise, Scott endorsed a full-scale research program, but he objected privately to Kenya's colonial secretary that "according to Dr. Gordon, I am engaged in preparing Africans for *dementia praecox.*"[136]

Among the authors Scott drew upon was the young psychologist, Richard Oliver, who had been in Kenya since February 1930 to conduct research on intelligence tests with a grant from the Carnegie Corporation. Oliver wrote a series of articles on the "Comparison of the Abilities of Races" for the *East African Medical Journal, Oversea Education* (Britain's "Journal of Educational Experiment and Research in Tropical and Subtropical Areas"), and the IIALC's periodical, *Africa.*[137] His work was supervised by James Dougall, director of the experimental Jeanes School, and had been arranged in consultation with Joseph Oldham in his capacity as member of the Colonial Office Advisory Committee on Education in the Colonies.[138] Dougall had summarized his own position on "racial psychology" in a 1932 article, "Characteristics of African Thought." In this piece, his main aim was to demonstrate that "the African thinks as we do, argues and makes deductions from commonly accepted premises . . . he is a reasonable being capable of logical thought, inference, and speculation."[139] Like Malinowski and Leakey, Dougall believed that whatever differences existed between Africans and Europeans did not result from disparities of "mind" but were rather "due to a difference in culture and civilization," which could best be studied by social anthropology.[140]

Under Dougall's tutelage, Richard Oliver began to develop intelligence tests that might assist educators in Kenya. He hoped they could help teachers select candidates for secondary school, classify students once admitted, guide students toward different jobs, and identify those who might "suffer from some mental or physical abnormality."[141] Oliver had been warned by the Advisory Committee on Education in the Colonies to proceed with caution. When the question of psychological tests first came up, Arthur Mayhew remarked, "Whether it is psychologically possible to isolate 'intelligence' and whether the operation of such 'intelligence' can be distinguished from reaction to teaching and environment, are questions to which dogmatic answers cannot at present be given."[142] When the committee decided to experiment with the tests in Kenya, the members asked Lord Passfield, the colonial secretary, to emphasize to all colonial governments "the danger of encouraging persons who were not trained psychologists to work with these tests, as too much importance might be attached to them, and only illusory results obtained." Passfield was only too happy to oblige.[143]

Oliver's articles describing his research challenged a number of Gordon's and Vint's underlying assumptions. He began by disclaiming any relationship between "innate racial ability" and "cultural achievement." Cultural advances, he argued, could not "be explained by biological principles."[144] Accepting the testimony of an American psychologist speaking on "Racial Differences in Mental Traits," Oliver acknowledged that "all [humans] have the same senses, the same instincts and emotions. All discriminate, compare, reason, and invent."[145] Perhaps most significant, Oliver challenged the assumption that "intelligence test results provide an accurate picture of the real state of affairs."[146] "It is of course impossible to measure [innate educable capacity] directly, detached from what has actually been learned; nature and nurture are inextricably mixed up in a man."[147]

For intelligence tests to be used for comparative purposes, Oliver believed their architects had to acknowledge inherent cultural biases that were difficult, if not impossible, to overcome. Even nonverbal "performance" tests of the kind he devised might contain visual cues that were culturally specific; they might also be administered in a language or setting that was unfamiliar to those being examined. The very idea of a test, which emphasized speed, efficiency, and performance, was hardly universal. Oliver concluded, "It is therefore probably safe to say that no test has yet been devised which draws equally upon the experience of advanced and primitive groups, and which is therefore a completely true measure of the intelligence of such groups."[148] While his research provided a start in this direction, he was just as interested in helping educators improve the way they approached and evaluated students from other cultural backgrounds, which is why he also suggested that psychologists cooperate with anthropologists.[149]

Oliver began his investigations by working with medium-size sample groups of 90 to 125 students at the "high school" level who ranged from fourteen to twenty years of age.[150] One of his initial challenges was to figure out how to correlate test questions with measurements of intelligence. His method reveals the extent to which attempts to measure intellect were inevitably relative and even self-referential. As he explained, before he conducted any test with students, he asked their teachers to give each student a numerical rank "in order of intelligence as they estimated it." Once he received these assessments, he averaged the scores, using a scale of zero to one hundred, and then divided the students into three groups, "comprising the best, the middle, and the poorest third of the pupils respectively." Only *after* he had ranked the students based on their abilities did he try out various test questions. "Any problem was considered a good test of intelligence if it was solved by more pupils in the best group than in the middle group, and by more pupils in the middle group than in the poorest group."[151] In other words, the tests were calibrated from the start to confirm teachers' preexisting appraisals. In one instance, the Porteus maze, anomalous results led him to omit the question.

Oliver's final test consisted of six elements, which were designed to stress nonverbal interpretations based on images, numbers, and letters: (1) picture numbering, in which numbered images were given to the pupil in a random order, and the pupil had to write the correct number, using a picture of the full sequence, below each image; (2) picture classification, in which four of five pictures had something in common and one was an anomaly; (3) picture completion, in which a student had to identify the missing element of a common object in each picture; (4) comparison, in which a student was presented with images that were either the same or different and had to identify them as one or the other; (5) number series, in which students had to complete a series of related numbers, that is, 2 4 6 8 __ ; and (6) picture absurdities, in which, for example, a goat was drawn with the tail of a cow and the student had to cross out the absurd part (figure 5.1).[152]

Although Oliver was initially asked to develop tests primarily for African students, it was almost inevitable that he would eventually use the tests to compare "European" and "African" subjects. What he noticed in their different reactions to the Porteus maze and how he chose to deal with the results reveal Oliver's own blind spots. That he reported these findings so openly indicates that he felt there was nothing exceptional about what he had done. The maze tests were designed to gauge a student's acumen in seeing and avoiding dead ends in a route from the outside to the center of a maze (figure 5.2). According to Oliver, "the mazes form a series, graded in difficulty, and constituting an age scale of intelligence. . . . If [the student] enters blind alleys [before reaching the center], he fails." When administering the test to European and African students, all boys, from two different schools, Oliver observed a pattern he did not expect.[153]

A European child, when he reaches a maze beyond his mental age, tends to enter a blind alley and explore it to the end, and then to retrace his path to the entrance of the blind alley and go on again. He penetrates to the centre of the maze quickly enough, but with many errors. The typical procedure of the Africans tested was different. The subject would study the maze for many minutes without making a move: then he would trace his path to the centre without hesitation or error. The test had to be abandoned as a test of intelligence, for even the most difficult mazes in the series were solved in this way by too many of the subjects.[154]

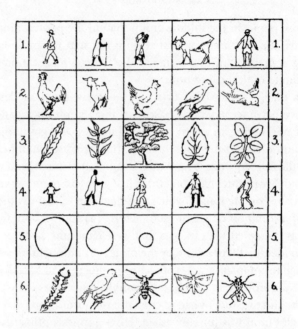

Figure 5.1. Picture classification questions from Richard
Oliver's Kenya Intelligence Test, 1932. Oral instructions (also
given in Swahili): "Look at the first row of pictures, row number 1. There are five pictures in the row—a man, another man,
another man, a cow, and another man. Four of these pictures
are like one another, one is different. The cow is different from
the men. So take your pencil and cross out the picture of the
cow." [*Assistants see that everyone makes a cross in the proper place,
and report all correct before examiner proceeds.*] "Look at the next
row of pictures, row number 2. You see a hen, a goat, another
hen, a bird, and another bird. Four of these pictures are like
one another, one picture is different. Which picture is different? [*Pause for correct answer.*] Yes, the goat is different. So everyone cross out the picture of the goat." [*Assistants supervise as
before.*] "When I say 'begin,' but not before, go on in the same
way yourselves. . . . Try to do all the rows on this page. Do not
turn over the page. Remember, in each row find the one picture
which is different from the other four pictures, and cross it
out with your pencil. In each row cross out one picture only.
Ready? Begin!" Sources: R. A. C. Oliver, *General Intelligence Test
for Africans: Manual of Instructions* (Nairobi: Kenya Colony,
1932); and R. A. C. Oliver, "Report on 'General Intelligence
Test for Africans,'" August 9, 1932, BY/26/7, KNA.

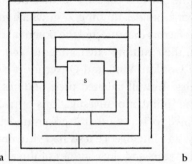 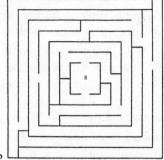

Figure 5.2. Sample Porteus maze tests, circa 1933. *a*, Age 12. *b*, Age 14.
Source: Stanley Porteus, *Porteus Maze Test: Fifty Years' Application*
(Palo Alto, CA: Pacific Books, 1965), 268–69.

Had Oliver not removed the Porteus maze from his tests, he might not have arrived at his result: "the average African score in the intelligence test is about 85 per cent of the average European score."[155] When he was comparing Africans among themselves, he might well have deemed the maze an ineffective tool for sorting and classifying students into different ability groups. But when comparing African to European students, the omission unquestionably skewed the results in Europeans' favor, assuming Oliver's observations were correct. This did not even seem to occur to him when he published his results. Even so, Oliver believed his "comparison of the abilities of races" was still preliminary and "of little value in itself."[156] In describing his two test groups, he was quick to concede that the "quality of the schooling has doubtless been superior in the case of Europeans" and noted they had the advantage of being given instructions in their "native language." In contrast to Gordon, Oliver remained a skeptic when it came to interpreting his own results. Indeed, he admitted that when asked, "'How do Africans compare with Europeans in intelligence?' he must at present answer that he does not know."[157]

Oliver's skepticism resonated with an officer in Kenya's medical service, Hugh Trowell, who was developing a training program for Africans. Kenya had the dubious distinction of being behind the Sudan and Uganda in its educational services, and both officials and Africans had begun to criticize the government openly for this failing.[158] The medical department in its memorandum on the subject placed responsibility squarely in the hands of the administration: the lack of "satisfactory" personnel resulted primarily from there being "as yet no organized mechanism by which African Staff may be trained in the highly technical duties they are required to perform."[159] Trowell wished to enter the "mental capacity" debate because it had a direct bearing on his own work. Besides funding, over which he admitted he had no control, the other factors that affected the potential of his recruits were their physical health, their cultural environments, and their inherited capacity. The writings of Gordon, Vint, and Oliver most clearly addressed the last of these variables. Trowell situated their findings in a wider geographical context since the "number of investigations carried out in East Africa is so limited." It seemed to Trowell that the Kenyan debates surrounding physiology, neurology, and intelligence tests had virtually "reenacted within the compass of a few years and the confines of three investigators, the history of racial comparison in North America: firstly, the measurement of cranial capacity, brain weights, and brain structure; secondly anthropometric measurements having significance for mental ability, and thirdly, intelligence tests. Each tool in its turn has been thought capable of effecting comparison, each tool has been found defective and has been superseded. The tool has never failed to record racial standards, it has always failed to record valid racial comparisons."[160] Like Oliver, Trowell thought the evidence that he reviewed, including a recent summary prepared for the International Congress of Psychology, was inconclusive.

Skepticism and Ambivalence: Reactions in the Colonial Office and the African Research Survey

While the specialists were wrestling to establish their positions, the British government needed to determine its own. Gordon's research plan would require government funds, and in the midst of the world economic depression it was already clear that the Kenya government had none to offer. The prospect of funding Gordon's and Vint's research had first been proposed in the fall of 1932. The Colonial Office's medical advisor, Dr. A. T. Stanton, initially declared the research "fanciful" and based on "scanty data [from which] rather sweeping conclusions had been reached." Indeed, he felt the investigation was "not really a medical one—it is an anthropological one directed towards issues which hardly fall within the medical sphere."[161] Gordon's suspicion that officials at the Colonial Office were unsupportive was on the whole accurate. Above all, the nonscientific staff was concerned that his research had "political" objectives and would stir up "anti-black" sentiments that were neither "justifiable" nor useful.[162] These reservations were applied more broadly to any scientific research that might undermine educational and social aims within the continent. Commenting on the danger of introducing psychological methods to Africa, one official wrote cynically: "I would rather give people firearms than intelligence tests to play with."[163] The general sentiment in the Colonial Office seemed to be that research in the human sciences would, at best, reinforce existing administrative efforts and, at worst, divert attention from these objectives and prompt unfounded and unnecessary controversies.

Much to Gordon's frustration, the Kenya government made no official request for funds until the summer of 1934.[164] In the intervening period, officials in Britain had decided it best to create a special subcommittee of the Economic Advisory Council that would consider "whether further research into this subject is desirable."[165] Doubts remained, particularly in the mind of the prime minister, Ramsay MacDonald, about whether such a topic should even be tackled by the government.[166] Although the new proposal had been broadened to include not just "the general question of heredity" but also "the special parts played by the environment" and "social heritage of the tribes," it was still considered controversial, not least because of the "propaganda" campaign that had been waged in the British newspapers.[167] The assistant secretary of state for East Africa wrote to Kenya's chief secretary, "There is definitely a risk of the uneducated charging into the field and saying that it is scientifically established that the black man of Kenya is of a lower order and mentality than the white, when nothing of the sort has been established at all, and there is the further danger that generalisations will be made from particular tribes in Kenya to 'the African' generally."[168] As misgivings on a number of fronts grew, the prime minister asked to consult with Joseph Oldham.[169]

Oldham was involved in a number of different research projects simultaneously: those of the International Institute of African Languages and Cultures (IIALC), the African Survey, and the efforts to coordinate human research in East Africa. In addition to helping facilitate Oliver's work in Kenya, he had been involved in approving the work of a social anthropologist, Dr. Yates, who had arrived in the colony late in 1932 with a partial fellowship from the IIALC.[170] Initially, Oldham seemed supportive of a coordinated project between Yates, Vint, and Oliver. After Gordon's visit to England, however, Oldham joined forces with his IIALC colleagues, Malinowski and Fortes, in "an emphatic protest" against Gordon's claims. "Nothing is to be more deprecated than that racial resentment should be aroused or that a race which is struggling to advance should be discouraged and that racial prejudices and pride, of which there is already too much, should be nourished by inferences which the facts as they are at present known do not in the slightest degree warrant."[171] By the end of his meeting with the prime minister, Oldham had convinced MacDonald to relegate the enquiry into Africans' mental capacity, not to a subcommittee of the Economic Advisory Council, but to "Sir Malcolm Hailey in connection with his projected survey of African conditions."[172]

By involving an "independent body" such as the African Survey, the Colonial Office and British government were able to distance themselves from Gordon's research program while simultaneously appearing to give credence to its importance. This stance did not stop Gordon, and a growing number of allies, from putting pressure on the government for direct funds. In 1934 and 1936, the East African Branches of the British Medical Association (a private body) offered a general endorsement of the proposal, although the East African Medical Service itself *did not*, despite the fact that it held inter-territorial research conferences in 1933 and 1936.[173] The project was reviewed positively in the *South African Medical Journal*, though the reviewer was careful to mention that environmental as well as biological factors would be taken into account.[174] According to East African officials, the Kitchener School of Medicine in the Anglo-Egyptian Sudan approached Gordon about doing "a parallel research in the Sudan," which Gordon's supporters mentioned with some frequency to the Colonial Office.[175] Yet when the Foreign Office corresponded with Sudan's officials on this subject, they replied that they were willing to submit information if a more general study were undertaken, but were unwilling to provide any financial contribution.[176] In England, there was also a steady trickle of letters to the *Times* and a series of questions on Gordon's proposals in Parliament.[177] None of these efforts translated into formal government support in either Kenya or Britain. As one official in the Colonial Office explained, when Gordon's supporters claimed the project was demanded by "the whole medical profession of East Africa," that "means the local branches of the British Medical Association and nothing else."[178]

In June 1935, Malcolm MacDonald was appointed secretary of state for the colonies.[179] Son of the former prime minister, MacDonald was a member of the African Research Survey's general committee; on receipt of his government post, he submitted his resignation. The following month, he was asked in the House of Commons whether the Colonial Office was "now in a position to recommend a grant" for the Kenyan investigations "into the causes of physical and mental backwardness in the natives. . . . An investigation," the questioner underscored, that had been "demanded on repeated occasions by scientific bodies in Kenya and in [Britain]." MacDonald replied in the negative, noting that he wished to await the issue of the African Survey's reports since these would highlight the "inter-relation of African problems" and help to determine "the course of future research."[180]

In private, the members of the African Survey continued to express doubt about the merits of Gordon's plans. The Oxford historian, Reginald Coupland, a member of the Survey's general committee, told Malcolm Hailey he thought the proposal was "amateurish" and "inadequate," especially on the "sociological side," a view Malinowski shared wholeheartedly.[181] Coupland was developing a research proposal on East Africa at the University of Oxford under the auspices of the newly founded Social Studies Research Committee.[182] While his proposal moved forward and was funded, the research plans in East Africa were held in check.[183] Even in 1935, the African Survey was already being positioned by the government as an arbiter of future African research.[184]

The period between 1934 and 1938, when the final reports for the Survey were published, was pivotal with respect to a number of scientific debates concerning African populations. During this time, the very concept of "race" came under heightened scrutiny, particularly by scientists who used the term to frame their research. This shift arose in part from concerns over "racial hygiene" in Nazi Germany, but it also stemmed directly from internal critiques of research methods in psychology, eugenics, and anthropology. In these fields there was an increasing recognition that scientific studies could incorporate and reproduce unfounded assumptions concerning "racial inferiority." Research methods were subjected to critical analysis. Gradually a community of scholars emerged who were willing to contest biases in their own fields. Siegfried Nadel, another psychologist turned social anthropologist and advisor to the African Survey, wrote of the "danger of [intelligence] test results being used as a social or political weapon, since they may be interpreted as demonstrating the constitutional inferiority and uneducability of racial or ethnic groups." It was important, he believed, to "formulate conclusions" from such tests in ways that would not "lend a scientific backing to popular prejudices . . . [because] even if our testing technique could counter the more conspicuous and tangible factors in the environment . . . it would still have failed to isolate the substratum of race and heredity."[185]

On a more fundamental level, scientists also began to challenge the biological criteria that sustained the concept of race. Even before the full scope of Nazi science was well known, this controversy concentrated on the question of pure racial types. As early as 1898 a social psychologist had observed that "the word 'race' is a vague formula to which nothing definite may be found to correspond. On the one hand, the original races can only be said to belong to paleontology, while the more limited groups, now called races, are nothing but peoples, or societies of peoples, brethren by civilization more than blood." Joseph Oldham seized upon this analysis in his book *Christianity and the Race Problem*.[186] If a race were constituted neither by language nor nationality, then it could only be characterized by physical features, but there was increasing scientific consensus that differences, whether in appearance or genetic make-up, were hardly rigid between so-called racial groups. "Let us remember," wrote Julian Huxley in 1931 in *Africa View*, "that 'race' is a mere convenience to help in pigeon-holing our knowledge of human diversity. The term is often used as if 'races' were definite biological entities, sharply marked off from each other. This is simply not true. In a community like the human species, in which inter-breeding is possible between each and every variety, and migration has been the rule since the earliest of times . . . such entities cannot exist."[187]

To dispel some of the "myths" concerning racial biology Huxley joined forces in 1935 with the anthropologist A. C. Haddon and the demographer Alexander Carr-Saunders to write "a survey of 'racial' problems." According to these authors, race was a "pseudo-scientific rather than a scientific term . . . it turns out to have no precise or definable meaning." Rather than continue in a state of confusion in which "race" could be "employed to bolster up the appeals to prejudice," they recommended abandoning the term altogether and substituting "ethnic group."[188] This two-pronged critique discredited the ideas of fixed racial differences and of pure racial types and drew attention to the abuses that arose precisely because the term was so ill-defined. For Haddon, in particular, these positions reflected a philosophical departure from his earlier standpoints. At the 1911 Universal Races Congress in London he had been reluctant to abandon racial nomenclature, arguing that it was still a useful zoological term; by 1935 he felt the term itself, and the biological assumptions that underpinned it, had become meaningless.[189]

As a consensus began to emerge that destabilized the biological suppositions of race, scientists began to shift their attention more confidently to the social arena. Raymond Firth, another of the social anthropologists who advised the African Survey, wrote: "When we speak of human race differences, we mean only a set of broad divisions, artificial, and not established on any very solid basis. . . . Where racial antagonism exists it is really a social antagonism. . . . It is founded not on difference of capacity of mentality between the races as such, but on differences of economic and other social interests."[190] Quoting an American "racial psychologist," Firth argued that it was "inexcusable" to withhold rights and privileges from particular social groups simply because they could be identified and isolated by certain biological traits. Such acts, he believed, stemmed from deep-seated "fear of competition," "jealousy," and the desire to "deliberately exploit."[191] Europeans, he felt, were guilty of denying "equality of rights" because this would prove "a threat to [their] social privilege and control." Firth regarded the Jim Crow laws of the United States and the color bar of South Africa as "set up in defence of vested interests."[192] Although his own area of expertise was in the Pacific, Firth felt compelled to refute Gordon's arguments: "Examination of the brains of different races tells us practically nothing about their mental character."[193]

An Official Verdict: The African Survey's Views on Race, Difference, Equality

This kind of candid criticism was hardly sufficient to alter the legal and political structures that sustained inequalities within racial states, but it did compel British officials to be more cautious about justifying policies of difference on biological grounds. The choice before the authors of the African Survey in evaluating Gordon's work was therefore momentous. The Survey was designed to provide an overview of existing conditions and policies in Africa, as well as to indicate the direction of future research. Receiving its stamp of approval or its censure could translate into long-term success or failure. How then was Gordon and Vint's research ultimately described in the final published volumes, *An African Survey* and *Science in Africa*?

In the first place, both texts acknowledged the "abundance of criticism" that surrounded the presentation of Gordon's results. They were quick to point out that the study of brain physiology was in its scientific infancy and that research along these lines had thus far demonstrated no proven correlation between "physical characteristics of the brain and intellectual or moral qualities."[194] "There would seem little advantage to be derived from the suggested inquiry into the physical characteristics of the African's brain as a basis of conclusions regarding his mental capacity. Certainly research of this type could not produce results of the social and political importance which some have expected from it." A key objection leveled against Gordon's work was his assumption that Africans were, in their normal state, inherently deficient. The tautology of his argument was precisely the problem. The *Survey* pointed out: "The abnormal condition of the Kenya native is taken for granted, yet it is difficult to see what is to be the criterion of normal 'self-development', or exactly how *bradyphysis* [backwardness] is to be diagnosed."[195]

Turning to intelligence tests, the *Survey* also cast considerable doubts. "At one time hopes were entertained that important conclusions regarding racial abilities could be drawn from tests of this kind, but as the methods of use came to be examined more critically it was realized that they were not so devised as to eliminate the effects of differences in environment and upbringing."[196] While investigations into Africans' physiology or psychology were "to be welcomed . . . it would not be reasonable to suggest support for [any] investigation based on assumptions such as those which characterize the theory of Kenya *bradyphysis*." The kinds of studies that would be most helpful "in formulating policy," wrote E. B. Worthington, were those that combined "knowledge of the mental and physical development [of Africans] . . . with a survey of diseases, of nutrition, of animal husbandry methods, and even botany, zoology, soils, etc. In fact, the ecological outlook would be all-important in such an inquiry in order to understand the several factors working on native development."[197]

These questions were discussed in the *Survey*'s chapter on "the African Peoples," in a section titled "the problems of ethnic classification." Drawing on the work of Hogben, Leakey, Huxley, Haddon, and the missionary ethnographer Edwin Smith, the *Survey* acknowledged that physical differences did exist among human populations, "but the use of physical characters, however carefully elaborated, will not in itself afford a basis of classification, in the sense of making it possible to delimit groups all members of which are marked by common and clearly defined characteristics." In rejecting both mental testing and brain analysis because neither would "form a safe basis for the determination of general administrative policy," the *Survey*'s authors were shifting emphasis away from racial and hereditary criteria in policy making and endorsing social criteria instead.[198] As racial science was increasingly discredited and marginalized, the conceptual foundations of racial states also came to be more unstable.

In the struggle to determine research agendas, it might seem that priorities in Britain superseded those developed in East Africa. However, an alternative interpretation turns on changing disciplinary and institutional parameters. Dr. Stanton, the Colonial Office's medical advisor, was first to suggest that Gordon's research had no place in colonial medical departments; anthropology was a more appropriate discipline to undertake such studies. Yet at the first International Congress of Anthropological and Ethnological Sciences held in London in 1934, the more than *one thousand* congress participants passed a resolution that "insisted on the importance of a profound study of methods of research into the mental aptitudes of the African peoples."[199] The operative word was "methods": the congress was recommending scrutiny not of "mental aptitudes" themselves, but of research methods used to determine them. In a draft resolution, the African section of the congress pronounced "any comparison between the mental aptitudes of the European and the African in our present state of knowledge [is] unscientific and unjust."[200] In the final plenary this language was toned down, and a clause calling on all scholars "to deprecate the drawing of conclusions from the inadequate data at present available" was also dropped, but the wider debate had lasting effects. Present at this meeting were E. B. Worthington, Audrey Richards, Lucy Mair, Dietrich Westermann, and Louis Leakey, all of whom later influenced the African Research Survey considerably. As British social anthropology defined its research program in the mid-1930s, many of its practitioners—including Raymond Firth, Siegfried Nadel, Bronislaw Malinowski, and Meyer Fortes—discredited such investigations for their racial and ideological biases.

Colonial medical departments also abandoned any systematic study of racial aptitudes or mental ability and concentrated instead on research into Africans' nutrition, diseases, physiology, and metabolism.[201] Likewise, discussion leaders at the newly inaugurated Summer Schools on Colonial Administration at Oxford University, which brought together more than 150 officials from across the empire with the largest delegations coming from tropical Africa, felt the need to discredit the implications of Gordon's and Vint's research. In the only lecture on "Race" during the 1938 summer school, W. E. Le Gros Clark made it clear that "there is no physical basis for the theory of racial superiority." "It has been affirmed by some medical investigators that as soon as the native is taken out of his ordinary environment, ridden of his parasites, and brought into good hygienic surroundings, he becomes mentally more alert." Citing the work of Worthington and Audrey Richards, he suggested that the study of diseases and nutrition would be more significant than "quite superficial" studies of anatomy and intellectual capacity.[202] Arthur Mayhew, who was present to speak on the topic of "Education in the Colonies," reinforced these points, stressing that "methods of imparting knowledge or skill must vary according to local conditions." While students' abilities would indeed differ across the empire, "there are no grounds . . . for assuming that there are permanent and ineradicable differences in educational capacity or for supposing that present differences are biological. These differences tend to disappear as education advances, and as environment is modified."[203]

The African Survey in effect closed the door on any program of research sanctioned by the imperial state that would pursue biological dimensions of "racial difference" in intelligence. Indeed, concerns about "racial discrimination" and "race prejudice" had become powerful enough in administrative and scientific circles relating to African affairs that they preempted lines of scientific inquiry that continued to flourish in other parts of the world. Ambiguities in the *Survey's* analysis remained, however, resting in particular on the distinction between inferiority and equality. While the authors made clear that they did not endorse the idea that Africans were inherently inferior, they also acknowledged that in scientific terms "universal equality of innate capacity" was still an unresolved question. The African Survey's advisors were unwilling, even unable, to put the racial genie back in the bottle. By discounting inferiority, they were neither embracing nor rejecting a principle of equality.[204] This question remained implicit and unanswered throughout the pages of *An African Survey*, and its openendedness had important consequences for future debates and policies regarding colonial rule. With respect to Gordon's research, however, the *Survey* was unequivocal. An official at the Colonial Office who was asked to summarize the history of the controversy wrote, "The view has been expressed that there is very little substance in Dr. Gordon's theories. This extract from Lord Hailey's *Survey* settles the matter finally."[205]

Conclusion: A Racial Laboratory?

A number of historians have critically examined the relationship of gender and reproduction to African colonial states.[206] Their studies reveal that the language of eugenics and demography, with their colloquial emphases on race, fertility, health, and improvement, had important effects on the conduct and legislation of colonial administrations. In the Belgian Congo, Kenya, French West Africa, and Uganda, among other places, regulating reproduction, controlling women's bodies, and managing population levels were always a high priority for officials. In this sense, eugenic and racial concerns, defined as population concerns, were pervasive across the continent. Yet racial research and eugenics societies developed beyond a "fledgling" form in only two territories of British Africa: Egypt and South Africa.[207] In both regions these efforts were initiated as much by individuals in educational and scientific institutions as by officials of the state. To sustain eugenics movements and racial science in these territories required at least two things: the existence of a critical mass of autonomous institutions of higher education, including medical associations, and comparatively large-scale capital investment in a scientific infrastructure, including national scientific societies. While the central problem in South Africa in the interwar period was "poor whites" and the pivotal issue in Egypt was birth control, in both societies eugenics activities and racial research functioned within a larger context of nation building.[208]

Colonial dependencies, in contrast, lacked the capital and people necessary to maintain an autonomous realm of scientific research and were hampered from pursuing nationalist ambitions. Kenya, with its settler population, could have been an exception to this rule. Gordon, and the handful of individuals who backed the African mental capacity proposals, tried to found a Society for the Study of Race Improvement in Nairobi in the 1930s. The preeminent reason they failed was the requirements of colonial administration, which placed a much higher premium on research into the colony's natural resources, its infectious diseases, and its inhabitants' nutritional and cultural status. Racial research was something colonial states could rarely afford.

Partly as a consequence of social anthropology's ascendancy over other human sciences and partly as a result of the new structures for interterritorial and inter-imperial coordination that emerged after the Second World War, researchers engaged in colonial development often tried to play down exoticizing and pathologizing tendencies in the scientific literature and play up their utilitarian and functional value. Long before African colonialism became publicly unfashionable, imperial administrators were pushed to avoid "racial prejudice" and "race discrimination."[209] They tended to eschew scientific studies that would uphold a strong interpretation of racial difference, that is, one that rested on the idea that differences were inherent and unchanging. Even in 1950, before any state in tropical Africa had achieved political independence, Lord Rosebery's question, "What is Empire but the Predominance of Race?" no longer had the self-evident ring it had in 1900.[210] The recalibration of racial politics around the world combined with the uncertainties surrounding racial science successfully undermined justifications of imperial rule on racial grounds.

CHAPTER SIX

An Anthropological Laboratory: Ethnographic Research, Imperial Administration, and Magical Knowledge

It is the business, and the practice, of any properly organized Government to maintain a continuous review of the social and economic circumstances of country and people. For some odd reason this is called "anthropology" in Africa.

—Philip Mitchell, *Secretary for Native Affairs, Tanganyika*, 1934

The African Survey now under way . . . promises to give an authoritative statement on the place of anthropology in our scientific knowledge of Africa.

—Bronislaw Malinowski, *"Anthropology," Encyclopedia Britannica*, 1938[1]

In late April 1932, the editor of the journal *Nature*, Richard Gregory, made a surprising decision.[2] After several weeks of intense public debate in Britain over a Kenyan Colony "witchcraft trial," he chose to run a lead article on "magic and administration in Africa," arguing that only the science of anthropology could adjudicate these matters. It was not good for officials and experts to disparage or ignore "tribal customs and beliefs" and to call them irrational or untrue. "From the point of view of the native . . . the law of the European administration is both illogical and unjust. . . . Unless, moreover, the law recognises the African point of view, no measure is likely to provide a solution." The "lessons" of anthropology "have been neglected time and again." Colonial interventions had often been pursued before officials understood the function or meaning of different social practices. The article endorsed further study of "the effect in the modification of native institutions of these varied influences brought into play by the impact of white civilisation"[3] and singled out the International Institute of African Languages and Cultures (IIALC) as doing the best research of this kind.

The court case that touched off the controversy had been settled nearly three months earlier.[4] It was not so much the verdict as the punishment that elicited a public outcry. In a village in Ukamba Province,[5] seventy young men, including ten under the age of sixteen, had been accused and convicted of murdering a woman. During testimony at the trial, one of the accused, Kumwaka Wa Mulumbi, explained that he believed the murdered woman had bewitched his wife, making her lose her ability to speak. He had organized the group of men to confront the "witch" and brought her back to his dwelling so she could remove the spell. When his wife seemed to recover partially, the woman tried to escape, but was tracked down by the men and beaten with sticks until her injuries were so severe that she died.[6] While various details relating to the case remained vague, none of the participants denied the death or the broad outlines of how it happened. When the trial came to a close, the chief justice, following Kenya's existing laws, sentenced sixty of the men to death and remanded those underage to custody. Within days of the verdict, the governor had cabled to the Colonial Office that he intended to "exercise [his] prerogative of mercy," but he could not do so officially or say so publicly until the East African Court of Appeals handed down the final sentence.[7] That process would take another two and a half months.

In the intervening period, individuals within the Colonial Office, the BBC, Parliament, the IIALC, Buckingham Palace, the Aborigines Protection Society, the League of Coloured Peoples, the Church Missionary Society, the Friends Society, the University of London, and the offices of *Nature* were all drawn into the fray.[8] When the editor of the *Times* learned in late March 1932 that the Court of Appeals had refused to alter the sentence, he wrote to the colonial secretary: "This sentence seems to me likely to create a disastrous impression in this country unless something is done rather quickly to make it clear that you are not actually going to execute sixty young men for the murder of one 'reputed witch'!"[9] A divinity professor from Edinburgh, who had recently edited a history of the Church of Scotland Mission to Kenya and had spent four months in the field in the mid-1920s, protested not just its disproportion but what it would signify to Africans: "I know sufficient of East Africa . . . to understand that the natives will suppose that such a sentence is merely part of a scheme for destroying them in order to get possession of their land. Besides they know that when a black man is killed by a white man, even one execution does not take place."[10] An MD from Liverpool was equally blunt: "The facts as read sound much more like mass murder than justice."[11] Even Kenya's chief native commissioner later called the episode "so tragic a farce as to

demonstrate the desirability of allowing to Judges some discretion in passing sentence."[12] Shortly after the appeal was declined the governor convened Kenya's Executive Council, which voted unanimously to commute
the death sentences to terms of hard labour, the "ringleader" for three years
and the rest for three to six months.[13] The *Times* and other papers reported
the decision the following day.

As so often happens in controversies of this sort, once the dust had settled what lingered in people's minds was not just a preoccupation with
"witchcraft" but also an enduring question about whether colonial states
were operating justly in situations where ruler and ruled held radically
different worldviews. The editorial in *Nature* placed these issues in sharp
relief and brought them within the realm of scientific debate. Several critics argued that whereas legal expertise seemed an obstacle to mutual understanding, anthropological science could hold the key for greater harmony. A strong proponent of this perspective was Frank Melland, a former
magistrate from Northern Rhodesia, who had himself been involved in a
controversial court case relating to an African-led "witchcraft eradication"
movement. For too long, Melland argued, lawyers had had the last word on
witchcraft, often taking uncompromising stands, "confusing witches and
'witch-doctors,'" and even going so far as to force administrators to deny
its existence.[14] If anthropologists were to undertake research on the question, he suggested, they would "deal with [the subject] sympathetically and
not scornfully: scientifically and not with *force majeure*." "Our attitude towards the whole problem has hitherto been wrong, and our law is wrong.
It is absurd to expect native chiefs and headmen (who are themselves often
witch-doctors) to administer the present law fairly and truly when it is so
contrary to their beliefs."[15]

One of Melland's correspondents on this subject was Frederick Lugard,
who had previously, in his capacity as governor of Nigeria, condemned "fetish worship and the hideous ordeals of witchcraft, human sacrifice, and
twin murder." By the early 1930s, however, Lugard had begun to see things
differently, in large part because of his association with the IIALC. As he
reported to the *Times*, the investigation of witchcraft beliefs was "a job the
International Institute of African Languages and Cultures, with the aid of
a psychologist and of anthropologists of international reputation, is now
engaged upon."[16] Melland, Lugard, and a number of anthropologists and
administrators discussed some of their results publicly during the eight
panels on Africa at the 1934 International Congress of Anthropological
and Ethnological Sciences in London. These speakers were often openly

critical of policies in British Africa and adopted resolutions supporting further research into Africans' conceptual and religious systems, including a special study of "witchcraft and colonial legislation."[17] A journalist reporting on the "witchcraft" panel for the *Daily Telegraph* explained that the congress was highlighting "the new spirit which anthropology is introducing into Colonial administration. The old attitude of complacent superiority is disappearing."[18]

Given the overlap in leadership between the IIALC and the African Research Survey, the IIALC's active involvement in debates over witchcraft and law virtually insured that these issues would eventually be addressed by the African Survey as well. Following the 1934 congress, the IIALC published a special issue of the journal *Africa* on the subject, which Lord Hailey consulted and even paraphrased. By the time the Survey published its reports in 1938, social anthropology had become central to the way its authors understood and defined specific problems, including "witchcraft." Even more important, the Survey recommended that anthropology occupy pride of place in future investigations in colonial Africa.[19] This recommendation proved influential: the Colonial Social Science Research Council, which was founded in 1944 as a direct offshoot of the Colonial Research Committee, allocated the vast majority of its funds in British Africa to ethnographic and anthropological studies.

This chapter considers, from a variety of vantage points, the changing attitudes of colonial officials to formal anthropological research and of anthropologists to empire and its apparatus of rule. The synergies and antagonisms of interactions between anthropologists and officials, especially in three institutions—the IIALC, the African Research Survey, and the Rhodes-Livingstone Institute in Northern Rhodesia—shed light on how social anthropology ultimately achieved a dominant position in research programs related to African development. This preeminence was not inevitable, nor was it greeted with unanimous approval. Skeptics' greatest criticism was that anthropologists appeared resistant to social change and wished to preserve African societies as static "museum specimens."[20] Advocates of anthropology, in contrast, often saw it as a means to reduce abuses of power and preempt cross-cultural misunderstandings. In their eyes, it was a tool to hold exploitative interventions at bay and to enable sympathy rather than contempt to inform policy making. The controversy over African witchcraft, therapeutics, and law considered at the end of the chapter illustrates the kinds of barriers that blocked anthropologists from having the reforming impact on colonial rule that they desired and highlights as well some of their own blind spots and biases.

Anthropological Training and Bronislaw Malinowski's Disciplinary Redefinitions

Although cadets bound for the Anglo-Egyptian Sudan had received training in anthropology since 1908 as a result of the efforts of its governor, Reginald Wingate, it took nearly two more decades before members of the colonial services in the rest of tropical Africa were provided with similar courses. The Colonial Office and the British government had been reluctant to support anthropology for some time.

Indeed, aware of this history, a prestigious group of administrators and anthropologists, including Alfred Haddon and Harry Johnston, asked in 1908 for a government grant of £500 a year in order to establish an Imperial Bureau of Anthropology in London that would undertake research projects on the empire's subject peoples and offer preparatory courses to officers bound for the colonies.[21] Yet, as with a similar project in France, the proposal came to naught.[22] Governmental inertia, treasury constraints, and the First World War proved to be the greatest stumbling blocks. After the war, Haddon at Cambridge and Robert Marrett and Henry Balfour at Oxford tried again, arranging meetings with the colonial secretary, Leopold Amery, and the recruitment officer, Ralph Furse, who was then designing the new training programs for the African services. In the minutes on the meetings officials expressed doubts about anthropology's relevance. They had no interest in "having our officers delivered to be prey of wild enthusiasts . . . you could not make every man, or one in a hundred, into an enthusiastic ethnologist. . . . As far as we are concerned ethnology is a side issue."[23]

Official attitudes began to change between 1922 and 1926, with the founding of the Advisory Committee on Education in Tropical Africa (1923) as well as the IIALC (1926). One of the key premises of the IIALC was that anthropological studies could be of use to colonial administrators by helping them avoid erroneous interpretations of Africans' cultures and customs. This rationale struck a chord with some officials whose first-hand experiences taught them the limits of administrators' knowledge.[24] It helped that Frederick Lugard had endorsed the project wholeheartedly; his reputation as a Nigerian governor still carried considerable weight. The single greatest ally of the IIALC in the Colonial Office during the 1920s was William Ormsby-Gore, whose interest in anthropology had been piqued during his tour of East Africa in 1924 and was reinforced while in West Africa in 1926.[25] Aware of the tensions between administrators and anthropologists Ormsby-Gore recommended that the two groups unite around the principles they shared.

If we are to succeed in our duties towards these peoples as rulers . . . we must study them objectively and base our policy on real understanding acquired not only from personal contact, but from scientific study of their mental and moral characteristics, of native law and customs, of native history, language and traditions. Native methods of agriculture, native arts and crafts, should be examined scientifically *before any attempt is made to supersede what we find existing*. Herein lies the importance of anthropological work, an importance which it is difficult to over-estimate. The wider the knowledge of the elements of anthropology amongst the administrative staff the better, and it is also essential that the trained anthropologist should work in closest possible touch with them.[26]

Thanks in large part to Ormsby-Gore's endorsement, the "Tropical African Services Courses" from 1926 onward included lectures by Robert Marrett on "social anthropology," Henry Balfour on "African arts and industries," and Reginald Coupland on colonial and African history, as well as lectures on "the races of Africa," "Islamic law, history, and institutions," and language training in Hausa, Yoruba, Swahili, and Chichewa.[27]

When Joseph Oldham took up these questions in 1927, as part of his preliminary research for the Hilton Young Commission on Closer Union in East Africa, Bronislaw Malinowski was among the people he consulted first. Malinowski relished the opportunity to expound upon his own priorities for anthropological research. "You want to point out [in your report] the need of a new research . . . old-fashioned antiquarian and historical anthropology is of little use and we need, within our science, a resetting of problems and a far greater interest in the working of human institutions, customs and ideas, rather than retrospective speculations about their origins and vicissitudes."[28] As several historians of anthropology have noted, Malinowski and A. R. Radcliffe-Brown played central roles in spearheading the new field of British social anthropology in the 1920s and 1930s.[29] For most of that time, however, Radcliffe-Brown was absent from Britain, working in Australia, South Africa, and the United States; only in 1937 did he return to the United Kingdom to take up the chair in social anthropology at Oxford. His absence meant that it was Malinowski and his students, most of whom became fellows of the IIALC, who shaped the African Research Survey's attitudes toward anthropology.

The concerns that Malinowski and Oldham emphasized in their discussions about the Hilton Young Commission were focused on the present rather than the past, and required different theories and methods than those of evolutionary anthropology. Malinowski shared with Oldham a confidential memorandum that he had written for the Rockefeller Foundation in March 1926 in which he justified these new methods in the context of "practical affairs." Underlying his arguments was an assumption that anthropology could be useful in providing direction to colonial and social policies. "Science gives us practical control of events through a theoretical insight into their nature. Social events do not differ in this respect from natural phenomena. Social science should aim at being put to the test of reality in the same manner by which natural science has vindicated its value i.e., practical application."[30] The Hilton Young Commission report endorsed further research in East Africa along anthropological lines.[31] Two years later, the Parliamentary Joint Committee on Closer Union, on which Lugard and Ormsby-Gore both served, reiterated these views.[32]

The question of anthropology's relevance gained prominence in the high-profile controversy about "practical anthropology" between Malinowski and the Tanganyikan secretary of native affairs, Philip Mitchell, which was carried on in the journal *Africa* between 1929 and 1930. Mitchell argued that anthropologists were ill-equipped to analyze the different facets of colonial administration, a job he believed was better left to district officers, while Malinowski contended they were eminently suited to this task because they were more likely than administrators to study social institutions and beliefs in their entirety. Shortly after he published his article, "Practical Anthropology," in 1929, Malinowski wrote a private memorandum for "Lionel Curtis and Co." on the "Teaching of Practical Anthropology in Connection with Colonial Studies." He hoped this lengthy essay would influence Curtis and Philip Kerr's plans for the Institute of Government at Rhodes House, which ultimately became the African Survey. Eager to persuade them that anthropology was "indispensable" to their wider goals, Malinowski argued that colonialism was in its essence "sociological and cultural engineering." Yet, "nothing serious" had been undertaken to ensure that British officials were properly trained in the tools they would need to be successful, such as "native languages, ethnology, customary law and comparative institutions." The "African Services" courses were insufficient, given the heterogeneity of the various colonies and the lack of funding for social science research in most of the dependent territories. How could Britain conceivably carry out its responsibilities, Malinowski asked, without an adequate understanding of these different societies' "daily life [and] forms of government"?

Malinowski encouraged colonial and African studies to shed common prejudices toward the societies the British governed. "We should not regard native forms as 'queer' or 'funny' types of superstition, but should study

how they work, how they function, and what they do morally and spiritu-
ally to help the native, and how amazingly consistent and sensible they are
in most cases." With regard to "witchcraft and magic," Malinowski argued,
it could "be regarded as a crude, often comical expression of 'savage superstition,'" or it could be understood through the "functional approach,"
which encouraged anthropologists to study "how magic, religion, and
knowledge work in primitive communities, what they contribute towards
the happiness of the individual, the cohesion of society and the general run
of practical affairs." Understanding the social functions of witchcraft and
magic seemed to Malinowski much more useful to "the practical man."
The "anthropological outlook" "consists above all in the respect for other
peoples' customs and social values. . . . Too much stress could never be
laid on the development of this tolerant, sympathetic, yet in a way, dispassionate attitude towards different cultures."[33] Malinowski's cultural relativity, however, was marked by hierarchies of its own. He continued to rank
societies into "lower and higher culture level[s]" and referred to anthropologists as those who dealt with "the simple mind" and helped to explain
"simple cultures."[34]

Drawing attention to colonized peoples' vulnerabilities and to the place
of anthropologists as cultural translators, Malinowski expressed his profound concern with "one of the greatest crises in human history, namely . . .
the westernization of the world."[35] The global advance of capitalist modes
of production and the "mechanical imperative of the age" were introduced
by Europe to other parts of the world through the vehicle of colonialism.
During a meeting to consider the roles of the IIALC and the proposed Institute of African Studies in 1930, Malinowski opened the discussion by
elaborating on this topic. The "whole trend of modern economic development was to stimulate increased production and consumption. The assumption underlying present economic conditions was that the greater the
volume of trade, the better it was for humanity." Africa was being drawn
into the world economy and Africans' standard of living was increasingly
linked to global markets. In this process, Malinowski suggested, "it was for
the anthropologist to see . . . that [the native] was not forced to labour on
products he did not wish to produce so that he might satisfy needs that he
did not wish to satisfy."[36]

What most disturbed Malinowski was that African colonization seemed
to be proceeding in an entirely irrational fashion, without clear principles
or plans. While championing his own form of expertise, he challenged critics of anthropology.[37] In his rejoinder to Philip Mitchell, Malinowski disputed the assertion that anthropologists were disconnected from the world,

more like secluded laboratory scientists than general practitioners whose role was to apply their knowledge.[38] To the contrary, Malinowski replied, anthropologists had to grapple constantly with the vagaries and complexity of human societies; their laboratory was "the surface of the globe." In an excursus into the philosophy of science, Malinowski sought to clarify the different ways scientific specialists produced knowledge. Whereas Mitchell drew a sharp line between laboratory research and practical problem solving, Malinowski preferred to see it as a blurred continuum.

> Even that ridiculed and maligned place, the laboratory, is not outside nature. It stands, in fact, in connexion with the whole world. It is an open piece of nature harnessed to the limited means of human observation. . . . In the present world of science it is very difficult to distinguish between the experimentalist, the theorist, the explorer, and the applied chemist or physicist. In natural science it is the give and take between theory and application which always leads to most fructifying progress, and exactly the same holds good of social science.[39]

The vitality of anthropology lay in the interplay between abstraction and observation in the field.

Anthropology potentially had the most direct, and perhaps unsettling, application in colonial Africa, in Malinowski's view, because it was beginning to explore the way colonial systems of rule affected African societies. "The functional anthropologist . . . studies the white savage side by side with the coloured, the world-wide scheme of European penetration and colonial economics . . . and he is prepared to face, as part of his problem, the turmoil of everyday life and even the chaos of maladministration and predatory politics."[40] Malinowski openly criticized the inconsistencies and contradictions of land policies in British Africa and advocated systematic studies. "Most of the blunders of the administration are due to lack of knowledge, which made determined constructive policy and a sound administration impossible. This weakness . . . created discontent and disorder and was responsible for half the troubles in Africa."[41] Colonial interventions that were not based on careful research and a comprehensive plan, he believed, had disastrous effects.

Although his solution to these problems was to try to reconcile anthropological research with colonial administration, Malinowski's faith in "scientific colonialism," even in the early 1930s, was hardly straightforward. Science, in its broadest definition, was in his opinion "the worst nuisance and the greatest calamity of our days." Echoing sentiments that were becoming increasingly common in Europe, North America, and India during this period, he wrote that "it has made us into robots, into standardized interchangeable parts of an enormous mechanism," a process that forced him to take an "extreme pessimistic view of 'progress' . . . [and of] the aimless drive of modern mechanization."[42] Anthropology, he admitted, would in some measure contribute to these trends. "As soon as the study of man becomes 'rationalized' it will proceed as ruthlessly to dehumanize human nature as science is even now obliterating the natural face of the inanimate world."[43] Although he believed the process of rationalization was "inevitable" in colonial Africa, he also held out the hope that anthropologists could temper its most egregious effects by studying Africans "in their full reality" without making them into "'primitives' or 'savages', into 'prelogical' beings, or into 'representatives of prehistoric times.'"[44] Ironically, Malinowski was as critical of the objectifying tendencies in science as critics of anthropology sometimes were.

What was new in Malinowski's views was neither cultural relativism nor the impulse to connect the discipline to British colonial administration but the methodological shift toward contemporary social issues. This shift entailed an emphasis not on such questions as cultural diffusion or racial improvement, but rather on "land tenure, primitive law, primitive economics, the changing native," and most important—and least defined—colonialism itself.[45] Malinowski's correspondence and articles express a general disillusionment with much of European culture, stemming not only from the horrors of the Great War but also from his amorphous sense that the modern world dehumanized "man" and de-animated "nature." During this period he was scathing in his criticisms of colonial administrators, even as he also expressed pejorative and paternalistic opinions concerning Africans and peoples of the Pacific. Whatever pragmatic impulses his research and proposals helped to satisfy, one thing seems certain: in seeking a place of relevance for anthropology within colonial states, he also sought insight into what he considered a common human condition.[46] In this respect, anthropology could help bring different societies closer together.

The Survey's Meeting of Anthropologists and Governors' Reactions

When the African Survey was first conceptualized in the years between 1929 and 1932, anthropology was presented as one of the four fields of knowledge that it would review, along with economics, natural science, and political science. When the project got underway in 1933, it was still unclear just whose anthropological expertise they would draw upon or how they would review the work of existing institutions. This uncertainty not only ruffled the egos of anthropologists, including Malinowski, but also set in motion a range of projects that affected imperial research activities both in England and across British Africa. The Survey's interest in reviewing scientific research, for instance, prompted the ethnographer, Edwin Smith, to devote his presidential address to the Royal Anthropological Institute in 1934 on the subject "Africa: What Do We Know of It?" Furthermore, had the governor of Northern Rhodesia, Hubert Young, not been disgruntled that the African Survey's leadership was being allowed to question colonial governments on how well they understood their subject populations, he might not have selected anthropological research as the Rhodes-Livingstone Institute's main priority.[47] British anthropologists' turn toward political questions at the end of the 1930s was also a direct response to behind-the-scenes conversations involving the African Survey and the IIALC.[48]

An interesting dimension to the story of anthropology's place in the African Survey is E. B. Worthington's role as he began to work actively on the early drafts of *Science in Africa*. Just a month after Worthington had accepted his assignment, Lord Lothian informed the general committee "that Mr. Worthington is not including anthropology proper in his survey" because they wanted to avoid duplicating work the IIALC was already undertaking.[49] Being a relative novice in the "science of man," however, Worthington still wanted to understand the discipline. He suggested that the executive committee sponsor a meeting of anthropologists to "discuss in the first place the points of contact between scientific research and anthropological research, and, in general, directions in which anthropologists might give help during the course of the enquiry."[50] There was, as yet, no consensus among the coordinators about where anthropology would fit in their research schema. While the study of African societies was certainly a prime concern, anthropology had no monopoly on the field; social history and human geography might also fill the bill. The South African historian William Macmillan had early in 1933 begun to formulate his plans for the book that eventually became *Africa Emergent*. This "comparative survey of social and economic questions throughout British Africa" was in Macmillan's eyes "almost precisely along the lines" of the African Survey's original plan of research.[51] Macmillan was among the skeptics with respect to anthropology's importance to colonial affairs.

Partly because of its loose and decentralized structure, the African Survey's leadership took its time deciding if Worthington should include anthropology within his mandate. His research was designed to illuminate relationships among scientific disciplines, fieldwork, and socioeconomic development. Yet it remained an open question whether anthropology was integral to or distinct from the other scientific disciplines he was considering. When the executive committee finally approved Worthington's proposal to hold a special meeting of anthropologists in March 1934, the decision took on added significance because it marked the *only* occasion when members of a single discipline were convened to discuss the relevance of their work to the African Survey.[52] Several months later, when Joseph Oldham asked Worthington to include a chapter in his book on "the almost sacred subject anthropology," Worthington consented, but acknowledged that he could be entering contested terrain: "You know how vehemently Malinowski's school feel that [natural] scientists cannot understand social anthropology."[53]

Indeed, so nebulous was the status of anthropology during this formative period (1933–34) that Edwin Smith, acting as both president of the Royal Anthropological Institute and as a member of the IIALC's executive committee, had the impression that "Anthropology was to be omitted" from the Survey's rubric of topics. It was for this reason that he decided to deliver his 1934 presidential address on the subject of Africa. His speech, which in published form ran to more than eighty pages, was premised on the idea that British administrators were still "groping in the dark" in terms of their understanding of African societies. Working from a "completely distorted and erroneous picture" of Africa could only be "misleading" and was no way to discharge imperial responsibilities. Smith declared, "The best [research] we have is not good enough: does not satisfy our thirst for full, precise, and accurate facts."[54] That the African Survey prompted him to address "the essential poverty of our knowledge" highlights its growing significance to wider academic and colonial pursuits and underscores the point that communication between the Survey and other institutions was not always clear or efficient.

One reason the executive committee gave for holding the meeting with anthropologists was that it would prevent them from thinking they were being "shut out" of the project.[55] Yet, absent from the list of individuals invited were many of the most prominent names in British anthropology, such as Charles Seligman, Malinowski's senior at the London School of Economics (LSE), and Robert Marrett, reader in social anthropology at Oxford. There were no representatives from Europe, although the co-directors of the IIALC, Westermann and the French ethnographer Henri Labouret, were already actively involved providing commentary, contacts, and sources of information.[56] Those who were invited—Malinowski, Jack Driberg, Louis Leakey, Raymond Firth, and Gregory Bateson—represented a mixture of junior and senior scholars, and as a group reflected a new set of concerns toward both anthropology and Africa.

The meeting was scheduled as a dinner and discussion on the evening of March 13, 1934. In addition to the anthropologists, there was a large contingent representing the African Survey: Joseph Oldham, Hilda Matheson, Lord Lugard, Lord Lothian, Lionel Curtis, E. B. Worthington, Ivison Macadam, and Margery Perham (in her capacity as author of a preliminary report for Hailey on colonial administration, which became her book *Native Administration in Nigeria*).[57] Summing up the evening to Lord Lothian, Hilda Matheson commented, "Personally I felt that it was well worth while having got these people together. They will not now feel shut out, and they will also realise that in this kind of enquiry all we want is to keep in touch with them and ask their advice as occasion arises. I am afraid Dr. Driberg may be a bit of a nuisance. He is very anxious, I think, to use us to press his claims to be sent on this two or three-year mission to Africa investigating records. My own impression was that he was the least able, as Malinowski was clearly the most able, of those present."[58] These were revealing words from the woman who was becoming central to the flow of information and to the development of the African Survey's research agenda. Of the five anthropologists, Driberg was the only individual who had spent a substantial part of his career as a colonial administrator, first in Uganda and then the Anglo-Egyptian Sudan; he was also the only other participant of similar age and professional status to Malinowski. Leakey, Firth, and Bateson were all considerably younger and in many ways under the tutelage of Malinowski himself.[59]

Driberg, who was then based at Cambridge University, had recently proposed to the Colonial Office that the records of each African territory be examined from the point of view of anthropological science. District notebooks and records, he claimed, generally included "notes on population and vital statistics, succession and inheritance, native beliefs and customs which are of importance to the local administration, health and sanitation, economics (pastoral, agricultural and manufacturing), labour and so forth."[60] These details, if analysed in the context of "the whole culture" and "with the requisite scientific knowledge" could immeasurably improve efforts to develop the territories. He concluded, "It is not too much to say that economic development cannot be isolated from other cultural activities, and only an appreciation of the unity of the whole culture will make real advance possible."[61] Driberg hoped that he would be selected to undertake just such a comparative investigation across the British territories over the course of the next several years.

Beyond its internal effects on the African Survey, the meeting of anthropologists catalyzed an important exchange between the Colonial Office and its African dependencies. Just weeks afterward, the assistant undersecretary of state for the colonies, Sir Cecil Bottomley, decided to provide the African governments with a more extended description of the Survey. Because he had also recently received Driberg's memorandum, Bottomley distributed the Survey's update simultaneously with a more general request that each territory consider both projects in light of anthropologists' increasing interest in the "information in District Offices." Bottomley wished to know what the different territories thought of "people outside the [Colonial] Service" having access to their records. He told the governors that this question "was likely to arise, with regard to anthropology and other subjects in connection with the African Research Survey. . . . It may be thought that the . . . Survey would be the most appropriate body [to have such access], especially as we might be able to secure that their anthropologist is not identified with any particular theories of anthropology." Based on the information he had received, Bottomley was under the mistaken impression that the African Survey might employ an anthropologist to work full-time alongside Hailey and Worthington. This proposal had actually been shelved, but he was correct that the Survey's leadership had a keen interest in district records and reports. He ended on a note of caution: "I personally am rather alarmed at the multiplicity of effort in regard to African Research and the resulting demands on the time of Government Officers and I think that it might be well if this sort of effort were led into one single channel."[62] That channel could be the African Survey.

Until this stage, the territories had received only one brief, "semi-formal" notice concerning the Survey.[63] To most of the governors, the project must have seemed inchoate. Driberg's memorandum further confused recipients of Bottomley's letter, which was also distributed in most territories to the secretaries for native affairs, because it was unclear how much Driberg's description of anthropological research corresponded to the kind of research the African Survey had in mind. In Tanganyika and Northern Rhodesia, colonial officials found much in the project descriptions that put them on their guard.

Responding to Bottomley's letter, Philip Mitchell was particularly offended by a quotation in the materials from the 1929 *Report on Closer Union . . . in Eastern and Central Africa*. The objectionable passage began,

"Anthropological science has shown the close inter-connection of all aspects of native life." Pronouncing this opinion "a simple platitude" that "is as true of Africans as of every other community that has emerged from the anthropoid," he embarked on a tirade against anthropology that concluded, "Except for the *obiter dicta* of the apostles of it, what makes anthropology a science?" Reiterating points he had made three years earlier in his exchange with Malinowski, Mitchell stressed that all governments, colonial or otherwise, are responsible for gathering information about their subjects. Anthropologists had no particular claim to special knowledge in this respect, and their approach could be a hindrance if they worked with flawed assumptions. Taking an indirect swipe at Malinowski, who had recently published a monograph, *The Sexual Life of Savages*, Mitchell continued: "Land tenure . . . is emphatically not a subject for an anthropologist, who may be a mine of information about the sexual oddities of savages and the marks with which they scarify the more intimate parts of their anatomies, but who almost for certain has no conception of land tenure and administration. Land is a highly technical and extremely complex subject and its administration is a profession in itself." The fundamental problem with anthropological projects, it appeared to Mitchell, was that they "assume a static society: a collation of facts such as would be made by a geographer, geologist, forest surveyor, and so on is of great value because generally speaking rivers, rocks, mountains and forests do not change appreciably, in time reckoned administratively. People, especially in new countries, do; and *customs and customary law are in a state of continuous development.*"[64]

The governor of Tanganyika had already drafted his own reply to the Colonial Office laying out similar concerns.[65] Harold Carmichael agreed with Bottomley that government officers were under increasing demands to complete "innumerable returns and statistics and digests," which prevented them from spending "the time they ought in administering their people." This difficulty could well be dealt with, he believed, if requests were centralized in the manner Bottomley suggested, and only those that were "really necessary" were placed before officers. To the possibility of Driberg or any other nonofficial being given access to "browse upon the files in our offices . . . I have . . . stronger objections. It seems to me quite out of the question that official records even if they were (which they are not) no more than a series of statements of facts, should be put at the disposal of people outside the Service."[66] As Mitchell later reminded Carmichael, "well vouched-for researchers (e.g. Margery Perham, Professor Macmillan, Gordon Brown) have for the last seven years been allowed free access to District Books and Native Court Books, but not of course to files."[67] The point was not simply "where to draw the line" in terms of who gained access to files, but how to control "the uses to which the information gleaned is put." Carmichael and Mitchell believed that colonial policies were best determined by administrators and not outside researchers. Anthropologists might develop their research in irrelevant directions at substantial cost; "the tendency of each such project gradually and insensibly to expand its scope are by no means trifling."[68]

After registering these general objections, Carmichael addressed a series of specific problems he had with Driberg's proposal, which "seem to have received little attention." Above all, he was insulted that Driberg suggested that "the average Administrative Officer and Departmental man fail to realise [the] truth" that development could only proceed through an understanding of the "unity of the whole culture." "One would suppose that such collation and co-ordination of facts with a view to economic development was not being done at present. . . . In actual fact all who are worth their salt put it into practice every day of their lives." Administrative officers were constantly assessing their "knowledge of the native's method of earning his living and the native's mental processes . . . and daily pooling the fruits of their experience and knowledge in discussion."[69] At issue for both Carmichael and Mitchell were the role of colonial administrators and the place of research in their official responsibilities. If anyone were to have authority for "knowing Africans," they argued, it should be administrators.

Mitchell's strident opposition to anthropology in print and private correspondence belies the fact that he made concessions to the new field and considered it an ally in the face of claims made by white settlers, and even Indians, to a permanently superior status over Africans.[70] In a proposal sent to Malinowski in 1931, he lamented that "the average member of the European or Indian community is astonishingly ignorant of the nature and composition of the population of the country, of how it lives and is administered, of what taxation it pays, of its social structure, and in fact of all the factors which make up the political, social, and economic life of the Territory in which he lives." Advocating a "comprehensive survey of the whole field, political, social, and economic," Mitchell named Margery Perham and Cecil Morison (the agricultural economist at Oxford) as potential candidates for political and economic research; "it would only remain to seek out a really competent social anthropologist" to round out the project.[71] While Mitchell clearly objected to exalted claims made in the name of social anthropology, he agreed that the field had its uses for administration. The research "experiment" he advocated combined the forces of Gordon Brown, an IIALC fellow based in Tanganyika, and Bruce Hutt, a district officer for Iringa, Tanganyika. The plan of research was designed by the district officer and implemented by the anthropologist. In the introduction to the published study, *Anthropology in Action*, he explained, "It must clearly remain the responsibility of the administrator to decide whether to act or intervene in consequence of information so obtained. But, he [the administrator] would be careful not to dispute, or be drawn into argument about [the information's] correctness, for that is the responsibility of the anthropologist."[72] Mitchell invoked a clear demarcation between the production of knowledge and its application, a division that was much less rigid in the minds of many anthropologists.

Northern Rhodesia's governor and administrators' responses were similar to Carmichael's and Mitchell's, but conveyed more suspicion of the African Survey itself. Officials worried that the "collection and preparation of information of this kind imposes a considerable burden on District Officers and . . . there is a real danger of the duplication of effort."[73] Northern Rhodesia had recently provided a "comprehensive response" to the British Association's special research project, the "Human Geography of Tropical Africa."[74] E. H. Jalland, the acting secretary for native affairs, replied that "a study of the Human Geographies . . . would prove of equal and probably greater value than an examination of books and records at the various bomas." Additional research should not burden the district officers with "extraneous tasks of this nature, demanding time and thought which would be more profitably expended on their ordinary duties."[75]

The replies from the governments of Northern Rhodesia and Tanganyika contain an essential contradiction: on the one hand, the respondents suggested that administrators were in a privileged positioned to speak to the socioeconomic and political conditions of the peoples under their jurisdiction; on the other, they recognized that these officials were overburdened by their core responsibilities and could not easily take time out to produce original studies relevant to African or wider colonial questions. This tension recurred throughout the decade, particularly in discussions concerning the implementation of indirect rule.

Sir Hubert Young, the governor of Northern Rhodesia, raised a question that was probably on the minds of several administrators across Africa. He wrote to Bottomley, "I cannot help asking myself what the attitude of the Government of India would be if some distinguished ex-Governor from the African territories—say myself, for example—were selected by Lord Lugard and Co., . . . to tour India with the object of making a general comparative survey of the chief problems of the Peninsula 'with a view to ascertaining how far the resources of modern knowledge are being utilised in dealing with them and what need there is for further research.'"[76] Young's main fear was that Hailey would consider his role that of an outside critic who would both detail and evaluate Northern Rhodesia's administrative policies. Worse still, Young could imagine Hailey in conversation with his "subordinates" across the territory giving them the "liberty to criticise" particular policies Young had implemented. "I may say frankly that I fully share your apprehension at the multiplicity of effort in regard to African Research and the resulting demands on the time of Government officers, and I cordially agree that it might be well if this sort of effort were led into one single channel, but so far as the territories under the control of the Secretary of State are concerned, I feel very strongly that this one channel should be the Colonial Office."[77]

In his reply, Bottomley attempted to assuage Young's "misgivings." He explained that the Colonial Office considered Hailey eminently qualified for the task of surveying the different African territories, but governors were not being asked to defend their policies to him. Bottomley had also discussed Young's concerns with Sir George Tomlinson, the Colonial Office's liaison to the Survey.

> Tomlinson indeed is pretty sure that your fears will be dispelled as soon as you make Hailey's acquaintance. And after all the African Dependencies have for years past been the scene of all sorts of problems—administrative, anthropological, economic and scientific. In a sense the Chatham House scheme will be the same sort of thing but on a bigger scale and conducted by a bigger man, and I personally hope that it will result in our being able to save Governors and their officials from a great part of the attentions of the lesser fry.[78]

This exchange may well have prompted Young to consider the need for locally controlled anthropological research in Northern Rhodesia: the very next month he proposed to establish a research institute there that would concentrate on "archaeology, geology, and particularly . . . the sociological side of anthropology."[79] This proposal, which emerged from Young's interest to commemorate the centenary of Livingstone's arrival in Africa, was to become the Rhodes-Livingstone Institute, the first social science institute established "on the spot" in British Africa.[80] Sometimes forgotten in discussions of its origin, the institute's history was inextricably bound up with debates concerning the African Survey, the International Institute of African Languages and Cultures, and the Colonial Office's plans for agricultural and medical research in the African territories.

Anthropological Auto-Critique and an "African Point of View"

Anthropologists themselves were raising many of these same points at precisely the same time. Although anthropological studies had always had a reflexive dimension, causing investigators to shine the spotlight on their own societies, this method was becoming more pronounced and explicit by the interwar period. British practitioners were looking inward and outward simultaneously. Some recognized that social categories they applied to non-European peoples might be just as relevant to European "civilizations." Malinowski wrote tongue in cheek in a private memorandum that the IIALC "understands also under anthropology the study of the 'detribalised tropical European', whether he develops the pathological forms of Kenya lunacy or only the ordinary idiocy of the average Colonial administrator or missionary."[81] In a similar, though less caustic vein, he wrote publicly:

> Anthropology is the science of the sense of humour. . . . For to see ourselves
> as others see us is but the reverse and the counterpart of the gift to see others
> as they really are and as they want to be. And this is the *métier* of the anthropologist. He has to break down the barriers of race and of cultural diversity;
> he has to find the human being in the savage; he has to discover the primitive in the highly sophisticated Westerner of to-day, and, perhaps, to see that
> the animal, and the divine as well, are to be found everywhere in man.[82]

This impulse toward reciprocal study led to repeated calls in the 1930s for Africans to become more involved in anthropological investigations. In its Five-Year Plan of Research, the IIALC declared, "Not only can they supply valuable information, but only from them can the African point of view be fully learned."[83] Malinowski is well known for supporting the work of Jomo Kenyatta during this decade,[84] but it was Margery Perham who first criticized the IIALC for being slow to move in this direction. "In any research that is attempted in the future, Africans should be brought in as recognised co-operators. I think we are far too prone in all our schemes for Africa to regard the natives as a passive, almost abstract element for whom everything must be done. . . . I should like to see researchers choose suitable natives to whom they could communicate the methods and objects of their work and so train them up as junior collaborators."[85] In his 1934 speech, Edwin Smith imagined a time when "African Fellows" of the Royal Anthropological Institute "will be . . . not only cooperating with us in the study of their own culture but also engaging in a study of European culture."[86] (The Nigerian, Nnamdi Azikiwe, soon obliged in several chapters of *Renascent Africa*.)

While such declarations can today be interpreted as savvy, even perhaps cynical, attempts to co-opt Africans into the colonial project and relegate their research activities to a subordinate role, they were intended to rectify what was then regarded as a serious inequality.[87] At this time, the production of knowledge about Africans was largely out of African hands. The recognition of the need to create conditions for a more reciprocal exchange of knowledge and information led Malinowski, Smith, Perham, and the IIALC Executive Council to seek to involve Africans in the research and publishing process. In 1928, the IIALC established its "Prize Competition" "to encourage Africans to write books in their own languages." The subject matter was open-ended, but it could include "stories, songs, dramas, riddles, proverbs, historical and other traditions, social institutions and customs, myths and religion in its every aspect."[88] Once translated into English, these texts would offer "a striking demonstration to [Europeans] of the flexibility and expressiveness of the language as used by an educated native."[89] Edwin Smith later declared that "no anthropologist can afford to ignore the books written by Africans about themselves," mentioning Thomas Mofolo's *Chaka*, Samuel Ntara's *Man of Africa*, and René Maran's *Batouala* and *Le livre de la brousse*. He might have added Akiki Nyabongo's *Africa Answers Back*, which was published the following year. (Nyabongo, who styled himself a "Prince" from the Toro kingdom, was just beginning a doctoral thesis in anthropology at Oxford on "Religious Practices and Beliefs in Uganda.")[90] Novels such as these, Smith believed, would help "redress the balance" in literature and remind anyone who had been "neglectful of the fact that Africans are flesh and blood and soul, even as we ourselves are."[91] Margery Perham had a similar intention when she edited the volume *Ten Africans*, which offered both biographies, often prepared by anthropologists, and autobiographies of ten individuals from different parts of the continent. In her introductory note, Perham observed that "knowledge of Africans as persons makes it impossible to dismiss them all as savage or backward" or "to regard them all as uniformly good, simple, unfortunate, or oppressed."[92]

Understanding cultural diversity and difference did not necessarily mean freezing African societies in some permanent condition of "custom," although a number of anthropologists acknowledged this pitfall both in their own studies and in indirect rule. Even Julian Huxley, who had become a strong advocate of technical officers embracing anthropological methods and who believed indirect rule was "a very interesting experiment," cautioned that it could also "become a fetish, a cast-iron thing, like everything else. You have to try and avoid fixing an institution which merely marks a stage [of change] . . . and saying, 'It is the native custom'. You have to allow for elasticity and expansion, and in some cases it looks as though sufficient allowance has not been made for this."[93] In their seminars on "Colonial Administration" and "Anthropology in Colonial Studies" at the LSE, both Bronislaw Malinowski and Lucy Mair tried to drive these points home to their graduate students who were asked to read critically not just anthropologists but also missionaries, administrators, and natural scientists, including Joseph Oldham and Julian Huxley. A reviewer for *Nature* in 1936 considered two books written in this tradition: Lucy Mair's *Native Policies in Africa* and Monica Hunter's *Reaction to Conquest* (on South Africa). In contrast to previous generations of anthropologists, the reviewer claimed, "they neither play upon the strings of philanthropy, nor as anthropologists enter unqualified protests against the modification of custom, arguing that if preserved it might serve as material for scientific investigation. Recognizing that Africa cannot be kept a museum piece, they accept the inevitability of change and record the facts and appraise the tendencies in relation to native institutions as social and economic phenomena *per se*."[94] The reviewer did not mention that both Mair and Hunter were critics of South Africa's segregation policies, which relied on the idea of custom to justify social discrimination. As Audrey Richards noted in her review, Hunter was among the first to undertake a thorough study of an "African community living in urban conditions. . . . [She] is one of the growing band of anthropologists who are determined to study the native as he is today, reaching out for European ways of living, and the trappings of Western civilisation."[95]

Officials' critique that anthropological research assumed stasis rather than change was increasingly incorrect. The primary purpose of "functional anthropology" was to understand and explain existing conditions in African societies based on a close analysis of their social networks and institutions. Those who promoted the study of "culture contact," especially the close associates and fellows of the IIALC, chose to examine the effects of the "rapid and forcible transformation" caused by the dual forces of colonialism and "westernization." In a 1934 article explaining these studies, Lucy Mair elaborated: "This does not mean that it is necessary to look for the 'original' native culture. Clearly, all such cultures have changed in the past, not only by a process of internal evolution but through the influence of external forces."[96] "Nor must we forget," reiterated Margery Perham during an address on "indirect rule" that same year, "that African institutions showed themselves to be remarkably plastic long before we came to

Africa, and are still showing it."[97] Social anthropologists emphasized that European colonialism, accompanied by new forms of economic production, catalyzed social changes "comparable only to the violent changes of a revolution." These "violent disturbance[s]" were different, both in degree and kind, from "the gradual, almost imperceptible, process of adaptation in which the normal evolution of human cultures consists."[98]

The force and power of colonialism and "modern" economic production, in these anthropologists' eyes, did not negate Africans' agency; to the contrary, "culture contact" studies aimed to address European interventions from "a native point of view." This endeavor would always be incomplete, because there was no single viewpoint to reconstruct and their authors inevitably remained cultural outsiders. Still, scholars pursued this goal valiantly. For example, Audrey Richards observed that administrators' and missionaries' contradictory policies had often led Africans to conclude that "the white man . . . is just mad . . . and must therefore be humoured . . . — and to 'humour' [they] of course deceive."[99] She elaborated, "On many so to speak controversial issues, the people have adopted as a protective measure different model answers suited to Government officials, to missionaries, and also probably by now to anthropologists!"[100] Margery Perham observed, when summarizing her assessment of the functional studies produced by "Dr. Richards, Dr. Mair, Dr. Hunter, and Dr. [Isaac] Schapera," that "they reveal Africans as no passive sufferers of our influence and all the effects it should logically produce in them, but as showing considerable initiative in taking what they want from us and keeping what they want of their own influences."[101] Change and flexibility rather than stasis were becoming the theoretical order of the day.

So, while officials critical of anthropology could at times deny that these studies were relevant to colonial administration, they shared several key assumptions with social anthropologists who were striving to establish a separate disciplinary identity for themselves. As the ethnographic interests of colonial states and the functional concerns of anthropologists began to converge, suspicion between the two sides endured. Besides questions of relevance and applicability, anthropologists' academic base was an important reason for official suspicion, as a review of Colonial Office minutes on the subject makes clear. Anthropologists were rarely direct employees of colonial states: fewer than ten anthropologists had been appointed in British tropical Africa between 1900 and 1940, and their studies only occasionally had a direct impact on policy making. Most of the professional anthropologists conducting fieldwork in British Africa in the 1930s were accountable not to the imperial states, but to their disciplinary and professional institutions, including the IIALC. No matter how keen anthropologists might be to produce studies useful for "native policy," they could also publish results that would "prove embarrassing to [the] Government."[102] As Africa became, in the words of Isaac Schapera, "a happy hunting ground for anthropologists," government officials became more sensitive to the role they might play as a destabilizing force.[103] When reviewing the work of Meyer Fortes among the Tallensi in the Gold Coast in mid-1935, the chief commissioner, W. J. A. Jones, commented to the Colonial Office that "while there is no evidence to suggest that Dr. Fortes is actively opposing Government policy, he certainly is not in sympathy with it, and may put wrong ideas into the heads of natives who are at present at a very impressionable stage."[104] Officials were quick to point out that anthropologists' relatively small numbers meant they were not a serious threat, but that did not mean they were considered natural allies in colonial development either.

The Rhodes-Livingstone Institute and the African Survey

Just how anthropologists became central to both development and "native administration" is intimately connected to the history of the African Survey and its role in deciding whether to sanction or reject Governor Hubert Young's proposal to establish a Rhodes-Livingstone Institute. When Northern Rhodesia's Legislative Council took up this question in 1937, its members spoke openly of the need for the government to retain total control. They insisted that the governor himself be responsible for appointing officers to the institute's board of directors and have veto power over staff appointments. The attorney general commented, "There is a danger that an institution of this nature might develop along the wrong lines and might conflict with researches for work done by Government, so that a safe-guard in that direction" is necessary. The chief secretary agreed: "In a thing like this, it is essential that it should not get out of line." Sir L. F. Moore, the member in whose district the institute would be established, cautioned: "If this Institute is to be anything at all, it will take a good deal of watching and managing."[105]

When Hubert Young first discussed founding an institute in his territory with the secretary of state for the colonies in mid-1934, Sir Philip Cunliffe-Lister asked him to consult with the Colonial Office's expert advisors, Sir Thomas Stanton and Sir Frank Stockdale, "about the desirability and practicability of any research institute and whether there would be overlapping with other places."[106] These questions were especially important because research funds were so scarce. Stockdale suggested that a more centralized research institute in the territory might become a headquarters for the "present ecological survey." Because Young had spoken of his institute as being of benefit to all of Central Africa, Stockdale pointed out that "similar ecological work was required in Nyasaland, and elsewhere in Africa, and was definitely linked with any study of the habits and customs of the African peoples, and of their development."[107] If Young were to add "ecological research" to his Central African institute, Stockdale felt, it would strengthen the appeal. Similarly, Stanton told Young that "from the point of view of the medical services we should welcome the establishment in Central Africa of a medical research institution." Because facilities in Northern Rhodesia were "very much behind" other territories, Stanton encouraged Young to avoid "highly specialized research" and consider instead "the possibility of making provision for the training of Africans as medical assistants."[108] Despite the generally positive tone of these meetings, which left Young with the impression that both advisors were "strongly in favour of" the proposal,[109] Cunliffe-Lister remained "doubtful about this project. It *must* not interfere with existing Research, and it can't be paid for by the territories."[110] Young planned to issue a public appeal in Britain for funds that he envisaged would raise an endowment for the institute. Cunliffe-Lister questioned whether Young's proposal had sufficient substance to warrant a public endorsement from the secretary of state for the colonies.[111] He wrote to Young, "I must tell you that I feel there is a danger that you are going too fast. . . . Before any appeal is made to the public, . . . I should like you to furnish much fuller details of the scheme than are given in your draft appeal."[112]

Meanwhile, word of Young's hostility to the African Survey had reached members of the Survey's staff and executive committee. Young had already registered certain complaints with Hilda Matheson, the Survey's secretary, who had in turn suggested to Lord Lothian that he try "to get him a bit tame." According to Matheson, Young "is very proud of his idea of a research museum for his own territory, but evidently extremely averse to any form of co-operation with anybody outside his own patch of Africa. He is enough of an autocrat, I gather, to feel that it is nobody's concern but his own to know what goes on in his territory, and that all these busybodies, like the Merle Davis people and others, are nothing but a nuisance." She advised that "the right way to deal with him would be to assume his friendly interest and thus make it difficult for him to be obstructive."[113] Matheson's impression of Young corresponds with Audrey Richards's recollections from the 1970s when she remarked that the institute "probably appealed to his vanity and self-importance."[114] His personality notwithstanding, Young *was* governor of a British territory and could create obstacles for the Survey should he wish to do so. For that reason, Lothian, Huxley, and Oldham all met with him in an effort to "win over Sir H. Young to the [Malcolm Hailey] scheme."[115] Oldham's involvement was once again pivotal to the Colonial Office's perception of Young's project.

In addition to meetings with members of the Survey's executive committee, Young also met with Audrey Richards and Major Archibald Church, who had been a member of the 1924 East Africa Commission and was a strong proponent of applied science in British Africa. In a characteristically optimistic manner, Young then wrote a detailed reply to Cunliffe-Lister in which he claimed that his research institute had "been welcomed by all whom I have consulted." Listing each individual by name, Young offered the impression that the new project summary reflected a consensus among those consulted. "The proposal is to start with applied anthropology and ecology, the museum . . . providing the necessary link with natural phenomena."[116] The research undertaken there would "help me and my successors, and possibly other authorities, by providing expert advice upon the potential economic and political future of the two communities," "white and black." In case the secretary of state had any residual misgivings concerning overlap, he also claimed that Oldham himself "guarantees the support of the African Institute and welcomes the proposals generally."[117]

Expecting that his response would satisfy Cunliffe-Lister, Young seemed entirely unaware of others' caution. After meeting with the secretary of state and Sir John Maffey (the permanent undersecretary of state for the colonies) on October 3, 1934, Cunliffe-Lister agreed to allow the Colonial Office two months to determine its position on "whether the appeal might go forward."[118] At this meeting, however, the Colonial Office learned that Young had already written to Godfrey Wilson, then undertaking anthropological fieldwork in Tanganyika, as a potential director of the institute.[119] This news placed Cunliffe-Lister in a difficult position, but the extra time would allow the Colonial Office staff to gather more information.[120] Specifically, officials wanted to understand how Young's proposed institute related to "the research now being carried on by the International Institute of African Languages and Cultures on the 'five year plan' . . . and [by] the African Research Survey."[121] The person who could answer both these questions was Joseph Oldham.

Compared to H. L. Gordon's proposed research on mental capacities, about which Oldham was advising the prime minister at almost precisely the same time, Young's project aroused few objections. Oldham's main goal in assisting Young, he told Sir George Tomlinson, had been "with a view to removing certain crudities from the proposals" and making Young aware of the difficulty of securing "a first rate man as Director." While Young seems to have exaggerated the degree of support Oldham had offered, Oldham felt there was little "opposition or conflict between the work of the [IIALC] and the projected Institute at Livingstone." His dilemma, however, was that the research undertaken for the IIALC had yet to be applied in practice. Especially in view of his skepticism that "the appeal was not likely to be a financial success," Oldham said he "would rather like to test the value of the work now being done . . . before setting on foot a new agency with similar objects in view."[122] Regarding any relation between the "Hailey Survey" and Young's institute, Oldham felt it was "impossible to forecast the nature of Sir M. Hailey's report . . . and was therefore impossible to say whether . . . [the institute] would be found to conflict or be in harmony with Sir M. Hailey's ideas." The best approach for the Colonial Office, Oldham thought, was for the secretary of state to take "the line that he would prefer to await the outcome of Sir M. Hailey's survey before giving his support to local schemes. In other words, it might be better to see the general plan and then see how local Institutes might be related to it."[123] Oldham had offered the same advice to Ramsay MacDonald with regard to Gordon's proposals.

Cunliffe-Lister followed Oldham's suggestions. After hearing Oldham's views, he privately disclosed that he had never had real faith "in either the ability or the success of Sir H. Young's proposal." Yet he had "great confidence in Sir Malcolm Hailey" and "intend[ed] to give him [his] full support." "I will do nothing to prejudice . . . his work," he declared.[124] In communicating his refusal to lend his name to Young's appeal, Cunliffe-Lister emphasized the need to coordinate research strategies and to utilize limited funding to maximum effect. Several institutions were already undertaking research in agriculture, medicine, anthropology, and linguistics. However, "none . . . can be said to have reached full fruition." What was really needed was "a larger plan to which existing and future schemes may be related."[125] This was what he believed the African Survey would provide. For that reason, he wished to wait until it was issued before drawing any conclusions about new proposals. Young seemed to take this rejection diplomatically and wrote that he would "possess my soul in patience until Malcolm Hailey has reported." In the meantime, he proposed to move forward with plans for a Livingstone Museum through a local appeal in Northern Rhodesia, which met with Cunliffe-Lister's approval.[126]

What makes these negotiations noteworthy to a history of scientific research in British Africa is the way the African Survey was used first as a gatekeeper for the Colonial Office, allowing it to hold various research proposals at bay, and second as a genuine means to inform its overall research strategies for its African territories. That social anthropology and ecological research were linked in Young's early proposals, as a result of his consultation with members of the Survey's executive committee and the expert advisors to the Colonial Office, supports the view that both disciplines were considered crucial to successful economic development. Although ecology as an explicit research field was later dropped when Young's proposal was resurrected in 1937, Sir Cecil Bottomley commented, "I don't know where ecology begins and applied anthropology ends."[127] The real question that bears asking with respect to the fate of Young's proposal is how it managed to overcome Colonial Office opposition and gain the stamp of approval of Malcolm Hailey *before* the final reports for the African Survey were even published. The answer should help explain the research priorities gaining credence within the inner circles of the African Survey's staff and advisors, which made it possible to reject the Kenya Colony proposal on mental capacities on the one hand and embrace Young's proposal on the other.

Much of the work conducted on behalf of the African Survey prior to February 1935 did not involve Malcolm Hailey directly.[128] The bulk of the preparatory research was coordinated by Hilda Matheson and Lord Lothian in conjunction with the Survey's various committees, specialist advisors, and members of its inner circle. Consequently, Hailey was unaware, until he returned to England from India in the winter of 1935, that the Survey had already been positioned by the Colonial Office as an arbiter concerning African research proposals, and specifically as "an excuse for withholding support to Sir Hubert Young's scheme for a Livingstone Memorial Institute [and] . . . the suggested investigation into the mental capacity of the East African native." This fact caused Hailey "dismay" for two reasons. First, he was "anxious not to find himself committed ahead [of time] to the considerations of specific questions or proposals, and [second] he felt that it would tend to embarrass his relations with the administrations if the Survey were to be used as a means by which their schemes might be indefinitely held up."[129] Above all, Hailey knew that if his Survey were to be a success, he needed to have smooth diplomatic relations with the African territories he planned to tour later that year. He understood that if he accepted the role of project evaluator, it would undermine his ability to be impartial. Nonetheless, Hailey's protests to the Colonial Office did not stop its senior staff from waiting for his "report" before making critical decisions concerning future research, nor did it relieve Hailey from being central both to the mental capacity debate and to the proposed Livingstone Memorial Institute. Even at this stage, the Survey had assumed a significance to African and imperial affairs far surpassing any other single project.

A number of individuals influenced Hailey's opinions concerning social anthropology during the years 1934 to 1938. Foremost among them were Joseph Oldham and Bronislaw Malinowski. Yet, Malinowski's relationship to the Survey was at times in question. In 1935, he even lamented to Reginald Coupland, "I am regarded as taboo!"[130] Describing this conversation to Oldham, Coupland wrote, "he clearly feels that Hailey should have gone and sat at his feet. He should have come, for instance, and been instructed about land-tenure. . . . I assured him that we had not consulted any other anthropologist, and did what I could to convince him that he was under a misapprehension. . . . Might not Hailey go consult him about the matter?"[131] While Malinowski's sense of his own marginality was exaggerated, other scholars, and in particular individuals such as Margery Perham and later Lucy Mair, were being called upon to bridge the gap between questions of colonial administration and social anthropology.[132] Perham and Mair, both fellows of the IIALC during its five-year plan, were deeply influenced by Malinowski's instruction and writing.[133] Yet, unlike many of their IIALC colleagues, who held posts in anthropology, Perham and Mair first taught in the field of "colonial administration."[134] This made their work on indirect rule and culture contact particularly relevant to the African Survey.

Just two weeks after she participated in the meeting of anthropologists for the Survey in March 1934, Perham delivered an important address to the Royal Society of Arts in which she discussed anthropological research in the context of "Some Problems of Indirect Rule in Africa."[135] Indirect rule, according to Perham, was not some vague and amorphous set of ideas, but an evolving political philosophy that rested on clear legal principles. It was *"a system by which the tutelary power recognizes existing African societies, and assists them to adapt themselves to the functions of local government. . . . The judicial aspect of indirect rule [was] not only the recognition of native law and custom, which, of course, obtains and must obtain even where there is direct rule, but the recognition of native law courts."*[136] Problems arose in indirect rule when recognition, in practice, was "faulty and incomplete," making "mutual comprehension and co-operation extremely difficult."[137] Responsibility for these deficiencies lay in the hands of political officers, who, according to Perham, "were not qualified to carry out what [was], after all, expert anthropological research." Not only were officers "often denied the time or the conditions to carry out their preliminary investigations," but their other responsibilities, such as collecting taxes, building roads, and supplying labor, often put them at odds with a genuine understanding of the societies over which they governed.[138] According to Perham, "Accurate information was not to be achieved by questioning a group of influential natives for a day or two and writing down their replies, but by patient research. . . . It is by winning the confidence of people who think they have only too much reason to distrust us that they can be induced to tell us, or, more usually, to show us what they really want." In Perham's view, "prolonged inquiries" conducted by "skilled anthropologist[s]" were required to help colonial administrations avoid the tendency of political officers to "misunderstand" and "mishandle native institutions," which allowed indirect rule "to deteriorate into a policy of mere preservation."[139] The prime reason Perham was so willing to criticize the "defects and difficulties" of indirect rule, she said, was that the "Colonial Service [was] in the peculiar position of a bureaucracy which can hardly be checked from above or below."[140] Lacking any form of political accountability raised the stakes of colonial administration: getting it wrong could lead to serious injustices.

Attending Perham's lecture were a number of individuals who later advised the African Survey: Lord Lugard, William Macmillan, William McGregor Ross, Thomas Drummond Shiels, C. K. Meek (a former anthropologist for Nigeria), and Sir James Currie. There were also several African participants, many of whom were in England working toward higher degrees.[141] Notable to the discussion afterward were the strong feelings Perham's comments about anthropology aroused among her audience and the harsh criticisms that the African members of the audience leveled at indirect rule. Macmillan felt Perham showed "excessive regard" for the new field to the detriment of her first calling, historical study. "Anthropology, he was afraid, was going to lead us astray in dealing with human material."[142] Stella Thomas, a young lawyer from Nigeria, after condemning indirect rule for "making puppets of African chiefs," announced that she "was not in favour of anthropologists. Africans were not curious to be studied in order to find out from where they came." What they really desired was "sound education" that could lead them toward "self-government."[143] Another graduate student from the Gold Coast, Modjaben Dowuona, who was preparing a paper on witchcraft and colonial legislation for the upcoming International Congress on Anthropological Sciences, endorsed Thomas's comments and spoke of the "irritating and discriminatory treatment" professional Africans received at the hands of administrators. Echoing comments Perham had herself made in other contexts, Dowuona continued with "a word about the anthropologists":

Professor Macmillan appeared to limit anthropology to the study of the so-called primitive peoples. He (the speaker) would extend that definition to include the study of the white races, whose manners, customs and institutions were not always easy for Africans to understand. . . . He would like to see, at no distant date, young Africans similarly trained, who would study the white peoples, especially the English, their customs and institutions, and interpret them to the world.[144]

Even the chair of the session, Thomas Drummond Shiels, said in closing that he "was a little suspicious of the anthropologists." While he acknowledged that "many of them were good and helpful both to people and government," he felt they "tended to look backwards rather than forwards." He also thought "there was rather too much feudalism and aristocratic privilege in [indirect rule] for his liking." He "had never been able to feel that there was such a cleavage between the essential outlook of Africans and ourselves as to make it necessary to delve deeply into local superstitions and folk-lore in order to devise proper lines of progress in government and administration."[145]

These speakers' ambivalence toward anthropology and anthropologists arose from several sources. First, they were referring to a variety of different experts when they used the term "anthropologist." With the exception of William Macmillan, who had taken part in IIALC work in the early 1930s, most of the other participants were not thinking of social or functional anthropology when they made their criticisms. Second, in much the same way that there was confusion and controversy over the fundamentals of indirect rule, there was also confusion over the objectives of "the new anthropology." By March 1934 few of the IIALC researchers had published their studies, and they were still quite marginal to the operations of colonial administrations. There were certain discrepancies between what its promoters promised and what they could actually deliver. Finally, there was a growing recognition that anthropological studies could be perceived to exacerbate or exploit existing inequalities. As Stella Thomas made clear, being an object of study was not a desirable end in itself: "There must be real co-operation and real understanding." At the crux of the matter was an imbalance of "power."[146]

In her response to the audience's comments, Perham felt compelled to "say a word in defense of anthropology." "The younger generation" of anthropologists had different goals in mind than those suggested by some of the speakers; they "were no longer looking at Africans as specimens, nor seeking to preserve the past for the convenience of their researches." Of course, she acknowledged, anthropology was not a panacea for the problems of colonial rule, but it could help to illuminate "existing conditions" and in that sense "would be of the greatest value to Africa." Above all, she agreed that anthropology should be a reciprocal endeavor. "It would be most interesting to carry out Mr. Dowuona's suggestion for anthropological research in this country by Africans."[147]

What might be overlooked in this exchange concerning the usefulness of anthropology was Perham's willingness to admit that British rule had, in many parts of Africa, already destroyed social structures that had operated in a profoundly "democratic spirit." That these institutions had gone unrecognized both in the early days of colonization and at the present time was, in her eyes, part of the travesty of European rule and the misguided actions of "impatient, revolutionary white men."[148] As she herself admitted, it was not a pretty picture she painted. Her willingness to support indirect rule as a means of appropriate governance rested on the assumption that it could only work if conducted as a "partnership."[149] To be genuine, such a partnership would require a much deeper understanding than existed of the complex, varied, and ever-changing social and economic relations in the British territories. The best means to this end, in her opinion, was social anthropology.

Perham's views were similar to Lucy Mair's. In her work on "native policies," Mair reprimanded colonial powers for their "complete failure to understand the nature of human society" and their "unquestioning belief in the inherent superiority" of their own modes of social organization.[150] That colonizers had been "indifferent to the nature of the societies" they were invading meant they had "attempted to mould a material of which they were ignorant into a form determined without reference to the ends which it would have to serve."[151] Like Perham, Mair acknowledged the pivotal role of Bronislaw Malinowski in drawing attention to the "function" of indigenous institutions in satisfying "social needs." The study of culture contact, spearheaded by social anthropologists, was concentrating on the extreme changes taking place in Africa without disparaging or ignoring Africans themselves. According to Mair, the field was "indispensable to persons who make the direction of such change their aim."[152] Tempering Malinowski's previous optimism regarding "social engineering," Mair concluded that the "rational control of human development," while it could be guided by a "dispassionate application of scientific principles," could never be brought entirely under control. Human affairs were simply "too intractable."[153]

Just a month after she completed her book, *Native Policies*, Malcolm Hailey "sounded Dr. Mair . . . about the possibility of . . . assisting [him] for the next nine or ten months with the preparation of [his] Report for the African Research Survey." Mair, he felt, possessed "special qualifications, including knowledge of the documentary material over a wide field, which would make her an invaluable assistant."[154] By the summer of 1936, Hailey was already being closely advised by Perham, Oldham, and Malinowski. Adding Mair as a "general drafter and sub-editor" stacked the deck in favor of social anthropology.[155] Any residual tensions between Malinowski and the Survey subsided at this time, thanks in large part to Hailey's correspondence and meetings with him. "Since your visit," Malinowski wrote in August 1936, "I have looked several times through your scheme of the report. It really seems to me that it covers all the points which I might have liked to suggest in a most satisfactory manner. . . . As far as the structure of the whole is concerned, I do not see any gaps, redundancies, nor indeed any points to cavil."[156] What probably appealed most to Malinowski was Hailey's emphasis on "the contribution of the trained and detached sociological worker" to the "type of knowledge or nature of studies that appears to be of most importance."[157]

Hailey had by now become a firm supporter of social anthropological research as an important dimension of colonial rule. When in January 1937 he was approached by Sir Hubert Young about the prospect of Young renewing his effort to establish a social research institute in Northern Rhodesia, Hailey "told him that he had no objection, and did not want it held up on account of his survey."[158] Tipping the scales in Young's favor, William Ormsby-Gore had become secretary of state for the colonies and wrote that he "certainly [did] not feel like opposing or putting any obstacle in the way of this project."[159] Not only did the Colonial Office pose no difficulties during this second round, but also Young was able to secure as signatories to his public appeal Ormsby-Gore, Lord Lugard (in his capacity as chair of the IIALC's Executive Committee), and Lord Hailey (in his capacity as director of the African Research Survey).[160] All three men were then involved in promoting the Survey.

In slightly later correspondence with Young, Hailey was more candid in expressing his views on social anthropology and the British government's lack of support for it. In June 1937, he wrote:

There is I think in the minds of those who approach this subject often some prejudice based on the fact that old-time anthropology was largely of an "antiquarian" interest, and sought to establish the origins of different races by indications of culture, physical characteristics and the like. Modern social anthropology is a study of their social organisation as we now find it, and of the changes which it is undergoing under the influence of new economic conditions. A study of this nature is necessary if we are to base on any sure foundation the improvement we are seeking to make in African conditions, whether it take the form of increase in agricultural production, or the betterment of diet and living conditions, or of health or of capacity for labour.[161]

Hailey lamented that so few institutions in Britain had actually undertaken such investigations; only the International Institute of African Languages and Cultures was conducting any systematic "inquiry into native conditions in Africa." Pointing out that no African government had, at that time, any official post of anthropologist, Hailey disclosed, "It is to my mind characteristic, I refrain from adding discreditable, that the main [financial] contribution to essential studies of this nature comes from America."[162] To remedy the situation, Hailey began to formulate his ideas concerning an African Bureau and Fund for research. Young, meanwhile, forged ahead with his plans for the Rhodes-Livingstone Institute, which by the early 1950s had become a model for anthropological research not just in Central Africa, where it was located, but also in East and West Africa where the British government sponsored two new centers for social and economic research.[163]

Before turning to an analysis of the African Survey's discussion of research into the "human factor," it will help to explore the history of the 1934 International Congress of Anthropological Sciences since it was there that anthropologists' criticisms received a much wider and more public airing. It was also there that questions of African magic, therapeutics, religion, and witchcraft once again took center stage.

Law, Witchcraft, and *Ngangas*: The 1934 International Congress on Anthropological Sciences

Scholarship on African magic, medicine, and witchcraft, since its inception in the late nineteenth and early twentieth centuries, tended to encapsulate three different kinds of concerns. The first, which was often brought to the fore by missionaries and administrators, focused on questions of religion and law: what kinds of spiritual beliefs prevailed in African societies and how were these maintained? Who adjudicated disputes and in what ways did they do so? What effects did new systems of religion and law, often introduced through imperial processes, have on African forms of jurisprudence and custom? Were these effects always positive, or could they be harmful? The second cluster of questions had to do with health and healing in African societies: what kinds of problems were brought before so-called witch doctors and how, if at all, did they handle individual and collective misfortune? Since physicians and colonial regimes were often unable to meet the health needs of African societies, particularly in rural areas, was it responsible to attempt to "eradicate" the practices of "witch doctors" with little or nothing to put in its place? Finally, the scholarship raised a number of questions about comparative rationality, causality, evidence, and logic: what kinds of conceptual frameworks and philosophies of nature underpinned African societies' belief in witchcraft and how did these attitudes compare to European frameworks? If African and European "therapeutic" systems were thought to be incompatible, or incommensurable, did that necessarily imply that one was true and the other false, that one was effective and the other ineffectual? Could these systems be allowed to coexist?

When critics today suggest that Europeans concerned with colonial Africa operated primarily through a dualism in which the rationality of science was on one side and the irrationality of magic or the supernatural on the other, they run the risk of ignoring significant consequences of the interplay between science and empire.[164] While such a dichotomy existed in the minds of many Europeans and even some African elites, disciplinary and legal changes in Britain itself, combined with anthropologists' skepticism about colonialism and industrialization, at times produced a complex and occasionally topsy-turvy outlook. Scientific research into auto-suggestion, hypnotism, and the role of "the mind as a therapeutic agent"; parliamentary discussions concerning Britain's own witchcraft laws and the need to pass a bill protecting those conducting "psychical research"; the growing social influence of ideas from psychoanalysis and psychology; and finally the ad hoc embrace of cultural relativism—all played a part in creating a scenario in which attitudes to witchcraft, psychic phenomena, therapeutics, and rationality were far more fluid and context-dependent than the concern about false dichotomies suggests.[165] This assertion is borne out in an examination of metropolitan debates on witchcraft in British Africa in the mid- to late 1930s.

In early August 1934, Britain's daily newspapers were littered with a range of provocative headlines: "1,200 Scientists Talk of Our Foibles," declared an article in the *Daily Herald*. "How Whites Harm Blacks," reported Newcastle's *Daily Chronicle*. "Licenses for Witches," announced the *Daily Telegraph*; "The Secret of Black Witchcraft," promised the *Sunday Dispatch*. The articles referred to the First International Congress of Anthropological and Ethnological Sciences, held at University College London from July

30 to August 4.[166] Attended by more than one thousand delegates from over fifty countries, including large contingents from Anglophone Africa as well as British India, the congress program offered eight panels devoted to African topics, more than any other region of the world. The series ended with a panel on "witchcraft and colonial legislation," which according to firsthand accounts attracted the largest audience.[167]

Plans for the congress had begun early in 1933 under the banner of the Royal Anthropological Institute. In their missive to the Colonial Office, the organizers noted that they were particularly interested in drawing attention to tropical Africa because policies there were still very much in formation.[168] After receiving input from anthropologists across Europe, India, North Africa, and the United States, the institute decided that it was best to combine anthropology and ethnology in order not to "offend our physical [anthropologist] colleagues and risk the failure of the Congress by excluding them."[169] The bulk of the papers and panels, however, ultimately addressed cultural and sociological topics: even the physical anthropologists present seemed more interested in discrediting rather than preserving various theories of racial supremacy.[170] As Grafton Elliot Smith, who chaired the congress's section on Anatomy and Physical Anthropology, was quoted as saying in a talk on the "Aryan fallacy," "We in this section are surely within our rights in criticising the fallacies that come into flagrant conflict with the generally recognised teaching of anthropological science."[171]

The African component of the congress was organized by Edwin Smith and a former government anthropologist from the Gold Coast, R. S. Rattray. In their announcement to prospective panelists, they wrote that "we wish to promote discussion that will illuminate those problems which seem more vital in Africa to-day. . . . We shall be particularly happy if Africans will come forward to take part in these discussions."[172] Among the topics they had in mind were "the essentials of indigenous African culture," "the validity of translations from native texts into European tongues and vice versa," "African marriage laws and customs and the effect upon them of contact with Western civilization," "the religious aspects of land tenure," "methods of investigating the mental aptitudes of African peoples," and "witchcraft and colonial legislation."[173]

Given his connections to various scientific and learned societies, including the IIALC, Smith was well positioned to recruit speakers on these topics. The most obvious choice was Edward Evans-Pritchard, who had recently published a series of articles on magic, "witch-doctors," and sorcery based on his fieldwork in the southern Sudan.[174] In 1932, Evans-Pritchard told the Royal Anthropological Institute that he undertook his research in order to help "white men whose work as missionaries, doctors, and administrators brings them into touch with Africans. These functionaries seldom regard the witchdoctor with favour and often enough are bitterly opposed to his activities. . . . I hope that this account will permit some one or two to see more clearly how easy it is to condemn and how hard to understand."[175]

When Evans-Pritchard began his fieldwork among the Azande in the late 1920s, he was drawing not only upon his own firsthand observations and interviews but also on the work of two missionaries based in the Belgian Congo. Robert Lagae, a Roman Catholic Bishop, had launched his career in the Congo at the turn of the twentieth century and, following the First World War, published scholarly work on Azande language as well as their religious and magical beliefs, including a major monograph in 1926, *Les Azande ou Niam-Niam: L'organisation d'Azande croyances religieuses et magiques, coutumes familiales*.[176] Lagae was elected a member of the IIALC in October 1930.[177] Father M. De Graer, also Catholic, had been involved in the northern Congo sleeping sickness campaigns from the turn of the century and published two detailed articles on the "art of healing" among the Azande in 1929.[178] It was from these articles that Evans-Pritchard derived much of his information about Azande therapeutics.[179]

Historians and anthropologists commonly associate discussions of African witchcraft and magic in the 1930s with the work of Evans-Pritchard. In disciplinary terms, there is no question that his research had far-reaching impact, culminating in his classic 1937 text, *Witchcraft, Oracles, and Magic among the Azande*. Yet Evans-Pritchard was actually based in Egypt between 1932 and 1934, teaching at the University of Cairo, and through an unexpected twist of fate was unable to attend the 1934 congress, despite having prepared a paper for the meeting. He was bitten by a rabid dog and had to undergo the long, painful course of preventive injections. Participants joked during the event that it was a case of witchcraft. Asked by a journalist to comment on the matter, Modjaben Dowuona, who was giving a paper on the subject, replied with amusement: "Why, because a mad-dog bites a scientist who is to speak on witchcraft, people should connect [the] two thing[s] and throw in also a touch of black magic and the supernatural, I cannot say. . . . I assure you the African witch doctors did not instruct the dog where to bite or how to go mad!"[180]

The most active behind-the-scenes organizer of the "witchcraft" panel was Frank Melland, whose 1932 article in the *Times* as well as a subsequent radio address for the BBC touched off the nationwide debate about African witchcraft. Melland had gone to Northern Rhodesia as a district officer in

1901 and was promoted to magistrate in the mid-1920s. After retiring from the Colonial Service, he continued to work as a consultant in the Office of the East African Dependencies for several more years.[181] Melland had two claims to fame: he was one of the first people in Anglophone Africa to write a semischolarly book on "witchcraft" beliefs, *In Witch-Bound Africa* (1923), which situated them in a wider context of religion and magic. (Charles Hobley was the other.)[182] By contemporary standards Melland's book appears sensationalist, but his intention was to dispel European misunderstandings about the nature of witchcraft. He told readers, "Further investigation might prove of unique anthropological value if we could only free our minds of prejudice and study the whole question scientifically."[183] As a result of these interests, in 1926 Melland was appointed by his administration to defend Tomo Nyirenda (also known as Mwana Lesa, "son of God"), one of the prime movers behind the Watchtower witch-cleansing baptismal movement that reached into Nyasaland, Tanganyika, the Congo, and Northern Rhodesia in the first decades of the twentieth century.[184] This movement resulted in more than two hundred people being drowned after witchcraft accusations. For his part in organizing these activities, Nyirenda was found guilty and hanged.

Melland's experience defending Nyirenda as well as trying a number of other witchcraft-related cases in his court ultimately had a radicalizing effect on his views. When he returned to London, he took his criticisms to a wider public through the press and public speeches as well as correspondence and meetings, reaching out to the secretary of state for the colonies and to Frederick Lugard in his capacity as representative to the League of Nations Mandates Commission and chair of the IIALC's executive council. Melland was the first ethnographer to comment publicly on the Kenyan witchcraft case, delivering an address on the subject in late February to the School of Oriental Studies as part of the Royal Anthropological Institute's 1932 winter lecture series.[185] As Melland wrote to Lugard later that spring, "I do not see how native courts can function adequately" in witchcraft accusation cases "until we give them a law which is adequate in native eyes. Our present laws in many parts (e.g. my own country, Northern Rhodesia, and Kenya) are hopelessly wrong" and "show abysmal ignorance on one point—confusing witches and 'witch-doctors.'"[186] Melland's experience in Northern Rhodesia had taught him that "witches" tended to be those who did harm, sometimes intentionally and at other times unintentionally, while witch doctors, or *ngangas* as he increasingly preferred to refer to them, tended to protect people from harm and restore order. If the law demonized both groups, where could people turn for help? He believed that one reason witch doctors' powers were increasing in various African territories was "because the natives feel our law is no help—it is even a hindrance—in their greatest need." "Our attitude on the subject appears to the natives as being unsympathetic, illogical (we teach them that Christ cast out devils, and dealt with those possessed), and—in plain language 'quite mad'. This is bound to make them suspect our other ideas for their welfare."[187] Distrust was partly responsible for what was happening among the Akamba in Kenya. Pointing to a recent report by the Native Affairs Department, Melland observed that some administrators were beginning to question the wisdom of rejecting the idea of witches or magic outright. According to a medical officer in the Ukamba Province, "To suggest . . . that witchcraft is powerless, is only to look absurd. . . . The only way to gain their confidence at all is to agree that witchcraft exists, and has very great power, but that white witchcraft can in many cases be proved to have even greater power."[188]

Shortly after Melland had published his own views on witchcraft, Clement Chesterman, a medical missionary, author of an *African Dispensary Handbook* (1929), and one of the lead physicians at the Baptist missionary station in Yakusu, Congo, elaborated on the intersections that could be drawn out between African understandings of healing and curing and those of Europeans. Chesterman asked: "Should we not welcome this intense, though misdirected, search for private and public health?"

> It only needs to be sublimated into useful channels to become a force of immense value. . . . This sublimation of witch-smelling into parasitology and of witch-killing into practical hygiene is so easy and fruitful by the simple process of cooperation with the educated youth of both sexes in the provision of modern medical and sanitary service for rural Africa. The charm to break the evil spell lies in the training and employment of the native African medical assistant.[189]

Chesterman gave a radio address on "The Curse and the Cure of Witchcraft in Africa" in July 1932, reiterating this point and continuing the public conversation on the subject.[190]

Melland thought both the medical officer and Chesterman had a point, but he was less cynical about ways to resolve the question of coexisting systems of knowledge. He insisted to Lugard, "We want (so I believe) to cooperate with the witch-doctors. (Some of the most able and astute headmen presiding over our courts *are* witchdoctors) but we want, by the provision of a really fair and adequate substitute, to remove from him one particular sphere of his activities." The substitute, presumably, was a more systematic legal system that would begin to reconcile British and African views of the law. "To do this we must think less of our pre-conceived ideas, in fact shed them altogether. Somehow or other we should make the law fit the people for whom it is made. It means a completely fresh orientation of our ideas."[191]

When Melland learned of the plans to consider African witchcraft and colonial legislation at the upcoming Congress of Anthropological Sciences, he was delighted. Finally, he thought, anthropologically minded experts and Africans themselves would have an opportunity to discuss the problems critically and openly. "What I readily want to emerge from this Congress," he wrote to one of his collaborators, Granville Orde-Browne, "is a UNITED EFFORT (not kudos to any individual) from which a case will be built up for reform of our official attitude, which will *command* attention by Government."[192] In January 1934 he had made his own views explicit in an article for the BBC journal *The Listener*.

> The *ng'anga* has been largely instrumental in maintaining order for centuries. . . . When [their] methods, like laws, no longer meet the case they are dropped or modified. These ng'angas are the ablest men in Africa: they wield tremendous power. . . . We register and license our own doctors and we legislate as to what they can do and what they cannot. I believe we could work on similar lines with the ng'anga and give a new and better lease of life to a wonderful profession, instead of delegating it to the extremely dangerous role of king of the underworld —using that term in the sense in which it is applied to our cities.[193]

As Orde-Browne observed, Melland's proposition was nothing less than a "revolutionary programme" which was likely to "meet with . . . much opposition from all quarters." Even if "it seems inevitable that we shall have to compromise," Orde-Browne agreed that this suggestion "will certainly provoke thought and criticism, which is what is chiefly wanted."[194]

Orde-Browne, who had worked as a district commissioner in Kenya and Tanganyika and was then serving as a "native labor" expert with the International Labor Office, was correct to anticipate opposition. Several prominent imperial administrators not only had a hard time imagining alternatives to existing laws but also voiced suspicions of anthropologists. During the 1932 debate, an administrator who had helped to pass one of British Africa's witchcraft ordinances wrote in response to Melland's article that in his "opinion, the less the 'man on the spot' is interfered with by anthropologists and others, the sooner will witchcraft die out."[195] Lord Olivier, who was an outspoken critic of racial prejudice and exploitation, wanted no part of an accommodating approach. The same day as Chesterman's radio address in July 1932, Olivier gave a speech in Liverpool in which he declared, "he was not at all happy about the idea of the present-day anthropologists who wished us to leave the superstitions of the black men alone so that they might study primitive man." Anticipating some of the criticisms later expressed during Margery Perham's 1934 speech on indirect rule, Olivier aligned himself implicitly with the views of Thomas Drummond Shiels and William Macmillan. He "realized the . . . mischief the belief in witchcraft and the evils associated with it too strongly," and argued that "you could not destroy those evils by tolerating them. The only method was the supplanting of witchcraft by the Christian religion."[196]

The problem with such definitive positions, according to the 1934 congress organizers, was that they overlooked existing flaws and contradictions in the laws of colonial states. In Melland's view, this was "a legal point of some magnitude." When he had worked in Northern Rhodesia, he had been instructed to ignore witchcraft beliefs: his court, he was informed, was not to take "cognizance of a non-existent thing." The leaders of "native" courts, he said, were likewise "forbidden to recognize it." "Therefore the native who believes in it . . . has no LEGAL REMEDY when he believes he is bewitched. He is therefore driven, by our acts, to have recourse to the witchdoctor, and him we proscribe. . . . This seems to me like saying to a man dying of thirst, I will not give you any water, and I will punish you if you get any from the only available tap. Is that morally or ethically defensible?"[197]

The panelists assembled for the discussion on "witchcraft and colonial legislation" offered a clear and uncompromising "no" in answer to Melland's question. Panelists included Evans-Pritchard, whose paper was read by Edwin Smith; Modjaben Dowuona, who spoke on "an African's perspective on witchcraft"; Louis Leakey, who was invited to address "what the African thinks of the way we are doing things"; Granville Orde-Browne and Clifton Roberts, who dealt with the "legal position"; and Melland himself, who promoted "registering and licensing" African healers.[198] As Orde-Browne explained to one of his magistrate informants in southern Nigeria, "In the group who are preparing the various papers on the subject, we hold strongly that the present attitude of the law is ignorant and wrong, and that to treat the whole subject from the crude standpoint that it is all [nonsense], is most dangerous."[199]

Orde-Browne was especially concerned to disaggregate the various African practitioners to whom "patients" turned, including those he referred to as priests, magicians, diviners, charm sellers, and "the much-disputed term 'witch doctor.'" He wrote to Melland, "The main and essential point, to my mind, is to differentiate between the witch, who does the harm, the patient, who suffers, and the witchdoctor, who defeats the witch and cures the patient or rights the damage. Our laws talk in a sloppy way about all three classes; we ought, to my mind, take a very different attitude towards the witchdoctor than that of the usual Ordinance."[200] While he admitted that "I may be wrong in my classification, and in any case I feel sure that it needs revision or recasting," he insisted that they "try to secure a more intelligent and sympathetic standpoint; the mere lumping together of all cases in which magic appears, as 'witchcraft' to be prosecuted as fraud, [was] the attitude that [he was] criticising."[201] Melland agreed emphatically about the need to correct "inaccurate terms and vague ideas." The subject was entangled with a belief in ancestors and cultural taboos; diviners, too, had multiple functions depending on where they lived and what services they were asked to perform. "Sometimes [herbalists] are quite separate from Diviners and Witchdoctors, but generally the witchdoctor is diviner and herbalist too."[202] Not only could anthropologists help tease out these various distinctions, but they could also establish a strong case that the laws themselves needed to be reformed. "In Africa," Melland insisted repeatedly, "it was first necessary to realise that the laws were made for the people, and not to act as if the people were made for the laws, as at present." "The British had grown up in the idea that British laws were perfect and sacrosanct, whereas they ought to be viewed primarily in the interests of the governed. The laws at present, especially in connection with witchcraft, were sometimes not in the interest of the governed, and they were certainly not imposed with the consent of the governed."[203] Seen from Africans' point of view, Melland thought British laws on witchcraft must appear "to be contrary to all reason, as well as a piece of injustice autocratically imposed by an alien ruling race."[204] He asked rhetorically, "Is it to be wondered at that natives think we are on the side of the witches?"[205]

Although Melland referred to himself as a "fanatic" on the subject of witchcraft and legislative reform, he was in good company during the 1934 congress. While not everyone was as openly critical, all the speakers drew attention to the inadequate ways administrations handled the question of "witchcraft." Evans-Pritchard's paper focused on definitions, helping to clarify relationships among "witchcraft, magic, witchdoctors, and oracles." "Notions of witchcraft comprise natural and moral philosophies. As a natural philosophy it reveals a theory of causation. Misfortune is due to witchcraft co-operating with natural forces. . . . As a system of moral philosophy notions of witchcraft define the moral sentiments and have great influence upon conduct."[206] Evans-Pritchard suggested that administrations rethink their approach not just to "witchcraft" but also to "African institutions in general." Other speakers expressed similar views during the congress's panel on African cultures. According to Dora Earthy, who had recently published a monograph on Mozambique with the support of the IIALC, it was necessary to recognize that "one of the essentials of indigenous African culture is the cult of health. This tends to assume a religious character, being closely connected with ritual, prayers to ancestral spirits, votive offerings to the spirits, magical medicines consumed or worn. It accounts for much of the influence of the native doctors. It permeates the whole life, especially the care of mothers for their children, indispensable links between venerated ancestors and their descendants."[207] Based on these views, the congress passed a resolution observing that "Africans' inherent religious and spiritual sense was, and should continue to be, regarded as 'essential' in African life and material progress."[208]

Because most colonial legislation still made no effort, as Clifton Roberts put it, to distinguish "between 'witch-doctor' and 'sorcerer', a distinction carefully maintained in Bantu society," and because legislation was never published in "native languages," it was easy to see why existing laws would seem so "unsatisfactory" to indigenous populations.[209] "The legal system we have imposed is entirely unsuitable," Roberts stressed during the discussion; "European conceptions of law and justice must be discarded as inapplicable."[210] Orde-Browne asserted during the discussion that one of the biggest problems was "the obvious lack everywhere of any Native co-operation in drafting the laws." He found "the super-imposition of the white man's laws and opinions without consultation with the Natives" "incomprehensible."[211]

Louis Leakey, in his presentation, deplored the miscarriages of justice that ensued when colonizers failed to take the time to understand the societies they governed. "What has always struck me most is that the African considers us to be the most illogical people living, and I must say he is not unreasonable in thinking so, for when the statements and actions of white men concerning witchcraft are looked at from the black man's point of view, nothing could be more absurd and illogical. . . . In short, to the African the attitude of white men to witchcraft is incomprehensible, illogical and selfish, besides being unutterably foolish."[212] Leakey saw more parallels than differences between magical and scientific knowledge. "English 'magic' included the clinical thermometer, the taking of blood tests and finger prints, gramophones, wireless and the camera." Europeans "use their own magic all the time," he insisted, which helped to explain why Africans were suspicious of colonial legislation that prohibited their own techniques. "It was naturally thought that we wanted a monopoly of magic for ourselves."[213]

Leakey's remarks illuminate an important subtext of the panel discussions, which focused on the reality of magic and the question of its compatibility with science and medicine. "Speaking as a scientist," Leakey told a reporter for *East Africa*, "I believe witchcraft does exist, and that laws based on the denial of its existence must necessarily be wrong. . . . There is no doubt . . . that the African can be killed by supernatural powers. Also such death can be prevented by calling in someone with superior supernatural powers." Others in the group expressed similar views. Edwin Smith acknowledged that "he saw nothing impossible in the transmission of power by thought."[214] Dauncey Tongue, a district commissioner who had spent twenty-one years in Uganda, told the audience, "Death by suggestion is fact. . . . I myself have seen no less than seven cases where men and women—otherwise sane and healthy—have simply withered away—because the native 'witch-doctor' has told them that they will die."[215]

Modjaben Dowuona from the Gold Coast, the only African student invited to give a paper, differentiated between witchcraft and healing. "Witchcraft is considered anti-social, whereas the practice of genuine African medicine is not so considered, and a genuine African 'medicine-man' does not practice witchcraft."[216] Dowuona called attention to the reality of the phenomena under discussion. He told a reporter, "Witchcraft . . . is a scientific fact, on a par with telepathy—all the laws of will, in fact, brought in to do harm or good. It certainly is not nonsense—as science has proved to its astonishment times without number. I would like to see more medico-psycho-analysts studying the problem, as this can no longer be dismissed as mere primitive belief."[217] Dowuona ended his paper with a call for more science education in Africa since he believed it would "enable Africans to develop a critical mind and give them confidence in themselves and ability to control their own environment; then they will be able to say, in due course, not that witchcraft is not marvelous, but that it is not so marvelous as the achievements of science."[218]

Jomo Kenyatta, who played an active role in the panel but did not give a formal paper, spoke openly of being the son of an Elder Kikuyu and the grandson of a "seer." His grandfather, he told the gathering, "knew healing and he could read—not only the things of the past, but the things of the future. And I can assure you that death can be willed on to one human being by another through suggestion and fear. I have seen cases with my own eyes. One cannot pretend to explain these things. But they are true, unalterable facts."[219] Like Dowuona, Kenyatta stressed that "Europeans confuse witchcraft with the skilled work of the medicine-man. African medicine is worth studying by European sciences. It is hard for Europeans to understand African magic. . . . Much harm is done by passing laws condemning all sorts of magic under the name of 'witchcraft.' . . . Europeans should consult Africans before they make laws about witchcraft, and train both Africans and Europeans in scientific medicine."[220] He endorsed the idea that governments consult with Africans before they revised laws relating to witchcraft, magic, and medicine.

In order to press the points arising from the discussion in a practical way, "the African section of the Congress" adopted a resolution urging "the appointment of a special advisory committee of experts (including Africans) on legislation affecting the native population in each territory under European rule, [which] would be of great value to the several governments in their task of assisting the progress of Africans."[221] Because the witchcraft panel was held on the final day, the text was presented too late to be endorsed by the entire congress, but the IIALC quickly took up the call, deciding within weeks that it would put out a special issue of *Africa* on the subject. Along with several of the presentations, they published new submissions by Edwin Smith, Siegfried Nadel (on Nigeria), Audrey Richards (on Northern Rhodesia), and Henri Labouret (on French West Africa). Given the congress's emphasis on African perspectives, Dietrich Westermann, the editor, also chose to solicit further material "from [IIALC] members in various parts of Africa," asking them "to consult experienced Africans in order to obtain their views on what they understand by witchcraft." They received and published twenty-one different submissions as a result.[222]

For many of the individuals who had taken part in the exchange, medical and philosophical questions were at the root of the issue. "If we are to understand the firm belief in witchcraft, we must first investigate the Africans' conception of 'medicine,'" Edwin Smith argued.[223] Even authors who focused primarily on legal and ethical questions, such as Orde-Browne, Melland, and Roberts, concurred. "The African witch-doctor must be accepted," insisted Roberts, and more attention had to be paid to "the medico-legal side of the case."[224] Coordinating the content of his paper with Melland's, Orde-Browne suggested, "might it not reinforce your case if you referred to the various forms of faith-healing which obtain in

European circles?" Perhaps thinking of the work of William H. R. Rivers, he continued, "The efficacy of firm belief on the part of the patient is of course generally admitted by the medical world, as is the existence of real ailments resulting from purely imaginary causes."[225] Looking back on the congress three years later, Orde-Browne commented: "The International Congress of Anthropology . . . was obviously quite opposed to the rigid skepticism of the established law. The discussions all served to indicate the desirability of further investigation and consideration of the whole subject."[226] Thanks largely to the close relationship that had developed between the leadership of the IIALC and the African Research Survey, Orde-Browne soon saw his wish fulfilled.

The African Survey's Synthesis on the Human Sciences and Its Verdict on Witchcraft

In its final report, the African Survey referred to witchcraft as "the outstanding problem of the lawgiver in Africa." Before considering this position in detail, we must first review its assessment of the kind of human research it deemed necessary for successful colonial governance. Anthropologists and their administrative allies played active roles in helping to draft the various chapters.[227] The starting point of the Survey's analysis was that, although Africans were culturally different from Europeans, they "individually show every variety of human feeling or ability." Difference in religion, custom, or "social observances, however marked, does not in itself convey any indication of an intrinsic divergence of character or capacity."[228] The primary objective of any study of "African peoples" should be to examine "the reactions of the African to the rapid changes in cultural and material environment to which he is being subjected."[229] Policies enacted in the absence of this kind of information were "apt to provoke unforeseen and unwelcome reactions."[230] "Knowledge," not uninformed "goodwill," was to be the "basis for administrative action." Otherwise, administrators risked perpetrating further "misinterpretation of native customs."[231]

The cultural expertise claimed by administrative officers and their objections to anthropological science were deprecated in *An African Survey*, as in the work of Perham, Mair, and Malinowski. The Survey dismissed the idea, expounded at times by Philip Mitchell and William Macmillan, that anthropological investigations of "native institutions" would inevitably "lead to an undue conservatism in maintaining custom which is either undesirable in itself or has outlived its use."[232] It reiterated the point that Hailey had made to Hubert Young, that African governments had "as a rule, given little direct support" or "encouragement to workers using the method employed by social anthropologists. Experience has shown that the reliance by governments on reports furnished by their own officers engaged in ordinary duties for certain types of information regarding African institutions may lead to serious misunderstanding."[233] The "administrative officer as an anthropological inquirer works under certain disadvantages," including his short term of service in a particular region, his position as government representative, and his lack of "long knowledge of any one locality." Even if an officer had adequate training and could overcome these other obstacles, "the professional anthropologist" would "provide a more complete picture of a native society."[234] Like Perham, the Survey stressed that officers' records, especially details of "cases in the high courts, magistrates' courts, and native courts," were important sources of information, in particular for "revealing native custom" and the "state of social institutions at moments of change." It recommended that these should be kept carefully, not only by administrators but also by technical officers and "literate Africans" who might wish "to place on record some of the traditions of their tribes."[235]

Governmental resistance to social anthropology warranted explanation. It was possible, the Survey conceded, that anthropology was still considered by some to be an "antiquarian" endeavor with little bearing on administrative activities. Other officials could be suspicious or "apprehensive" that "non-officials" would begin to play too large a role in the "day-to-day decisions of the administration."[236] These anxieties would need to be addressed explicitly, the Survey believed, if "governments [were] to take a more direct part in supporting sociological inquiry in Africa." The most obvious tension existed around research priorities and whether anthropologists or administrators would set them. Governments would have particular, and often relatively urgent, questions that concentrated on "aspects of native life which present immediate problems of policy." But thorough anthropological inquiry required a variety of preliminary, intensive, and general studies before it could undertake "any special investigation" that would address specific policies. Striking a balance between research that focused on the context and investigations into a particular problem would, the Survey acknowledged, be a constant challenge. "While no government could reasonably seek to prescribe the precise range which an investigator's inquiries should take, it may, with some reason, require that he should endeavor to throw all possible light on the problems which it selects as being of practical importance."[237]

The issues the Survey suggested anthropologists work on were those already being promoted by Malinowski with his students at the LSE and with the fellows of the IIALC: diet, land tenure, social customs, religious beliefs, marriage and kinship, law and justice, political organization, and economic relations. These matters were considered so closely linked that they could only be properly explored through "joint investigation" in which anthropologists worked collaboratively with other experts.[238] "Perhaps no area in the world offers more conspicuous proofs than are to be found in Africa of the necessity for recognizing the close interdependence of all scientific effort."[239] The Survey envisaged a matrix of environmental, medical, and anthropological research to address the most pressing problems of development in Africa: malnutrition, disease, demographic change, agricultural production, labor migration, land rights, and rapid social transformation.[240] Among the human sciences, social anthropology was singled out for investigating "the African peoples," having moved over the course of the decade from the margins of African administration to the center.

On the question of witchcraft and law, the Survey's authors—and here we know Hailey himself was directly involved[241]—adopted a position that resonated with the views of several of its advisors, including Malinowski, Orde-Browne, and by extension Frank Melland. "It would seem reasonable that the existing legislation on the subject should now be reviewed, with the purpose of bringing under the penal law only that use of wizardry or magic which has, or is intended to have, harmful consequences."[242] Orde-Browne had commented in 1934 that if Melland hoped to "*command* attention from Government, a Mills bomb is what you need."[243] He was joking, but his point was apt. Several participants at the International Congress had lamented that the Colonial Office sent no representative and seemed to take little interest in the outcome. As soon as the Survey took up the question, however, the Colonial Office almost immediately began a review of "witchcraft legislation."[244] The African Survey had literally become the backdoor to official policy making, as colonial critics hoped would be the case. Paraphrasing Orde-Browne's article on the subject, the Survey pronounced:

> The intentions of these laws, reflecting an attitude which regards magic and witchcraft as a form of fraud, are not readily understood by people whom large numbers regard the operations of magic as normal events of everyday life. . . . There is no clear-cut solution of the difficulties arising out of the divergence of European and African conceptions of law; all that is possible is to secure the utmost acceptance for the law by the fullest discussion before it is made, and by a careful study in advance of its probable reactions on African society.[245]

This perspective was not lost on the book's reviewers. The *Manchester Guardian* noted that "The 'Survey' indeed suggests that we should do well . . . to deal more sympathetically with African belief in magic."[246] Administrators and magistrates chimed in, calling for revision of the laws. In Nyasaland, Musgrave Thomas, Judge of its High Court, wrote to the Colonial Office's legal expert, Sir Grattan Bushe, urging that they undertake a "proper inquiry into the question of witchcraft and witchcraft laws, and I do not think the question can be put off much longer, especially in view of the general policy to grant natives much greater powers of controlling their own affairs." In Thomas's view, "The Witchcraft Laws in the African Territories make punishable acts which nothing can convince the natives are wrong. . . . To punish people for what to the native is sanctioned by custom creates a thorough distrust of our system of justice."[247]

Conclusion: Anthropology among the Human Sciences

Unfortunately for those who had campaigned long and hard to transform colonial laws, the Second World War curtailed their efforts. In the words of an influential player in the Colonial Office, the reform of witchcraft legislation was not a "pressing" issue. "It occasions great stress of mind to individual magistrates and other officers who come up against the problem of black magic, but it does not embarrass governments on the political plane. This being so, I think we should be wise to leave it for the present. What with the inquiries into land policy, native law, native administration and the future of legislative councils (not to mention the war), invitations to investigate equally complex but less urgent problems are likely to do more harm than good."[248]

While proponents of legislative reform were disappointed, shifts across British Africa took at least some of their views into account. These changes owed a great deal to the continuing influence of anthropological studies. When Audrey Richards began teaching at the University of Witwatersrand in the late 1930s, she included court cases on witchcraft from Northern Rhodesia where she had done her fieldwork. In sending her one of these, Thomas Fox-Pitt, a midlevel administrator, prefaced it with the comment, "You must tell your students that until there are better medical facilities in N. Rhodesia we don't like disturbing the people's trust in the only doctors they have, at least I don't. . . . Your class will observe that although the witch diviner may not be the ideal instrument of justice, the High Court can also make justice impossible."[249] Max Gluckman, who became director of the Rhodes-Livingstone Institute in 1942, wrote openly of "African science" in his review of Evans-Pritchard's *Witchcraft, Oracles, and Magic.* "The African has a wide technical knowledge which is accurate and scientific. . . . They are also knowledgeable companions, for every African knows much about his own laws, politics, history, art, medicine, so that conversation with them often becomes general and philosophical."[250] Had he lived long enough, Frank Melland would have been pleased to learn that by 1959, according to Richards, nearly 10,000 African healers were registered to practice in Ghana alone.[251] Even so, these results were not the revolutionary legal transformations that participants in the International Congress had envisaged.

Social anthropologists were more successful in securing a prominent place among the disciplines. As Hailey and the Survey committee moved forward with various plans for African research between 1938 and 1939, their priorities included not just the natural sciences, economics, and administration, but also anthropology. They envisaged a scheme that would combine the best of the IIALC with a more comprehensive "African Bureau."[252] Although even these plans had to be scaled back as a result of the war, with the founding of the Colonial Social Science Research Council (CSSRC) in 1944, funding became available for a far more concerted effort in the human sciences than had yet been undertaken.[253] Combined with the creation of the Scientific Council of Africa South of the Sahara in 1950, the CSSRC helped cement a transdisciplinary approach to human problems, with anthropology often at the helm.

Historians of colonial Africa have paid a fair amount of attention to the outlook and research of a handful of human scientists in the postwar period, particularly psychologists and psychiatrists.[254] More significant, but less noticed, are the several hundred studies funded by Britain's Colonial Development and Welfare Act based on the advice of the CSSRC.[255] (Table 6.1 provides a summary of the first eight years.) Lyn Schumaker, David Mills, Benoît de l'Estoile, and Frederick Cooper have considered different dimensions of these studies, often by taking up a specific institute, discipline, or set of questions.[256] The cumulative impact of CDWF-sponsored research, however, remains underappreciated, as do its multidisciplinary preoccupations. Given that the CSSRC's definition of "social science" included anthropology, sociology, linguistics, history, nutrition, colonial administration, human geography, demography, economics, and psychology, it seems worthwhile to consider the broader effects these studies had.

One striking feature of this research, which is noticeable more from

Table 6.1   Research in the human sciences: Colonial development and welfare grants, 1945–53

Territory	No. of projects	Amount (£s)
Africa (General)	11	79,830
Basutoland	1	2,985
Swaziland	1	850
East Africa (General)	8	297,046
Kenya	14	40,554
Tanganyika	1	4,185
Uganda (excluding research institutes)	8	17,845
Zanzibar	1	4,185
Somaliland	1	2,466
Central Africa (General)	4	128,750
Northern Rhodesia	2	7,943
Nyasaland	1	740
West Africa (General)	7	170,305
Gambia	3	6,696
Gold Coast	12	35,578
Nigeria	15	34,325
Sierra Leone	4	6,724
Total	94	843,822

Source: CSA, *Research in the Social Sciences in Africa South of the Sahara* (Bukavu: CSA, 1954), 44.
Note: Between 1945 and 1954 approximately £450,000 of the £843,822 went to three institutions: the Rhodes-Livingstone Institute (Northern Rhodesia, founded 1937), the East African Institute of Social Research (Uganda, founded 1950), and the West African Institute of Social and Economic Research (Nigeria, founded 1950).

its absence than its presence, was its lack of emphasis on the kinds of racial comparisons that preoccupied certain psychiatrists. When the Scientific Council for Africa South of the Sahara convened to review research in the human sciences across all of sub-Saharan Africa in February 1953 and again in September 1955, none of the seventy topics it identified for further study dealt explicitly with hereditary or racial questions construed in terms of inherent qualities. The vast majority of their recommendations— sixty-five of the seventy topics—dealt with economic, sociological, historical, and cultural concerns, including land use, urbanization, economic development, subsistence economies, regional histories, inter-territorial migration, the effects of large hydroelectric schemes, linguistics, African arts and literature, African therapeutics, and cross-disciplinary studies of "native laws and customs."[257] While they did advocate developing psychological tests for "educational and vocational guidance," this suggestion was pursued only on a very modest scale in Nigeria and the Gold Coast and was prefaced by a range of caveats about the limits of all intelligence tests.[258]

Their overall priorities were focused on sociological and cultural interpretations rather than biological and physical ones. Perhaps more significant, these projects increasingly drew attention to the effects of colonial rule. Development funds were being used not just to grease the wheel of empire but also to slowly erode its foundations. Anthropologists, who had worked for so long to stake out a claim to be included among the empire's experts, used their tools to chip away at it once on the inside.

CHAPTER SEVEN

A Living Laboratory: Ethnosciences, Field Sciences, and the Problem of Epistemic Pluralism

Africa is, as Lord Hailey points out, a vast living laboratory of biological and social experiments. . . . But, as [he] constantly reiterates, our scientific and sociological knowledge of Africa is extremely inadequate.

—Meyer Fortes, *Man*, 1939

While it is understood that scientific research must use universally accepted criteria, there is a general wish to bring scientific research closer to African problems
and traditions. Such an "Africanisation" of scientific research does correspond in
no small measure to many recent findings in the field of agriculture, nutrition,
medicine, and others, which tend to prove that certain traditional methods were,
in fact, perfectly adapted to the local environment. The growing importance of
synthetic traditions like ecology makes the scientists more ready to accept [this]
notion.

—Scientific Council for Africa South of the Sahara, 1957

The juridical conquest of tropical Africa at the end of the nineteenth century was a watershed moment both for geopolitics and for knowledge. It
transformed sub-Saharan Africa into an imperial laboratory where political, economic, and scientific experiments could be pursued with relative
impunity. These experiments had a lasting impact both within and beyond
the continent. Could we explain how international law or multinational
institutions of governance rose to prominence without being forced to include some account of the effects of Africa's partition?[1] Would it be possible to tell the story of the ways in which tropical medicine and social
anthropology were codified unless we considered how their practitioners
used Africa to justify their endeavors?[2] Certainly, no analysis of the history

of racial theories or of human origins would be complete without describing the effects of archaeological and disease research in colonial Africa.[3] African experiences and precedents loom large in any book on the history of international conservation efforts.[4] It would be difficult to evaluate the legacies of the colonial state and of colonial development if we ignored African cases.[5] In all of these examples, the continent of Africa and its peoples have been far more than an incidental backdrop: they provided the bricks and mortar of disciplines, theories, institutions, and even laws. As we unpack this history it helps us to see both how European rule affected tropical Africa and how African experiences shaped key elements of the modern world.[6]

Decentering Europe and Africanizing Science

Enrique Dussel has recently argued that "modernity is not a phenomenon of Europe as an *independent* system, but of Europe as center."[7] Recent imperial history has taken this point even further: key elements of ostensibly "modern" phenomena originated not in Europe but elsewhere, so much so that a number of scholars have taken to seeing colonies as "laboratories of modernity." This new angle of analysis has stemmed from a desire to unsettle Eurocentric biases in earlier historiography and has challenged narratives that see European developments spreading from center to periphery. Colonial cities, for instance, were a venue to work out new techniques of surveillance and hygiene; tropical islands helped cement new kinds of commodification and conservation; territorial conquest generated the phenomena of "concentration camps" and genocide; the tenets of liberalism and nationalism could not be imagined without experiences in extra-European contexts; imperial expansion enabled new understandings of self and identity to emerge—all this evidence forces scholars to resist arguments that assume that European countries were hermetically sealed.[8] To understand the interplay between metropole and colonies, we must study not only connections and networks that have tied European nations to the rest of the world but also those circulations and developments outside Europe that differed from and influenced trajectories within Europe itself.

Vernacular and Patriotic Science

This kind of close examination reveals four trends that this book has explored at some length and that each played a role in the process of *African-*

izing science. First, we find that the very people engaged in creating and maintaining structures of imperial domination in Africa were, ironically, among those who shared with postcolonial scholars a desire to "provincialize Europe."[9] In spite of their different motives and ideologies, it was they who began to question Europe's epistemic authority, challenging truth claims that accepted European examples and standards as the norm. E. B. Worthington's book, *Science in Africa*, was in part grounded in this premise: as he put it, Africa had much to teach Europe. Indeed, the push to decenter Europe was an enduring feature of overseas empire building and often included a turn toward *patriotic* and *vernacular* sciences.[10] The former suggests connections to state building while the latter implies an emphasis on local knowledge and cultural interaction. To pursue vernacular science was to emphasize ethnography and the significance of Africans' own natural and technical knowledge. Its proponents sought in their research to connect everyday forms of expertise, especially orally transmitted knowledge, to formal scientific systems.

By the early twentieth century, theorists' loyalties to the sites in which they produced knowledge could make them insist that Africa, that "enchanting abstraction,"[11] was a unique and important place in its own right. Some of these patriots could be exclusionary: settlers and officers who wanted to use the material of science to help their communities cohere, usually around a presumed racial identity. Others were nascent cosmopolitans who wished to ensure that European perspectives and evidence were not given unwarranted privilege. Many displayed a mixture of these qualities. One of the most overt expressions of this kind of patriotism was made by South African statesman and naturalist, Jan Smuts, who in 1925 gave the presidential address to the South African Association for the Advancement of Science. Smuts spoke of the dangers of developing scientific theories only in Europe: "The European situation is best known, it is the classic ground of science. No wonder that it has come to be considered the centre of the world." As scientists undertook research in other continents, Smuts felt sure the picture would change.

> While for the statesmen the problems of the African continent may become all-important during [the twentieth] century, it is more than probable that for the scientist also this continent will assume a position of quite outstanding importance. From many points of view, Africa occupies a key position among the continents of the world. . . . In many ways Africa is the great "scientific divide" . . . where future prospectors of science may yet find the most precious and richest veins of knowledge.[12]

Jan Hofmeyr reiterated these provincializing ambitions in his 1929 address to the British and South African Associations when he put out a call to "*Africanise*" science,[13] anticipating remarks made by Thabo Mbeki nearly seventy years later when he was president of South Africa.[14] Not only did Smuts and Hofmeyr, and many of their contemporaries, wish to reposition Europe, they also wished to challenge the supremacy of scientific perspectives developed in the Northern Hemisphere, a critique that a number of technical officers and fieldworkers in colonial Africa reinforced in the inter-war period.

Both an interest in indigenous knowledge and patriotic tendencies were evident in the archaeologist Louis Leakey's research and writings from the 1920s and 1930s. Joining a number of scientists who had firsthand experience in the field, Leakey emphasized vernacular science when he chose to align himself with defenders of Africans' "magical" and agricultural knowledge.[15] Not everything European scientists wished to introduce to Kenya was accurate or sound, Leakey argued; Africans had their own systems of knowledge that were worthy of defense. Leakey also enjoyed turning the tables on his European audiences, calling himself "more a Kikuyu than an Englishman," since he was born and raised in Kenya. He emphasized that colonial laws and administration were at times "irrational" and "grossly unjust and unfair."[16] Indeed, his fieldwork in East Africa led him to put forward the provocative hypothesis that Africa might be "the cradle of modern man."[17] Nothing could subvert Europeans' complacent sense of their own history so profoundly as suggesting that *Homo sapiens* originated in Africa. Leakey's counterpart in South Africa, Raymond Dart, also a patriot, expressed similar frustration with people's "false love of European literature, culture, prehistory and politics. The consequences of this misplaced policy have been fatal to African philology, African ethnology, African archaeology and African anthropology."[18]

Patriotic and vernacular sciences could cut many ways, undermining European hegemony, reinforcing "white" control, and even promoting "indigenous" perspectives. As African nationalists and social critics joined debates about the substance and consequences of scientific theories and redirected their arguments away from European and toward African audiences, they could produce patriotic science of their own, such as Jomo Kenyatta's ethnographic defense of "African medicine" and Nnamdi Azikiwe's critical analysis of "African super-science" in the 1930s. Should Africans' knowledge be called "superstition," or might it be better labeled "superscience"? Azikiwe himself was hardly sure. "If an African would only study the science of the West, and try to correlate the same with African science, an important contribution could be made to the studies of science."[19] Both men called for more rigorous investigations of Africans' theoretical and practical knowledge and both had formal affiliations with anthropologists in Britain. Kenyatta later became the first president of Kenya, Azikiwe the first president of Nigeria. If they shared nothing else, Smuts, Hofmeyr, Kenyatta, and Azikiwe all considered African experiences central to the process of knowledge production and state building.

Complexity and Interdependence

An analysis of networks and intellectual exchange during the colonial period reveals a second, equally important pattern: tropical Africa has served as a key site in which to work out a scientific discourse of complexity, interrelations, and interdependence, concepts that were at the heart of governmental and development interventions. This emphasis emerged as much from the transnational task of managing colonial states and directing the flow of information within and across African territories as it did from the interplay between field and laboratory sciences. The reason scholars have largely missed these patterns is that they are most visible only when one examines the interstitial spaces that linked Africa and Europe through the apparatus of science and empire. What might seem marginal to a single territory, such as Kenya, Nigeria, or Tanzania, appears central when examined in the context of inter-territorial networks that were attempting to coordinate the circulation of ideas and methods. The sciences of geography, anthropology, and ecology were most significant to this process, but so too were field epidemiology, tropical medicine, nutritional science, psychology, demography, and even archaeology.

By 1900, the African continent was the largest *colonial* landmass in the world, and its tropical states were also, on average, the youngest. While some colonial sites, such as British India or the Dutch East Indies, had comparatively well-established networks of laboratory facilities, British tropical Africa did not. That meant that field and laboratory sciences existed on a par and, in many locations, field sciences were conceptually more important to scientific research. In epistemic terms, because field sciences and "teamwork" were so central to imperial coordination, approaches that stressed interactions and integrated analyses occupied a more prominent, even dominant place. In metropolitan centers, meanwhile, researchers increasingly had to confront reductive tendencies in laboratory methods and scientific reasoning, which privileged parts over wholes and drew conclusions about organisms in the absence of an analysis of wider interactions.

We should avoid drawing too sharp a dichotomy between laboratory and field methods, since many disciplines relied on both techniques, but in British tropical Africa, no matter which method individuals identified with more, the sites in which they produced knowledge went well beyond the boundaries of any institutional lab. The "field" was essential to everyone.

By the end of the twentieth century, with the introduction of computer modeling, global information systems, and more thorough aerial surveys, distinctions between field and laboratory sciences were being broken down and reconfigured yet again. These techniques made it easier to study African topics theoretically and from a distance. The decades of "foot safari" that technical officers during the colonial period pursued have fewer parallels in the present than we might expect, even among scientists based permanently in tropical Africa. Colonial field officers' proximity to the land and constant contact with peoples in their environs enabled at least some of them to produce a kind of vernacular science that is increasingly difficult to imagine, much less produce, today. Although their rhetoric tended to outpace their performance and their ambitions were sometimes at odds with one another, their methods relied upon conceptual tools that stressed the interdependence of human and nonhuman nature. When administrative control of Africa merged with scientific field research in an effort to bolster the interventionist power of colonial states, it inadvertently helped to generate epistemic communities that thought about African development in terms of complexity and local specificity.[20] This was not a critique from the margins of power, but a view that emerged within the epicenters of colonial and metropolitan control. Thinking like an empire generated—at least in the case of British tropical Africa—an emphasis on research into interdependent problems.

Localizing Knowledge and the Problem of Authenticity

A third pattern this book has explored is the imperial imperative to *localize* knowledge. Soils, deserts, forests, diseases, climate, species, and even witchcraft beliefs all underwent scientific scrutiny during the colonial period. While the research might have left a light footprint in terms of its intensity—in 1937, for instance, there were more scientists active in South Africa than there were across *all* of British tropical Africa—it still had lasting effects on how people thought about these physical features. "Forests in Africa," Julian Huxley wrote in a report prepared for UNESCO in 1961, "should be conserved not merely for timber-production and watershed protection, but as being among the chief attractions of a National Park system, as well as providing natural laboratories for ecological study."[21] In much the same vein, geological formations including the Rift valleys and Mount Kilimanjaro not only became tourist destinations but also points of scientific interest. Harry Johnston's and John Gregory's research in these areas in the 1880s and 1890s, respectively, started a long-standing tradition of scientific fieldwork.[22] Kilimanjaro, and the arc of mountains to which it is connected, was recently deemed a "biological hot spot" with "the highest density of endangered animals anywhere on earth," while the Rift System was described as "Africa's most interesting continental-scale land-form" where "research methods . . . [and] modern team projects are multidisciplinary in their approach."[23] With the ascendancy of plate tectonic theories, which had originally inspired Jan Smuts to call Africa the great continental divide, "Africa has become something of a test-bed for tectonic and geomorphic models over the past decade or so."[24] Even its human populations, considered en masse, encouraged demographers to question the underlying methods and assumptions of their discipline in new ways. The historical demographer, John Caldwell, observed when he reviewed the effects colonialism had on population levels: "From the mid-1950s large-scale demographic surveys were carried out in greater numbers in Africa than anywhere else in the world . . . to an extent that the challenge of African data has revolutionized methodology in demography."[25] The same challenges, especially in terms of oral evidence, have arguably revolutionized historical methods.

The act of localizing knowledge meant that many field analysts were careful to distinguish the particular from the universal, or the local from the general. The site specificity of phenomena forced this task upon them. Both field scientists and laboratory experts were evaluated by their peers in relation to how well they were able to disaggregate specifics of place and people; most could not afford the oversimplified cognitive frameworks that scholars sometimes attribute to them. Their emphasis on site specificity did not mean that their interpretations were correct or even good in a normative sense. Nor did it mean that they easily accommodated competing epistemologies. Yet we should not assign all their "mistakes" to an ostensibly narrow and exclusive definition of science's universality, because this argument rarely holds up under scrutiny.

Yet the process of localizing knowledge was paradoxical: as insights derived from African experiences were folded into the fabric of scientific disciplines, as well as the policies of colonial states, Africans themselves were rarely at the helm of decision making. While there was much give and take in epistemic terms, there was little social parity. This meant that while colonial states and scientific projects might privilege "indigenous knowledge," often calling into question any simple dichotomy between "Western" and non-Western science, empires in Africa could not entirely escape this dichotomy. Lurking in the background were always other questions: could science be Africanized without African scientists?[26] Just what counted as science, and who would decide?

The same year *Science in Africa* was published (1938), the Indian Science Congress Association produced a review of *The Progress of Science in India during the Past Twenty-Five Years*. More than two-thirds of the authors were South Asian; so, too, was the editor, Baini Prashad, who then directed the Zoological Survey of India.[27] All the Indian contributors held senior scientific posts directing research institutes or university departments. In the entire history of scientific research in colonial sub-Saharan Africa, considered across *all* the European regimes, Africans were never present in such numbers and never had the autonomy to define research questions or set scientific priorities. Thus, even as colonial administrators and technical officers attempted to supersede Africans' systems of land management, healing, disease control, and social organization, they occasionally tried to understand these dynamics through scientific analysis. The imperial emphasis on Africans as ethnographic subjects produced, ironically, some of the most influential studies of subalterns' ideas and practices. As various specialists came to terms with local specificity and local knowledge, they helped to codify a range of vernacular approaches to research, which also unexpectedly destabilized the foundations of imperial rule.

The relative paucity of scientific experts, combined with the weakness of colonial states, produced an enduring ambivalence in the minds of many— African and non-African—about what would be considered authentic and legitimate forms of knowledge within the continent. The proliferation of studies of "indigenous" and "endogenous" knowledge systems since independence reflects one part of this legacy.[28] A tacit reluctance to accept African scientists as viable and powerful actors in their own right forms another part.[29] On the one hand, anyone who unproblematically promotes "science," at least in humanist as opposed to policy circles, runs the risk of being characterized as naive and unsophisticated at best and neocolonial at worst. On the other hand, advocates of science in policy circles, while sometimes sensitive to questions of "indigenous knowledge," rarely grapple explicitly with the challenges of a pluralistic approach.[30]

For those who remain dubious about modes of reasoning and interventions that they label "science," one of their main alternatives is to suggest that other kinds of knowledge, experience, and logic are more legitimately *African*. The fact that currently "there are more African scientists and engineers working in the U.S.A. than there are in Africa" only feeds into these patterns, since the "brain drain" itself deprives the continent of a critical mass of scholars.[31] Taken to extremes, an emphasis on "ethnoscience," which is, in part, a construct of colonial relations, can lead to a cul de sac where people search in vain for essential insights that only Africans might possess.

These disputes over authenticity and representation come across vividly in the behind-the-scenes controversy touched off by a 1943 film script for *Men of Two Worlds*, inspired by Joyce Cary's novel, *The African Witch* (1936), and sanctioned by Britain's Colonial Office. The script focused on a Tanzanian protagonist, who had been educated in England and had decided to return to his native country as a member of the colonial service. Fittingly, it depicted a sleeping sickness epidemic and the colonial state's "development scheme" to combat its spread. The central conflict in the film is between the colonial biomedical model and, as the film puts it, an African "witch doctor" and his followers.[32] One of the more vocal critics of the film was the London-based West African Student Union (WASU), which the film's producers approached early in the process to help identify potential actors for the cast. WASU objected in the strongest possible terms to the film's depiction of witchcraft. Its Nigerian secretary general, Ladipo Solanke, remarked pointedly: "The fact that the script makes no reference of any kind to, or, denies, in a way, the existence, among Africans of any organised indigenous system of curing diseases and of doing any purely African social and health services, at once, makes the whole representation unrealistic, and harmful."[33] Solanke was a strong proponent of ethnographic studies, writing in WASU's journal in 1926 that "it is the duty of Africans to investigate and give to the world in suitable literary form, an account of their history, laws, customs, institutions and languages. Without such material it would be impossible for us to know what lines our development should take."[34] To gain a better understanding of African healing, WASU's executive council suggested several African-authored texts and urged that "the promoters of this film . . . go and read very carefully here the whole of a book written by George Way Harley, M.D., entitled *Native African Medicine*."[35] Harley, an American medical missionary who had been active in Liberia since 1925, began and concluded his book with summaries of a range of British anthropological and ethnographic literature on the subject.[36] In fact, he drew upon the work of key protagonists affiliated with the International Institute of African Languages and Cultures in London.

What WASU members were never told was that the public relations staff in the Colonial Office arrived independently at an equally interesting point. Writing to the director of the film, Thorold Dickinson, an official conveyed his colleagues' opinions at precisely the same time WASU put forward its own objections. "The Department concerned here has seen" the script, the official wrote, "and the only comment they have is that . . . they do not think Kijana [the lead protagonist] would have referred to witch-craft as 'nonsense' unless he were trying to put across a superior attitude, which is not in keeping with his character as portrayed. They think that his education in England should have taught him, as it was supposed to teach us, not to regard witchcraft, of which we cannot understand the origin or mechanics, as nonsense."[37] That the lead actors in the film were from the Caribbean, Nigeria, and Uganda and at least one, Robert Adams, was friends with the actor and political activist, Paul Robeson, should help us appreciate that disputes over epistemology in colonial Africa were anything but simple controversies.

Epistemic Decolonization and Auto-Critique

Finally, the process of producing new knowledge and synthesizing its results often had the unexpected and unintended effect of prompting epistemic *decolonization*. Scientific research could subvert imperial ideologies and practices in unpredictable ways. This trend should not be elided with political change: however weak colonial states were, they clung to their existence powerfully and, when necessary, with brutal force. Yet epistemic decolonization, combined with the kinds of auto-critique scientists periodically expressed, weakened the rationale for empire and had lasting indirect effects on the political will to maintain colonial structures of rule. Indeed, at least some of the research sanctioned by Britain's "imperial organism" and its subsidiary colonial states following the Second World War bears surprising resemblance to existing research priorities in African studies in European and North American institutions today.[38] If we focus less on anomalies, exceptions, and egregious examples—studies that admittedly stand out and demand attention—and more on the quotidian and mundane priorities of British Africa's research institutes and technical departments, a different picture emerges. This comprehensive view forces us to acknowledge decolonizing impulses that were present in tropical Africa long before the political "wind of change" swept across the continent in the 1960s. Serious analysis of egregious examples has much to teach us about the nature of colonial power and its attendant ideologies.[39] It also reveals important insights about the objectifying drive that underpins many scientific disciplines and activities. Yet these examples should be situated alongside the vast body of scientific literature produced during the colonial period in and on Africa that did *not* fall into these patterns. Only then can we fully appreciate the norms and standards, both explicit and unstated, that guided research in the human, environmental, and medical sciences in tropical Africa. All of this research is open to critical scrutiny, but if we wish to make claims about its specifically *colonial* nature we need a sound overview of what projects were supported and how influential they were. The challenge that scholars and social critics face is to explain the coexistence of radically different points of view within scientific debates, some of which fed into a colonial status quo, while others transformed and undermined it. Indeed, one of the questions *Africa as a Living Laboratory* raises for further debate is the extent to which scientific knowledge and its production played a role not just in the *construction* of empires but also in their *dismantling*.

Development and the End of Empire

The Second World War was a tumultuous turning point in African colonial history.[40] By 1945, many European powers had become much more vulnerable economically and had begun to emphasize anew tropical Africa's commercial potential. At a conference of African governors sponsored by the Colonial Office in November 1947, Sir Stafford Cripps, then British minister for economic affairs, made the point explicit: "We have for a long time talked about the development of Africa, but I do not believe that we have realized how, from the point of view of world economy, that development is absolutely vital. . . . In Africa, indeed, is to be found a great potential for new strength and vigour in the Western European economy, and the stronger that economy becomes the better Africa itself will fare."[41] The continent of Africa continued to be thought of as a laboratory for scientific research as well as a place to experiment with different kinds of development plans.[42] Yet economic imperatives accompanied by changing institutional and disciplinary arrangements tended to raise the stakes of these interventions and open the door for new approaches to development that were radical departures from previous models worked out in British colonial Africa. Cripps remarked to the governors, it is "essential that we should increase out of all recognition the tempo of African economic development," and for that purpose "we must be prepared to change our outlook and our habits."[43]

Scientific Expertise and Pan-African Coordination

At the time of Cripps's speech, several African territories had already drafted ten-year development plans in order to qualify for the newly available funds from the 1940 and 1945 Colonial Development and Welfare Acts. Most continued to grapple with the issue of scientific expertise, considering the question at the 1946 British Commonwealth Scientific Conference in London, which was preceded by a three-week-long Empire Scientific Conference sponsored by the Royal Society. At both gatherings, the only region of the world to warrant a special committee on "fundamental scientific research" was Africa.[44] Endorsing a recommendation originally made by the African Research Survey, these committees concluded that the time had come to establish an inter-imperial African Research Council and agreed that this body might best be launched at a conference in South Africa, following up on the various pan-African scientific conferences held there in the interwar period.[45] Out of this recommendation came the 1949 African Regional Scientific Conference in Johannesburg, which endorsed creating the Scientific Council for Africa South of the Sahara (CSA) and helped to establish its long-term research priorities.[46]

It took delicate negotiations to ensure that the CSA's headquarters were situated not in South Africa, as scientists and officials there had hoped, but in the Belgian Congo. During the 1949 conference British officials privately expressed serious concerns with the new policies of apartheid and asked the South African government to give them "unqualified assurance that should any of our African Governments wish to send African representatives to attend any meetings of [a] continuing organisation they would be treated on *exactly* [the] same footing, both for official and social purposes, as European members of delegations."[47] Rather than concede, South African officials struck a compromise in terms of the CSA's structure. "The choice of the Belgian Congo was really part of a bargain," explained Britain's assistant undersecretary of the Colonial Office, "whereby the Secretary-General of the [Scientific] Council (key post) was British, the Chairman South African, the Vice-Chairman, French, the site of the Secretariat, Belgian."[48] Between 1950 and 1960, CSA and its governmental counterpart, the Commission for Technical Cooperation in Africa (CCTA), sponsored more than two hundred conferences on various specialist subjects.[49]

Epistemological and Methodological Continuities and Discontinuities

The period following the Second World War involved key continuities with the interwar period, especially in scientific fieldwork. Imperial coordinating bodies and a number of colonial states continued to embrace transdisciplinary and inter-territorial approaches to research and development in which ecological interdependence and sociological complexity remained underlying suppositions. This pattern held not just for projects sponsored by the Scientific Council for Africa South of the Sahara but also for several small-scale development projects and nutritional studies, including the Anchau settlement scheme in northern Nigeria, the Azande development scheme in southern Sudan, the Nyasaland Nutritional Survey, and Tanganyika's Bukoba Nutritional Survey, all of which were launched in the late 1930s.[50] These projects, which were often considered models for other territories, integrated key dimensions of scientific fieldwork that had been pioneered in the interwar period. The Azande scheme, for instance, not only began with an ecological survey, but included a nutritional survey and was also the subject matter of one of Edward Evans-Pritchard's own PhD candidates; ecology, nutrition, and anthropology intersected actively in the field.[51] Perhaps more significant, the same patterns held for territorial development plans, including those for Uganda and Northern Rhodesia and for inter-territorial training programs and research institutes in which ecologists and anthropologists often played central roles. We can also find continuities in postwar journals founded to guide colonial policy making, including the short-lived journal *Farm and Forest* for West Africa.

The discontinuities are even more important, however, because they signaled changing disciplinary patterns, including the ascendancy of economists and engineering experts as advisors to development plans. Some of these changes were by design, since metropolitan leaders began to foster a new kind of rhetoric that urged colonial states, to quote Cripps again, to "go ahead with as many large-scale experimental schemes as possible."[52] Other changes were by default and resulted as much as anything else from the increased professionalization of key disciplines, including anthropology and ecology, which led specialists to pull back from applied research.[53] In much the same way as the African Survey had done, those in charge of coordinating research after the Second World War struggled to synthesize insights across disciplines and among different experts. Their successes were usually partial and incomplete.

Endogenous Models of Development and the Problem of Scale

The push to promote large-scale development efforts did not always sit well with those who had pioneered the bottom-up and small-scale development strategies in the interwar period. Colin Trapnell, who had spent nearly twenty years in Northern Rhodesia working on its territory-wide Ecological Survey, remembered butting heads with the Colonial Office's new agricultural advisor, Sir Geoffrey Clay. Clay was interested in promoting centralized and mechanized agricultural projects, which fundamentally undermined what Northern Rhodesia's leadership had pursued since the early 1930s. Trapnell believed he was transferred out of Northern Rhodesia in 1950 because he disagreed so strongly with Clay's approach.[54]

Yet many technical officers and scientists embraced the postwar ethos, agreeing that the time had come to make a more concerted effort to scale up development. Officials had a new sense of economic urgency. What opponents and advocates of massive development schemes had to consider was *how* to accomplish them. Would models essentially derived from British tropical Africa serve as templates, or would approaches drawn largely from experiences in other parts of the world, including the United States, South Africa, and India, become the framework? Even with the emphasis on scaling up development and with economists playing much more significant roles, environmental, medical, and human scientists in specific African territories as well as in metropolitan institutions were expected to weigh in on territorial plans. That an ecologist, E. B. Worthington, was selected to prepare Uganda's first ten-year development plan in 1944 underlines this point.[55] He had influential counterparts in the Gold Coast (John Phillips), Nigeria (Thomas Nash), Northern Rhodesia (Colin Trapnell), and Southern Rhodesia (John Ford).[56]

These sorts of choices would have been much less imaginable in most of Europe, North America, Latin America, and Asia. We need only compare the secretary of state for the colonies' 1945 dispatch on "Colonial Development and Welfare" with the official report of the inaugural meeting of the World Bank and the International Monetary Fund the following year to see some of the stark differences in their approaches to expertise and development.[57] While the colonial secretary continued to emphasize broad and interdisciplinary scientific research, the individuals attending the World Bank meeting stressed economic and financial knowledge and looked no further.[58] The British delegation to the World Bank meeting was led by John Maynard Keynes as well as three representatives of the Treasury and two representatives of the Bank of England. A clearer endorsement of the importance of economic knowledge could hardly be found.[59] Yet during the colonial period in tropical Africa, there was no such economic hegemony: states and their imperial coordinators relied on a much wider array of experts and explicitly endorsed a teamwork approach to problem solving. Only in the last decade has the World Bank begun to reinvent itself as a "Knowledge Bank," traversing some of the same epistemic ground as European empires almost a century ago.[60]

On the eve of political independence, British colonial states in tropical Africa oscillated between supporting small-scale and bottom-up strategies in which subaltern knowledge and practice played an important role and large-scale interventions that gave little credence to Africans' ideas or methods. Although the larger projects attracted more attention and required considerably more resources, they were never the sole model. Still, we must not lose sight of the fact that both approaches were externally imposed and allowed individual Africans at most a minor role in decision making. Charles Wilcocks, an epidemiologist who worked in Tanganyika between 1927 and 1937, offered in his memoirs a disarming statement of what he saw as colonialism's fatal flaws. "We, the colonizers, dominated the indigenous people. This is difficult to justify, just as it is morally indefensible for one person to dominate another to the extent of making him lose self-respect through servitude." He acknowledged that many Europeans, himself included, had "committed the fault of confusing technical, mechanical knowledge with intelligence. Because we had motor cars, aeroplanes, medical knowledge and perhaps above all a written literature, we considered ourselves much superior to the Africans."[61] Seventy years earlier John Hobson had made a similar point about colonial power relations, observing that the "vice" of empire became visible in "the autocratic idea and temper which [it] commonly assumes and by a certain spurious temporary strength which emanates from military organization."[62] Given these hierarchical arrangements, many African nationalist leaders and their allies, including Fabian socialists, were often ignorant of the extent and longevity of transdisciplinary debates about African development. Nor were they aware of the social and ethnographic interactions that research officers pursued in their long-term field studies. This gap goes some way to explaining why few African elites tried to adopt what can rightfully be called *endogenous* strategies of development for modern African states. Even technical officers themselves were often oblivious to the heterogeneity of approaches since their spheres of expertise and their administrative status often kept them only partially informed.

Elite versus Vernacular Knowledge

Political decolonization entailed another contradiction for scientists in terms of the roles that elite and subaltern knowledge would play in the future of African state building. This point is illuminated by a declaration made at the annual meeting of the Scientific Council for Africa South of the Sahara in 1957, the year the Gold Coast (Ghana) achieved political independence. The council recorded, without irony, that it was now in a "race against time." African territories were rapidly achieving "complete independence," interrupting the council's efforts at the "Africanisation of scientific research." In the eyes of the CSA's leadership, Africanizing science meant ensuring that scientific research conducted *within* the continent would be given a privileged position over research on Africa produced in other parts of the world. "This, of course, is a different position than the one prevailing until very recently, when scientific research in Africa was but a subsidiary branch of scientific research carried out in Europe or in America." They hoped to ensure that "the governments which are based in the African region will have a natural tendency to consider research on Africa undertaken outside Africa as an addition to their own research *actually based in the continent* and submitted to its day-to-day conditioning to the natural factors of Africa."[63] Their goal to reposition African institutions and scholarship and to decenter knowledge produced elsewhere led them to conclude that recent research had shown that Africans' "traditional methods" of "agriculture, nutrition, [and] medicine" "were . . . perfectly adapted to the local environment."

The CSA's council was correct to recognize that political transitions were likely to undermine existing scientific structures and insights. As E. B. Worthington pointed out in the mid-1990s, European technical officers were often unable to train their successors, much less ensure that incoming political leaders understood the role and function of different institutes and coordinating conferences, before they had to leave Africa.[64] Above all, what was lost sight of in the political struggles and recalibrations of the 1960s and 1970s was the extent to which various traditions of fieldwork within the continent during the colonial period had attempted to achieve a rapprochement between indigenous, orally transmitted expertise and field research. Although this point was not always easy for scientists or administrators to admit, learning to accommodate and build upon Africans' vernacular knowledge could hold the key to successful development. Inadvertently, field scientists, aided by weak colonial states, constructed a space for epistemic pluralism to persist. In the process, they helped to codify the conceptual categories—indigenous, local, traditional knowledge—that would sit uneasily side by side with the other kinds of scientific expertise. Whether and how to sanction pluralism officially were questions they left by and large to state leaders, scholars, and social critics in the postcolonial period.

Many critics of the relationship between science and empire in colonial Africa tacitly assume that international development agencies and even independent African states were the repository of colonial ways of thinking about African environments and peoples. This is partly true, but it misses another important pattern. After the Second World War, many innovative ideas and methods, which had developed with the sanction and funding of imperial governments, found refuge within programs of "African studies." At the same time, scholars and fieldworkers distanced their perspectives from colonial power and ideologies of domination. They also occasionally constructed mythological histories of "official colonial doctrine," as we saw, for instance, in the work of John Ford.

Why have scholars interested in African imperial history so often misunderstood these patterns? In the decades immediately following independence, intellectuals in Africa and elsewhere were understandably concerned to show the destructive effects colonialism had on African societies and self-perceptions. A rich literature in environmental, social, and medical history attempted to reconstruct Africans' "survival strategies" and methods of healing and land management. Numerous scholars highlighted asymmetries in the production of knowledge, which they argued made it easier for colonizers to dismiss Africans' own knowledge claims. For these reasons, many historians and anthropologists from the 1970s through the 1990s spent a great deal of energy both trying to recover what they believed colonizers overlooked and to correct what they got wrong. *Africa as a Living Laboratory* reevaluates this work, particularly in terms of the history of ideas, and points to the colonial origins of a range of critiques that scholars in African studies have long suggested are products of postcolonial thinking.

Reflections on an Intellectual Journey

I have written this book in dialogue with several rather distinct audiences: development experts and social critics who rarely get a chance to consider the past; scientists who care about the nature and history of expertise; and scholars, students, and citizens who share an interest in exploring the dialectical relationship between African colonial conquest and the production of knowledge. Astute readers may notice a certain tension that is inherent in my own interpretations of the evidence. The book's opening epigraphs make it obvious where my fundamental sympathies lie, yet I am more interested in stimulating questions than in providing definitive answers. Must scientific truth always be paradox? Is any society exempt from living in a "system of approximations" with respect to the natural world? What does it mean for an "outdoor laboratory" on the scale of the African continent to be of its "own local kind"? Did "scientific knowledge of the facts" and anthropological perspectives really generate a "new power of self-criticism" among experts?

Some readers may find my emphasis on scientific epistemologies unsettling, in particular because, in the words of the colonial critic Norman Leys after he read *An African Survey*, it is difficult to find much "mention of what Africans themselves think or wish."[65] Others may feel that the picture I paint of scientists' theories and criticisms is overly coherent when, in fact, their perspectives were far more fragmentary and impressionistic. Still others may believe that there is too much attention to rhetoric and too little attention to practice or implementation. All these critiques are, to some degree, fair.

Nonetheless, the research that forms the core of this book has enabled me to answer questions that hitherto I could only wonder about. It has also helped me find new intellectual currents and conceptual tools that will allow me to redress some of these imbalances in future studies. Most important, this book addresses a set of heated debates in African studies that will likely remain unresolved for some time to come. These center on the effects and legacies of "colonial science" (a formulation that I hope I have demonstrated is untenable), the construction and relevance of indigenous knowledge and ethnosciences, the role and methodologies of economic development plans, the centrality of environmental change to human livelihoods, the agency of disease in delimiting and constraining sociocultural organization, and the nature of medical and epistemic pluralism.

If I seem to be a defender of "science," there is some truth in that perception. Both in my career as an environmental and social justice organizer and in my academic pursuits, I have been struck by the popularity of scientific critiques. The patterns that these studies have revealed around colonialism, power, domination, hegemony, and violence are extremely important and should never be discounted. Several protagonists in my research had their own insights along these lines, and I admire them for it.

Rarely, however, do we find a study that explores the question of scientific contributions to liberation or human freedom. We have moved beyond a time when scientific knowledge, facts, or rationality could be discussed as unproblematic contributions to "human progress." Yet, the longer I consider questions of social change and human emancipation, the more I am convinced that truth matters and cannot be "relativized" out of existence. There is something beautiful in truth, even if that truth is approximate, paradoxical, and ephemeral. There is also something mocking and unpredictable in the way both the nonhuman and human worlds react to our attempts to control them. These lessons have been learned in epochs before ours, in many other cultures, as well as in the "modern West." Critiques of science, at least in fields outside the history of science, sometimes tend to disregard the multifaceted and complex histories of thought in Europe itself.

The anthropologist Tom O'Meara recently drew attention to several hazards associated with extreme epistemological relativism. Such a standpoint, he argued, could become

> an instrument of subjugation, not of liberation. No matter how righteous the cause, it is dangerous as well as false to claim any "special way of knowing" about the physical world that produces "knowledge" which is immune to empirical testing and logical contradiction. [66]

This book should attest to the fact that I am well aware of the potential for knowledge to be used in ways that subjugate. Truth is a highly contested issue; determining who is right and wrong in circumstances where competing epistemologies exist, each with its own domain of accuracy and authority, is difficult indeed. Central to these issues is the question of power: who has it, who wields it, and who benefits from it. Yet, even with all these attendant problems, the idea of truth cannot easily be forsaken.

Africa as a Living Laboratory opens up a range of new questions about the history of science in colonial Africa and about the relationship between ethnosciences and field sciences. In both legal and epistemological terms, it seems too late to put the genie of "traditional knowledge" and its conceptual cognates back in the bottle. One need only read through the publications on this subject by special committees of the member states of the African Union, the World Health Organization, or the World Intellectual Property Organization, among others, to understand how pervasive the concept has become. This raises the question, born in part out of the end of empires, whether this knowledge is or ought to be commensurate with

APPENDIX

African Colonial Service Employment, 1913–51

The Colonial Office Lists

These tables were compiled using the annual publication *Colonial Office List*, which was a detailed summary of every employee appointed to serve in one of the territories overseen by the Colonial Office.[1] This excluded India, which had its own administrative structure managed by the India Office as well as Egypt and the Anglo-Egyptian Sudan, which were supervised through the Foreign Office. Because the British South Africa Company had its own bureaucratic structure and because in 1923, Southern Rhodesia was granted attenuated self-governance, the *Colonial Office Lists* never included a full summary of Southern Rhodesia's personnel. I have therefore elected to omit Southern Rhodesia from the tables since its figures would have been misleading. I have also omitted the territories overseen by South Africa's High Commissioner for similar reasons. That leaves the colonial dependencies.

Statistics are always relational, and scholars who would like to use these numbers for their own interpretive purposes should do so with care, particularly if they wish to draw conclusions about a single territory or a single element of the service (i.e., "Medical" personnel). (Unless scholars knew, for instance, that personnel for the railways and certain public works projects in Uganda and Kenya were often listed only under "Kenya" it might lead them to draw flawed conclusions about Kenya's higher numbers under "Infrastructure.") These totals include *both* personnel appointed by the Colonial Office (usually with the nominal approval of the secretary of state for the colonies) *and* certain select staff hired within the territory. That means it is more inclusive than a simple listing of "officers." It also means the totals are actually *larger* than they would be if I had omitted clerks, female nurses, and subordinate staff in each of the three categories, Administrative, Technical Services, and Infrastructure. Depending on the territory, most of these subordinates were originally from Europe or hired from within settler populations; in East Africa there were also a number of South Asian subordinate staff members. In West Africa, there were occasionally African staff on the employment rolls as well, but these numbers were relatively low and were not always labeled as such. (Sometimes an individual is identified as "African Medical Officer," or "African Surveyor," but I have also found listings for individuals who I know to be from the region, but who were listed with no special designation.)

While the totals for each year are complete in terms of the *Colonial Office List*, that does not mean they are entirely accurate. For instance, when Britain took possession of Tanganyika (German East Africa) during the First World War, it took two full years before the Colonial Office began to report personnel for Tanganyika. British personnel were on the ground before this, but their numbers were not reported in any detail until 1920. The same applies to Northern Rhodesia. The *List* began to include personnel figures for Northern Rhodesia before the First World War, but even after it was transferred from the British South Africa Company to the Colonial Office in 1924, its totals seem to be incomplete, especially for the technical services. Not until the early 1930s does the list appear to reflect the employment figures more accurately. When military personnel were included in the *Colonial Office List*, I have added them to the totals under Administration, but the vast majority of territorial staff was civil rather than military. (As a number of imperial historians have pointed out, this was a key difference between the French and the British in tropical Africa.) These lists do not include totals for the West African Frontier Force or for the King's African Rifles since such totals were never reported in the *Colonial Office Lists*.

It is sometimes easy to forget, when studying the history of a single region in tropical Africa, just how minimal the total personnel was for each territory. Colonial conquest and rule had such dramatic effects that it can be tempting to ascribe more power and coherence to the whole process than may be warranted. One of the striking patterns that comes across from this overview is that the scope of the totals had a surprisingly small spread. The year 1913 represented a capstone of pre–World War I growth, adding up to 4,039 personnel. And the year 1931 was the capstone of the growth in the 1920s, totaling 7,447 individuals, a figure that was not to be matched again even in the years between 1946 and 1951. (The last year in which the *Colonial Office List* published complete totals was 1951; after that, it provided only summaries. There were no *Lists* published between 1941 and 1945 or in 1947.)

If there was a Second Colonial Occupation of British tropical Africa following the Second World War—to use Anthony Low and John Lonsdale's
phrase—it had less to do with sheer personnel numbers (at least *official*
numbers) and more to do with the kinds of resource extraction and development plans introduced in this period. It is possible that personnel totals
increased dramatically between 1952 and 1957 (the year Ghana achieved
independence), but given that the Recruitment Officer, Ralph Furse, was already signaling the end of empire in 1951, that possibility seems unlikely.[2]
Even if the 1951 totals doubled over the next five years (which would have
had *no* historical precedent for the personnel in tropical Africa), that would
still have put the total at only 14,000 individuals.

The Categories and Their Totals

In order to smooth out idiosyncrasies in the way numbers for different territories were reported—since each territory had its own bureaucratic structure, which developed in an ad hoc fashion—I have elected to report the
figures in three broad categories: Administration, Technical Services, and
Infrastructure. This allows me to give a rough breakdown of the staff in
each territory and the different state functions they were employed to address. There is nothing hard and fast in these categories although in the
aggregate they do tell us something important about imperial priorities as
a whole. For instance, by the interwar period Administrative personnel accounted for roughly 50 percent of the totals—on average—and the Technical Services and Infrastructure staff were roughly 25 percent each.

In my doctoral research I used recruitment totals—compiled by Anthony Kirk-Greene in his excellent statistical and analytical summary of
the Colonial Service, *On Crown Service*—in order to determine the relative
importance of the technical services. It turns out that recruitment totals are
actually somewhat misleading since they make it seem as if the technical
services accounted for more than 40 percent of all appointments. The *Colonial Office Lists*, I believe, give a more accurate thumbnail sketch of the
importance of the technical services.

In terms of the categories they include the following:

§ Administration: executive branch, district and provincial officers, judicial
staff, the police force, education, and customs.

† Technical Services: medicine (including public health and sanitation),
agriculture, veterinary and game, geological, and surveys and land.

¶ Infrastructure: public works, railways, harbors and marine/ports, mines, and posts and telegraph.

Anyone interested in studying the history of engineering in British tropical Africa would do well to look more closely at the personnel in the Infrastructure category. There were hundreds of engineers employed across the different territories, but ironically they rarely had the impact—in epistemic terms—that researchers in the technical services did. (Nor did their professional societies get involved in tropical African affairs to the extent that other scientific societies did.) Those personnel who were given responsibility for producing *new* knowledge about tropical Africa were employed in the technical services, not in public works, railways, or posts and telegraph. Even within the category "technical services" there was a division of labor between those hired largely to intervene—physicians, public health experts, agricultural demonstrators, soil erosion officers, foresters—and those hired to study and represent—ecologists, soil scientists, entomologists, bacteriologists, epidemiologists, and occasionally demographers, psychologists, and anthropologists. Although many technical personnel often represented *and* intervened (undertaking agricultural or disease control campaigns and also publishing certain results in scientific journals) there was only a limited number of technical staff who were hired *explicitly* to represent. Despite their relatively small numbers, these individuals often had a disproportionate impact on inter-territorial and inter-imperial thinking.

By the interwar period, several territories included research institutes that existed to serve more than one territory. Their personnel are included in the territory in which they were based. I have chosen not to break out these numbers since they could actually be somewhat misleading. When I provide figures for research officers in chapter 3 (see table 3.2), I am drawing upon another set of lists published by the Colonial Office specifically for research officers who were conducting entomological, agricultural, botanical, veterinary, and forestry investigations. There was no equivalent list for medical research officers, unfortunately, although the Colonial Office did publish four reports on medical research (in 1928, 1929, 1930, and 1947), which included such lists as well as a bibliography of studies published by Colonial Service personnel (in 1947).

Two idiosyncrasies within the technical services also need to be explained. First, most territories had personnel devoted to tsetse and trypanosomiasis control and research by the interwar period. These projects were often funded through "public health" grants, which would place them in the "medical" category, but because of their mixed functions, I have elected to place a number of the research officers working specifically on tsetse flies in the veterinary/game category. These totals never amount to more than two-dozen individuals and have the greatest impact on the figures for Tanganyika (where the Tsetse Research Department was based). Because the Colonial Office considered these public health activities, I discuss their work in the "Medical Laboratory" chapter.

Second, because of the way each territory listed personnel responsible for land, it is often difficult to distinguish between those staff conducting topographical and cartographic research and those undertaking more administrative and quotidian functions. That makes the "Surveys and Land" category anomalous since it is obvious that the majority of those listed were not surveyors, using this term in its technical sense. The Colonial Office published periodic reports on its surveying activities in tropical Africa, starting in 1906; it also sponsored conferences of Survey Officers in 1931 and 1935 (see table 3.4). Anyone interested in the history of geodetic and cartographic surveying would do well to begin with these as well as territory specific reports. From a summary of the Ugandan surveying efforts, for instance, it becomes obvious that African surveyors played an important role.

Totals for each category are listed in ITALICS and are CENTERED.

The subtotals for the Technical Services are INDENTED to the right and are listed in PLAIN text.

Table A.1   Colonial Office List, 1913

Category	Nigeria[a]	Gold Coast[b]	Sierra Leone	Gambia	East Africa Protectorate	Uganda	Nyasaland	Totals
Administration§	649	1,208	230	61	223	158	90	2,619
Tech Services‡	182	190	60	9	89	66	30	626
Medical	143	129	54	9	37	30	17	419
Ag/Vet/Game	11	29	3	0	25	15	12*	95
Forestry	14	7	3	0	2	3	1	30
Surveys	14	25	0	0	25	18	0	82
Infrastructure¶	204	324	101	19	70	55	21	794
Totals	1,035	1,722	391	89	382	279	141	4,039

[a]Nigeria was not amalgamated officially until 1914; these numbers combine Northern and Southern Nigerian totals.
[b]Gold Coast figures have been double-checked; this is one of the only years when its personnel dramatically exceeded Nigeria's.
* This figure includes 5 staff for a one-time scientific commission to the territory to investigate correlations between game, tsetse flies, and trypanosomiasis.

Table A.2 Colonial Office List, 1918

Category	Nigeria	Gold Coast	Sierra Leone	Gambia	Kenya	Tanganyika	Uganda	Northern Rhodesia	Nyasaland	Totals
Administration§	*501*	*355*	*315*	*87*	*252*	*0*	*142*	*0*	*83*	*1,735*
Tech Services†	*231*	*83*	*99*	*23*	*121*	*0*	*77*	*0*	*30*	*664*
Medical	165	66	90	20	45		37		16	439
Agriculture	18	11	4	0	13		17		7	70
Veterinary & Game	2	0	0		21		0		3	26
Forestry	24	3	5		6		2		1	41
Geology & Labs	1									1
Surveys & Land	21	3		3	36		21		3	87
Infrastructure¶	*342*	*189*	*157*	*21*	*100*	*0*	*43*	*0*	*20*	*872*
Totals	*1,074*	*627*	*571*	*131*	*473*	*0*	*262*	*0*	*133*	*3,271*

Table A.3 Colonial Office List, 1919

Category	Nigeria	Gold Coast	Sierra Leone	Gambia	Kenya	Tanganyika	Uganda	Northern Rhodesia	Nyasaland	Totals
Administration§	*434*	*342*	*291*	*73*	*279*	*0*	*126*	*0*	*78*	*1,623*
Tech Services†	*195*	*73*	*98*	*25*	*116*	*0*	*75*	*0*	*22*	*604*
Medical	135	56	89	22	40		35		12	389
Agriculture	16	11	4	0	13		16		4	64
Veterinary & Game	2	1	0	0	20		0		3	26
Forestry	21	3	5	0	7		2		1	39
Geology & Labs	1	0	0	0	0		0		0	1
Surveys & Land	20	2		3	36		22		2	85
Infrastructure¶	*315*	*173*	*136*	*22*	*103*	*0*	*42*	*0*	*21*	*812*
Totals	*944*	*588*	*525*	*120*	*498*	*0*	*243*	*0*	*121*	*3,039*

Table A.4 Colonial Office List, 1920

Category	Nigeria	Gold Coast	Sierra Leone	Gambia	Kenya	Tanganyika	Uganda	Northern Rhodesia	Nyasaland	Totals
Administration§	600	150	291	75	82	131	113	107	83	1632
Tech Services†	184	89	96	24	156	33	83	3	25	693
Medical	116	70	87	21	59	31	38	1	13	436
Agriculture	15	10	4	0	14	1	19	0	6	69
Veterinary & Game	0	1	0	0	21	1	0	1	2	26
Forestry	19	3	5	0	23	0	4	0	1	55
Geology & Labs	0	0	0	0	5	0	2	0	0	7
Surveys & Land	34	5	0	3	34	0	20	1	3	100
Infrastructure¶	334	311	73	7	98	55	51	2	17	948
Totals	1,118	550	460	106	336	219	247	112	125	3,273

Table A.5 Colonial Office List, 1921

Category	Nigeria	Gold Coast	Sierra Leone	Gambia	Kenya	Tanganyika	Uganda	Northern Rhodesia	Nyasaland	Totals
Administration§	*738*	*294*	*141*	*86*	*338*	*229*	*139*	*105*	*109*	*2179*
Tech Services†	*180*	*131*	*36*	*26*	*196*	*81*	*107*	*2*	*33*	*792*
Medical	111	78	30	21	75	51	36	1	15	418
Agriculture	15	14	4	0	28	4	17	0	9	91
Veterinary & Game	3	8	0	0	58	22	28	1	3	123
Forestry	35	5	2	0	25	2	4	0	2	75
Geology & Labs	0	5	0	0	6	0	2	0	1	14
Surveys & Land	16	21	0	5	4	2	20	0	3	71
Infrastructure¶	*263*	*315*	*55*	*17*	*122*	*88*	*55*	*0*	*6*	*921*
Totals	*1,181*	*740*	*232*	*129*	*656*	*398*	*301*	*107*	*148*	*3,892*

Table A.6 Colonial Office List, 1922

Category	Nigeria	Gold Coast	Sierra Leone	Gambia	Kenya	Tanganyika	Uganda	Northern Rhodesia	Nyasaland	Totals
Administration§	739	358	134	102	399	264	160	72	118	2,346
Tech Services†	186	208	38	33	270	124	116	3	42	1,020
Medical	120	94	30	28	89	49	44	1	20	475
Agriculture	19	33	4	0	36	16	21	0	10	139
Veterinary & Game	3	6	0	0	57	34	11	1	4	116
Forestry	26	13	4	0	25	11	6	0	3	88
Geology & Labs	0	7	0	0	9	0	5	0	0	21
Surveys & Land	18	55	0	5	54	14	29	1	5	181
Infrastructure¶	340	327	83	21	155	116	75	2	23	1,142
Totals	1,265	893	255	156	824	504	351	77	183	4,508

Table A.7 Colonial Office List, 1923

Category	Nigeria	Gold Coast	Sierra Leone	Gambia	Kenya	Tanganyika	Uganda	Northern Rhodesia	Nyasaland	Totals
Administration§	748	302	132	93	345	255	151	68	117	2,211
Tech Services‡	225	193	34	29	231	141	122	3	49	1,027
Medical	142	84	27	24	92	51	54	1	23	498
Agriculture	22	31	4	0	32	16	19	0	11	135
Veterinary & Game	3	5	0	0	55	33	11	1	4	112
Forestry	29	15	3	0	22	12	7	0	4	92
Geology & Labs	5	7	0	0	4	0	4	0	0	20
Surveys & Land	24	51	0	5	26	29	27	1	7	170
Infrastructure¶	625	339	45	33	139	111	74	0	24	1,390
Totals	1,598	834	211	155	715	507	347	71	190	4,628

Table A.8 Colonial Office List, 1924

Category	Nigeria	Gold Coast	Sierra Leone	Gambia	Kenya	Tanganyika	Uganda	Northern Rhodesia	Nyasaland	Totals
Administration§	637	282	90	97	326	199	158	72	103	1,964
Tech Services†	241	190	36	33	191	145	126	3	41	1,006
Medical	150	87	27	28	71	56	54	1	17	491
Agriculture	27	32	4	0	24	14	22	0	8	131
Veterinary & Game	3	5	0	0	50	35	15	1	4	113
Forestry	31	15	3	0	20	11	6	0	4	90
Geology & Labs	6	7	0	0	4	0	3	0	2	21
Surveys & Land	25	44	2	5	22	29	26	1	6	160
Infrastructure¶	602	234	47	36	115	122	74	0	25	1,255
Totals	1,481	706	173	166	632	466	358	75	169	4,226

Table A.9 Colonial Office List, 1925

Category	Nigeria	Gold Coast	Sierra Leone	Gambia	Kenya	Tanganyika	Uganda	Northern Rhodesia	Nyasaland	Totals
Administration§	*733*	*316*	*139*	*91*	*346*	*269*	*133*	*218*	*105*	*2,350*
Tech Services†	*257*	*190*	*43*	*37*	*201*	*161*	*148*	*48*	*48*	*1,133*
Medical	147	98	29	29	79	60	69	29	23	563
Agriculture	30	27	7	3	27	16	27	5	8	150
Veterinary & Game	7	6	0	0	52	42	18	7	4	136
Forestry	36	16	3	0	22	13	7	0	5	102
Geology & Labs	5	7	4	0	0	0	3	0	2	21
Surveys & Land	32	36	0	5	21	30	24	7	6	161
Infrastructure¶	*638*	*212*	*50*	*45*	*119*	*118*	*84*	*34*	*25*	*1,325*
Totals	*1,628*	*718*	*232*	*173*	*666*	*548*	*365*	*300*	*178*	*4,808*

Table A.10 Colonial Office List, 1926

Category	Nigeria	Gold Coast	Sierra Leone	Gambia	Kenya	Tanganyika	Uganda	Northern Rhodesia	Nyasaland	Totals
Administration§	740	342	129	78	383	265	158	281	119	2,495
Tech Services†	297	233	50	31	238	174	165	63	49	1,300
Medical	178	116	32	21	103	71	75	30	24	650
Agriculture	31	48	11	5	32	19	32	5	8	191
Veterinary & Game	11	5	0	0	58	43	21	20	5	163
Forestry	38	16	5	0	23	13	6	0	5	106
Geology & Labs	5	8	0	0	0	0	4	0	2	19
Surveys & Land	34	40	2	5	22	28	27	8	5	171
Infrastructure¶	859	289	63	25	145	122	77	13	19	1,612
Totals	1,896	864	242	134	766	561	400	357	187	5,407

Table A.11 Colonial Office List, 1927

Category	Nigeria	Gold Coast	Sierra Leone	Gambia	Kenya	Tanganyika	Uganda	Northern Rhodesia	Nyasaland	Totals
Administration§	740	356	139	99	435	275	174	246	112	2,576
Tech Services†	338	254	51	24	267	243	183	55	46	1,461
Medical	198	140	31	16	116	113	88	32	21	755
Agriculture	44	45	12	5	45	25	31	4	8	219
Veterinary & Game	14	6	0	0	61	59	27	10	5	182
Forestry	37	19	5	5	22	14	5	5	5	107
Geology & Labs	4	7	0	0	0	4	5	0	2	22
Surveys & Land	41	37	3	3	23	28	27	9	5	176
Infrastructure¶	879	288	68	18	149	178	110	21	22	1,733
Totals	1,957	898	258	141	851	696	467	322	180	5,770

Table A.12 Colonial Office List, 1928

Category	Nigeria	Gold Coast	Sierra Leone	Gambia	Kenya	Tanganyika	Uganda	Northern Rhodesia	Nyasaland	Totals
Administration§	863	411	142	52	512	386	175	249	113	2,903
Tech Services†	361	251	54	22	295	251	197	77	46	1,554
Medical	207	145	35	15	143	115	100	37	22	819
Agriculture	41	41	15	5	37	29	30	10	8	216
Veterinary & Game	19	5	0	0	65	59	28	21	5	202
Forestry	39	19	4		25	14	6		5	112
Geology & Labs	4	6	0	0	0	4	5	0	2	21
Surveys & Land	51	35	–	2	25	30	28	9	4	184
Infrastructure¶	499	355	85	18	193	189	111	34	42	1,526
Totals	1,723	1,017	281	92	1,000	826	483	360	201	5,983

Table A.13 Colonial Office List, 1929

Category	Nigeria	Gold Coast	Sierra Leone	Gambia	Kenya	Tanganyika	Uganda	Northern Rhodesia	Nyasaland	Totals
Administration§	885	400	148	50	470	349	180	262	103	2,847
Tech Services†	448	277	64	21	327	295	185	86	46	1,749
Medical	241	166	40	14	191	124	89	39	22	926
Agriculture	69	42	16	4	40	43	33	11	6	264
Veterinary & Game	31	6	0	0	69	71	26	23	5	231
Forestry	45	21	5	0	27	14	7	1	5	125
Geology & Labs	5	5	0	0	0	6	4	0	3	23
Surveys & Land	57	37	3	3	–	37	26	12	5	180
Infrastructure¶	1,023	344	89	18	265	244	94	48	52	2,177
Totals	2,356	1,021	301	89	1,062	888	459	396	201	6,773

Table A.14 Colonial Office List, 1930

Category	Nigeria	Gold Coast	Sierra Leone	Gambia	Kenya	Tanganyika	Uganda	Northern Rhodesia	Nyasaland	Totals
Administration§	*940*	*442*	*154*	*48*	*431*	*420*	*193*	*327*	*121*	*3,076*
Tech Services†	*472*	*313*	*69*	*23*	*345*	*332*	*192*	*94*	*53*	*1,893*
Medical	240	188	45	15	179	139	96	43	26	971
Agriculture	72	45	12	5	37	53	31	15	8	278
Veterinary & Game	37	6	0	0	75	73	26	23	6	246
Forestry	47	20	4		28	20	8	2	5	134
Geology & Labs	4	6			0	9	6		3	28
Surveys & Land	72	48	8	3	26	38	25	11	5	236
Infrastructure¶	*986*	*499*	*100*	*17*	*279*	*264*	*92*	*93*	*48*	*2,378*
Totals	*2,398*	*1,254*	*323*	*88*	*1,055*	*1,016*	*477*	*514*	*222*	*7,347*

Table A.15   Colonial Office List, 1931

Category	Nigeria	Gold Coast	Sierra Leone	Gambia	Kenya	Tanganyika	Uganda	Northern Rhodesia	Nyasaland	Totals
Administration§	964	419	153	37	532	437	180	243	125	3,090
Tech Services†	452	308	66	17	289	482	208	120	55	1,997
Medical	265	190	47	10	122	250	98	51	28	1061
Agriculture	80	48	12	5	42	75	34	17	8	321
Veterinary & Game	35	7	0	0	67	85	34	31	6	265
Forestry	–	19	4	0	29	21	11	3	5	92
Geology & Labs	5	6	–	0	0	10	5	0	3	29
Surveys & Land	67	38	3	2	29	41	26	18	5	229
Infrastructure¶	1,218	373	100	20	210	148	114	128	49	2.360
Totals	2,634	1,100	319	74	1,031	1,067	502	491	229	7,447

Table A.16 Colonial Office List, 1932

Category	Nigeria	Gold Coast	Sierra Leone	Gambia	Kenya	Tanganyika	Uganda	Northern Rhodesia	Nyasaland	Totals
Administration§	963	330	94	40	399	418	202	457	127	3,030
Tech Services†	373	231	45	16	288	354	207	150	62	1,726
Medical	180	109	29	9	117	148	97	64	32	785
Agriculture	76	55	10	5	47	68	39	26	9	335
Veterinary & Game	16	8	0	0	62	64	31	37	7	225
Forestry	51	13	3	0	31	19	11	4	5	137
Geology & Labs	5	6	0	0	0	13	5	0	4	33
Surveys & Land	45	40	3	2	31	42	24	19	5	211
Infrastructure¶	515	194	34	19	273	227	78	146	54	1,540
Totals	1,851	755	173	75	960	999	487	753	243	6,296

Table A.17   Colonial Office List, 1933

Category	Nigeria	Gold Coast	Sierra Leone	Gambia	Kenya	Tanganyika	Uganda	Northern Rhodesia	Nyasaland	Totals
Administration§	*844*	*249*	*87*	*37*	*374*	*378*	*191*	*227*	*114*	*2,501*
Tech Services†	*336*	*169*	*44*	*15*	*266*	*270*	*174*	*70*	*59*	*1,403*
Medical	165	83	27	9	100	80	68	26	30	588
Agriculture	62	44	10	4	48	68	40	14	9	299
Veterinary & Game	17	7	0	0	64	58	29	12	6	193
Forestry	45	15	4	0	27	17	9	3	5	125
Geology & Labs	5	5	0	0	0	10	4	0	4	28
Surveys & Land	42	15	3	2	27	37	24	15	5	170
Infrastructure¶	*414*	*174*	*28*	*16*	*220*	*174*	*68*	*84*	*48*	*1,226*
Totals	*1,594*	*592*	*159*	*68*	*860*	*822*	*433*	*381*	*221*	*5,130*

Table A.18   Colonial Office List, 1934

Category	Nigeria	Gold Coast	Sierra Leone	Gambia	Kenya	Tanganyika	Uganda	Northern Rhodesia	Nyasaland	Totals
Administration§	774	233	88	37	360	359	200	191	119	2,361
Tech Services†	317	183	40	17	256	274	164	54	62	1,367
Medical	150	95	25	10	97	98	64	21	32	592
Agriculture	60	46	9	4	48	70	38	10	11	296
Veterinary & Game	16	7	0	0	60	53	27	9	6	178
Forestry	47	15	3		25	17	9		5	121
Geology & Labs	5	4	0	0	0	9	4	0	4	26
Surveys & Land	39	16	3	3	26	27	22	14	4	154
Infrastructure¶	402	159	29	16	233	114	60	27	53	1,093
Totals	1,493	575	157	70	849	747	424	272	234	4,821

Table A.19 Colonial Office List, 1935

Category	Nigeria	Gold Coast	Sierra Leone	Gambia	Kenya	Tanganyika	Uganda	Northern Rhodesia	Nyasaland	Totals
Administration§	805	270	91	37	356	364	199	183	117	2,422
Tech Services†	297	174	41	16	226	270	197	53	65	1,339
Medical	145	82	25	10	79	93	87	21	34	576
Agriculture	57	48	9	3	44	70	46	10	12	299
Veterinary & Game	16	6	0	0	57	55	25	9	5	173
Forestry	44	18	4		25	16	9		5	121
Geology & Labs	5	5	0	0	3	9	6	0	5	33
Surveys & Land	30	15	3	3	18	27	24	13	4	137
Infrastructure¶	379	153	26	16	238	110	64	58	52	1,096
Totals	1,481	597	158	69	820	744	460	294	234	4,857

Table A.20 Colonial Office List, 1936

Category	Nigeria	Gold Coast	Sierra Leone	Gambia	Kenya	Tanganyika	Uganda	Northern Rhodesia	Nyasaland	Totals
Administration§	777	235	84	35	357	368	196	178	116	2,346
Tech Services†	291	214	44	16	226	279	185	53	66	1,374
Medical	137	87	28	11	91	94	74	21	35	578
Agriculture	60	49	9	2	40	78	44	10	12	304
Veterinary & Game	15	6	0	0	56	56	25	9	5	172
Forestry	43	17	4		23	17	10		5	119
Geology & Labs	5	40	1	0	2	7	7	0	3	65
Surveys & Land	31	15	2	3	14	27	25	13	6	136
Infrastructure¶	340	145	29	14	248	123	63	60	50	1,072
Totals	1,408	594	157	65	831	770	444	291	232	4,792

Table A.21 Colonial Office List, 1937

Category	Nigeria	Gold Coast	Sierra Leone	Gambia	Kenya	Tanganyika	Uganda	Northern Rhodesia	Nyasaland	Totals
Administration§	824	287	99	33	298	369	206	192	132	2,440
Tech Services‡	310	180	49	20	124	303	182	55	65	1,288
Medical	144	88	29	14	57	93	74	22	33	554
Agriculture	68	47	10	3	34	77	44	11	13	307
Veterinary & Game	16	7	0	0	17	60	22	9	6	137
Forestry	48	21	4	0	10	17	11	0	6	117
Geology & Labs	5	4	4	0	6	0	6	0	7	32
Surveys & Land	29	13	2	3	0	56	25	13	0	141
Infrastructure¶	320	149	26	15	127	102	60	17	37	853
Totals	1,454	616	174	68	549	774	448	264	234	4,581

Table A.22 Colonial Office List, 1938

Category	Nigeria	Gold Coast	Sierra Leone	Gambia	Kenya	Tanganyika	Uganda	Northern Rhodesia	Nyasaland	Totals
Administration§	823	247	102	35	291	437	166	181	121	2,403
Tech Services†	331	229	51	19	130	285	187	57	68	1,357
Medical	148	96	31	13	61	93	73	24	34	573
Agriculture	72	48	12	3	31	77	49	8	14	314
Veterinary & Game	18	7	0	0	17	57	24	9	6	138
Forestry	51	21	5		9	17	12	3	6	124
Geology & Labs	12	42	1	0	6	0	5	0	5	71
Surveys & Land	30	15	2	3	6	41	24	13	3	137
Infrastructure¶	303	148	33	17	135	105	77	60	47	925
Totals	1,457	624	186	71	556	827	430	298	236	4,685

Table A.23 Colonial Office List, 1939

Category	Nigeria	Gold Coast	Sierra Leone	Gambia	Kenya	Tanganyika	Uganda	Northern Rhodesia	Nyasaland	Totals
Administration§	863	299	98	37	289	455	217	199	129	2,586
Tech Services†	341	206	53	19	140	310	233	61	73	1,436
Medical	147	95	32	13	55	96	109	28	36	611
Agriculture	79	54	13	3	35	75	50	8	13	330
Veterinary & Game	18	8	0	0	17	61	22	10	6	142
Forestry	50	28	4	0	9	17	14	3	6	131
Geology & Labs	13	7	1	0	6	0	14	0	7	48
Surveys & Land	34	14	3	3	18	61	24	12	5	174
Infrastructure¶	387	149	36	17	142	102	52	75	44	1,004
Totals	1,591	654	187	73	571	867	502	335	246	5,026

Table A.24   Colonial Office List, 1940

Category	Nigeria	Gold Coast	Sierra Leone	Gambia	Kenya	Tanganyika	Uganda	Northern Rhodesia	Nyasaland	Totals
Administration§	*758*	*274*	*87*	*30*	*253*	*389*	*215*	*200*	*129*	*2,335*
Tech Services†	*354*	*216*	*45*	*16*	*158*	*300*	*231*	*68*	*70*	*1,458*
Medical	151	98	35	11	64	92	110	28	33	622
Agriculture	80	58	0	2	33	80	48	8	15	324
Veterinary & Game	20	8	0	0	25	65	21	11	6	156
Forestry	48	27	5	0	9	2	14	4	5	114
Geology & Labs	15	8	1	0	6	0	14	0	9	53
Surveys & Land	40	17	4	3	21	61	24	17	2	189
Infrastructure¶	*404*	*148*	*37*	*14*	*137*	*97*	*48*	*30*	*49*	*964*
Totals	*1,516*	*638*	*169*	*60*	*548*	*786*	*494*	*298*	*248*	*4,757*

Table A.25 Colonial Office List, 1946

Category	Nigeria	Gold Coast	Sierra Leone	Gambia	Kenya	Tanganyika	Uganda	Northern Rhodesia	Nyasaland	Totals
Administration§	749	320	113	33	346	422	162	305	125	2,575
Tech Services‡	384	249	82	21	170	293	221	105	53	1,578
Medical	164	132	47	12	62	106	130	58	19	730
Agriculture	99	58	18	5	43	68	37	13	21	362
Veterinary & Game	27	8	4	3	34	60	16	13	5	170
Forestry	45	29	8	0	13	16	9	9	4	133
Geology & Labs	17	4	1	0	0	0	9	0	1	32
Surveys & Land	32	18	4	1	18	43	20	12	3	151
Infrastructure¶	384	149	47	7	175	96	47	78	18	1,001
Totals	1,517	718	242	61	691	811	430	488	196	5,154

Table A.26 Colonial Office List, 1948

Category	Nigeria	Gold Coast	Sierra Leone	Gambia	Kenya	Tanganyika	Uganda	Northern Rhodesia	Nyasaland	Totals
Administration§	*1077*	*424*	*121*	*30*	*297*	*489*	*209*	*288*	*128*	*3,063*
Tech Services†	*413*	*288*	*94*	*22*	*253*	*466*	*268*	*88*	*81*	*1,973*
Medical	204	156	55	14	135	182	143	50	46	985
Agriculture	94	74	20	4	53	101	42	12	23	423
Veterinary & Game	32	5	2	2	38	92	21	10	4	206
Forestry	42	35	10	0	12	23	16	4	5	147
Geology & Labs	18	6	1	0	0	0	23	0	1	49
Surveys & Land	23	12	6	2	15	68	23	12	2	163
Infrastructure¶	*408*	*195*	*81*	*6*	*194*	*95*	*55*	*41*	*25*	*1,100*
Totals	*1,898*	*907*	*296*	*58*	*744*	*1,050*	*532*	*417*	*234*	*6,136*

Table A.27 Colonial Office List, 1949

Category	Nigeria	Gold Coast	Sierra Leone	Gambia	Kenya	Tanganyika	Uganda	E.A.H.C.*	Northern Rhodesia	Nyasaland	Totals
Administration§	*869*	*391*	*122*	*42*	*353*	*327*	*204*	*41*	*305*	*125*	*2,779*
Tech Services‡	*378*	*200*	*75*	*23*	*244*	*207*	*212*	*30*	*105*	*53*	*1,527*
Medical	208	77	39	15	144	79	89	0	58	19	728
Agriculture	62	68	18	4	51	42	41	10	13	21	330
Veterinary & Game	26	3	2	2	25	29	16	19	13	5	140
Forestry	45	33	9	0	13	11	17		9	4	141
Geology & Labs	17	7	1	0	0	1	25	0	0	1	52
Surveys & Land	20	12	6	2	11	45	24	1	12	3	136
Infrastructure¶	*431*	*141*	*47*	*8*	*47*	*31*	*64*	*76*	*78*	*18*	*941*
Totals	*1,678*	*732*	*244*	*73*	*644*	*565*	*480*	*147*	*488*	*196*	*5,247*

* The East Africa High Commission—employment figures were separated out for the EAHC from 1949 onward.

Table A.28   Colonial Office List, 1950

Category	Nigeria	Gold Coast	Sierra Leone	Gambia	Kenya	Tanganyika	Uganda	E.A.H.C.*	Northern Rhodesia	Nyasaland	Totals
Administration§	*1,093*	*397*	*128*	*45*	*365*	*338*	*260*	*77*	*347*	*153*	*3,203*
Tech Services†	*411*	*331*	*82*	*22*	*251*	*187*	*216*	*60*	*112*	*56*	*1,728*
Medical	210	162	43	14	144	54	93	3	63	17	803
Agriculture	64	103	19	3	49	43	43	19	15	26	384
Veterinary & Game	28	8	2	3	26	33	16	29	13	6	164
Forestry	62	34	10	0	8	17	19		9	3	162
Geology & Labs	20	8	1	0	0	3	21	0	0	1	54
Surveys & Land	27	16	7	2	24	37	24	9	12	3	161
Infrastructure¶	*429*	*369*	*44*	*8*	*40*	*35*	*58*	*178*	*97*	*22*	*1,280*
Totals	*1,933*	*1,097*	*254*	*75*	*656*	*560*	*534*	*315*	*556*	*231*	*6,211*

* The East Africa High Commission—employment figures were separated out for the EAHC from 1949 onward.

Table A.29 Colonial Office List, 1951

Category	Nigeria	Gold Coast	Sierra Leone	Gambia	Kenya	Tanganyika	Uganda	E.A.H.C.*	Northern Rhodesia	Nyasaland	Totals
Administration§	1,186	553	156	45	407	385	290	87	424	167	3,700
Tech Services†	468	418	84	26	281	213	222	66	138	68	1,984
Medical	225	150	39	17	161	67	83	8	75	26	851
Agriculture	77	184	20	4	49	47	44	22	18	28	493
Veterinary & Game	53	13	2	3	39	38	26	28	13	0	215
Forestry	60	36	13	0	11	18	21		13	3	175
Geology & Labs	22	10	5	0	0	19	25	0	0	6	87
Surveys & Land	31	25	5	2	21	24	23	8	19	5	163
Infrastructure¶	438	412	63	6	95	57	61	202	106	26	1,466
Totals	2,092	1,383	303	77	783	655	573	355	668	261	7,150

* The East Africa High Commission—employment figures were separated out for the EAHC from 1949 onward.

NOTES

INTRODUCTION

1. See Dubow, "A Commonwealth of Science," 67–99; and Worboys, "The British Association and Empire." Additional details are derived from the London *Times* and the National Archives of Kenya and Britain.

2. Hofmeyr, "Africa and Science," 11, 18.

3. Ibid., 13–14.

4. Ibid., 13.

5. Ibid., 17.

6. Ibid., 19–21.

7. See Lord Lothian's foreword in Hailey, *An African Survey*, v.

8. For evidence, see Hailey, *An African Survey*, 1956 rev. ed.; Scientific Council for Africa South of the Sahara, *Inter-African Technical and Scientific Co-Operation*; Scientific Council for Africa South of the Sahara, *Eighth Meeting of the Scientific Council*; and Worthington, *Science in the Development of Africa*.

9. In their volume on the pertinence of Antonio Gramsci's concept of hegemony to colonial settings in India and Africa, Dagmar Engels and Shula Marks stipulate that hegemony "suggests the ways in which colonial ideology served the ruling class by helping it to make their rule appear natural and legitimate; . . . hegemonic colonial ideologies reflected the material and cultural conditions of both the dominant and dominated classes. To be successful, imperial hegemony had to come to terms with, incorporate and transform Indian and African values." Engels and Marks, *Contesting Colonial Hegemony*, 3.

10. Hailey, *An African Survey*, xxiv–xv, emphasis added.

11. For a lengthier definition, see Thackray, "History of Science," 4–5.

12. Here I list the name of the independent country first and the name of the colonial dependency, if it was different, in parentheses; in the rest of the book, I use only the historical designations.

13. Weindling, "Social Medicine at the League of Nations Health Organisation," 134–53, esp. 136–37.

14. Drayton, "Science, Medicine, and the British Empire," 264.

15. Huxley, *Africa View*, 378–79.

16. Crawford, Shinn, and Sörlin, "The Nationalization and Denationalization of the Sciences," 2.

17. Schroeder-Gudehus, "Nationalism and Internationalism," 909.
18. See also Sörlin, "National and International Aspects of Cross-Boundary Science," and Jamison, "National Political Cultures and the Exchange of Knowledge"; Chambers, "Period and Process in Colonial and National Science," and Sheets-Pyenson, "Civilizing by Nature's Example"; and Crane, "Transnational Networks in Basic Science."
19. This phrase is used not only by historians but also by sociologists, anthropologists, and critics of development.
20. Cohn, *Colonialism and Its Forms of Knowledge*; Marks, "What Is Colonial about Colonial Medicine?"
21. The classic statement is Basalla, "The Spread of Western Science." Use of this phrase is not limited to Africanists. In the 2005 special issue of *Isis* on "Colonial Science," contributors use the phrase but often adroitly sidestep many of its limitations. See also Cañizares-Esguerra, *Nature, Empire, and Nation*; Philip, "English Mud"; Osborne, "Acclimatizing the World"; Chambers and Gillispie, "Locality in the History of Science"; Conte, "Colonial Science and Ecological Change"; Gilfoyle, "Veterinary Immunology as Colonial Science"; Raina, "Beyond the Diffusionist History of Colonial Science"; and Kumar, "Patterns of Colonial Science in India."
22. A number of authors writing on science and empire have avoided this phrase, including Roy MacLeod, David Livingstone, and Warwick Anderson.
23. I am leaving "science" in the singular here, but readers should bear in mind that we are talking about sciences in the plural. See Livingstone, "The Spaces of Knowledge"; and Shapin, "Placing the View from Nowhere."
24. See Ferguson, *Global Shadows*.
25. One of the most interesting statements on the lack of a "system" of knowledge is Last, "The Importance of Knowing about Not Knowing."
26. See Headrick, *The Tentacles of Progress*. Headrick does not discuss the introduction of the airplane, which began to feature with some regularity in British colonial scientific circles by the interwar period.
27. On other colonial regions and periods, see Cañizares-Esguerra, *Nature, Empire, and Nation*; Ballantyne, *Science, Empire, and the European Exploration of the Pacific*; Arnold, *Science, Technology, and Medicine in Colonial India*; Drayton, *Nature's Government*; MacLeod and Rehbock, *Darwin's Laboratory*; McClellan, *Colonialism and Science*; Pyenson, *Empire of Reason*; Pyenson, *Civilizing Mission*; Burnett, *Masters of All They Surveyed*; Edney, *Mapping an Empire*; Miller and Reill, *Visions of Empire*; and Grove, *Green Imperialism*.
28. I must set aside the question of precolonial Islamic sciences, though it would repay closer examination.
29. On northern Africa, see Keller, *Colonial Madness*; El Shakry, *The Great Social Laboratory*; Pelis, *Charles Nicolle, Pasteur's Imperial Missionary*; Osborne, *Nature, the Exotic and the Science of French Colonialism*; Jagailloux, *La medicalisation de l'Egypt*; Gallagher, *Medicine and Power in Tunisia*; and Gallagher, *Egypt's Other Wars*. Two recent books on southern Africa illuminate trends there: Dubow, *A Commonwealth of Knowledge*, and Harries, *Butterflies and Barbarians*. The concept of a "center of calculation" comes from Latour, *Science in Action*.
30. See Collins, "The British Association as Public Apologist for Science"; Adas, *Machines as the Measure of Men*; MacLeod and MacLeod, "The Social Relations of Science and Technology"; Owen, "Critics of Empire in Britain"; Wylie, "Norman Leys and McGregor Ross"; Louis, *Imperialism at Bay*; and Constantine, *The Making of British Colonial Development Policy*.

31. For more on these themes, see MacLeod, *Nature and Empire*; Bonneuil, "Mettre en ordre et discipliner les tropiques"; Carroll, "Engineering Ireland"; Smith and Agar, *Making Space for Science*; Wright, "Tradition in the Service of Modernity"; Livingstone, "The Spaces of Knowledge"; Kuklick and Kohler, *Science in the Field*; Findlen, *Possessing Nature*; Cittadino, *Nature as the Laboratory*; Comaroff and Comaroff, "Medicine, Colonialism, and the Black Body"; and Schumaker, *Africanizing Anthropology*.

32. See, for instance, Roll-Hansen, "Studying Nature without Nature?"; and Gooday, "'Nature' in the Lab."

33. James, *The Development of the Laboratory*; Cunningham and Williams, *The Laboratory Revolution in Medicine*; and Fox and Guagnini, *Laboratories, Workshops, and Sites*.

34. See also Anderson, *Colonial Pathologies*; Anker, *Imperial Ecology*; Haynes, *Imperial Medicine*; Nandy, *The Savage Freud*; Kuklick, *The Savage Within*; and Asad, *Anthropology and the Colonial Encounter*.

35. Kuklick and Kohler, *Science in the Field*; and Kohler, *Landscapes and Labscapes*.

36. See, for instance, Müller-Hill, *Murderous Science*; Proctor, *Racial Hygiene*; Bock, "Sterilization and 'Medical' Massacres in National Socialist Germany"; and Baumslag, *Murderous Medicine*.

37. Warren T. Reich, ed., *Encyclopedia of Bioethics*, 2nd ed., Appendix, 2763–64; Schmidt, *Justice at Nuremberg*; Weindling, *Nazi Medicine and the Nuremberg Trials*; and Freyhofer, *The Nuremberg Medical Trial*.

38. On Japan's history of human experimentation during the Second World War, which did not result in a similar trial, see Baader, Lederer, Low, Schmaltz, Schwerin, "Pathways to Human Experimentation"; and Harris, *Factories of Death*.

39. Hooper, *The River*; Cohen, "Vaccine Theory of AIDS Origins Disputed at Royal Society"; and special issue on "Origins of HIV and the AIDS Epidemic," *Philosophical Transactions of the Royal Society Biological Sciences* 356 (2001): 777–977. These issues were central to the accusations made by Patrick Tierney on the role of scientific researchers among the Yanomamo: Tierney, *Darkness in El Dorado*.

40. For an expression of this view, see Smith, "Medicine in Africa as I Have Seen It," 28: "It is the almost unlimited field that Africa offers for clinical research that I find so enthralling."

41. Waller and Homewood, "Elders and Experts."

42. Vaughan, *Curing Their Ills*; Hunt, *A Colonial Lexicon*; and White, *Speaking with Vampires*; Livingston, *Debility and the Moral Imagination in Botswana*.

43. Vaughan, "Healing and Curing," 288.

44. Ranger, "Plagues of Beast and Men"; and Feierman, "Healing as Social Criticism in the Time of Colonial Conquest."

45. Janzen, "Toward a Historical Perspective on African Medicine and Health," 104.

46. Schumaker, *Africanizing Anthropology*; Jezequel, "Itinéraire lettres sous la colonisation"; Jezequel, "Voices of Their Own?" and Pugach, "Of Conjunctions, Comportment, and Clothing"; and West, "Inverting the Camel's Hump." See also Sanjek, "Anthropology's Hidden Colonialism."

47. See, for instance, Shapiro, "Doctors or Medical Aids"; Lyons, "The Power to Heal"; Iliffe, *East African Doctors*; and Bell, *Frontiers of Medicine*, esp. chaps. 2 and 7.

48. The most provocative discussion of this point remains Horton, "African Traditional Thought and Western Science, Part I," and Horton, "African Traditional Thought and Western Science, Part II."

49. Ian Hacking, quoted in Keller, *Secrets of Life, Secrets of Death*, 2.

50. Feierman, "Struggles for Control," 80.

51. Hacking, *Representing and Intervening*, 130–31.
52. Aidan Southall quoted in Vincent, "Sovereignty, Legitimacy, and Power," 151.
53. Comaroff, "Reflections on the Colonial State," 18.
54. Statistics include Bechuanaland Protectorate and Somaliland.
55. See Koponen, *Development for Exploitation*.
56. Colonial Survey Committee, *Surveys and Explorations of British Africa*, 8.
57. This claim is borne out from even a cursory examination of governors' and commissioners' reports in the early years of state building.
58. Lonsdale and Berman, "Coping with Contradictions," 489. I cannot agree with Crawford Young's statement that colonial states "had little to do with 'building capitalism,'" since state actors clearly saw this as one of their functions. The interests of the state and of capital interests were not one and the same, but an interest in capital development preoccupied state leaders. See Young, "Country Report," 104.
59. By comparison, sixty years into its existence, the World Bank employed 13,000 individuals globally. Cassidy, "The Next Crusade," 36.
60. Taylor, "Technocratic Optimism." For an earlier groundbreaking analysis, see Leiss, *The Domination of Nature*.
61. Scott, *Seeing like a State*, 89–90.
62. Herbert Simon quoted in ibid., 45.
63. Ibid., 311.
64. Ibid., 279.
65. Haldane, "The Constitution of the Empire and the Development of Its Council," 17–18.
66. "Report of Committee on Colonial Scientific and Research Services," in *Summary of First Colonial Office Conference*, 35.
67. Ibid., 39.
68. Jeffries, *The Colonial Empire and Its Civil Service*, xvi.
69. Kirk-Greene, "The Thin White Line"; and Berry, "Hegemony on a Shoestring."
70. The imperial bureaus covered such subjects as entomology, forestry, agriculture, and disease; likewise, the imperial and pan-African coordinating conferences focused on soil science, forestry, botany, sleeping sickness, agriculture, medicine, and public health. Inter-imperial bodies, such as the International Institute for African Languages and Cultures (founded in 1926), were clearinghouses for anthropological research.
71. Sleeping sickness, or African trypanosomiasis, is a vector-borne disease conveyed by the tsetse fly, which inevitably leads to death in humans unless treated. The equatorial epidemics at the turn of the twentieth century (circa 1902–14) resulted in the death of hundreds of thousands of people in Uganda and the Congo and many more thousands in French, German, and Portuguese territories.
72. For a discussion of nineteenth-century infrastructures prior to the era of "high imperialism," see Laidlaw, *Colonial Connections*.
73. Smith, "Africa: What Do We Know of It?" 78.
74. Dover, the foreword to his collection of poems, *Brown Phoenix*; and Dover, *Half-Caste*.
75. Arnold, *Colonizing the Body*, 15.
76. Bayly, *Empire and Information*, 25.
77. Trigger, "The History of African Archaeology in World Perspective."
78. Sears, "Ecology—A Subversive Subject," 12.
79. Lewis, *Inventing Global Ecology*; Anker, *Imperial Ecology*; Anker, "The Politics of Ecology in South Africa on the Radical Left"; Dunlap, *Nature and the English Diaspora*;

Bocking, *Ecologists and Environmental Politics*; Mitman, *The State of Nature*; and Cittadino, *Nature as the Laboratory*.

80. Here I am thinking of James Fairhead and Melissa Leach's concern to "replace" humans in ecological science in their conclusion to *Misreading the African Landscape*, 283–87.

81. Mackenzie, *Land, Ecology, and Resistance in Kenya*, 104.

82. For a more detailed discussion of the construction of vernacular science as a global process, see Tilley, "Global Histories, Vernacular Science, and African Genealogies." The following paragraph is taken from that article.

83. International Institute of African Languages and Cultures, 1935 renewal appeal to Rockefeller Foundation [no date, but February 1935], GD40/17/126, Lord Lothian Papers (LP), National Archives of Scotland.

84. Curtin, *The Image of Africa*, 70, and chaps. 3 and 4.

85. Saunders, "On the Present Aspects of Africa," 261.

86. Worthington, *Science in Africa*, 8.

87. Hailey, *An African Survey*, 1115.

88. Goldberg, *The Racial State*.

89. See Hallen, *African Philosophy*; Ayittey, *Indigenous African Institutions*; and Ferguson, *Global Shadows*.

CHAPTER ONE

1. Green, "Annual General Meeting," 156.

2. See Yearwood, "Great Britain and the Repartition of Africa"; and Smith, "The British Government and the Disposition of the German Colonies in Africa."

3. Green, "Annual General Meeting," 156. Also see Shelford, "The Late Mrs. J. R. Green," 413–14.

4. Fabian, *Out of Our Minds*, 16 and 183.

5. See Helly, "'Informed Opinion' on Tropical Africa in Great Britain"; and Helly, "British Attitudes towards Tropical Africa."

6. Bridges, "The R.G.S. and the African Exploration Fund."

7. Ravenstein, "The Climate of Tropical Africa."

8. John Kirk to Earl Granville, May 5, 1884, "East Africa, Kilimanjaro Expedition," Royal Botanic Gardens—Kew Archives, London.

9. Comments by Sydney Buxton in "The African Society Inaugural Meeting," xx. *West African Studies* is the title of Kingsley's second book, published in 1899; she died in South Africa in 1900, just before the Royal African Society was founded. Biological scientists had some influence in African affairs during this period, and a vast body of popular and scholarly literature on African topics circulated across Europe. But the organized efforts of scientific societies were sporadic, especially in comparison to the projects they undertook after the turn of the twentieth century.

10. Farr, "Inaugural Address [November 19, 1872]," 425 (emphasis in original).

11. Kennedy, "Review: The Theory and Practice of Imperialism," 762 and 764.

12. Bayly, *Birth of the Modern World*, 176.

13. Lieven, "Dilemmas of Empire 1850–1918."

14. Cain and Hopkins, *British Imperialism*, chap. 11; and Munro, *Africa and the International Economy*, chap. 3.

15. Lloyd, "Africa and Hobson's Imperialism," 151.

16. Dunn, "For God, Emperor, Country!"; and Akpan, "Liberia and Ethiopia 1880–1914."

17. On "African initiatives and resistance," see Boahen, *Africa under Colonial Domination*, chaps. 3 through 9.
18. Ranger, "Plagues of Beast and Men," 252.
19. Ravenstein, "Notes on the Maps," in Silva White, *The Development of Africa*, 323.
20. Koskenniemi, *The Gentle Civilizer of Nations*, 126, and chap. 2.
21. Grovogui, *Sovereigns, Quasi-Sovereigns, and Africans*, 77, and chap. 3.
22. Cohen, "Malaria and French Imperialism." A definitive comparative study of quinine use remains to be done; the French took up quinine more slowly than the British, and even some British failed to rely on it systematically.
23. Curtin, "The End of the 'White Man's Grave'?"
24. For early debates on firearms and related questions, see *Journal of African History* 12, nos. 2 and 4 (1971). Recent studies on southern Africa include Mavhunga, "Firearms Diffusion, Exotic and Indigenous Knowledge Systems"; and Storey, "Guns, Race, and Skill in Nineteenth-Century Southern Africa." On railways and steam technologies, see Headrick, *The Tentacles of Progress*. There is a pressing need to evaluate the effects of telegraphs, railways, and roads in African colonial history.
25. The French were particularly prone to grandiose visions; see Soleillet, *Avenir de la France en Afrique*; Heffernan, "The Limits of Utopia"; Carrière, "Le Transsaharien"; and Broc, "Les Français face à l'inconnue saharienne."
26. "The Congo," *Times (London)*, August 28, 1884, 2, col. A.
27. Galton, *The Art of Travel*; British Association, *Notes and Queries on Anthropology*.
28. These figures may not be comprehensive, as they are derived from three sources. Figures up to 1869 can be found in Wright, "The Field of the Geographical Society," 548. Numbers and geographical distribution for 1870–90 come from comparing Sparn, "Cronología, diferenciación," esp. 324–26, with Schneider's "Geographical Reform," 92.
29. Depping, "Sociétés de géographie," 5878. Also see McKay, "Colonialism and the French Geographical Movement."
30. Galton, "Presidential Address: Geography," 199.
31. Ibid., 200.
32. The list of 123 questions proposed by the conference organizers can be found in *Congrès International des Sciences Géographiques*, vol. 1, lvii–lxviii. Also see Delagrange, "Rapport sur la colonisation"; and Ravenstein, "Statistics at the Paris Geographical Congress," esp. 426–27.
33. For discussions about African expeditions, see *Congrès International des Sciences Géographiques*, vol. 1, 509–12, 577–99, 601–2; and Rohlfs, "Projet de voyage en Afrique," 612–14.
34. Comments by Rohlfs, *Congrès International des Sciences Géographiques*, vol. 1, 577.
35. Grandidier, "Méthode pratiques à employer," vol. 1, 620–21, and discussion, 598–99.
36. Schweinfurth, "Discours prononcé au Caire," 114 and 122.
37. Isma'il to Baker, February 1872, quoted in Gray, "Ismail Pasha and Sir Samuel Baker," 204.
38. Schweinfurth, "Discours prononcé au Caire," 114. The society was founded by decree in May, and Schweinfurth was appointed in June. See Reid, "The Egyptian Geographical Society."
39. Léopold is listed among the participants, but it is not clear whether he actually attended. He certainly gained information from the head of Belgium's delegation, E. de Borchgrave, with whom he corresponded; see Roeykens, *Léopold II et l'Afrique 1855–1880*, 50.

40. On the German society, see Schneider, "Geographical Reform," 113, and the review of Von Adolf Bastian, *Die Deutsche Expedition au der Loanga Kuste nebst alteren Nachrichten uber die zu erfoschenden Lander* (Jena: Trübner and Co., 1874) in the *Geographical Magazine* 1 (1874): 115–16. On plans for de Brazza's expedition, including an account of de Brazza's intention to carry an electric battery in his pocket to shock and inspire those "Chiefs" he might negotiate with on his journey, see "Central Africa," *Times (London)*, 6. De Brazza's expedition set off in the autumn of 1875 and returned in 1878.

41. Roeykens, *Léopold II et l'Afrique*, 53; the letter was written on August 22.

42. See "Cinquantième Anniversaire." On Léopold's offers of support to the RGS, see Roeykens, *Léopold II et la Conférence Géographique de Bruxelles*, 19; and Rawlinson, "Address to the Royal Geographical Society," 438.

43. Hochschild, *King Leopold's Ghost*, 43; Vandewoude, "De Aardrijkskundige Conferentie (1976) [sic]," esp. 416–18. On Frere's meeting with Léopold, see Bartle Frere to Lord Carnarvon, December 10, 1876, PRO 30/6/4, Lord Carnarvon Papers, BNA, London. The King's plans were already known in England as early as April; see "Arrival of Lieutenant Cameron," 104.

44. Léopold to Jean-Baptiste Nothomb, June 14, 1876, quoted in Stengers, "Introduction," vii.

45. Roeykens, *Léopold II et la Conférence Géographique de Bruxelles*, 83. The president of Rome's geographical society was also invited, but did not attend. See also Bridges, "The First Conference of Experts on Africa," 14–15; and the minutes, *Conférence Géographique de Bruxelles*.

46. Baron Lambermont, August 16, 1876, quoted in Roeykens, *Léopold II et l'Afrique*, 79. Lambermont later attended the Berlin West Africa Conference as a Belgian delegate.

47. The first of these objectives was articulated by members of the Belgian planning committee; the other objectives are stated in Léopold's letter of invitation, July 1876, published in full in Roeykens, *Léopold II et la Conférence Géographique de Bruxelles*, 28–29. Léopold repeated much the same thing in his opening address, reprinted in Banning, *Africa and the Brussels Geographical Conference*, 152–54.

48. Banning, *Africa and the Brussels Geographical Conference*, 149–51.

49. Lambermont, from first meeting on August 16, 1876, quoted in Roeykens, *Léopold II et la Conférence Géographique de Bruxelles*, 79. The second meeting took place on September 9.

50. Banning, *Africa and the Brussels Geographical Conference*, xiv. I have used the official English translation, but Banning used *peuples*, not *races*, in his original text, which avoids an explicit racial connotation. Banning's French manuscript was sent to both the Foreign and Colonial Offices in Britain before it was published.

51. See Johnston, *Africa*.

52. Roeykens, "Banning et la Conférence Géographique," 227–28. Cameron's first reports were published on January 12, and Banning's first article was published on January 17.

53. Frere, "Memorandum of Instructions for the Livingstone East Coast Expedition."

54. Summary of figures in "Astronomical and other Observations by Lieut. L. V. Cameron, R.N. Across Central Africa—A brief summary, continued from Map Room Report of January 10, 1876, brought up to date of June 3, 1876," File 5. Full figures in "Reduction of Astronomical Observations made by Commander Cameron in the years 1874 and 1875 in his journey across Africa," File 7. Verney Lovett Cameron Papers (hereafter VLC Papers), Royal Geographical Society Archives (RGS), London.

55. Rawlinson comments in Cameron, "Lieut. Cameron's Letters," 133.

56. "Extract from Mr. Ellis's last report on Lt. Cameron's Observations dated May 27th 1876," File 5, VLC Papers, RGS.

57. Rawlinson comments in Cameron, "Lieut. Cameron's Letters," 133.

58. V. L. Cameron, "Report on the work of Messrs. Strachan and Ellis," February 1877, File 5, VLC Papers, RGS.

59. Conference minutes, *Conférence Géographique de Bruxelles*, 16.

60. *Conférence Géographique de Bruxelles*, 14.

61. Ibid., 19.

62. Ibid., 15.

63. Sanford, "Report on the Annual Meeting of the African International Association," 104. Sanford served as Frere's replacement on the executive committee of the IAA. Although this report is from the following year, it captures the broader discussion about the stations' functions during the original meeting.

64. Banning, *Africa and the Brussels Geographical Conference*, 113.

65. Frere to Carnarvan, December 4, 1876, and Frere to Carnarvan, December 10, 1876, PRO 30/6/4, "Correspondence—the Prince of Wales, 1874–1878," BNA.

66. *Conférence Géographique de Bruxelles*, 30–31.

67. Afterward, many participants spoke in glowing terms of the event publicly. Some of these comments are quoted verbatim in Rutherford Alcock's address at the first meeting of the RGS for the 1876–77 session; see Alcock, "Address," [November 13, 1876], esp. 16–20.

68. Cameron, *Across Africa*, 479–80. Also see Cameron, "On Proposed Stations in Central Africa," 141–42; and "British Association for the Advancement of Science—Report—Geography Section," 409.

69. De Courcel, "The Berlin Act of 26 February 1885," 247; and Gründer, "Christian Missionary Activities in Africa," 92. For excellent interpretations of Léopold's changing motives and priorities, see Stengers, "Léopold II and the *Association Internationale du Congo*"; Stengers, "Leopold II et Brazza en 1882"; and Stengers, "King Leopold and Anglo-French Rivalry."

70. Some of these are described in Fabian's *Out of Our Minds*. The full scope of the activities can be found in *L'Afrique Explorée et Civilisée* (1879–94); *Exploration: Journal des conquêtes de la civilization sur tous les points du globe* (1876–84); and *Mittheilungen der Afrikanische Gesellschaft in Deutschland* (1878–89).

71. "Summary of conversation with B. Frere by M.H.O. for Lord Carnarvon on 8 December 1876," PRO 30/6/4, BNA.

72. Nowell, "Portugal and the Partition of Africa," 8. The society presented its petition on October 16, 1876.

73. Sociedade de Geografia de Lisboa, July 1876, quoted in Axelson, *Portugal and the Scramble for Africa*, 47.

74. Carnarvan to Frere, December 12, 1876, PRO 30/6/4, emphasis in original. Parts of this quotation appear in numerous histories of the partition of Africa, but most authors often fail to mention the context that generated it, and none ever notes the opening and closing sentences, which connect Carnarvan's concerns to geographical and scientific undertakings.

75. These details can be found in FO 84/1463 and PRO 30/6/4.

76. "African International Association—Memorandum," December 21, 1876, PRO 30/6/4.

77. "African International Association—Opinion—Henry Thring, December 18, 1876," PRO 30/6/4.

78. John Kirk to the Earl of Derby, March 7, 1877, FO 881/3183. This letter is also partially quoted in Coupland, *The Exploitation of East Africa*, 330–31.

79. Mr. Malcolm, memo to Lord Carnarvan, December 15, 1876, PRO 30/6/4.

80. Elton to Principal Secretary of State for Foreign Affairs [Carnarvan], March 1, 1877, FO 84/1479.

81. Thring to Ommanney, January 23, 1877, PRO 30/6/4.

82. Circular, "African Exploration Fund," no date [mid-January 1877], PRO 30/6/4. For Frere's role in asking Thring to draft the document, see Thring to Ommanney, January 23, 1877.

83. Carnarvan to Thring, February 2, 1877, PRO 30/6/4.

84. Keith Johnston to John Scott Keltie, November 30, 1875, Corr. Block 1871–1880, RGS archives.

85. "Meeting of the Subscribers to the African Exploration Fund," *PRGS* 22 (1877): 463–76, on 467. This report includes a brief discussion of Henry Morton Stanley's proposal, which was rejected, and a more detailed description of Keith Johnston's proposal, which was accepted.

86. Reeves, *The International Beginnings of the Congo Free State*, 17–26; and Stengers, "Léopold II and the *Association Internationale du Congo*."

87. Sanderson, "The European Partition of Africa," 116.

88. Robinson, Gallagher, with Denny, *Africa and the Victorians*, 169–73; and Stengers, "Leopold II et Brazza en 1882."

89. Quoted in Reeves, *International Beginnings of the Congo Free State*, 24.

90. Excerpts from Stanley's speech, *The Manchester Examiner and Times*, 19 September 1884, FO 84/1679, "Slave Trade. Zanzibar. From Sir J. Kirk. Nos. 127–176. October to December 1884."

91. Reeves, *International Beginnings of the Congo Free State*, 20.

92. Robinson, Gallagher, with Denny, *Africa and the Victorians*, 169.

93. The Swiss established the Comité National Suisse-Africain shortly after the 1876 conference; independently, another group founded the monthly journal *L'Afrique Explorée et Civilisée*, intended to complement the journals and publications produced by the British and German geographical societies. See "A nos lectures" and "L'exploration moderne de l'Afrique."

94. Besides Germany, the host, Britain, Belgium, Holland, Denmark, Norway, Sweden, Austria-Hungary, Italy, Russia, Spain, Turkey, Portugal, and the United States attended the Berlin Conference. See Johnston, *Sovereignty and Protection*, chap. 6. "The Berlin Conference and European Protectorates," esp. 168–69.

95. Rawson, "The Territorial Partition of the Coast of Africa." Also see "The Partition of Africa," *Times (London)*, November 7, 1884, 2, col. E.

96. "The Scramble for Africa."

97. See Fisch, "Africa as Terra Nullius."

98. From Article 35 of Berlin Act, *Protocols and General Act of W. African* [C. 4361].

99. Report on Stanley's speech in "Latest Intelligence."

100. Count Knapist, in *Protocols and General Act of W. African Conference* [C. 4361], 183.

101. Count de Launay, November 19 and 27, 1884, in *Protocols and General Act of W. African Conference*, 23 and 45. In both sets of comments, de Launay also mentioned the need to protect missionaries, who were included in Article 6 as well.

102. November 1889, quoted in Roberts, *Salisbury*, 517.

103. Hertslet was the Foreign Office (FO) librarian, a position that required him to serve as the institutional memory of the FO and during his career he wrote approximately 3,000 memoranda dealing with the historical and legal backgrounds of particular policy issues. He was elected a fellow of the RGS in 1858. *DNB*. The first edition of the *Map of Africa by Treaty* appeared in two volumes in 1894.

104. John Kirk, 1885, quoted in Miller, *The Lunatic Express*, 169.

105. Horace Waller, letter to the editor, *Times (London)*, January 7, 1887, 7, col. D.

106. Cust, "The Ethics of African Geographical Explory [*sic*]." On Cust's thirteen-year service on the council (1877–96), see "Obituary: Robert Needham Cust"; also see Cust, "The Scramble for Africa," in Cust, *Linguistic and Oriental Essays*, 449–53.

107. For an analysis of this phrase, see Driver, *Geography Militant*.

108. V. Lovett Cameron, letter to the editor, *Times (London)*, November 25, 1884, 4, col. A.

109. Thomson, "East Central Africa," 66. Thomson was referring primarily to German emigrants, who were most encouraged by the Berlin Conference to look to Africa.

110. For an analysis of some of this rhetoric, see Bendikat, "The Berlin Conference."

111. Greswell, "Europe and Africa," 843–44 and 852. Greswell was the author of *Our South African Empire*, 2 vols. (London: Chapman and Hall, 1885) and of *Geography of Africa South of the Zambesi* (Oxford: Clarendon Press, 1892). He was elected a fellow of the RGS in 1893.

112. Thomson, "East Central Africa," 66.

113. Ibid., 73.

114. Ibid., 67.

115. Silva White, *Development of Africa*, vi.

116. John and Alice Stopford Green, *A Short Geography of the British Isles* (London: Macmillan, 1880), preface, quoted in Freshfield, "The Place of Geography in Education," 700.

117. Wise, "The Scott Keltie Report 1885," 375.

118. H. J. R., "Proposals for the Training of Colonial Medicine Officers," June 1, 1898, FO 2/890, "London and Liverpool Schools of Tropical Medicine, 1898–1904."

119. Patrick Manson, comments in "Dinner to Dr. Andrew Balfour . . . December 8th 1902 before Dr. Balfour left to take up his appointment as Director of the Wellcome Tropical Research Laboratories, Khartoum" [66 pp.], Balfour Dinner, WA/HSW/PE/A.7, Wellcome Library for the History and Understanding of Medicine Archives, London. Also see FO 2/890, "London and Liverpool Schools of Tropical Medicine, 1898–1904," BNA.

120. A. Lane Fox comments in Cameron, "On the Anthropology of Africa," 178.

121. Galton, "Opening Remarks by the President," 337–38.

122. Galton, "Address Delivered to the Anniversary Meeting," 391.

123. Macalister, "Presidential Address," 467–68.

124. Kingsley, *Travels in West Africa*, 435.

125. See her talk, "African Therapeutics from a Witch Doctor's Point of View," delivered on November 19, 1896; details found in RAMC 801/6, Wellcome Library, London; she later wrote this up as two chapters on "African Medicine" and "Witchdoctors" in *West African Studies*.

126. Ravenstein, "Lands of the Globe Available for European Settlement," 27; his remarks are nearly identical to those of Sir William Farr expressed twenty years earlier.

127. For earlier perspectives, see Curtin, *The Image of Africa*.

128. These maps were not unique to this period, but they circulated in far greater numbers and with far more new empirical evidence in this era. Improvements in German cartography had a significant influence on British developments as well.

129. Felkin, "On the Geographical Distribution of Tropical Diseases in Africa," 416; for a summary of the Francophone and Germanic literature, with some discussion of British Africa as well, see Poskin *L'Afrique equatoriale*.

130. Keltie, *Applied Geography*, 15; also see Silva White, *The Development of Africa*, v.

131. Silva White, *The Development of Africa*, v–vi.

132. As the editor of the *Statesman's Yearbook* since 1883, Keltie would have known these statistics better than anyone in Britain. He began his career as a staff member of the newly founded journal *Nature*.

133. Keltie, *Applied Geography*, 36–37; also see 55.

134. Silva White, *Development of Africa*, 3.

135. Ibid., 97.

136. Ibid., 96.

137. Keltie, *Applied Geography*, 80.

138. Silva White, *Development of Africa*, 98.

139. Ibid., 86. Also see Silva White, "On the Comparative Value of African Lands," 193, for his six categories, three "natural" and three "social," which he uses to calculate the value of specific African regions.

140. Keltie, *Applied Geography*, 31.

141. Ibid., 22.

142. Silva White, *Development of Africa*, 51.

143. This tradition began in the 1920s; see McPhee, *The Economic Revolution in British West Africa*; Forman, "Science for Empire"; Semmel, *Imperialism and Social Reform*; Kubicek, *The Administration of Imperialism*; Gann and Duignan, *The Rulers of British Africa*, 57–62; Worboys, "Science and British Colonial Imperialism"; Havinden and Meredith, *Colonialism and Development*; and Hodge, *Triumph of the Expert*.

144. Although Lord Milner coined the phrase, Chamberlain is widely seen as the first person to put it into effect; see Saul, "The Economic Significance of 'Constructive Imperialism.'"

145. Dumett, "Joseph Chamberlain, Imperial Finance and Railway Policy"; Storey, "Plants, Power, and Development"; and Drayton, *Nature's Government*, esp. chap. 7.

146. *Report from the Select Committee on Colonization and Settlement (India)*, iii.

147. "Royal Geographical Society . . . ," *Times (London)*, May 25, 1869, 8, col. D.

148. Because their books were reviewed "as books of the week" in early December, I surmise that they appeared in November. Ravenstein, "The Climate of Tropical Africa." The first report appeared as "Report of the Committee . . . appointed to inquire into the Climatological and Hydrographical conditions of Tropical Africa," 36[7]–68.

149. Reported in "British Association," 29 August 29, 1891, *Times (London)*, 13, col. A.

150. For a summary of the discussion, see "British Association," *Times (London)*, September 9, 1890, 4, col. A.

151. Ravenstein, "Lands of the Globe Available for European Settlement," 27.

152. *Hints to Meteorological Observers* went through three editions before the First World War (in 1892, 1902, and 1907). The last version is the most easily available: Meteorological Office (London), *Hints to Meteorological Observers in Tropical Africa*. Also see Ravenstein, *Climatological Observations at Colonial and Foreign Stations—Tropical Africa*.

153. Ravenstein, "Tables of African Climatology," in Murray, *How to Live in Tropical Africa*, 28. The committee's annual reports mentioned more than eighty stations sending in statistics, but not all followed the suggested guideline of making at least one measurement per day.

154. This was the subtitle on the panel provided by the *Times*'s correspondent in "International Geographical Congress."

155. Ravenstein comments in *Report of the Sixth International Geographical Congress*, 547.

156. Silva White, "To What Extent Is Tropical Africa Suited for Development," 550–51.

157. Christy, "Sleeping Sickness," 3.

158. Cain, "Economics and Empire," 47.

159. Curtin, Feierman, Thompson, and Vansina, *African History from Earliest Times to Independence*, 316.

160. For instance, in 1901 there were approximately 11,070 Europeans in Rhodesia (mostly residing in what became Southern Rhodesia); in 1905–06 there were only 1,813 in the East Africa Protectorate (Kenya). Figures from Kennedy, *Islands of White*, 197.

161. Blyden, *The African Society and Miss Mary H. Kingsley* (quotations in order of appearance) 7–8, 20, 14, 24, and 22.

CHAPTER TWO

1. Quotation from a 1935 renewal appeal to the Rockefeller Foundation [no date, but February 1935], GD40/17/126, Lord Lothian Papers (LP), National Archives of Scotland.

2. Quoted in Roberts, "The Imperial Mind, 1905–1940," 48.

3. Members of the Kindergarten included Robert Brand, Lionel Curtis, John Dove, Patrick Duncan, Richard Feetham, Lionel Hichens, Philip Kerr, Dougal Malcolm, Peter Perry, Geoffrey Robinson (later Dawson), and Hugh Wyndham.

4. [Kerr], "Editorial Manifesto in the First Number, November 1910," 381.

5. *Eighth Annual Report*, RIIA, 1928, quoted in Lavin, "Lionel Curtis and the Founding of Chatham House," 69; and Lavin, *From Empire to International Commonwealth*, 165–66.

6. This building was the former residence of prime ministers, including Pitt, Earl of Chatham.

7. On the "African Group" at Chatham House, see the Oldham Papers, Box 2, File 1; the description of the group is taken from a letter in that file from Oldham to W. H. Himbury (British Cotton Growing Association), December 16, 1924.

8. Smith, "The Story of the Institute," 1 and 4. The IIALC's constitution listed eight "major objects" related to research, publishing, international cooperation, and practical affairs. Despite its desire to bridge the gap between practical affairs and scientific research, it also described itself as "entirely non-political" and declared that it would "not concern itself with matters of policy or administration."

9. Description of the IIALC's function in "Resolutions and Recommendations of the Conference on the Christian Mission in Africa Held at Le Zoute, Belgium, September 14th to 21st, 1926," in R856/Health/126/56397/2625, League of Nations Archives, Geneva, Switzerland; also see "Minutes of a Meeting Held at the School of Oriental Studies (University of London) on Sept. 21st and 22nd, 1925," File 1/3, International African Institute Archives, LSE.

10. Lugard was in the chair for the founding meeting; see his remarks in "Minutes of the First General Meeting of the International Institute for the Study of African Lan-

guages and Cultures held at the School of Oriental Studies London, on June 29th and June 30th, 1926," File 1/3, International African Institute Archives, LSE.

11. Draft memo by Philip Kerr and sent to Reginald Coupland, Lionel Curtis, Geoffrey Dawson, and R. H. Brand, no date [but before April 12, 1929]; a second and revised draft was produced on July 1, 1929, "Suggestions for an Institute of Government at Oxford," both in RTA, File 2792 (1)—Oxford University Institute of Government.

12. Quotations from both documents in RTA, File 2792 (1).

13. Kerr to Coupland, October 3, 1929, LP, GD40/17/234.

14. Cooper, "Modernizing Bureaucrats, Backward Africans," 64.

15. Low and Lonsdale, "Introduction: Towards the New Order, 1945–1963," 12–16.

16. Smuts's activities have been most thoroughly documented by Dubow, *Scientific Racism in Modern South Africa*, and by Anker, *Imperial Ecology*.

17. Smuts, "African Settlement. Part I," 127; also see Smuts, "British African Conference [Smuts Lecture]."

18. MacDonald resigned from the general committee of the survey upon his appointment as secretary of state for the Dominions in 1935; between the years 1935 and 1940 MacDonald alternated between the two positions of secretary of state for the Dominions and secretary for the colonies. Ormsby-Gore was a member of both the general and the scientific committees of the survey; he resigned from these posts in 1936 when he was appointed secretary of state for the colonies.

19. Several of the individuals involved in the African Research Survey wrote autobiographies or had biographies written about them: Perham, *Lugard*; Clements, *Faith on the Frontier*; Cell, *Hailey: A Study in British Imperialism*; Butler, *Lord Lothian*; Lavin, *From Empire to International Commonwealth*; Armytage, *Sir Richard Gregory*; Huxley, *Memories I & II*; Waters and Helden, *Julian Huxley*; Sanger, *Malcolm MacDonald*; Orr, *As I Recall*; and Carney, *Stoker*. Hilda Matheson served as the secretary to the African Research Survey from 1933 to 1938.

20. Lord Lothian to Edmund Day, March 30, 1931, Rockefeller Foundation Archives, Tarrytown, New York.

21. Recent studies include Hodge, *Triumph of the Expert*; Rajan, *Modernizing Nature*; Tignor, *W. Arthur Lewis*; Feinstein, *An Economic History of South Africa*; Zachariah, *Developing India*; Cooper and Packard, *International Development and the Social Sciences*; Cowen and Shenton, *Doctrines of Development*; Havinden and Meredith, *Colonialism and Development*; and Arndt, *Economic Development*.

22. See, for instance, Hetherington, *British Paternalism and Africa*, esp. chaps. 3 and 6, "The Meaning of Colonial Trusteeship," and "Development and Research."

23. Hailey, *An African Survey*, 1662.

24. Ibid.; Worthington, *Science in Africa*; and Frankel, *Capital Investment in Africa*. Other volumes that in part owed their origins to the Survey include Perham, *Native Administration in Nigeria*; Macmillan, *Africa Emergent*; and Kuczynski, *The Cameroons and Togoland*.

25. The budgets and expenses for the African Survey can be found in Lord Lugard's Papers (LLP), Box 149, File 4, Rhodes House, Oxford. At the end of 1936, E. B. Worthington had received £1,490 for his services; Kuczynski had received £236, Perham had received £200, and Frankel had received £150. These figures do not include additional expenses for Worthington's tour of sub-Saharan Africa or the fees provided to the scientific advisors for their draft reports and comments. The print runs for Worthington's and Frankel's volumes were 2,500 while Perham's was only 1,000; all were published by Oxford University Press (OUP). Perham's and Kuczynski's vol-

umes were published outside the auspices of the Survey though they were used as preliminary reports. Publishing figures from out-of-print archives, OUP, Oxford.

26. "Notes of an informal meeting held at Chatham House on Sept 20th [1933]," [present were Lord Hailey, Julian Huxley, Joseph Oldham, and Hilda Matheson], Folder 1 16/26, African Survey, Royal Institute of International Affairs; and Minutes of the 2nd Meeting of the General Committee, October 31, 1933, "Appointment of Experts—Scientific," LLP, Box 149, File 4. The full quotation reads, "It has been found on enquiry that this [the scientific research] constituted almost a separate task, necessitating the services of a trained scientist to collect and evaluate data."

27. An important exception is Anker's *Imperial Ecology*, which focuses most attention on relationships between Britain and South Africa. For a review of the literature in the history of ecology, see chapter three. John Cell's work remains the most extensive overview in "Lord Hailey and the Making of the African Survey," and in his chapter in *Hailey: A Study in British Imperialism*. Also see Lee, *Colonial Development and Good Government*; Hargreaves, "History: African and Contemporary," 3–8; Robinson, "Experts, Colonialists, and Africanists, 1895–1960"; Pearce, *The Turning Point in Africa*, esp. chap. 3, "Lord Hailey and Colonial Office Thought on African Policy"; Hetherington, *British Paternalism and Africa*; and more peripherally, Lavin, *From Empire to International Commonwealth*; and Constantine, *The Making of British Colonial Development Policy*.

28. For some of the discussions leading up to the establishment of the fund, and the transition from Africa to all the colonies, see the materials in CO 847/13/13, BNA.

29. See, for instance, Worthington, "Organization of Science in East Africa," 451–53; Worthington, *A Survey of Research and Scientific Services in East Africa*; and CO 927/124/4 "Regional Co-Ordination of Research: East Africa Research Priorities, 1950," BNA.

30. Like the African Survey itself, the Scientific Council for Africa South of the Sahara had multifaceted origins, but an important strand can be traced back to the Survey's call to establish a permanent African Research Bureau. These connections are spelled out most clearly in Audrey Richards to Charles Jeffries, February 25, 1942, CO 847/23/4, BNA. The Colonial Office and the British government more broadly were able to support the Scientific Council because it was seen as an initiative in line with the African Survey's findings.

31. Hailey, *An African Survey*, 1662.

32. Figures for *An African Survey* from Oxford University Press out-of-print archives and in CO 323/1524/8, BNA; figures for *The Dual Mandate* from Perham, *Lugard*, vol. 2, 645.

33. John W. Gardner to Mr. Morris Hadley [Trustee], Esq., October 25, 1957, Box 314, Carnegie Corporation Archives, Columbia University.

34. Hargreaves, "History: African and Contemporary," 7.

35. [H. A. L. Fisher], Chair, Warden of New College, "Minutes of a Conference on Africa at Rhodes House, 9–10 & 16 November, 1929," LP, GD40/17/120. It should be noted that these are not verbatim transcripts, but brief summaries of each speaker's comments.

36. The heads of the colleges were Fisher (New College), W. G. S. Adams (All Souls), Walter Buchanan-Riddell (Hertford), Michael Sadler (University), and Alexander Lindsay (Balliol); the natural scientists were John S. Haldane (physiology), Arthur Tansley (botany/plant ecology), Edward Poulton Hope (entomology/zoology), Robert Troup (forestry), George Dreyer (pathology), and James Watson (agricul-

ture); the social scientists included Gilbert Murray (Greek), Robert Marrett (social anthropology), James Brierly (international law), and Henry Harrod (economics); other Oxford participants included Sir Francis Wylie, the Oxford secretary to the Rhodes Trust, and four individuals identified only by their surnames: Mr. Rian, Mr. Gillett, Mr. Harlow, and Mr. Manning.

37. Constantine, *The Making of British Colonial Development Policy*, chap. 7, "The Colonial Development Act, 1929." Also see, more recently, Hodge, *Triumph of the Expert*.

38. Leopold Amery, in *Times (London)*, January 30, 1925, also partially quoted in MacLeod and Andrews, "The Committee of Civil Research," 683; MacLeod and Andrews list the year as 1924 instead of 1925.

39. Unless otherwise noted, all quotations from the conferences come from "Minutes of a Conference on Africa at Rhodes House, 9–10 & 16 November, 1929," LP, GD40/17/120.

40. Given Kerr's later descriptions of Smuts's comments, it seems likely that it was at this point in the discussion that Smuts spoke of the need for a comparative study of the different colonial powers in Africa "as a whole." This was the summary Kerr provided in his letter to Frank Aydelotte, president of Swarthmore College and U.S. secretary for the Rhodes Trustees; see Kerr to Aydelotte, November 14, 1929, RTA, File 2792.

41. Coupland to Oldham, October 15, 1929, OP, Box 3, File 4.

42. Joseph Oldham, statement on "The Aim of Christian Missions," February 1920, quoted in Clements, *Faith on the Frontier*, 252; also see Oldham, *Christianity and the Race Problem*. See chapter five for an extended discussion on racial politics.

43. *Report of the Commission on Closer Union of the Dependencies in Eastern and Central Africa*.

44. Oldham, letter to unnamed individual, January 31, 1926, quoted in Clements, *Faith on the Frontier*, 233.

45. Oldham to Lord Archbishop of Canterbury, November 13, 1929, OP, Box 4, File 1.

46. *Memorandum Relating to Indians in Kenya*; this was also known as the Devonshire declaration.

47. Barnes and Nicholson, *The Empire at Bay*, 53.

48. Ibid., 53–54.

49. Oldham to William Temple, Lord Archbishop of York, November 11, 1929, OP, Box 4, File 3.

50. Murray, "At Home in the Modern World," 34.

51. Oldham to William Temple, November 11, 1929, OP, Box 4, File 3.

52. Barnes and Nicholson, *The Empire at Bay*, 53–54.

53. See Philip Kerr, "The African Highlands, 25 February 1927," LP, GD40/17/83; Kerr's population estimate was much higher than official reports. In his cover letter Kerr explained his purpose: "The attached Memorandum is an attempt to set forth some of the problems which are beginning to confront that part of Central and Southern Africa which Mr. Cecil Rhodes worked so hard to include in the British Empire, in whose development and problems he always showed so much interest and in which I feel sure he would have wished his Trustees to take some special concern."

54. I have taken the second half of this quotation from his published rejoinder to Smuts; he said something similar at the conference, but his expression here is more succinct; Oldham, *White and Black in Africa*, 70.

55. Emphasis in original.

56. These are discussed in chapter four.

57. Joseph Oldham to Sir Walter Fletcher [of the Medical Research Council], September 8, 1925, OP, Box 2, File 3.
58. Committee of Civil Research, Minutes of a Meeting of the Sub-Committee held in Confidence, January 10, 1927, 9, LP, GD40/17/93. Among those present at his testimony were Thomas Jones, Sir Frank Heath, secretary, Department of Scientific and Industrial Research; Charles Strachey, asst. undersecretary of state, Colonial Office; Dr. Andrew Balfour, director, London School of Hygiene and Tropical Medicine; Edwin Smith, British Foreign and Bible Society; Sir Robert Greig, chair, Board of Agriculture for Scotland; Sir James Currie, director, Empire Cotton Growing Corporation; Professor E. B. Poulton, Hope Professor of Zoology, Oxford; and E. N. Fallaize, secretary, Royal Anthropological Institute.
59. Oldham was appointed to the commission in November of 1927; on his early connection with Malinowski, see his "Notes of Conversation with Professor Malinowski, 2nd December 1927," and B. Malinowski, "Memorandum on Colonial Research," December 1927, (presumably produced for Oldham), OP, Box 2, File 2; also see "Notes of Conversation with J. H. Driberg, 5th December 1927."
60. Malinowski's draft article on "Practical Anthropology" was first discussed by the IIALC's executive council during its 6th Meeting on December 19–20, 1928; File 1/3 "Executive Council Minutes, 1925–1931," International African Institute Archives, LSE.
61. Bronislaw Malinowski, "Re: Discussion with Philip Kerr," March 1930, MP, Africa I, #8; the research priorities of Malinowski and the IIALC are explored at greater length in chapter six.
62. See Oldham to Kerr, October 14, 1929, OP, Box 4, File 2; Kerr to Westermann, November 14, 1929; and Westermann to Kerr, November 19, 1929, RTA, File 2792. The minutes of the conference do not indicate which participants attended for which days so I must go by the transcripts of who spoke as well by the evidence I have located for particular individuals that allows me to conclude they missed the final session.
63. Onderstepoort had become a world center for veterinary research by the 1920s. See Brown, "Tropical Medicine and Animal Diseases"; and Gilfoyle, "Veterinary Immunology as Colonial Science."
64. In response, Thomas Jones recommended that Troup write up his recommendations and submit them to the Committee on Civil Research and to the Colonial Development Fund, which "contained special mention of surveys and scientific research as coming within its purposes."
65. This point was made explicitly by Walter Buchanan-Riddell, Master of Hertford College.
66. Kerr to Felix Frankfurter, Harvard Law School, July 16, 1929; also see Coupland to Kerr, June 6, 1929 and Kerr to Robert Brand, October 8, 1929, in LP, GD40/17/235.
67. Oldham to Coupland, November 12, 1929, OP, Box 3, File 4.
68. Westermann to Kerr, November 19, 1929, RTA, File 2792.
69. Coupland to Kerr, November 11, 1929, also Coupland to Kerr, November 11, 1929 (two written on the same day), and Coupland to Kerr, November 12, 1929, in LP, GD40/17/234; emphasis in original.
70. Oldham to Lord Lugard, November 12, 1929, OP, Box 3, File 1.
71. [Julian Huxley], "Memorandum on the Possibility of Establishing an Ecological Department as Part of the Proposed Oxford School of African Studies," and Kerr to Fisher, February 19, 1930, in which Kerr says he is returning Julian Hux-

ley's memorandum, RTA, File 2792; and Coupland to Kerr, February 25, 1930, LP, GD40/17/239.

72. The members and their respective fields were Arthur Tansley (botany/plant ecology), Robert Marrett (applied/social anthropology), Robert Troup and John Burtt Davy (forestry), Charles Elton (animal ecology), Edward Poulton (entomology), Cecil Morison (agriculture), and Beckit (geography).

73. All quotations in "Report of the Science Committee of the Conference on African Problems," February 28, 1930, RTA, File 2792.

74. "Memorandum Prepared by Professor Julian Huxley and Dr. W. K. Spencer for the Advisory Committee on Education in Tropical Africa," 1928, and Huxley, *Biology and Its Place in Native Education in East Africa.*

75. "Memorandum on the Possibility of Establishing an Ecological Department as Part of the Proposed Oxford School of African Studies," RTA, File 2792.

76. Kerr to Curtis, February 28, 1930, RTA, File 2792.

77. Philip Kerr, "Future of Rhodes House: Proposed Institute of Government," [February 1930], LP, Box 149, File 1.

78. Oldham, *White and Black in Africa.* This book was published before Smuts's lectures appeared in book form, but they had already been printed almost in full in the *Times (London)* and in various journals, including the *Journal of the Royal African Society.* Oldham discussed his concerns about the parliamentary committee in a letter: Oldham to Archbishop of Canterbury, November 28, 1929, OP, Box 4, File 3.

79. Kerr to Oldham, February 12, 1930, LP, GD40/17/243.

80. Oldham to Kerr, February 19, 1930, RTA, File 2792.

81. Lugard to Wilson, February 25, 1930, LLP, Box 149, File 1 and CO 323/1069/6.

82. Passfield to Sir Cecil Bottomley and Sir Samuel Wilson, March 4, 1930, CO 323/1069/6. Lothian seems to have dramatically misread Passfield and Wilson's attitude; Kerr to Fisher, March 4, 1930, RTA, File 2792.

83. All quotations from an undated and unsigned memorandum that Lugard wrote following the meeting in early March, "Commentary on what a memorandum should contain," LLP, Box 149, File 1. Also see Lugard to Wilson, March 10, 1930, in which he comments, "Each side will now draw up a Memorandum, stating as tersely as possible its own point of view."

84. Bronislaw Malinowski, "Items of Expenditure under the Million Dollar Interlocking Scheme for African Research," March 1930, Malinowski Papers, Africa I, #8, London University, London School of Economics, London. Also cited in Cell, "Lord Hailey and the Making of the African Survey," 487.

85. Stocking, *After Tylor,* 400. Also see Cell, "Lord Hailey and the Making of the African Survey," 487: "Malinowski used his strong influence with Rockefeller to reinforce the Foundation's preference for London."

86. All quotations from "A Proposed Institute of African Studies at Oxford University," submitted to the Rockefeller Foundation on March 28, 1930, RTA, File 2792; in the same file is a copy of the IIALC's "Memorial Presented to the Rockefeller Foundation."

87. The seriousness with which the Oxford institute pursued anthropology is attested in the correspondence regarding a possible director; see Marett to Fisher, February 27, 1930, and A. C. Haddon to Fisher, April 23, 1930, RTA, Miscellaneous File; Huxley to Coupland, March 27, 1930, LP, GD40/17/239. Potential directors for the anthropology section included Captain Rattray, anthropologist for the Gold Coast, A. R. Radcliffe-Brown, who had just completed several years at the University of Cape

Town as professor of social anthropology; and Jack H. Driberg, a retired officer from Kenya who had just written *The Savage as He Really Is*.

88. "Report of the Science Committee of the Conference on African Problems," February 28, 1930, RTA, File 2792; this phrase was provided by Charles Elton under the heading "animal ecology."

89. "Memorandum Re International Institute of African Languages and Cultures: Interview between JVS and Malinowski," no date [after March 1930], Rockefeller Foundation Archives (RF), Tarrytown, New York; and Oldham to Coupland and Kerr, June 5, 1930, RTA, Box 218, File 2792.

90. Kerr to Robert Brand, October 30, 1929, LP, GD40/17/234.

91. Curtis to R. Feetham, October 14, 1930, quoted in Lavin, *From Empire to International Commonwealth*, 234.

92. Oldham to Tom Jones, February 14, 1929, OP, Box 4, File 2. Also see the review, "The Native Problem in Africa," 93–94.

93. The bureau had been established in 1924 and was supported by a grant of £10,000 per year for five years from the Laura Spelman Rockefeller Memorial; *Reports of the President and Treasurer of Harvard College, 1925–26*, 27.

94. "Memorandum on the Financial Needs of Oxford, by Mr. Lionel Curtis Submitted to the Hebdomadal Council by the Vice-Chancellor," December 16, 1930, LP, GD40/17/247. The first version was submitted in November; this is the final, revised version.

95. Minutes by A. Bevir, January 24, 1930; Lord Passfield, March 4, 1930, and C. Bottomley, April 24, 1930, CO 323/1069/6. Bottomley served as the assistant undersecretary of state for the colonies for the African territories from 1930 to 1938.

96. For the rumors regarding Smuts and Curtis, see Passfield's minutes, February 10, 1930 and March 4, 1930, and E. J. Harding to S. Wilson, May 2, 1930, CO 323/1069/6.

97. Minute by A. Bevir, April 23, 1930, CO 323/1069/6.

98. R. V. Vernon, January 25, 1930, CO 323/1069/6. Vernon was an assistant secretary of state without a geographical emphasis, but responsible for "general" issues.

99. C. Bottomley, May 5, 1931, CO 323/1115/18; and various minutes in CO 323/1166/14 and CO 847/2/1.

100. A. Bevir, April 23, 1930, CO 323/1069/6.

101. Bottomley, March 22, 1933, CO 847/2/1. A similar assessment was made in an unsigned handwritten comment, March 22, 1933, CO 847/2/1.

102. A. Fiddian, January 16, 1933, CO 847/2/1.

103. R. V. Vernon, May 19, 1931, CO 323/1115/18.

104. The results of these various committees can be found in *Report of Lovat Committee on Agricultural Research and Administration*, undertaken by Lords Milner and Lovat; *Summary of First Colonial Office Conference*, includes brief report on possible Scientific and Research Service; *Report of Lovat Committee on Agricultural Research*; and *Report of Lovat Committee on Veterinary Services*.

105. Alter, *The Reluctant Patron*.

106. The medical advisor was appointed based on recommendations made in the 1920 report on the Medical Service (Cmd. 939, 1920). The Advisory Medical Committee was founded in 1909 to cover matters relating to tropical Africa and was expanded in 1922 to cover the colonial empire (again as a result of the 1920 report). Jeffries, *The Colonial Office and Its Civil Service*, 157. The eight Agricultural Bureaus included

soil science, animal health, animal nutrition, plant genetics (2 sites), fruit production, animal genetics, and agricultural parasitology.

107. R. V. Vernon, May 19, 1931, CO 323/1115/18.
108. "Memorandum on the Possibility of Establishing an Ecological Department as Part of the Proposed Oxford School of African Studies," RTA, File 2792, emphasis in original; and the revised version, Julian Huxley, "A Proposal for a Bureau of Ecology for Africa," April 29, 1931, "African Survey Correspondence," File 3, 16/26d, Royal Institute of International Affairs, London.
109. Huxley, "Appendix No. 36 Memorandum," 261.
110. G. J. F. Tomlinson, May 5, 1931, CO 323/1115/18; Tomlinson was assistant undersecretary of state for personnel between 1931 and 1939.
111. R. V. Vernon, January 25, 1930, Bevir, January 24, 1930, G. G., January 27, 1930, and P[assfield], February 10, 1930, CO 323/1069/6.
112. W. C. B., January 27, 1930, CO 323/1069/6.
113. Sir Basil Blackett to H. A. L. Fisher (as chair of committee), December 11, 1929, RTA, File 2792.
114. Bulmer and Bulmer, "Philanthropy and Social Science in the 1920s," 387.
115. The two were the Rockefeller Foundation and the Laura Spelman Rockefeller Memorial Fund.
116. "Work of the Rockefeller Foundation," 438–40.
117. Havinden and Meredith, *Colonialism and Development*, 164.
118. Worboys, "Science and British Colonial Imperialism," 280.
119. Fisher, "The Role of Philanthropic Foundations in the Reproduction and Production of Hegemony"; Bulmer, "Philanthropic Foundations and the Development of the Social Sciences in the Early Twentieth Century"; Fisher, "The Rockefeller Foundation and the Development of Scientific Medicine in Great Britain"; and Fisher, "Rockefeller Philanthropy and the British Empire."
120. Henry Clay, September 27 1930, in Selskar M. Gunn's diary, "Memoranda of Conversations Held in London in Connection with Various African Projects," RF.
121. "Resolutions from 15 April 1931 for the IIALC and the Institute of African Studies at Oxford," RF. The sum to the IIALC was in the form of an unconditional grant of $125,000 to be dispersed in annual installments of $25,000 over five years and $125,000 in matching grants to be dispersed up to $25,000 per year at a ratio of $1 from the foundation for every $2 raised by the IIALC.
122. "Institute of African Studies at Oxford," Officers' Resolutions, April 15, 1931, RF.
123. R. V. Vernon, May 19, 1931, CO 323/1115/18.
124. Oldham to Coupland and Lothian, June 5, 1930, RTA, File 2792, and "Memorandum Concerning the Proposed Institute for African Studies at Oxford University," submitted by S. M. Gunn, March 16, 1931, RF.
125. Oldham to Curtis, January 8, 1931, LP, GD40/17/247. Edmund Day to Raymond Fosdick, January 27, 1930, RF, expresses the same sentiment.
126. "Memorandum Concerning the Proposed Institute for African Studies at Oxford University," submitted by S. M. Gunn, March 16, 1931, RF.
127. Ibid.
128. Oldham to Curtis, January 8, 1931, LP, GD40/17/247.
129. Curtis to Kerr, January 19, 1931, LP, GD40/17/247. Also see Flexner to Lothian, June 20, 1930, and Lothian to Flexner, June 24 1930, GD/40/17/248.
130. Lothian to Curtis, January 20, 1931, LP, GD40/17/247.

131. Lothian to Day, March 30, 1931, RF. Also see Fisher to Lothian, February 12, 1930, RTA, File 2792: "I think that the University ought out of its own resources, or out of the resources specially contributed by this country, to provide for general political studies. A thoroughly good exploration of the African difficulty could not I think under any possibility be undertaken by the University without support from outside."

132. Lothian to Oldham, April 20, 1931, and Oldham to Lothian, April 22, 1931; and Frederick Keppel, president of the Carnegie Corporation, to Curtis, June 30, 1931, LP, GD40/17/256.

133. Gunn, "Memorandum concerning the Proposed Institute for African Studies at Oxford University," March 16, 1931, RF.

134. Coupland to Lothian, March 1, 1931, RTA, File 2792.

135. Coupland to Lothian, April 27, 1931, which includes the invitation letter that Lothian was to send out, and Coupland to Lothian, March 31, 1931, in which Coupland originally proposes the meeting on May 17 and names the individuals who should be invited, RTA, File 2792; and Julian Huxley, "A Proposal for a Bureau of Ecology for Africa," April 29, 1931, "African Survey Correspondence," File 3, 16/26d, RIIA.

136. Tomlinson, May 18, 1931, CO 323/1115/18. Mitchell and Morison only attended the latter half of the meeting. Mitchell was an active participant in debates over the relevance of anthropology to colonial administration, having taken a skeptical view of Malinowski's argument that they could be easily combined. See Mitchell, "The Anthropologist and the Practical Man"; for more on Mitchell see chapter six.

137. Passfield minute, May 20, 1931, CO 323/1115/18, summarizing a conversation with Oldham. John Cell has conflated many of these negotiations and mistakenly concluded that at this time, "Oldham now took the matter in hand, drawing up a plan not for a half-baked centre but for something recognizably like the eventual African Survey: a comparative investigation of the major problems, current research, and future needs." Cell, "Lord Hailey and the Making of the African Survey," 489.

138. [J. H. Oldham,] "African Research," May 18/19, 1931, LP, GD40/17/121, and Oldham to Lothian, May 19, 1931, LP, GD40/17/259, in which he wrote, "I enclose a memorandum in which I have tried to embody the results of the conversation at Oxford."

139. Tomlinson, May 18, 1931, CO 323/1115/18.

140. [J. H. Oldham,] "African Research," May 18/19, 1931, LP, GD40/17/121.

141. Minutes of a luncheon for Dr. Frederick Keppel and Mr. Arthur Page of the Carnegie Corporation of New York, June 27, 1939, at the Royal Institute of International Affairs, Box 315, Carnegie Corporation Archives (CC), Columbia University, New York. Most of the records regarding the corporation's grant-making activities in its early decades were destroyed, but a handful of files of Keppel's annual London luncheons on the British Empire survive.

142. Curtis to Justice Feetham, June 8, 1931, LP, GD40/17/256, and Curtis to Keppel, July 20, 1931, LP, GD40/17/121.

143. For the foundation's history, see Lagemann, *The Politics of Knowledge*; on its later interests in Africa, see Murphy, *Creative Philanthropy*.

144. Lester, *Summary of Grants from the British Dominions and Colonies Fund, 1911–1935*, and "Luncheon for Dr. Frederick Keppel and Mr. Arthur Page of the Carnegie Corporation of New York," June 27, 1939, at the Royal Institute of International Affairs, Box 314, CCA.

145. Keppel to Curtis, June 30, 1931, LP, GD40/17/256.

146. Ibid., and Keppel to Curtis, July 29, 1931, GD40/17/256.

147. Curtis to Keppel, July 20, 1931, LP, GD40/17/121; Keppel to Curtis, July 29, 1931, LP, GD40/17/256; and Keppel to Curtis, October 17, 1931, LP, GD/40/17/120.

148. In attendance were Lothian, Oldham, Huxley, Coupland, Lugard, Basil Blackett, William Ormsby-Gore, Lord Snell, Malcolm MacDonald, and Ivison Macadam (of the RIIA). Both Lionel Curtis and Leopold Amery had been invited, but each had prior engagements. See Macadam to Lothian, November 12, 1931, LP, GD40/17/120, and "Extract from the Standing Orders Committee Minutes, Dec 2nd 1931," Folder 1 16/26, "African Survey," RIIA.

149. Coupland to Lothian, May 19, 1931, RTA, File 2792.

150. Keppel to Curtis, June 30, 1931, LP, GD40/17/256.

151. All quotations from Curtis to Lord Lothian, May 20, 1932, LP, GD40/17/120. Curtis described their criteria in this letter with respect to a candidate then under review, Sir William Marris, a South African barrister.

152. The number comes from Lavin, From Empire to International Commonwealth, 235. When Hailey's name first arose for consideration, Lord Lugard wrote to his nephew in India to sound him out on his qualities; Lugard to Frank Brayne (commissioner of Lahore), no date, and Frank Brayne to Lugard, July 28, 1932, LLP, Box 8, File 4. Although Brayne called Hailey "a very able man . . . with imagination, a large outlook and lots of ideas," he also wrote: "His trouble is jealousy, he cannot abide other people or their ideas getting the lime-light. If he can absorb and put them out as his own, well and good, but otherwise he will not encourage them." There is no other evidence that Hailey was considered by the committee in 1932, so it seems possible that the date on the letter is incorrect.

153. Lionel Curtis to Alexander Carr-Saunders (copied to Lord Lothian), June 29, 1933, LP, GD40/17/121.

154. Rich, Race and Empire in British Politics, 147.

155. On Grigg's candidacy, see Jan Smuts to Lothian, March 6, 1930, and Lothian to Smuts, March 28, 1930, in File 2792, RTA.

156. Cell, Hailey: A Study in British Imperialism, chap. 15, "Surveyor of Africa, 1935–1939," and Cell, "Lord Hailey and the Making of the African Survey."

157. Hailey to Lothian, April 30, 1934, LP, GD40/17/124; Hailey to Lothian, August 16, 1934; Hailey to Lothian October 1, 1934; Hailey to Lothian, December 12, 1934, LP, GD40/17/125; and Macadam to Hailey, June 6, 1935, LP, GD40/17/126.

158. Matheson began part-time in July 1933 and moved to full-time in February 1934. The executive committee consisted of Huxley, Oldham, Lothian, Curtis, and Ivison Macadam (as liaison for Chatham House); see Lothian to Hailey, January 31, 1934, LP, GD40/17/123, and Hailey to Lothian, February 22, 1934, LP, GD40/17/124.

159. Hailey and Malcolm departed on August 15, 1935 and journeyed through South and East Africa before being joined by Worthington in West Africa in late March 1936; the journey was complete by June. Malcolm had been selected in part because of his "knowledge of scientific work as well as administration"; he had a bachelor of science in agriculture. See Hailey to P. E. Mitchell, May 22, 1935, and Matheson to Lothian, May 24, 1935, LP, GD40/17/126.

160. Executive Committee Statement to Carnegie Corporation, no date, and Lothian to Keppel, December 16, 1936, GD40/17/128.

161. Tomlinson minute, July 18, 1933, CO 847/2/1/4204, in which he reported Cunliffe-Lister's views; and Macadam to Keppel, June 26, 1933, GD40/17/121, summarizing

statements at the Savoy Hotel luncheon on June 23 to discuss the African Research Survey.

162. The Kenya proposal is dealt with in chapter five; the Rhodes-Livingstone Institute in chapter six.

163. Comments of Julian Huxley in Minutes of Meeting of African Research Survey, July 15–16, 1933.

164. "Notes of discussion at Blickling on Sunday, July 30th 1933," Folder 1 16/26, RIIA; present at this informal meeting were Lothian, Curtis, Oldham, Hailey, Matheson, and Margery Perham.

165. Matheson to Lothian, August 22, 1933, Matheson to Lothian, August 14, 1933, and Huxley to Matheson, August 4, 1933, LP, GD40/17/122.

166. "Notes of an informal meeting held at Chatham House on Sept 20th [1933] at 4:30 pm," Folder 1 16/26, RIIA. These results were also reported to Lionel Curtis, who responded that he felt Worthington should be secured "for the duration of the Survey."

167. Huxley in Minutes of Meeting of African Research Survey, July 15 & 16 1933, LLP, Box 149, File 4.

168. "Notes of discussion at Blickling on Sunday, July 30th 1933," Folder 1 16/26, RIIA.

169. Minutes of Executive Committee meeting at Blicking, October 14–15, 1933, and "Scientific Survey" [from Hailey and the Executive Committee to Worthington, no date], LP, GD40/17/123.

170. Worthington, *A Report on the Fishing Survey of Lakes Albert and Kioga*, esp. appendix V, "Ecology and General Natural History," 114–28; Worthington, *A Report on the Fisheries of Uganda*; Worthington, "Scientific Results of the Cambridge Expedition to the East African Lakes, 1930–1." For an interpretation of Worthington's work in this period in the context of "human ecology," see Anker, *Imperial Ecology*.

171. E. B. and S. Worthington, *Inland Waters*, 243.

172. Worthington, "The Lakes of Kenya and Uganda," 294.

173. On the idea of local knowledge as a "gift," see Bravo, "Ethnological Encounters."

174. Worthington, *Ecological Century*, 8; on p. 27 he continued, "study of the fishing methods in these African lakes, of canoes, and of the men and women who use them, gave me a respect for the diversity and adaptability of African peoples. Their attributes in perception, adaptation, and inventiveness gave great hopes for the future."

175. E. B. Worthington, unpublished manuscript, chapter three, "Lake Victoria," 9.

176. Werskey, "*Nature* and Politics between the Wars," 462.

177. Letters to Lugard from Matheson, April 10, 1934 and October 12, 1934, LLP, Box 149, File 4.

178. Worthington, *Ecological Century*, 30–32; personal interview, November 16, 1996; letter from Matheson to Tomlinson, December 14, 1933; and minute by Frank Stockdale, January 9, 1934, CO 847/2/1/4204.

179. Matheson to Lothian, January 3, 1934, LP, GD40/17/123, and Minutes of Executive Committee Meeting, February 22 and March 6, 1934, LP, GD40/17/124. Matheson wrote on April 10: "Work in Progress—Scientific: special sections of the field are being covered by supplementary memoranda; e.g. Professor Troup of Oxford is preparing a report on forest policy and work in British Africa; Mr. Bonacina of the Royal Geographical Society is providing material on meteorology; Professor Tansley of Oxford on plant ecology; Dr. Hirsch, Director of the Physical Sciences in Egypt, on the upper reaches of the Nile; Sir David Chadwick has furnished introductions

to the Imperial Agricultural Bureaux. Arrangements are now being made, with Lord Lugard's help, to secure introductions to similar sources of information in France and Belgium."

180. Worthington to Malinowski, April 28, 1934, and Malinowski to Worthington, May 4, 1934, File Correspondence U to W, MP.

181. The twelve chapters consisted of "Surveys and Mapping," "Geology," "Meteorology," "Soil Science," "Botany," "Zoology," "Entomology," "Agriculture," "Forestry," "Fisheries," "Food Preservation and Stored Products Research," and "Health and Medicine." The complete set can be found in Folder 4 16/26F, RIIA.

182. Lothian to Hailey, November 15, 1934, LP, GD40/17/125.

183. Matheson to Lothian, November 14, 1934, LP, GD40/17/125.

184. Matheson to Hailey, January 15, 1935, LP, GD40/17/126.

185. Lothian to John Boyd Orr, 18 March 1935, LP, GD40/17/126.

186. General and Executive Committee Meeting Minutes, March 9–10, 1935, LP, GD40/17/126.

187. Tomlinson to Bottomley, February 4, 1935, CO 847/4/2/47002.

188. The meeting between the five men took place on February 15, 1935; Tomlinson minute, February 18, 1935, CO 847/4/2.

189. Hemming to Hailey, December 10, 1935, CO 847/4/2; Hemming was a specialist in entomology. A more disparaging comment was made by J. E. Flood, assistant secretary of state for East Africa, who remarked, "The thing so far as I can judge is a sheer waste and adds nothing to anyone's knowledge. It isn't even accurate. I certainly don't want to be mixed up in it." October 21, 1935, CO 847/4/2.

190. Tomlinson, February 18, 1935, CO 847/4/2.

191. Tomlinson, March 4, 1935, CO 847/4/2; according to Tomlinson, Ormsby-Gore had begun to read Worthington's chapters and "had already formed the view that the trees obscured the wood."

192. Gregory to Lothian, April 9, 1935, LP, GD40/17/126.

193. Werskey, The Visible College, 30.

194. Kay, "John Boyd Orr," 60; also see Orr, "Problems of African Native Diet: Foreword."

195. Minutes of Scientific Sub-Committee, 1st Meeting, April 30, 1935, LP, GD40/17/126. Ormsby-Gore was unable to attend this meeting, but was considered a member of the committee.

196. Ibid.

197. Ibid.

198. Minutes of Executive Committee Meeting, March 9–10, 1935, LP, GD40/17/126.

199. The first was completed in 1934, the second in 1935, and the third in 1937–38; a complete set of the second draft is located in CO 847/4/1, BNA.

200. Minutes of 2nd Scientific Sub-Committee, December 1935, LP, GD40/17/127. This number was ultimately increased to 275; see Worthington, Science in Africa, 615–25.

201. Malcolm MacDonald to Sir Hubert Young, Governor Northern Rhodesia (Zambia), October 8, 1935, SEC1/1727, "African Research Survey," Zambian National Archives (ZNA), Lusaka, Zambia.

202. Malcolm had been an assistant district officer in Tanganyika and would be appointed later in 1936 to conduct a survey of land tenure in Sukumaland, Tanganyika.

203. African Research Survey, File 21984, Folders 1, 5, 6, 7, and 8, Tanzanian National Archives (TNA), Dar es Salaam, Tanzania. Less extensive comments were also received

from the Tsetse Research Department (Folder 4) and the Fisheries Service (Folder 2). Comments from the Department of Veterinary Science and Animal Husbandry (Folder 3) are unfortunately missing. The individuals were as follows: game warden, S. P. Teare; Agriculture, E. Harrison; Forestry, W. F. Baldock; Lands and Mines, H. P. Rowe and F. B. Wade; and Medical, R. R. Scott and B. Gaffney.

204. Unfortunately only excerpts of most of these letters survive; these can be found in LP, GD40/17/127. Colonial Office and Economic Advisory Council replies from Frank Stockdale (Agriculture and Animal Husbandry), Francis Hemming (Entomology), and Thomas Stanton (Medicine) can be found in CO 847/4/2, BNA.

205. Hailey to Lothian, September 6, 1935, LP, GD40/17/127.

206. Minutes of 2nd Scientific Sub-Committee [December 11, 1935], LP, GD40/17/127.

207. Minutes of Executive Committee, July 20, 1935, and Minutes of 2nd Scientific Sub-Committee [December 11, 1935], LP, GD40/17/127.

208. Hailey to Lothian, April 24, 1936, LP, GD40/17/128.

209. See his contribution, Worthington, "On the Food and Nutrition of African Natives." Other members included K. Abrahams, dietician to University College Hospital; D. G. Brackett, secretary to the IIALC; J. L. Gilks, former director of Kenya Medical Service; and F. Kelly, assistant director, Imperial Bureau of Animal Nutrition.

210. Personal interview, June 22, 1998. This is borne out by an examination of his diary of the journey, which Worthington kindly allowed me to read. Most of the entries deal with specific departments, laboratories, dispensaries, hospitals, and research stations. Two of the more general entries read: April 10, "Back at the rest house . . . we [Hailey, Malcolm and Worthington] set to chatting about the better organisation of scientific departments, the possibilities of making annual reports of value instead of keeping them quite useless, as at present, and of mixed farming, animal diseases, and so forth." On May 1: "A long discussion after dinner about administration, native law, anthropology, and so forth."

211. Journal entry, April 7, 1936.

212. Lord Hailey, "A Note on the Scheme for an African Bureau," December 14, 1938, and Hilda Matheson to the president, Rockefeller Foundation, December 15, 1938, Box 204, File B, International Missionary Council (IMC), School of Oriental and African Studies Library, London; also Matheson, "Notes on Follow-Up of African Survey Report," January 21, 1937, LP, GD40/17/128.

213. Lord Lothian to Frederick Keppel (Carnegie Corporation), December 16, 1936, LP, GD40/17/128; emphasis in original.

214. D[aryll] F[orde], "Note on the Position and Prospects of the International African Institute – Confidential," April 27, 1944, in IMC, Box 204, File on "International Africa Institute," SOAS.

215. See the materials in CO 847/13/13, BNA; and also Havinden and Meredith, *Colonialism and Development*, 197–202.

CHAPTER THREE

1. Worthington, *The Ecological Century*, 36.

2. Hilda Matheson to Lothian, February 23, 1938, LP, GD40/17/129; and Hailey, *An African Survey*, 1624.

3. Haddon, "Presidential Address: Section H Anthropology," 512. Also see Haddon, "Anthropology, Its Position and Needs"; he discusses "anthropological oecology" and the "oecology of nature folk" on pp. 12–13.

4. George Taubman Goldie was best known as the founder of the Royal Niger Company and had been a representative at the Berlin Conference. See his comments in the discussion section following Elliot, "Suggestions for an Inquiry into the Resources of the Empire," 559.

5. Goldie, "Geographical Ideals [November 1906, address delivered to the Scottish Geographical Society]," 10, 14.

6. Agricultural activities tended to include veterinary staff as well, but I have separated these out in the appendix.

7. "Report of Committee on Colonial Scientific and Research Services," 27.

8. I take the phrase from Fairhead and Leach, *Misreading the African Landscape*. Landmark publications that have shaped these debates outside those I discuss in this section include Richards, "Ecological Change and the Politics of African Land Use"; Anderson, "Depression, Dust Bowl, Demography, and Drought"; Beinart, "Soil Erosion, Conservation, and Ideas about Development"; and Anderson and Grove, *Conservation in Africa*.

9. Ford, *The Role of the Trypanosomiases in African Ecology*. Scientists today are increasingly spelling it "trypanosomosis"; throughout this book, I have retained the former spelling for consistency.

10. Not coincidentally, most other authors who had worked on the history of tsetses and trypanosomes in African history were technical officers; see, for instance, Lambrecht (epidemiologist in the Belgian Congo, 1945–59), "Aspects of the Evolution and Ecology of Tsetse Flies and Trypanosomiases in Prehistoric African Environment."

11. Ford, *The Role of Trypanosomiases in African Ecology*, v, emphasis added. More recently, a group of ecologists and medical entomologists, several of whom are based at Oxford University, published an account of trypanosomiasis control in Africa, and they opened their book with this excerpt from Ford, under the heading "Déjà Vu?"; see Bourn, Reid, Rogers, Snow, and Wint, *Environmental Change and the Autonomous Control of Tsetse and Trypanosomosis in Sub-Saharan Africa*.

12. Ford, *The Role of Trypanosomiases in African Ecology*, 9 and 494, emphasis added.

13. All quotations, ibid., 8–11.

14. Ford, "African Trypanosomiasis," 72.

15. Ford, *The Role of Trypanosomiases in African Ecology*, 7.

16. Worthington, "East Africa High Commission: Research and Scientific Services—First Progress Report for the Period April to August 1948," BV/16/162, KNA.

17. Ibid.; and Ford, Whiteside, and Culwick, "The Trypanosomiasis Problem."

18. For a fascinating early example of Ford's work with oral interviews and historical reconstruction, including a discussion of rinderpest and trypanosomiasis, see Ford and Hall, "The History of Karagwe (Bukoba District)."

19. See Kjekshus, *Ecology Control and Economic Development in East African History*; Richards, *Indigenous Agricultural Revolution*; Vail, "Ecology and History"; Lyons, *The Colonial Disease*; and Giblin, "Trypanosomiasis Control in African History"; also see Johnson and Anderson, *The Ecology of Survival*.

20. Interest in this far-reaching epidemic, especially in southern Africa, has since produced a range of fascinating studies including a handful that address scientific research. See, for instance, Ballard, "The Repercussions of Rinderpest"; Phoofolo, "Epidemics and Revolutions"; Phoofolo, "Face to Face with Famine"; and Gilfoyle, "Veterinary Research and the African Rinderpest Epizootic." Charles van Onselen and Terence Ranger have also written on the subject.

21. Both quotations, Kjekshus, *Ecology Control*, 175 and **185**.
22. On "folk ecology," see Richards, "'Alternative' Strategies for the African Environment."
23. Richards, *Indigenous Agricultural Revolution*, 9.
24. Ibid., 159.
25. Ibid., 29.
26. Leach and Mearns, "Environmental Change and Policy," 27.
27. Ibid., 18–19.
28. Ibid., 27; also see Fairhead and Leach, "Rethinking Forest-Savanna Mosaic," 118–19.
29. Leach and Mearns, "Environmental Change and Policy," 18.
30. Fairhead and Leach, "Rethinking Forest-Savanna Mosaic," 114.
31. Ibid., 121.
32. Fairhead and Leach, *Misreading the African Landscape*, 283, 287.
33. Worthington, *The Ecological Century*, 42.
34. Adams, "Foundations of Economic Progress in Tropical Africa," 754.
35. "Mr. Churchill on African Affairs." Also see an address he gave to the African Society before his departure: Churchill, "The Development of Africa"; and Churchill, *My African Journey*.
36. For details, see Duignan and Gann, *Colonialism in Africa 1870–1960*, vol. 4, *The Economics of Colonialism*, esp. introduction, chaps. 3, 8, and 9.
37. Munro, *Africa and the International Economy*, chap. 4, "Colonial Foundations, 1896–1914."
38. See Havinden and Meredith, *Colonialism and Development*, tables 6.4 and 7.2: "Imports, exports and balance of trade, selected colonies, 1911–1921," and "Imports, exports and balance of trade, selected colonies, 1920–30," 118 and 154.
39. In 1913, the British tropical African territories, including both Rhodesias and the Anglo-Egyptian Sudan, had a total foreign trade of £41 million, while South Africa had just under £105 million; Munro, *Africa and the International Economy*, "Appendix I: Total Foreign Trade of Subsaharan Africa," 217–19.
40. Details can be found in the archives and publications at Kew Gardens and in the Colonial Office's confidential reports on agricultural and forestry correspondence in the CO 879 series.
41. Johnston, *British Central Africa*, includes chapters on physical geography, botany, zoology, "the natives of British Central Africa," and their languages.
42. Whyte was sixty years old when he began work in Nyasaland, after working as a botanist and planter in Ceylon (Sri Lanka); he was also a fellow of the Zoological Society. See Baker, *Johnston's Administration*, 25–26.
43. H. H. Johnston to Mr. Freshfield, RGS, March 4, 1891, Sir H. H. Johnston RGS Corr. Block 1883–1910, RGS Archives.
44. T. V. Lister for Lord Salisbury to Director, Royal Gardens Kew, March 21, 1891, reprinted in *"Diagnoses Africanae, I,"* 18.
45. Thiselton-Dyer to Colonial Office, February 21, 1900, CO 879/65/1.
46. Thiselton-Dyer to Colonial Office, November 4, 1901, CO 879/69.
47. Johnston, *Report by His Majesty's Special Commissioner*, 9 and 19.
48. Johnston to Marquess Lansdowne, January 24, 1901, *Despatch from His Majesty's Special Commissioner*, 2–3.
49. Lugard, *Northern Nigeria Report for 1901*, 19–20; Frederick Lugard to Joseph Chamberlain, July 22, 1902, CO 879/69, "Further Correspondence relating to Botanical

and Forestry Matters in West Africa, March 1901 to December, 1904," African (West) No. 661, BNA, London.

50. Lugard to Chamberlain, July 22 1902, CO 879/69.

51. Chamberlain to Lugard, December 22, 1902; and Chamberlain to Lugard, February 13, 1903; also see Lugard to Chamberlain, January 28, 1903; all in CO 879/69.

52. Alfred Maloney, "Memorandum on Establishing a Botanic Station at Lagos," reprinted in "Botanical Station at Lagos," 152.

53. W. R. Elliott, "Report on the Forestry and Agriculture of Northern Nigeria," in Lugard, *Northern Nigeria: Report for 1904*, 132. In another letter, he describes people in the area as "very intelligent in agriculture"; W. R. Elliott to Secretary, Northern Nigeria, November 24, 1903, CO 879/69. Elliott's other reports include: Elliott to Secretary, Northern Nigeria, August 26, 1903; Elliott to Secretary, Northern Nigeria, April 4, 1904; Elliott to Secretary, Northern Nigeria, July 20, 1904, all in CO 879/69.

54. Morel, *Nigeria: Its People and Its Problems*, 115.

55. Alfred Lyttelton to Frederick Lugard, March 23, 1904, CO 879/69, emphasis in original.

56. Lyttelton to Lugard, March 23, 1904, CO 879/69.

57. "German Colonies in Tropical Africa"; "German Colonies in Tropical Africa and the Pacific"; and Wright, "German Methods of Development in Africa."

58. This paragraph is based, in part, on a paragraph that also appears in my introduction to the reprint of William Allan's *The African Husbandman*.

59. Yves Henry, agricultural report for 1906, quoted in Bonneuil and Kleiche, *Du jardin d'essais colonial*, 40 and 60.

60. Auguste Chevalier, "Les cultures indigènes dans l'Afrique-Occidentale française," *Revue des cultures coloniales* 6 (1900): 257, quoted in Bonneuil and Kleiche, *Du jardin d'essais colonial*, 61.

61. *Summary of the First Colonial Office Conference*; and CO 323/972/8, "Proposal for the Establishment of an Imperial Science Service (1927)."

62. *Imperial Agricultural Research Conference, 1927*, 8–13 and 223–26.

63. "Report on the Re-Establishment of a Research Institute at Amani, in the Tanganyika Territory, to Serve the East African Dependencies," December 1921, from the Directors of the Tanganyikan, Kenyan, Ugandan, and Zanzibar Departments of Agriculture, BV/22/85, KNA.

64. *East African Agricultural Research Station, Amani First Annual Report, 1928–29*, 12; on the tour, see pp. 3–7.

65. Nowell, "The Work of the Amani Institute [talk on October 28, 1927]," 123.

66. Ormsby-Gore, "Agricultural and Veterinary Research in the Colonies, Protectorates, and Mandated Territories," 55.

67. Director of Agriculture, Tanganyika [E. Harrison] to Lord Passfield, April 14, 1931; and G. Seel, minute, May 15, 1931, CO 323/1137/10. "Agricultural and Veterinary: Pan-African Agricultural and Veterinary Congress, 1929, Reports and Resolutions," 1931, BNA.

68. Charlotte Leubeuscher, "Memorandum on the Economic and Social Problems of Africa," July 1934, LP, GD40/17/125. Leubeuscher was an important advisor to the project between 1933 and 1935; this memo was written as an overview for Hailey and the general committee.

69. Between 1929 and 1931, total export revenues fell in West Africa (including the Gambia, the Gold Coast, Nigeria, and Sierra Leone) from £31.6 to £26 and then to £16.5 million, and in East Africa (including Kenya, Uganda, Tanganyika, and Zan-

zibar) from a high of £14.5 to £11 and then to £8 million. In Central Africa (Northern Rhodesia and Nyasaland), export revenues rose from £1.5 million in 1929 to £1.6 million in 1931. All figures are from Havinden and Meredith, *Colonialism and Development*, 176, 180.

70. The figures are: Nyasaland, 42 percent; Kenya, 32 percent; Nigeria, 28 percent; Tanganyika and the Gold Coast, 24 percent. Havinden and Meredith, *Colonialism and Development*, 175–76.

71. Brown, *The Economies of Africa and Asia*.

72. Articles in Brown's edited collection that reinforce these arguments include Anderson and Throup, "The Agrarian Economy of Central Province, Kenya, 1918 to 1939," 8–28; Döpke, "Magomo's Maize," 29–58; and Clarence-Smith, "The Effects of the Great Depression on Industrialisation in Equatorial and Central Africa," 170–202. For evidence supporting an alternative interpretation, that the economic depression slowly and unevenly locked African producers into a cash economy beyond their control, see Martin, "The Long Depression," 74–94.

73. One of the most prolific advocates of this synthesis, Girolamo Azzi, began publishing on the subject in the 1920s and was appointed chair of Agricultural Ecology at the Instituto Superiore Agrario, Perugia, Italy, in 1921. This was probably the first chair in the subject anywhere. Azzi published his first textbook in 1928, a second edition in 1944, and an English edition in 1956. He helped spread these ideas through his affiliation with the International Agricultural Institute in Rome and his travels abroad. However, Azzi paid little attention to rural cultivation practices and was more interested in studying crops. See Azzi, *Agricultural Ecology*.

74. India had a marvelous tradition of research into peasant cultivation, but it seems not to have been linked so explicitly to developments in ecological and anthropological sciences. The best known of these researchers, Albert Howard, made little use of ecological concepts or anthropological analysis. See Howard and Howard, *The Development of Indian Agriculture*; and Howard, *An Agricultural Testament*.

75. In the United States, the term had its origins in the 1940s and 1950s; see Madison, "Potatoes Made of Oil." According to the OED, "agroecology" was first used in 1930 in the *International Agricultural Review*.

76. On Shantz as Clements's "pupil," see T[ansley], "Two Recent Ecological Papers," 219. A brief background to the journey can be found in Wagner, "Notes on the Shantz Collection, Tucson, Arizona."

77. For details of the expedition, see McKinley, *The Lure of Africa*, 127–31; specimen details on p. 131.

78. Shantz and Marbut, *The Vegetation and Soils of Africa*; the quotation is the title to chapter 10.

79. See Shantz, "Urundi, Territory and People," on agriculture, see 348–53. He left a 600-page manuscript about the first half of his journey: Homer Shantz, "Travel Notes on a Trip to Africa from Cape to Cairo," Boxes 5 and 6, Homer Shantz Collection MS 030, University of Arizona Library Special Collections.

80. The Phelps Stokes Fund, an American foundation established in 1911, was primarily concerned with the educational uplift of African Americans, Africans, and Native Americans. Its 1911 critique of racially segregated school systems in the U.S. South put it in the forefront of progressive thinking; so did its 1928 critique of the public education of American Indians. Members of the Phelps Stokes family had a longstanding interest in the American Colonization Society, and the fund was active in Liberia from the 1920s. In 1929, at the invitation of Liberia's king, it founded

and funded the Booker Washington Institute, the country's first and still its largest agricultural and technical vocational school, which was modeled on the Tuskegee Institute in Alabama.

81. *Report of the East Africa Commission*, 6.
82. Shantz, "Agriculture in East Africa," supplementary chapter in Jones, *Education in East Africa*, 358. The nine territories were Kenya, Uganda, Tanganyika, Northern Rhodesia, Southern Rhodesia, Zanzibar, Portuguese East Africa, Ethiopia, and Bechuanaland.
83. Shantz, "Agriculture in East Africa," tables 2 and 4, 357 and 359.
84. Ibid., 361 and 363. Shantz often used the male pronoun, but his work includes a fascinating analysis of "women agriculturalists."
85. Ibid., 364–65.
86. Ibid., 366–67.
87. Ibid., 369.
88. Ibid., 367. He included a brief discussion of plow-driven agriculture in Ethiopia, which he noted was an interesting exception. See McCann, *People of the Plow*.
89. Shantz, "Agriculture in East Africa," 365.
90. Ibid., 366.
91. Ibid., 370.
92. Ibid., 371.
93. Church, *East Africa, a New Dominion*, 114. For Joseph Oldham's approval of Shantz's suggestion for closer study of native agriculture, see J. H. Oldham to W. Ormsby-Gore, May 6, 1925, "A Note on the Report of the East Africa Commission," reprinted in Cell, *By Kenya Possessed*, full letter 295–316, key passage on 308–9.
94. Faulkner, "The Aims and Objects of the Agricultural Department in Nigeria," 5–7. Faulkner had joined the department after working in India, which Paul Richards has argued helped encourage his interests in peasant production. There is no evidence that Shantz had communicated with Faulkner before writing his own report.
95. *Report of the Commission on Closer Union of the Dependencies in Eastern and Central Africa*, 21; and *Report of a Committee on Agricultural Research and Administration in the Non-Self-Governing Dependencies*, 73.
96. Northern Rhodesia, *Department of Agriculture Annual Report for Year 1928*, 4; in 1924 revenue from maize trade was only £50,000 and for tobacco it was only £91,000. See *Report of a Committee on Agricultural Research and Administration*, 73. Smith was a veterinarian by training; when he retired from Northern Rhodesia in 1933 he joined the Colonial Office's Advisory Committee on Agriculture and Animal Health.
97. Northern Rhodesia, *Annual Report for Year 1927, Department of Agriculture*, 3.
98. *Report of the Commission Appointed to Enquire into the Financial and Economic Position of Northern Rhodesia*, appendix XXI, "Statistics Relating to District Administration (as at May, 1937)," 388; the breeds were then known as Barotse, Ila, and Ngoni. For an anthropological analysis, see Colson, *The Plateau Tonga of Northern Rhodesia*, chap. 5, "The Role of Cattle among the Plateau Tonga of Mazabuka District."
99. Northern Rhodesia, *Annual Report for the Year 1929, Department of Agriculture*, 4.
100. Northern Rhodesia, *Annual Report for the Year 1927, Department of Agriculture*, 4.
101. J. Smith to Empire Marketing Board, September 6, 1927; J. Smith, "Memorandum Showing the Necessity for a Botanical Survey of Northern Rhodesia," [September 1927], esp. 7–8; T. McEwan, "Survey of Agricultural Research Work and Problems

of Northern Rhodesia, for the Agricultural Conference, October 4th, 1927," June 11, 1927, CO 758/53/1, esp. p. 1.

102. These were planned in 1927 and inaugurated in 1928.

103. Acting Secretary for Agriculture, John Smith, to Chief Secretary, November 25, 1929; and memorandum by H. C. Sampson, "Vegetation Survey of Northern Rhodesia, 28 June 1928," MAG 2/9/1, Zambian National Archives.

104. Moore, "Some Indigenous Crops of Northern Rhodesia," 35.

105. Northern Rhodesia, *Annual Report for the Year 1928, Department of Agriculture*, 4.

106. Ibid., 5.

107. On the "serious effect" of the "exodus of youths and men from the villages," see Northern Rhodesia, *Annual Report for the Year 1929, Department of Agriculture*, 5.

108. H. C. Sampson, Memorandum, June 5, 1929, CO 758/53/1. In Fort Jameson, the village chief expelled seventy women who were involved in a religious revival led by the American-based Watchtower Movement and sent them to the provincial magistrate.

109. Moffat, "Native Agriculture in the Abercorn District," 55–62, on 55.

110. The extensive assistance he obtained from local inhabitants is not made clear in Moffat's 1932 report, but it is evident in Colin Trapnell's field notebooks during the Ecological Survey.

111. Bemba and Mambwe peoples overlap geographically, and their languages have much in common. For an excellent longitudinal study of social changes in the area in which Moffat worked that includes a chapter on the "colonial construction of knowledge," see Moore and Vaughan, *Cutting Down Trees*.

112. On finger millet's importance to contemporary agriculture, see *Lost Crops of Africa*, vol. 1, *Grains*, 39–58.

113. In these experiments the yield for citemene averaged 1,800 pounds per acre while that for mound cropping averaged 1,250 pounds.

114. Stent, "Observations on the Fertilizer Effect of Wood Burning in the 'Chitemene' System," 49.

115. According to the *Annual Bulletin* of Nigeria's Agriculture Department, Lewin joined the department in 1923 as a "superintendent of agriculture" and by 1924 had been made a botanical research officer.

116. "Editorial," *Second Annual Bulletin—Agricultural Department of Northern Rhodesia*, 3–4. Lewin identified himself as the author of the editorial in the first report of the Ecological Survey.

117. F. A. Stockdale, "Memorandum," June 25, 1929, CO 758/53/1. By that time, Ormsby-Gore, one of the few officials sympathetic to "native agriculture," had been replaced as assistant secretary of state.

118. The origins of the Ecological Survey are sometimes attributed to Bourne's recommendations, but he arrived in Northern Rhodesia after McEwan had already put forward his initial proposal. In fact, Smith wrote up his proposal to the Colonial Office while Bourne and McEwan were doing joint fieldwork. Bourne, *Aerial Survey in Relation to the Economic Development of New Countries*, 8.

119. For details of the follow-up ecological survey by air, with notes by Colin Trapnell in the appendix, see Robbins, "Northern Rhodesia."

120. Bourne, *Aerial Survey in Relation to the Economic Development of New Countries*, 6–7.

121. Ibid., 8 and 11. See also Bourne, "Some Ecological Conceptions."

122. [H. C.] Sampson, "Vegetation Survey in Northern Rhodesia Memorandum," June 28, 1928, MAG 2/9/1, ZNA.

123. *First Interim Report of the Colonial Development Advisory Committee*, 25–27. Far larger grants for other purposes, especially railways, harbors, and bridges, went to Tanganyika, Nyasaland, Uganda, and Sierra Leone.

124. Acheson, "Northern Rhodesia: Proposed Ecological Survey and the Question of Ambit," May 28, 1929; and Acheson, "Northern Rhodesia and the Question of Ambit," April 11, 1929, CO 758/53/1.

125. The fieldwork of the Ecological Survey was conducted between 1932 and 1940; the first report was published in 1937 and the second report in 1943, a final soil-vegetation map, plus a memorandum, was published in 1947.

126. Smith, *Ecological Survey of Zambia*, vol. 1, 8.

127. Wyndham, "The African Labourer and West African Agriculture," 824.

128. Northern Rhodesia, *The Agricultural Survey Commission Report, 1930–1932*. The commission consisted of three members: "an experienced technical agriculturalist, a practical farmer of over twenty years' experience in the Territory, and a [land] surveyor."

129. Baldwin, *Economic Development and Export Growth*, 149 and chap. 6.

130. Northern Rhodesia, *Annual Report for the Year 1929, Department of Agriculture*; and *Annual Report for the Year 1932, Department of Agriculture*.

131. See "Farmers Criticise Director—Said to Adopt Unsympathetic Attitude to Europeans—Separation of Departments Asked For," *Bulawayo Chronicle*, April 1941, in SEC1/65 "Criticisms of Agriculture Department," ZNA.

132. Author's interview with Colin Trapnell, November 12, 1998, at his home in Bristol, England.

133. The Oxford Exploration Club was "inaugurated in Max Nicholson's rooms . . . on the 8th December, 1927. Present: Charles Elton, Colin Trapnell, Bill Crouch, Mike Corbett and Max and Basil Nicholson." Elton became chair, Nicholson vice-chair, and Trapnell secretary. "Oxford Expedition Club [Constitution papers]." Trapnell allowed me to copy these papers during the interview on November 12, 1998.

134. Tom Harrisson's introduction, in Ford, Hartley, et al., *Borneo Jungle*.

135. In addition to our interview, Trapnell gave me an autobiographical summary, by year, that included a full list of his publications and a list of thirteen research institutes in South Africa that he visited in 1931.

136. Their fieldwork covered only portions of these areas, but their surveying techniques were designed to reveal representative patterns.

137. Trapnell and Clothier, "Report of the Ecological Survey of 1933," 23.

138. Trapnell worked with one of the translators, Jean Baptiste, who was from Kasama (adjacent to Abercorn), for several years.

139. Colin Trapnell, interview, November 12, 1998.

140. On Allan's work in Northern Rhodesia, see my introduction to the International African Institute's reprint of his 1965 book, Allan, *The African Husbandman*.

141. Trapnell and Clothier, "Report of the Ecological Survey of 1933," 23.

142. Trapnell, "Ecological Methods in the Study of Native Agriculture," 2.

143. Trapnell and Clothier, "Report of the Ecological Survey of 1933," 23: "When he has migrated of his own accord to a fresh environment he is capable of adjusting his methods to it. But when he comes under the influence of European contact this adjustment to the natural environment is lost, and many problems of native areas have their origin here."

144. Trapnell and Clothier, "Report of the Ecological Survey of 1934," 26 and 28.

145. Hubert Young to Cunliffe Lister, September 29, 1934, and Hubert Young, "Livingstone Memorial Institute," CO 795/72/2, BNA.

146. Richards revised her dissertation following her first fifteen months of research in Northern Rhodesia; see Richards, *Hunger and Work in a Savage Tribe*, vii and 1.

147. Reginald Coupland to Philip Kerr (Lord Lothian), November 5, 1929, File 2791, "Anthropological Research," Rhodes Trust Archives; Coupland was summarizing Richard's proposal to the trust.

148. Ibid.

149. On the IIALC's interest in studying labor migration, wage earning, economic conditions, and land tenure in a single "tribe," see Margery Perham, "Notes upon Research in Africa," n.d. [June 1931], and Joseph Oldham to Bronislaw Malinowski, June 10, 1931, File #62, Bronislaw Malinowski Papers, London School of Economics.

150. Richards, *Land, Labour and Diet*, ix.

151. Ibid., 277. This question corresponds to research interests Bronislaw Malinowski had been advocating since the mid-1920s.

152. Richards, *Land, Labour, and Diet*, 279.

153. Ibid., 283–87.

154. Moore and Vaughan, *Cutting Down Trees*, 21.

155. Ibid., 42.

156. Trapnell and Clothier, *The Soils, Vegetation and Agricultural Systems of North-Western Rhodesia*, ix, emphasis added.

157. Jack Morrell has questioned whether Tansley actually coined the term; see Morrell, *Science at Oxford*, 241.

158. Tansley, "The Use and Abuse of Vegetational Concepts and Terms," 300. See also Tansley, "The Classification of Vegetation and the Concept of Development."

159. Tansley, "Use and Abuse of Vegetational Concepts," 301–2.

160. Ibid., 303.

161. Ibid., 291.

162. C. J. Lewin to Chief Secretary (with notes by Trapnell), no date 1937, MAG 2/9/7; "Ecological Survey General, 1937–1949"; and Northern Rhodesia, *Annual Report for the Year 1940, Department of Agriculture.*

163. Gluckman, "The Rhodes-Livingstone Institute and Museum."

164. Winterbottom, "The Ecology of Man and Plants in Northern Rhodesia," 38.

165. Gillman, "Review: *The Soils, Vegetation, and Agriculture of North-Eastern Rhodesia*," 61–62.

166. Faulkner and Mackie, *West African Agriculture*, 3–4, 7, 41–42.

167. Faulkner and Mackie, "The Introduction of Mixed Farming in Northern Nigeria," 89. They made an almost identical statement in *West African Agriculture*, but there they referred to "English" rather than "European" farmers.

168. Faulkner and Mackie, *West African Agriculture*, 7.

169. Jones, *The Earth Goddess*, v.

170. Ibid., 35.

171. Ibid., 43–44.

172. Ibid., vii.

173. Ibid., vi–vii.

174. Anderson, "Depression, Dust Bowl, Demography, and Drought"; Beinart, "Soil Erosion, Conservation, and Ideas about Development"; Beinart, "Soil Erosion, Animals, and Pastures over the Longer Term"; Showers, "Soil Erosion in the Kingdom of Lesotho"; and Stocking, "Soil Conservation Policy in Colonial Africa."

175. Mackenzie, *Land, Ecology, and Resistance in Kenya*, 104.

176. Colin Maher, who features in the work of both Mackenzie and David Anderson as

the agricultural officer who assumed responsibility for Kenya's soil erosion control and produced numerous reports during the 1930s, was not a part of this network and attended none of the gatherings of soil scientists.

177. Nowell, "The Work of the Amani Institute," 123.

178. The International Congress included soil scientists from around the world. The greatest numbers came from the United States (forty-one) and Great Britain (ninety-four). The forty-five participants from the British dominions and colonies were listed separately; twenty-three came from Africa: Egypt and South Africa each had five, Sudan and Gold Coast each sent three, and Kenya, Nigeria, Nyasaland, Southern Rhodesia, Tanganyika, Uganda, and Zanzibar each had one. Significantly, the Belgian, Portuguese, and French colonies in Africa sent no representatives. The delegates from Africa met during the conference to discuss producing a "Soil Map of British Africa" for which the East African map would serve as a model. *Transactions of the Third International Congress of Soil Science*, vol. 2, xi–xx, 258–59.

179. Quotations from William Nowell and from conference resolutions, in *Proceedings of a Conference of East African Soil Chemists . . . 1932*, 13, 17. Present were representatives from Kenya (Beckley), Uganda, Zanzibar, and Tanganyika, including Amani's director, Nowell, and soil chemist, Geoffrey Milne. For the genealogy of proposals for research on shifting cultivation, see *Proceedings of the Second Conference of East African Agricultural and Soil Chemists . . . 1934*; and *Proceedings of Agricultural Research Conference Held . . . 1931*.

180. Milne in collaboration with Beckley, Jones, Martin, Griffith, and Raymond, "Provisional Soil Map of East Africa," and Milne et al., "A Short Geographical Account of the Soils of East Africa," 266–70, 270–74.

181. See Krupenikov, *History of Soil Science*, which is heavily slanted toward the Russian and American contributions to soil science; Yaalon and Berkowicz, *History of Soil Science*; and International Society of Soil Science, *Fifty Years Progress in Soil Science*.

182. For a closely related point in the French context, see Bonneuil, "Crafting and Disciplining the Tropics."

183. Nowell, "The Agricultural Research Station at Amani."

184. Ibid., 8. In many parts of Tanganyika both Europeans and Africans were engaged in coffee production; in the Mount Kilimanjaro region in 1934, Chaga cultivators produced 1,530 tons of arabica coffee while European planters produced only 1,361 tons. See Wakefield [Dep. Director of Ag.], "Native Production of Coffee on Kilimanjaro."

185. Nowell, "Agricultural Research Station at Amani," 6.

186. Ibid., 8; Nutman, "The Root-System of *Coffea arabica*, Parts 1 and 2," and Nutman, "The Root-System of *Coffea arabica*, Part 3."

187. Kirkpatrick, *Studies on the Ecology of Coffee Plantations in East Africa I*; and Kirkpatrick, *Studies on the Ecology of Coffee Plantations in East Africa II*.

188. Alex Holm, comments during discussion, in Nowell, "The Agricultural Research Station at Amani," 16–17.

189. This statement was made by the German soil scientist, Paul Vageler, in his 1935 volume (in German) *Basic Concepts of Soils of the Subtropical and Tropical Countries* and quoted in Krupenikov, *History of Soil Science*, 250. Also see Vageler, *An Introduction to Tropical Soils*. Vageler worked in German East Africa before the First World War and is acknowledged by Milne et al., in *Provisional Soil Map of East Africa*, 8.

190. Milne, "Normal Erosion as a Factor in Soil Profile Development," 549; also see

Milne, "A Soil Reconnaissance Journey through Parts of Tanganyika Territory," esp. 241–44.

191. Hartley, "An Indigenous System of Soil Protection," 66. For other contributions from Tanganyika, see Thornton and Rounce, "Ukara Island and the Agricultural Practices of the Wakara"; Smith, "The Sukuma System of Grazing Rights"; and Rounce, "The Ridge in Native Cultivation."

192. *Report of the Proceedings of the Conference of Colonial Directors of Agriculture*, 42; also see comments by S. Milligan (p. 46), "As Mr. Lewin has pointed out, soil erosion is due to recent economic development."

193. Hornby, the director of the Veterinary and Animal Husbandry Services of Tanganyika, wrote on the question of overstocking and soil erosion, "[I] must point out that it is wrong to refer to soil erosion as though it were practically synonymous with over-stocking. . . . If we interfere [in native practices] we must do so judiciously." Quoted in Harrison, *Soil Erosion—A Memorandum*, 8–9.

194. Tothill et al., *A Report on Nineteen Surveys*; the entire "questionnaire" is on pp. 30–31.

195. The acreage devoted to cotton had risen from approximately 92,000 in 1916 to nearly 1.5 million by 1936; that to cassava had risen from roughly 6,100 to 296,294 during the same period. Tothill et al., *Report on Nineteen Surveys*, 5. Also see Hansford (departmental mycologist), "Some Effects of the Development of the Cotton Industry on Native Agriculture in Uganda."

196. See the discussion of the densely settled areas in Masaka District in Tothill et al., *Report on Nineteen Surveys*, 12.

197. Tothill et al., *Report on Nineteen Surveys*, 5 and 10.

198. Hayes, "Agricultural Surveys in the Eastern Province, Uganda," 211.

199. Ireland, Hosking, and Loewenthal, *An Investigation into Health and Agriculture in Teso, Uganda*.

200. Hosking, "The Improvement of Native Food Crop Production," 84.

201. Tothill, *Agriculture in Uganda*.

202. Carswell, "Soil Conservation Policies in Colonial Kigezi, Uganda," 138.

203. "Copies of the 1937 Ecological Survey," August 20, 1937, "Ecological Survey General," MAG2/9/7, ZNA.

204. Worthington, draft of chapter 7, "Zoology," for *Science in Africa* (London: African Research Survey, n.d. [August/September 1935]), 74.

205. Joseph Oldham [in consultation with Julian Huxley and William Macmillan] to Sir Malcolm Hailey, July 19, 1933, OP, Box 1, File 3.

206. Phillips, "Ecological Investigation in South, Central, and East Africa," 474, 477, and 482.

207. Francis Hemming to E. B. Worthington, February 12, 1935, CO 847/4/2.

208. E. B. Worthington to Hemming, February 15, 1935, CO 847/4/2.

209. On the conceptual debates lying behind "the politics of holism," see Anker, *Imperial Ecology*, esp. chaps. 2, 4, and 5; on the idea of "super-coordination," see Charles Elton to Julian Huxley, March 9, 1932, OP, Box 2, File 4.

210. Ford, "The Ecological Method," a review of J. W. Bews, *Life as a Whole*, 1937, 157. Ford was commenting both on the philosophy of holism, which was the subject of Bews's book, and on other biological sciences that shared ecology's "method of thought," including the work of Ludwig van Bertalanffy, Joseph Needham, and Joseph Woodger.

211. Worthington, second draft of chapter 5, "Botany," for *Science in Africa* (London: African Research Survey, n.d.), 95.

212. Ibid., 99.

213. Pim, *Financial and Economic Position of Basutoland*, quoted in Staples and Hudson, *An Ecological Survey of the Mountain Area of Basutoland*, 1. Staples was a plant ecologist loaned from the Tanganyika government.

214. Their model was Northern Rhodesia; see *Report of the Proceedings of the Conference of Colonial Directors of Agriculture*, 8 and 47, for C. T. Morison's discussion of Trapnell's work.

215. Sampson and Crowther, "Crop Production and Soil Fertility Problems," 53, and "Ecological Surveys," 52–54. Sampson was the final editor of Worthington's chapters in *Science in Africa* on "agriculture," "crop-plants," and "plant industry."

216. L. B. Freeston to Governments of Kenya, Uganda, Tanganyika, Nyasaland, Northern Rhodesia, and Zanzibar, September 24, 1936, CO 822/74/9, BNA; also "Conference of Ecological Workers, 1936–38," File 24327, TNA.

217. H. L. Gurney, Colonial Secretary for Kenya, to L. B. Freeston, Secretary to East African Governor's Conference, November 11, 1936, CO 822/74/9.

218. Ogilvie, "'Science in Africa' by E. B. Worthington," 233.

219. Prof. E. J. Salisbury, letter prepared for 2nd Scientific Sub-Committee meeting on December 11, 1935, GD40/17/127.

220. Worthington, *Science in Africa*, 3.

221. Hailey, *An African Survey*, 1313.

222. Worthington, *Science in Africa*, "The Changing Environment," 4–15, on 15. There are interesting parallels between Worthington's analysis and vocabulary and that of Continental theorists concerned with "internal and external nature"; see Denham (now Tilley), "The Cunning of Unreason and Nature's Revolt."

223. Worthington, *Science in Africa*, 4.

224. Ibid., 144.

225. Ibid., 7–8, emphasis added. Similar views were expressed in *An African Survey*: "the African tradition of cultivation . . . teaches methods of rotation of crops, of usage of soils, and means of fertilization and even sometimes anti-erosion measures, which, though they may not be suited to modern demands on the soil, are often well adapted to the prevailing conditions of labour and climate" (881).

226. Ibid., 423, 337, 22, 15.

227. Ibid., 420, 423, 425.

228. Ibid., 446, 420–27. "It is certain that many distinct strains of trypanosomes exist, and that [cattle] immunity to one does not involve immunity to another" (451).

229. Ibid., 237.

230. Ibid., 399.

231. Ibid., 336.

232. Ibid., 306. Mixed cropping should not be confused with mixed farming, which is the integration of animal husbandry with horticulture.

233. Ibid., 124, emphasis added.

234. Ibid., 404–5.

235. Ibid., 302. This distinction was first made by E. Harrison, the director of agriculture in Tanganyika, in his comments on Worthington's drafts: E Harrison to Chief Secretary, November 26, 1935, "African Research Survey-Agriculture," File 21984/1, TNA.

236. Worthington, *Science in Africa*, 23–24.

237. Ibid., 17.

CHAPTER FOUR

1. Colonial Office, *A Note on Some of the Scientific Studies*, 3; unnamed author in reaction to the 1935 Annual Medical Report of the Kenya Colony, quoted with approval by A. R. Paterson, director of medical services, in "Note on Considerations arising out of the Question 'What trained Africans the Medical Department of Kenya is likely to require?' Including a Statement in Reply to the question 'What employment the Medical Department could give to trained Africans if such are provided?'" January 20, 1937 (47 pp. typed manuscript), Published Monographs, Kenya National Archives.

2. Champion Russell, "Memorandum on Native Health in Africa," [December 17, 1938], emphasis in original, CO 847/13/19, "Note by Mr. Champion on Native Health, 1938," BNA; all quotations in this paragraph from this memo.

3. Mitchell, "The Uganda Medical School"; and Williams, "The History of Mulago Hospital."

4. Hailey, *An African Survey*, 1654.

5. Mr. Freeston, on behalf of Malcolm MacDonald, to Secretary of East African Governors Conference, February 17, 1939 and file minutes, CO 847/13/19.

6. A. R. Paterson to Secretary of East African Governors Conference, April 13, 1939, CO 847/16/2; the six territories included Kenya, Uganda, Tanganyika, Nyasaland, Northern Rhodesia, and Zanzibar.

7. R. R. Scott, "Proposed Experiment in the Prevention of Disease at a Selected Place in Tanganyika, Dependent on a Gift by Mr. Champion Russell of £1,000," May 19, 1939, CO 847/16/2.

8. [W. H. Kauntze], "Offer by Mr. Champion to Contribute £1,000 towards the cost of an Experiment for the Prevention of Disease in a Selected District in Africa, Memorandum by the Director of the Medical Services, Uganda," June 14, 1939, CO 847/16/2.

9. On Kauntze's role, see "African Research Survey, Progress Report—October 28, 1937," LP, GD40/17/129; and Lord Hailey, *An African Survey*, 1198; also W. H. Kauntze, "Memorandum on Departmental Policy," 1935, Uganda Medical Department, Entebbe.

10. Worthington, *Science in Africa*, 461. Worthington lifted this language from letters submitted by medical directors and departmental experts; see R. R. Scott and Burke Gaffney, "Science in Africa—XII. Health and Medicine—Suggested Alterations and Additions," November 25, 1935, File 21984/7, African Research Survey—Medical, TNA.

11. H. S. de Boer, Director of Medical Service, Nyasaland, to the Secretary of the Governors' Conference, April 10, 1939, CO 847/16/2, emphasis added.

12. Paterson to Secretary of East African Governors Conference, April 13, 1939, CO 847/16/2.

13. Arnold, "Medicine and Colonialism," 1393.

14. H. S. de Boer to the Secretary of the Governors' Conference, April 10, 1939, CO 847/16/2.

15. He was referring to the epidemiological survey and medical "campaign" conducted between 1927 and 1931 in Kahama, Tanganyika, an area of approximately 8,000 square miles with 76,000 people, which attempted to address chronic diseases, sanitary conditions, and even research into "African medicine," but which was severely disrupted when a sleeping sickness epidemic broke out. Similar districtwide medi-

cal surveys and sanitary campaigns were conducted in Kenya, Uganda, Nyasaland, Nigeria, and the Gold Coast in the 1930s. On Kahama, see *Papers Relating to the Health and Progress of Native . . .* , 54 and 79. Also see the *Tanganyika Annual Medical Report of 1928*, 105–15.

16. The importance of medical issues to Malcolm MacDonald's thinking on development is apparent in Malcolm MacDonald, "Colonial Development: Note for the Chancellor of the Exchequer," October 11, 1939, CO 847/15/9, BNA.

17. Colonial Office, *A Note on Some of the Scientific Studies*, 3; the bulk of the research described in this document, which covers nutrition and a host of different diseases, was conducted in British colonial Africa.

18. Brantley, "Kikuyu-Maasai Nutrition and Colonial Science"; McCulloch, "An Enquiry into the Dietaries of the Hausa and Town Fulani," part I, 8–22, part II, 36–47, and part III, 62–73.

19. Committee of Civil Research, "Interim Report of the Sub-Committee on Dietetics July 23, 1926," and "Minutes from comments of J. L. Gilks on 15 June 1928 in the Kenya Legislative Council," in CO 758/19/6 "Dietetics Research, 1926," BNA.

20. Scott, *A History of Tropical Medicine*, vol. 1, 519.

21. On the British antecedents to this work, see the reports of Great Britain, *Yellow Fever Commission (West Africa)*.

22. Major Hon. E. F. Wood, 13th of June, comments in "Conference between the Colonial Office and Representatives of the Rockefeller Foundation, 13, 16, and 18 June 1921," CO 323/874.

23. The administrative district was Kota Kota; see Berry and Petty, *The Nyasaland Survey Papers*; and Brantley, *Feeding Families*.

24. On tuberculosis research, see Wilcocks, *Tuberculosis in Tanganyika Territory*.

25. These figures are compiled from the ten *Interim Reports of the Colonial Development Fund*, 1929–39; Command Papers 3540, 3876, 4079, 4316, 4634, 4916, 5202, 5537, 5789, 6062.

26. *Proceedings of the First International Conference on Sleeping Sickness*.

27. A more complete publication list can be found in League of Nations, "Bibliography of the Technical Work of the Health Organisation," 190–93.

28. See Ludwik Rajchman, "Memorandum of Conversation with Mr. Ormsby-Gore," May 6, 1925, File 41373, "Sleeping Sickness, Arrangements for a Conference of Powers Possessing Territories in Equatorial Africa," Box R855, LNA; and W. E. Rappard, Director of the Mandates Section to President of the Health Committee [Dr. Th. Madsen], August 12, 1922, File 26254x, "Tropical Diseases in Africa Various Correspondence," Box R854, LNA; this letter conveys the text of Ormsby-Gore's resolutions on "the prevention, incidence and treatment of disease in the mandated areas."

29. Ormsby-Gore, "Opening Address," *Report of the Second International Conference on Sleeping Sickness*, 14–19, quotations on pp. 15, 18–19.

30. These causal links are most clearly spelled out in "Health Committee, 14th Session, May 2nd 1929, Sleeping Sickness," (notes on meeting agenda prepared by Medical Director) 1st April 1929, C. H. 778; also see, "Health Committee Resolution, Sixteenth Session, October 1930," File 26316, "Sleeping Sickness—Meeting of Medical Officers of African Powers, Cape Town, November 1932, Representatives of Union of South Africa," Box R5920, Series 23393, LNA.

31. "Report on Rural Hygiene," in "Report of the International Conference . . . November 15th to 25th, 1932," 104, and more generally, 101–7; also see "Rural Hygiene

and Medical Services in Africa," in "Report of the Pan-African Health Conference, Johannesburg, November 20–30, 1935," 198–200.

32. Farley, *Bilharzia*, 266 and 275.

33. *The Colonial Territories (1948–1949)*, 92. Similar views were expressed by several contributors to *Conference on Co-Ordination of Tsetse and Trypanosomiasis Research and Control in East Africa* (1943).

34. Nash, "The Anchau Settlement Scheme."

35. This attitude was particularly prevalent among those who worked to develop new chemotherapies; the term *magic bullet* was coined and popularized by Paul Ehrlich. I have left largely to one side the history of drug testing and experimentation, although I occasionally discuss the botanical side of drug prospecting.

36. Lyons, "Diseases of Sub-Saharan Africa," 301–2; similar views were also expressed in Prins, "But What Was the Disease?" 159–79.

37. Ford, *The Role of Trypanosomiases in African Ecology*, 8–11.

38. Packard, "Visions of Postwar Health and Development," 96. But also see Packard, *White Plague, Black Labor*, and Packard, "The Invention of the 'Tropical Worker.'"

39. Marks, "What Is Colonial about Colonial Medicine?" Also see Roemer, "Internationalism in Medicine and Public Health," 1417–35; although Roemer begins with the question of colonialism, he drops this topic entirely in his discussion of the League of Nations.

40. Bell, *Frontiers of Medicine*, 233. One reason Bell overlooks connections between colonial and international medicine is because the Sudan's representatives rarely took part in various regional and pan-African gatherings.

41. Walter Fletcher to Rajchman, May 20, 1925, File 44260, "Conference on Sleeping Sickness: London, May 1925," Box R855, LNA.

42. In Africa, the Rockefeller Foundation was most active in yellow fever research. In 1925, the foundation's International Health Division (IHD) sponsored a Yellow Fever Commission to West Africa; between 1927 and 1934, it operated a yellow fever research laboratory in Lagos, Nigeria. In 1936, the IHD opened an East African Yellow Fever Research Institute at Entebbe, Uganda, following the closure of the Human Trypanosomiasis Institute. See Farley, "The International Health Division of the Rockefeller Foundation"; and Bell, *Frontiers of Medicine*, chap. 6, "The International Construction of Yellow Fever," esp. 167–68, 181–82.

43. Huntington Gilchrist, "International Campaign Against Sleeping Sickness," April 25, 1928, File 687, Box R5854, "Sleeping Sickness—General, 1928–1938," LNA. For debate on this subject, see *Report of the Second International Conference on Sleeping Sickness*, 8 and 10–13.

44. For early bibliographical references on this topic, see Harley, *Native African Medicine*.

45. E. B. Worthington, first draft chapter 12, "Health and Medicine," 194; Folder 4 16/26f, RIIA; Burke Gaffney, "Science in Africa—XII. Health and Medicine—Suggested Alterations and Additions," November 25, 1935, File 21984/7, African Research Survey—Medical, TNA; while the review was written by Burke Gaffney, it came with a cover letter from R. R. Scott, the director of the medical department, in which he stated that he agreed with Burke Gaffney's views.

46. Hailey, *An African Survey*, 1198. On Kauntze's favorable views of augmenting the medical services with African recruits trained at Makerere and Mulago, see Iliffe, *East African Doctors*, 38 and 47.

47. Hailey, *An African Survey*, 1198, emphasis added.

48. Janzen, *Quest for Therapy*; and Livingston, *Debility and the Moral Imagination in Botswana*.
49. Feierman, "Struggles for Control," 74.
50. Quoted in Iliffe, *East African Doctors*, 29. Janzen makes a similar point for the Belgian Congo, see *Quest for Therapy*, 51–52.
51. *Papers Relating to the Health and Progress of Native Populations in Certain Parts of the Empire*, 33.
52. See Lord Crewe, Secretary of State for the Colonies, to Governors and Acting Governors, August 6, 1909, CO 879/102/7, "Africa: Papers June, 1906 to December, 1910 relating to Medical and Sanitary Matters in Tropical Africa," BNA.
53. Mendelsohn, "From Eradication to Equilibrium," 303–4.
54. Ibid., 304.
55. See Anderson, "Natural Histories of Infectious Disease."
56. Mendelsohn, "From Eradication to Equilibrium," 323; I have amalgamated the terms into one quotation for convenience.
57. See, for instance, Gill, *The Genesis of Epidemics and the Natural History of Disease*, esp. part IV, "The Bionomics of Disease"; Dudley, "Can Yellow Fever Spread into Asia?"; Dudley, "The Ecological Outlook on Epidemiology—President's Address"; Strong, "The Importance of Ecology in Relation to Disease"; Jelliffe, "The Ecological Principle in Medicine"; Hackett, *Malaria in Europe*; Burnet, "Inapparent Virus Infections"; Burnet, *Biological Aspects of Infectious Disease*, esp. chap. 1, "The Ecological Point of View"; Burnet, *Virus as Organism*; Reed, "Environmental Medicine"; and Livingstone, "Environments."
58. Greenwood, "The Epidemiological Point of View," 405; Greenwood was the first professor of epidemiology and vital statistics at the London School of Hygiene and Tropical Medicine. Also partially quoted in Mendelsohn, "From Eradication to Equilibrium," 321.
59. Worboys, "Manson, Ross and Colonial Medical Policy," 22.
60. Balfour, Campenhout, Martin, and Bagshawe, *Interim Report on Tuberculosis and Sleeping-Sickness in Equatorial Africa*, 7.
61. Ross, "The Progress of Tropical Medicine," 271–72; and Li, "Natural History of Parasitic Disease."
62. From obituary for Sir William Leishman, member, U.K. Medical Research Council, 1925, quoted in Thomason, *Half a Century of Medical Research*, vol. 2, 190. Also see Worboys, "Tropical Diseases," esp. 523–24.
63. Huxley, *Biology and Its Place in Native Education in East Africa*, 6.
64. See Wilkinson, "Epidemiology," 1262–82, esp. 1278–79.
65. Soper, "Rehabilitation of the Eradication Concept"; and Cueto, "The Cycles of Eradication."
66. Swynnerton, "An Examination of the Tsetse Problem in North Mossurise."
67. Swynnerton, "Some Factors in the Replacement of the Ancient East African Forest," 493–518; on medicinal plants, see 509. It was an elderly African "doctor" to whom he turned for help identifying these plants. He had also undertaken a range of studies on birds and mimicry.
68. "C. F. M. Swynnerton."
69. Weinmann, *Agricultural Research and Development in Southern Rhodesia*, 15.
70. See, for instance, Swynnerton, "Some Factors in the Replacement of the Ancient East African Forest," 496, 501–2, and 509.
71. Swynnerton, "An Examination of the Tsetse Problem in North Mossurise," 316.

72. Charles Swynnerton, "Game Circular: Tsetse Control," August 7, 1921, reprinted in Glasgow, "Shinyanga," 22–34, Swynnerton circular on 31–34, quotation on 32.
73. Swynnerton, "An Examination of the Tsetse Problem in North Mossurise," 325.
74. Ibid., 332–33.
75. Ibid., 332.
76. Ibid., 376 and 384.
77. Swynnerton, comments in *Report on the Second Imperial Entomological Conference, June 1925*, 33.
78. Swynnerton, "An Experiment in Control of Tsetse-Flies at Shinyanga," 315.
79. Swynnerton referred to the railway plans in several of his articles; also see *Report of the East African Guaranteed Loan Committee, 1926–1929*, 5. This railway was built by 1930.
80. Swynnerton, "An Examination of the Tsetse Problem in North Mossurise," 317; and Swynnerton, *The Tsetse Flies of East Africa*, 10. Also see MacKenzie, "Experts and Amateurs."
81. By 1918 Swynnerton knew of the ecological studies of John Bews, a Scotsman employed in South Africa, and was also friendly with Guy Marshall, the director of the Imperial Entomological Bureau, and Arthur Bagshawe, the director of the Bureau of Tropical Diseases. Marshall also encouraged ecological studies. On Bews's networks, see Anker, *Imperial Ecology*.
82. Swynnerton, "Fauna in Relation to Flora," 10–15.
83. Swynnerton, "An Examination of the Tsetse Problem in North Mossurise," 381.
84. Johnston, "The Preservation of the African Fauna and Its Relation to Tropical Diseases"; Johnston is reviewing *Further Correspondence Relating to the Preservation of Wild Animals in Africa*.
85. Mutwira, "Southern Rhodesia Wild Life Policy (1890–1953)." Mutwira states that Swynnerton endorsed game slaughter. It would be more accurate to say that he was willing to test this strategy on a small scale, as he did in the Tsetse Research Department in Shinyanga, but he opposed radical culling.
86. Swynnerton, "An Examination of the Tsetse Problem in North Mossurise," 324.
87. Charles Swynnerton, comments in *Report on the Second Imperial Entomological Conference*, 31.
88. Austen, *Northwest Tanzania under German and British Rule*, 197.
89. Specialists interested in controlling other infectious diseases could find this frustrating; see Summers, "Intimate Colonialism," 790.
90. Zeller, "The Establishment of Western Medicine in Buganda," 162–78.
91. Mitchell, "The Uganda Medical School," 85–91; also see Iliffe, *East African Doctors*.
92. See Carpenter, "Progress Report on Investigations into the Bionomics of *Glossina palpalis*"; Carpenter, "Second Report on the Bionomics of *Glossina fuscipes (palpalis)* of Uganda"; and Carpenter, "Third, Fourth, and Fifth Reports on the Bionomics of *Glossina palpalis* on Lake Victoria."
93. Carpenter, *A Naturalist on Lake Victoria*, x.
94. Bruce, "Sleeping Sickness in Uganda"; and Fraser and Duke, "Duration of the Infectivity of the *Glossina palpalis*.
95. Carpenter, *A Naturalist on Lake Victoria*, 26.
96. Ibid., 60–61.
97. *Minutes of Evidence Taken by the Departmental Committee on Sleeping Sickness*, 51.
98. *Reports of the Sleeping Sickness Commission of the Royal Society* n. 16 (1915), 7. These comments indicate the lack of a uniform approach to tsetse or trypanosomiasis

control in British Africa and highlight the need to revisit Worboys, "The Comparative History of Sleeping Sickness."

99. Professor E. A. Minchin, testimony, *Minutes of Evidence*, 69; Minchin was chair of protozoology at the University of London and had been a member of the Royal Society's Sleeping Sickness Commission in Uganda in 1905.

100. Da Costa, *Sleeping Sickness*; and Da Costa, *Sleeping Sickness in the Island of Principe*.

101. *Report of the East Africa Commission*, 76.

102. *Report of the Inter-Departmental Committee on Sleeping Sickness*, 14. The Portuguese suspected that trypanosomiasis had originally been imported to the island in the early nineteenth century (circa 1825); according to interviews with "native" informants it seemed that epidemic sleeping sickness became a problem only late in the nineteenth century, circa 1893–94. See Da Costa, *Sleeping Sickness in the Island of Principe*, 1–2.

103. *Report of the East Africa Commission*, 79.

104. *Report of the Joint Committee on Closer Union in East Africa*, 816.

105. Few studies discuss the history of bionomics explicitly. An article on its development in the American context reinforces the idea that bionomics and ecology shared an interest in seeing "interactions in the natural world as complex and multi-causal"; see Largent, "Bionomics," 482.

106. Minute from Flood to D. O'Brien, 23rd October 1936, CO 822/74/9, "Co-Ordination of Research—Ecological," BNA.

107. The committee was created at the first meeting of the Civil Research Committee on June 18, 1925; see CO 533/589, "Tsetse Fly Sub-Committee of Civil Research Committee, 1925–26," BNA; veterinary research was meant to be concentrated at Kabete, Kenya, as well.

108. This work was reviewed over the course of three years; see Colonial Medical Research Committee, *Memorandum on Medical Research in the Colonies* [1928, 102 pp.]; *Medical Research in the Colonies, Protectorates, and Mandated Territories* [1929, 238 pp.]; and *Medical Research in the Colonies, Protectorates, and Mandated Territories* [1930, 208 pp.], London School of Hygiene and Tropical Medicine Library.

109. Both quotations from Walter Fletcher, testimony before the East Africa Joint Committee, *Report of the Joint Committee on Closer Union in East Africa* (1931), [156], 807.

110. The Depression had a much more severe effect on the East African Human Trypanosomiasis Research Institute in Uganda, which was closed in 1936; on the multiyear debates about its function and purpose, see CO 822/45/6, BNA.

111. He also did a documentary on the Maji Maji uprising during the German period in Tanganyika. "Film Production in East Africa by Thorold Dickinson," May 28, 1943, CO 875/17/6, "Two Cities Film Unit in Africa: 'Men of Two Worlds,' 1943," BNA.

112. A thorough history of the Tsetse Research Department has yet to be written. I remain indebted to Luise White, whose research initially alerted me to some of the overlap between the Ecological Survey of Zambia and the Tsetse Research Department: White, "Tsetse Visions."

113. Swynnerton, "A Late Dry-Season Investigation of the Tsetse Problem in the North of the Abercorn District," October 29, 1935, SEC 3/525, vol. 1, ZNA; and "Report on the Investigation of the Tsetse Problem in N. Rhodesia by Swynnerton," vol. 1, 1935–38, File 23098, TNA.

114. For details of the department's field surveys and their lead coordinators, see Tanganyika Territory, *Tsetse Research Report, 1935–1938*, 72–76.

115. Duke, "An Inquiry into an Outbreak of Human Trypanosomiasis"; and Swynnerton, "Entomological Aspects of an Outbreak of Sleeping Sickness."

116. Swynnerton, *The Tsetse Flies of East Africa*, 10.

117. Phillips, "The Application of Ecological Research Methods," 713. Also see *Tsetse Research Department Annual Report, 1928–1929.*

118. Staff numbers were reduced to 15 in 1934; figures from Swynnerton, *The Tsetse Flies of East Africa*, 17–18.

119. Tanganyika Territory, *Tsetse Research Report, 1935–1938*, 8–9.

120. Swynnerton, *The Tsetse Flies of East Africa*, 26, 128–29, 440–41.

121. Tanganyika Territory, *Tsetse Research Report, 1935–1938*, 9.

122. Swynnerton, "An Experiment in Control of Tsetse-Flies at Shinyanga, Tanganyika," 322.

123. Chief Makwaia testimony with P. E. Mitchell translating, *Report of the Joint Committee on Closer Union in East Africa*, 492.

124. He relied at times, for instance, on the natural knowledge of Clement Gillman, who was one of the few engineers in British Africa to engage actively in ecological debates. Gillman, "East African Vegetation Types"; and Gillman, "Review: *The Soils, Vegetation and Agriculture of North-Eastern Rhodesia.*"

125. The term *boy* was used both for young people and adults, a label that could be pejorative and paternalistic. Photographs from the Tsetse Research Department, however, show that many of the "fly boys" were indeed boys. There is a fuzzy picture of Abdallah in Swynnerton, *The Tsetse Flies of East Africa* (p. 41), but it is difficult to determine his age, except to say that he appears young, perhaps in his twenties.

126. Swynnerton, *The Tsetse Flies of East Africa*, 9 and 308.

127. Ibid., 272.

128. Ibid., 26.

129. Tanganyika Territory, *Tsetse Research Report, 1935–1938*, 21.

130. Phillips, "Who Dare Be So Bold?" 19; this is an autobiographical sketch.

131. "Report by Swynnerton on the Tsetse Flies of East Africa, 1932–37," File 25162, TNA.

132. Kirkpatrick, "East Africa and the Tsetse Fly," 412–13.

133. Buxton, "Tsetse Flies of East Africa," 93; Buxton had worked in Nigeria and was a member of the British Ecological Society.

134. Johnson, *Notes upon a Journey.*

135. Ibid., 38; emphasis in original.

136. Nash, *A Zoo without Bars*, 75 and 147; also see Nash, *Africa's Bane.*

137. Nash, "The Anchau Settlement Scheme."

138. Their critique of Swynnerton derives from John MacKenzie (1990); see Mustafa and Meager, "Agrarian Production, Public Policy and the State," 23–25. My thanks to Raufu Mustafa for sharing this with me.

139. While I am sympathetic with their desire to explore the ramifications of such projects for Nigerian development following independence, their assessment of the Anchau scheme overlooks its epistemic and sociological significance.

140. The American anthropologist, Horace Miner, undertook two stints of fieldwork in Anchau in 1957–58 and 1970–71; to the best of my knowledge he never wrote up the results of his second trip. See Miner, "Culture Change under Pressure."

141. *The Colonial Territories (1948–1949)*, 92.

142. See, for instance, Davis, "Social Medicine as a Field for Social Research"; and Weindling, "Social Medicine at the League of Nations Health Organisation."

143. Edge, "The Incidence and Distribution of Human Trypanosomiasis"; this is a review of medical department reports for 1936.

144. Swynnerton, *The Tsetse Flies of East Africa*, 6.

145. Swynnerton, "The Ecological Study and Control of the East African Species of Tsetse Fly," no date, but circulated in January 1936, CO 691/149/9, "Tsetse Fly Research and Control, Part 1, 1936," BNA 7–9. Much of this forty-six-page memorandum was published verbatim in *The Tsetse Flies of East Africa* later in the year.

146. Pearce, *The Treatment of Human Trypanosomiasis with Tryparsamide*.

147. Carpenter quoted in *Uganda News*, November 8, 1929; entry in Carpenter's diary for October 21, 1929, Geoffrey Hale Carpenter Papers, London School of Hygiene and Tropical Medicine Archives, London. The sequence of events and the governor's use of his report in a speech are in the diary.

148. See, as well, Carpenter, "Sleeping Sickness," esp. 141–42.

149. On the Oxford expeditions to Spitsbergen, see Anker, *Imperial Ecology*, 89–97.

150. See Jackson, "Some New Methods in the Study of *Glossina morsitans*."

151. Huxley, *Essays of a Biologist*; and Huxley, *Essays in Popular Science*.

152. On Lankester's influence on Wells and their collaboration on Wells's *Outline of History*, see Peter Bowler's entry on Lankester in the *Oxford Dictionary of National Biography*. For Lankester's views on sleeping sickness and nature's agency, see Lankester, *The Kingdom of Man*, chap. 3, "Nature's Revenges: the Sleeping Sickness."

153. The review was written in 1923 and republished as Huxley, "Biology in Utopia," in *Essays in Popular Science*, 65; also cited in Anker, *Imperial Ecology*, 111. Anker offers an analysis of the philosophical and ideological underpinnings of Wells and Huxley's collaborative writings.

154. Huxley, "Biology in Utopia," 65.

155. Also expressed by Huxley in University of Oxford, ed., *Scientific Results of the First Oxford University Expedition to Spitsbergen*, v.

156. All quotations from Wells, Huxley, and Wells, *The Science of Life*, vol. 3, 664, 677, 679.

157. Burnet, *Changing Patterns*, 23. A deeper analysis of Burnet's ecological approach to disease can be found in Anderson, "Natural Histories of Infectious Disease."

158. Burnet, "Inapparent Virus Infections," 100.

159. Burnet wrote in the preface that he "should like to think that this book expresses the same general point of view that runs through Wells, Huxley, and Wells's *Science of Life*." Burnet, *Biological Aspects of Infectious Disease*, ix.

160. Elton, *Animal Ecology*, 101.

161. Elton, "The Study of Epidemic Diseases," 436; also see Elton, "Plague and the Regulation of Numbers," 138.

162. Elton, "The Study of Epidemic Diseases," 436.

163. Also see Elton's monograph, *The Ecology of Animals*, 78–79.

164. Huxley, "Editor's Introduction," in Elton, *Animal Ecology*, xiv–xv, emphasis added.

165. Julian Huxley, "A Proposal for a Bureau of Ecology for Africa," April 29, 1931, Folder 3 16/26d, Royal Institute of International Affairs, London; Charles Elton in "Report of the Science Committee on African Problems," February 28, 1930, File 2792, Rhodes Trust Archives, Oxford.

166. Huxley, "Aspects of Africa, I. Some First Impressions, II. White and Black, III. The Tsetse Fly, IV. The Linked Tale," *Times (London)*, 6, 7, 8, and 9 January 1930; and Huxley, *Africa View*.

167. Huxley, "Editor's Introduction," xiv.

168. Huxley, "A Proposal for a Bureau of Ecology for Africa."
169. Charles Elton to Julian Huxley, March 9, 1932, Box 2, File 4, Joseph Oldham Papers, Rhodes House, Oxford.
170. See Huxley's rather confident assertion that "there is no reason why diseases like malaria and plague should not disappear from Africa as thoroughly as they have from northern Europe . . . the prime cause of malaria's diminution has always been and will continue to be the general raising of the human standard of life, which results in better houses, less contact with mosquitoes, more drainage, greater resistance, readiness to call in a doctor, and readiness to take more trouble about sanitary matters in general." Huxley, *Africa View*, 289–90.
171. Wells, Huxley, and Wells, *The Science of Life*, 694–95.
172. The Tsetse Research Department in Tanganyika offers an excellent example; see Swynnerton, *Tsetse Flies of East Africa*, 28–30.
173. *Report of the East Africa Sub-Committee of the Tsetse Fly Committee*, 19.
174. *Report of the Colonial Development Fund Advisory Committee Ninth Annual Report, 1937–1938*, 10.
175. On Barrett's work in Uganda, see Holden, "Doctors and Other Medical Personnel in the Public Health Services in Africa, 1930–1965," 132 and 137–43.
176. Barrett, "Notes on the Epidemiology of Sleeping Sickness," 20, emphasis added. Barrett cites Swynnerton and Carpenter.
177. Wilcocks, Corson, and Sheppard, *A Survey of Recent Work on Trypanosomiasis and Tsetse Flies*, esp. 54–69; and Colonial Office, *A Note on Some of the Scientific Studies*.
178. *Conference on Co-Ordination of Tsetse and Trypanosomiasis Research and Control in East Africa*, 3.
179. Historians have recently taken up this question, including Deborah Neill, Karen Brown, and Kirk Hoppe. While I do not share many of Hoppe's conclusions regarding "colonial science," I do support his effort to integrate social history and the history of science. See Hoppe, *Lords of the Fly*.
180. Quoted in A. R. Paterson, "Note on Considerations arising out of the Question 'What trained Africans the Medical Department of Kenya is likely to require?' Including a Statement in Reply to the question 'What employment the Medical Department could give to trained Africans if such are provided?'" January 20, 1937 [47 pp. typed manuscript], Published Monographs, KNA.
181. *Conference on Tsetse and Trypanosomiasis (Animal and Human) Research Entebbe, 22nd to 25th November 1933*, 5–6.
182. Paterson, "The Provision of Medical and Sanitary Services for Natives in Rural Africa," 440.
183. The phrase comes from a discussion heading reported in *Conference on Co-Ordination of General Medical Research in East African Territories, Entebbe, 27th to 29th November 1933*.
184. Quoted in "Conference on Co-Ordination of Medical Research in East African Territories," 75.
185. Paterson, "Some Observations on General Medical Research in East Africa," in *Conference on Co-Ordination of General Medical Research*, 57.
186. Ibid., 61.
187. "Malaria under African Conditions—Report of the Committee," in "Report of the Pan-African Health Conference," 110. The rapporteur for malaria was A. R. Paterson, director of Kenya's medical services; other members of the committee were Sir

Spencer Lister, G. Clarebout (Belgian Congo), G. Girard (Madagascar), J. M. Mackay (Nigeria), J. A. McGregor (Northern Rhodesia), G. A. Park Ross (South Africa), A. M. Wilson Rae (Gambia), P. S. Selwyn-Clarke (Gold Coast), and Fred Soper (Rockefeller Foundation).

188. See the two League of Nations reports, *Report on Principles and Methods of Anti-Malarial Measures in Europe*, and *Report on the Therapeutics of Malaria*; also pertinent are the 1934 Heath Clark Lectures delivered by Louis W. Hackett, *Malaria in Europe*.

189. See James, *Report on a Visit to Kenya and Uganda*; Swellengrebel, "Report on Investigation into Malaria in the Union of South Africa, 1930–31," esp. 443–56; Ross, "Insecticide as a Major Measure in Control of Malaria," 114–33; and Kohn, *Encyclopedia of Plague and Pestilence*, 294–95. G. A. Park Ross, participant in the 1935 conference and member of the Committee on Yellow Fever, Plague and Malaria, was responsible for coordinating control of the Natal epidemic.

190. "Malaria under African Conditions—Report of the Committee," in "Report of the Pan-African Health Conference," 110.

191. Ibid., 112.

192. Ross, "Insecticide as a Major Measure in Control of Malaria," 116.

193. An interesting story could be told about researchers' use of African assistants and the reaction of local populations during the control efforts in the early 1930s in Zululand; Park Ross, for instance, recorded: "It was evident that the native malaria assistants were gaining the confidence of their own people, and objections to quinine were disappearing. . . . In the native reserves, the position became grave, especially in newly infected areas, where the herbalists were still against us. In one of these, the hot, heavily infected valley of the Mfongosi, a trial was made with sulphur fumigation of huts. . . . While the fumigation did little good, the procedure, savouring of witchcraft, was enthusiastically taken up by the native populations." Ross, "Insecticide as a Major Measure of Control," 120–21.

194. "Malaria under African Conditions," in "Report of the Pan-African Health Conference," 112–13.

195. Worthington, *Science in Africa*, 461.

196. Ibid., 462.

197. Ibid., 14.

198. Ibid., 569.

199. Ibid., 517 and 519.

200. Ibid., 520.

201. Here Worthington is quoting from the League of Nations, "Report on Therapeutics of Malaria" (1933), *Science in Africa*, 520.

202. Worthington, *Science in Africa*, 578.

203. Ibid., 580. For an interesting recent study that provides references to other sources, see Sidibé et al., "Baobab: Homegrown Vitamin C for Africa."

204. Worthington, *Science in Africa*, 209; also 165–66 on "toxicology and medicine."

205. Ibid., 583.

206. Sir Robert McCarrison to Scientific Sub-Committee, no date [October–November 1935], LP, GD40/17/127.

207. Paterson, "Considerations arising out of the question . . . ," January 1937, KNA.

208. Worthington, *Science in Africa*, 505.

209. Comments by Hailey in conversation with A. T. Stanton at the Colonial Office, October 10, 1938, CO 847/11/1.

210. Hailey, *An African Survey*, 1182–93, quotation on 1182.
211. See details in CO 859/14/1, "Pan-African Health Conference—Nairobi, March 1940," BNA.
212. Hailey, *An African Survey*, 1183.
213. For one of the more complete syntheses, see Dubos, *Man Adapting*.
214. Thornton and Orenstein, "Co-Ordination of Health Work in Africa," 208–9; on Orenstein as an "opponent to the use of silver bullet approaches to health problems and an advocate of environmental reform," see Packard, "The Invention of the 'Tropical Worker,'" 288–89.
215. Doyal, *The Political Economy of Health*, 256.
216. Hailey, *An African Survey*, 1193.

CHAPTER FIVE

1. B. Malinowski to Editor, January 8, 1934; J. Webb, Editor's Secretary, to B. Malinowski, January 8, 1934, in MP, Africa I, #13.
2. Governor Brigadier General J. Byrne to Colonial Secretary Sir Philip Cunliffe-Lister, July 5, 1934, BY/26/7, Kenya National Archives, Nairobi, hereafter KNA.
3. H. L. Gordon, "Problems of African Education," January 22, 1934, *Times (London)*, in CO 822/55/1, and "Investigation of African Mental Development," first proposal to Colonial Development Fund, no date [likely 1931], BY/26/7, KNA.
4. L. B. Freeston to Francis Hemming, January 14, 1935, CO 822/61/14, "East African Native: Brain structure and mental capacity, 1934–35," British National Archives, London, hereafter BNA.
5. African Research Survey, Executive Committee Meeting Minutes, July 20, 1935, Lord Lothian Papers, GD40/17/127, National Archives of Scotland, hereafter LP.
6. A noteworthy exception is Dubow, *Scientific Racism in Modern South Africa*; for a perceptive account of racial thinking, see Glassman, "Slower Than a Massacre."
7. For a review of some of this literature, see Graham, *The Idea of Race in Latin America*.
8. For a few examples, see Hoffman, "The Race Traits and Tendencies among the American Negro"; and "A Selected Bibliography on the Physical and Mental Abilities of the American Negro," 548–64. This latter bibliography runs to 319 entries, which was meant to accompany the fifteen articles in the special issue of the *Journal* devoted to "a critical summary of the studies on the relative physical and mental abilities of the American Negro."
9. Goldberg, *Racial State*, 195–96. His emphasis on "whiteness" aside, Goldberg's more general definitions could apply to several other contexts than the ones he has in mind; for example, they could be extended to Japan, China, and Egypt.
10. See, for instance, Eluwa, "Background to the Emergence of the National Congress of British West Africa."
11. Brubaker and Cooper make a similar distinction in their analysis of the different ways "identity" is used in the secondary literature, which applies equally to the past; Brubaker and Cooper, "Beyond 'Identity.'" Also see Young, "Nationalism, Ethnicity, and Class in Africa."
12. On miscegenation in particular and sexual relations more broadly, see Barrera, "Dangerous Liaisons"; Mandaza, *Race, Colour, and Class in Southern Africa*; White, *Children of the French Empire*; Lee, "Colonial Kinships"; and Lee, "The 'Native' Undefined." For one of the first studies of miscegenation in South Africa, by a journalist, see Findlay, *Miscegenation*.

13. Darwin, *The Descent of Man*, 151.

14. See, for instance, Robertshaw, *A History of African Archaeology*.

15. Pearson, *The Grammar of Science*, 369; also partially quoted in Bowler, *Biology and Social Thought*, 70.

16. Cust, "On the Attitude of the White Man to His Coloured Fellow-Creature All Over the World" [1902], 20–21.

17. For a review of some of these perspectives, see Roberts, "The Racial Interpretation of History and Politics"; Cowen, "Race Prejudice"; Howsin, "Race and Colour Prejudice"; Scott, "A New Colour Bar"; Bacillus [pseud.], "Colour Prejudice"; Koyaji, "Colour Prejudice in the British Colonies"; and Singh, "Asiatic Emigration." For an essay that draws attention to the need for a more systematic discussion of racial issues at the Colonial Conference, see Wybergh, "Imperial Organisation and the Colour Question," 695–05 and 805–15.

18. Blyden, "West Africa before Europe," 363; the exact date is reported in Blyden, *West Africa Before Europe and Other Addresses*.

19. Blyden, "Africa and the Africans," 181.

20. The ordinance was ultimately passed in 1904 and Milner was one of its supporters; see Richardson, *Chinese Mine Labour in the Transvaal*, esp. 27–29.

21. Governor Milner to Joseph Chamberlain June 6, 1903, including enclosure of report from the *Rand Daily Mail*, in *Further Correspondence Relating to the Transvaal and Orange River Colony*, 40.

22. On the Gold Coast, see Akurang-Parry, "We Cast about for a Remedy." Akurang-Parry argues that the Transvaal and Gold Coast cases actually differed, but he bases this claim on a secondary source that does not analyze the objections of the "White League," which bore a striking resemblance to the Gold Coast protagonists.

23. Ibid., 378.

24. Also see Godard, *Racial Supremacy*. Ramsay MacDonald also pursued these themes in his work with Fabian Socialists and the Labour Party.

25. Wells, *A Modern Utopia*, chap. 10; and Wells, "Race Prejudice." On Wells's broader sympathies, see Partington, *Building Cosmopolis*.

26. Finot, *Les préjugé des races*, translated into English in 1906. On Finot's wider context, see Hecht, "The Solvency of Metaphysics."

27. Royce, "Race Questions and Prejudices"; Royce first delivered this address in 1905 to the Chicago Ethical Society. Also see Thomas, "The Psychology of Race Prejudice." Wells drew upon Thomas's work.

28. Hobson, *Imperialism: A Study*, 198–200 and, more broadly, chap. 2, "The Scientific Defense of Imperialism." Hobson was often reacting explicitly to Karl Pearson.

29. Hobson, *Imperialism: A Study*, 184, 198–200, 247, 295–96.

30. Finot, *Race Prejudice*, ix and 318–19.

31. Royce, "Race Questions and Prejudices," 268 and 285.

32. Wells, *A Modern Utopia*, 186. Wells was drawing on the work of Joseph Deniker, *The Races of Man* (1900) and H. N. Hutchinson, J. W. Gregory, and R. Lydekker, *The Living Races of Mankind* (1902).

33. This worry is expressed in these terms in H. S. Goldsmith, Acting Governor, Northern Nigeria, to Secretary of State, April 29, 1912, CO 879/109; these positions correspond to some of the patterns described in Füredi, *The Silent War*.

34. M. Cameron Blair, Senior Sanitary Officer to Principal Medical Officer, Zungeru, December 6, 1911, CO 879/109; "Further Correspondence relating to Medical and Sanitary Matters in Tropical Africa, January to June 1912," BNA.

35. See Simpson, *Report by Professor W. J. Simpson on Sanitary Matters in Various West African Colonies.*

36. Hugh Clifford to Colonial Secretary of State, March 8, 1913, CO 879/112 "Africa. Further Correspondence relating to Medical and Sanitary Matters in Tropical Africa, January to June 1913," BNA.

37. Hesketh Bell, response to segregation proposals, February 20, 1912, CO 879/109; for Lugard's objections, see "Minutes of the 40th meeting of Advisory Medical and Sanitary Committee for Tropical Africa July 2, 1912," CO 879/110 "Africa. Further Correspondence relating to Medical and Sanitary Matters in Tropical Africa, July to December 1912," BNA. For an overview on the history of segregation theory and practice, see Curtin, "Medical Knowledge and Urban Planning in Tropical Africa"; and Cell, "Anglo-Indian Medical Theory and the Origins of Segregation in West Africa." I am more sympathetic to Cell's interpretation, although neither author recognizes Lugard's about-face on the question of segregation in 1912.

38. Sambon, "Sleeping Sickness in the Light of Recent Knowledge," 201.

39. C. Jenkins Lumpkin, Ogurtola Sapara, W. A. Cole, and C. C. A. Jones, "Memorandum from 'native' Medical Officers in West Africa on the subject of the 4 paragraphs in the Report on the Departmental Committee on the West African Medical Staff," November 22, 1909; and Editorial, *Sierra Leone Weekly News*, November 6, 1909, both in CO 879/102/7 "Africa: Papers June, 1906 to December, 1910 relating to Medical and Sanitary Matters in Tropical Africa," BNA.

40. A. F. "Memorandum on the Employment of Native Medical Officers in West Africa," January 30, 1911, CO 879/107, BNA.

41. Goldberg, *Racial State*, 109–10.

42. Thiselton-Dyer to Colonial Office, April 23, 1900, CO 879/65/1.

43. For historical background, see Patton, *Physicians, Colonial Racism, and Diaspora in West Africa.*

44. Hugh Clifford to Secretary of State for the Colonies, March 3, 1913, CO 879/112.

45. Hugh Clifford to Secretary of State, April 10, 1913, CO 879/112.

46. Mr. Beilby Alston, Chargé d'Affaires, Tokyo to Lord Earl Curzon, Foreign Secretary, April 10, 1919, and general correspondence and minutes in FO 608/243/12. Also see Macmillan, *Peacemakers*, chap. 23; and Shimazu, *Japan, Race, and Equality.*

47. On India's uncertain relationship to the Colonial and Imperial Conferences before the First World War, see Kendle, *The Colonial and Imperial Conferences 1887–1911.*

48. This language appears in Government of India to His Majesty's Secretary of State for India, October 21, 1920, in *Correspondence Regarding the Position of Indians in East Africa*, 3–6; I am amalgamating the three concerns here, which are discussed separately in the letter.

49. According to the 1921 census, there were 9,651 "Europeans," 22,822 "Indians," and 10,102 "Arabs," while the African population was estimated at 2.5 million; reported in *Memorandum Relating to Indians in Kenya*, 9–10.

50. Quoted in Government of India to His Majesty's Secretary of State for India, October 21, 1920, in *Correspondence Regarding the Position of Indians in East Africa*, 2; emphasis added.

51. Quoted in *Memorandum Relating to Indians in Kenya*, 6.

52. Ibid., 10.

53. Maxon, "The Devonshire Declaration"; and Maxon, *Struggle for Kenya.*

54. Sir James Masterton-Smith, 1923, quoted in Robinson, "The Moral Disarmament of African Empire," 92.

55. William Simpson report quoted in *Correspondence Regarding the Position of Indians in East Africa*, 4–5.

56. Alfred Milner quoted in ibid., 5.

57. All quotations in Government of India to His Majesty's Secretary of State for India, October 21, 1920, *Correspondence Regarding the Position of Indians in East Africa*, 5–6. Uganda's Development Commission ruled out racial segregation in commercial areas, but left open the possibility of racial segregation in urban residential areas; see *Report of the Uganda Development Commission, 1920*, 33.

58. *Memorandum Relating to Indians in Kenya*, 8, emphasis added.

59. *Report of the East Africa Commission*, 23.

60. Oldham to Curtis, May 18, 1925, OP, Box 2, File 3.

61. J. H. O[ldham], "Research in East Africa," [no date, but sandwiched between documents dated 1925 and 1926], OP, Box 1, File 5. Also see Oldham to Sir Edward Grigg, December 3, 1925, OP, Box 6, File 1.

62. All quotations from Joseph Oldham, "Research into Native Welfare in East Africa," to William Ormsby-Gore, September 10, 1926, CO 533/648, "Kenya Native Welfare Research Organisation, 1926," BNA.

63. Leys to Oldham, October 6, 1921, reprinted in Cell, *By Kenya Possessed*, 194; Leys also became an avid reader of Du Bois's journal *The Crisis*.

64. See especially Murray, "At Home in the Modern World," 30–38.

65. Oldham to Reverend Donald Fraser, Church of Scotland Mission in Nyasaland, January 3, 1924, reprinted in Cell, *By Kenya Possessed*, 221.

66. See his outline for the book, Joseph Oldham "Inter-Racial Relationships," November 3, 1921; for Leys's response, see Leys to Oldham, November 14, 1921, both reprinted in Cell, *By Kenya Possessed*, 197–201 and 202–5.

67. Oldham, *Christianity and the Race Problem*, 67.

68. Both quotations in "Church Congress: Question of Race Equality."

69. "Summary of Proceedings of the Conference of Governors of the East African Dependencies, 16 April 1926," CO 533/589, "Tsetse Fly Sub-Committee of Civil Research Committee, 1925–26," BNA; and quotation in "The Mind of the Native," *East African Standard*, October 7, 1926, in CO 533/648, BNA.

70. CO 533/584 "International Commission on Sleeping Sickness, 1926," and CO 822/15/10, "Sleeping Sickness Research Station, Entebbe, Future Arrangements of the Station, 1929," BNA. The Human Trypanosomiasis Institute was officially established in June 1927.

71. Committee of Civil Research, "Interim Report of the Sub-Committee on Dietetics July 23, 1926," CO 758/19/6 "Dietetics Research, 1926," BNA; also see Brantley, "Kikuyu-Maasai Nutrition and Colonial Science," 49–86.

72. H. L. Gordon, "The Mind of the Native: Scientific Study Recommended by a Psychologist—Mental Processes," *East African Standard*, part I on October 7, 1926 and part 2 on October 8, 1926. These articles include many verbatim excerpts from Gordon's lecture; unless otherwise noted, all quotations in this paragraph are from these articles.

73. The proposal was circulated to the Imperial Conference in 1926, but was considered premature and unwieldy; see CO 323/965/16 "Imperial Conference, 1926: Imperial Co-Operation in Research, 1926," BNA.

74. "Henry Laing Gordon"; also see "Dr. H. L. Gordon," 109. My thanks to Sloan Mahone for this reference. Gordon had previously worked in the neurological division of a London hospital.

75. See *Report of the Commission on Closer Union*; and Malinowski, "Practical Anthropology."

76. *Report of the President and the Secretary as to an Educational Program in Africa*, 14.

77. Vint had been appointed by the Colonial Office, but Gordon had been given his post through the Kenya government; Vint was one of the Medical Service's research officers, while Gordon was not.

78. Gordon, "Relation of Malaria to the Alleged Rarity of Neurosyphilis," 227 and 229. Sloan Mahone has undertaken a detailed history of psychiatry and its intersections with neurology and psychology in the context of mental hospitals in East Africa in her 2004 doctoral thesis, which she is preparing for publication as a book.

79. Gordon, "A Note on Diagnosis of Amentia," 208.

80. For a more detailed history of eugenics in Britain and Kenya, which draws upon several of the same sources relating to H. L. Gordon's proposal, see Campbell, *Race and Empire*. Campbell is less interested in research priorities and debates about "racial prejudice" and more interested in popular support for eugenics in Britain and Kenya. Our different emphases lead to different arguments about the significance of racial science to colonial state building in British tropical Africa.

81. Much of the background material for this section can be found in BY/26/7, "Publications 1931–50, Mental Capacity of Natives," KNA, Nairobi. Parts of this story have also been analyzed by Vaughan, *Curing Their Ills*, 110–11; Dubow, *Scientific Racism in Modern South Africa*, 201–2; McCulloch, *Colonial Psychiatry*, 46–49; and Olumwullah, *Dis-Ease in the Colonial State*, 217–22.

82. British Medical Association, "Brain and Mind in East Africa."

83. Gordon, "Amentia in the East African." The talk was delivered on November 7; Gordon arrived in England on October 4 and departed on November 30.

84. Gordon, "The Native Brain"; for Vint's work, see, "A Preliminary Note on the Cell Content." In addition to his talk to the Eugenics Society, Gordon also delivered lectures to the Royal Society of Medicine, the African Circle of the Royal Institute of International Affairs, the Royal Medico-Psychological Association, and the Empire Parliamentary Association; see H. L. Gordon to Dr. A. R. Paterson, December 12, 1933, BY/26/7, KNA.

85. Gordon, "The Native Brain."

86. On the history of eugenics in Britain, see Searle, *Eugenics and Politics in Britain*; Stepan, *The Idea of Race in Science*, chap. 5, "Eugenics and Race"; Kevles, *In the Name of Eugenics*; Soloway, *Demography and Degeneration*; and Barkan, *The Retreat of Scientific Racism*.

87. H. L. Gordon, "Eugenics and the Truth about Ourselves in Kenya," delivered on March 18, 1933 to the Kenya and Uganda Natural History Society and "published to assist formation of a Society for Study of Race Improvement."

88. Ibid., 2, 3.

89. Ibid., 2.

90. Gordon, "Amentia in the East African," 229, 234.

91. See, for example, Chamberlin and Gilman, *Degeneration*; and Pick, *Faces of Degeneration*.

92. See Pearson, *The Life of Francis Galton*, vol. 2, chap. 9, esp. "The Passage from Geography to Anthropology and Race-Improvement," 70–87.

93. See Vaughan, *Curing Their Ills*, chap. 6, "Syphilis and Sexuality: The Limits of Colonial Medical Power"; and Hunt, "Colonial Medical Anthropology and the Making of the Central African Infertility Belt," in Tilley with Gordon, *Ordering Africa*.

94. For an overview of the historical evidence, see Caldwell, "The Social Repercussions of Colonial Rule," 458–86.
95. Also see Kennedy, *Islands of White.*
96. MacKinnon, "Education in Its Relation to the Physical and Mental Development of European Children"; Boer, "A Survey of European Children of School-Going Age." Also see "Physique and Health in Kenya"; and "Development of School Children in Kenya."
97. Burkitt, "The Medical Aspect of Closer Settlement of Europeans," 188.
98. See, for instance, Balfour and Scott, *Health Problems of the Empire*, 104–5.
99. Smuts, "African Settlement. Part I," 127; also see Smuts, "British African Conference [Smuts Lecture]"; this sentence also appears in Smuts's published version of his lectures.
100. Remarks of Leopold Amery and Dietrich Westermann, "Minutes of Conference on African Problems," LP, GD40/17/120.
101. Fortes, "Perceptual Tests of 'General Intelligence' for Inter-Racial Use."
102. See Stepan, *The Idea of Race in Science*, 139; on Huxley's shifting views on race, see chapters 4 and 5 of Barkan, *The Retreat of Scientific Racism.*
103. Leakey, "The Native Brain."
104. See his comments in the introduction to Kitson, "A Study of the Negro Skull."
105. Louis Leakey, "Plans for a Projected East African Scientific Research Institute," and L. Leakey to B. Malinowski, April 14, 1930, Bronislaw Malinowski Papers, Africa I, Mal 589; and Malinowski to Louis Leakey, April 28, 1930, MP, Africa I, 5, London School of Economics, hereafter MP.
106. Harry Leakey, "Appendix No. 6, Memorandum from the Rev. Canon Harry Leakey," 21. His father's views on the "fallacy" of the idea of "detribalisation" are in this memorandum.
107. Also see the letter Leakey signed objecting to the "prejudice" "coloured medical students" experienced during their studies in Britain and calling on the secretary of state for the colonies to intervene; he was joined by a prestigious cast of characters including Lord Olivier; the legal expert, A. Berriedale Keith; the psychologist, W. B. Mumford; the South African historian, William Macmillan; the former Kenya administrator, William McGregor Ross; and Thomas Drummond Shiels (most of whom also advised the African Survey); Leakey et al., "Coloured Medical Students."
108. Huxley, "The Native Brain."
109. Burt, "The Native Brain."
110. M. Fortes, unpublished letter to the *Times*, in MP, Africa I, #13; also see, Fortes, "A New Application of the Theory of Neogenesis to the Problem of Mental Testing. (Perceptual tests of 'g'.)"; and Fortes, "Review of S. Porteus, *The Psychology of a Primitive People*," in which he was critical of several of Porteus's "naive fallacies." Fortes was already a sharp critic of loose generalizations about racial difference.
111. Hogben, *Genetic Principles in Medicine and Social Science*; and Hogben, *Nature and Nurture.*
112. Haldane, "The Native Brain."
113. Hogben, "Letter to the Editor," 432.
114. Letter to the BBC from Clifton Roberts, Lancelot Hogben, Winifred Holtby, J. F. Harrabin, Lucy Johnstone Scott, Norman Leys, F. S. Livie Noble, W. H. C. Malton, Lord Olivier, V. S. S. Sastri, Georgiana M. Solomon, G. L. Steer, Josiah Wedgwood, November 14, 1930, LP, GD40/17/246. One of the signatories, Clifton Roberts, would go on to author a critical review of colonial law; see Roberts, *Tangled Justice.*

115. Huxley, *Africa View*, 394.
116. Eugenics Society to Under Secretary of State for the Colonies, November 21, 1933, BY/26/7, KNA. The letter's signatories, besides Huxley, were the physical anthropologists, Arthur Keith and Grafton Elliot Smith; the society's president, Humphry Rolleston; the former director of Medical and Sanitary Services for Kenya, John Gilks; the animal geneticist, Francis A. Crew; and the medical doctors, Lord Dawson and Lord Horder. A similar letter to the *Times* appeared on November 25, 1933, with the same signatories except Huxley, Crew, and Gilks.
117. Huxley, *Africa View*, 406.
118. Huxley, *Science and Social Needs*, 241.
119. Leakey, Huxley, and Burt all prefaced their letters with comments on the wider importance of the questions Gordon's research raised regarding the administration of territories in Africa; it was Burt who used the phrase *racial prejudice* in his letter.
120. Malinowski to Editor, January 8, 1934; this letter went unpublished because "so many letters had already appeared on the subject." J. Webb, Editor's Secretary, to B. Malinowski, January 8, 1934, in MP, Africa I, #13.
121. B. Malinowski, "Memorandum on Colonial Research," December 1927, OP, Box 2, File 2; also see Malinowski, "New and Old Anthropology."
122. Malinowski, "Memorandum for the Rockefeller Foundation," 1926, OP, Box 2, File 3.
123. B. Malinowski, "The Teaching of Practical Anthropology in Connection with Colonial Studies," 15 pp. [written by hand, "Memo drafted for Lionel Curtis & Co.", no date, but likely between January and June 1929], MP, Africa I, #2, Mal 541. Following Elazar Barkan, Barbara Bush argues that "British anthropologists were not directly concerned with refuting scientific racism and physical anthropology." The evidence contradicts this assertion. Bush, *Imperialism, Race, and Resistance*, 37.
124. Compare, for instance, Huxley and De Beer, *The Elements of Experimental Embryology* and Malinowski, *Culture as a Determinant of Behavior*.
125. B. Malinowski to Sir James Frazer, October 25, 1917, Sir James Frazer MSS, Add. Ms.b.36, Trinity College, Cambridge University.
126. Malinowski, "Anthropology," *Encyclopedia Britannica*, 1938.
127. Shiels, "The Native Brain."
128. Shiels in Minutes of meeting [unmarked and undated, but March 18, 1930], in MP, Africa I, #9, Temp 589.
129. H. L. Gordon to Dr. A. R. Paterson, December 12, 1933, BY/26/7, KNA.
130. Comments by H. L. Gordon, in Vint, "A Preliminary Note on the Cell Content of the Prefrontal Cortex of the East African Native," 50.
131. Vint did say that he thought the East African native's cerebral development only reached that "of the average European boy of between 7 and 8 years of age," but this did not preclude the idea that Africans' nerve cells might mature differently under different "conditions of life and education." Vint, "A Preliminary Note on the Cell Content," 48–49.
132. Scott, "A Note on the Educable Capacity of the African," 99.
133. Ibid., 104.
134. H. S. Scott, "Memorandum in Regard to Education of Africans" [19 pp.], enclosed in Scott to Acting Colonial Secretary, Nairobi, April 18, 1929, CO 533/388/11 "Re-Organisation of African Education—Kenya, 1929," BNA.
135. Scott, "Note on Educable Capacity," 108.
136. Minute by Scott to Colonial Secretary, February 3, 1934, BY/26/7, KNA.

137. Oliver, "The Comparison of the Abilities of Races"; Oliver, "The Adaptation of Intelligence Tests to Tropical Africa"; Oliver, "The Adaptation of Intelligence Tests to Tropical Africa, [Part] II"; and Oliver, "Mental Tests in the Study of the African."

138. On Oldham's role in Oliver's research, see Box 223, "Africa General-Education," file on "Psychology Tests," and Box 243 "East Africa, Kenya—Education," file on "Dr. Oliver's Intelligence Tests," in the International Missionary Council Papers, School of Oriental and African Studies, London; also see Oldham to Gilks, October 8, 1932 and Gilks to Oldham December 9, 1932, BY/26/7, KNA.

139. Dougall, "Characteristics of African Thought," 256.

140. Ibid., 259.

141. Oliver, "The Adaptation of Intelligence Tests to Tropical Africa," 186–88; also see Oliver, *General Intelligence Tests for Africans: Manual of Directions*.

142. Arthur Mayhew, "A note on Psychological Tests of Educable Capacity and on the possibility of their use in the Colonies," no date [circulated to Advisory Committee on Education in the Colonies for the July 18, 1929 meeting], CO 323/1036/20, "Education—Psychological Tests of Educable Capacity, 1929," BNA.

143. Minutes of 8th meeting of Advisory Committee on Education in the Colonies, September 26, 1929, "Psychological Tests of Educable Capacity"; and Circular from Lord Passfield to all Colonies and Protectorates (except Malta, Somaliland, Tanganyika, Sarawak, and Palestine), December 12, 1929, CO 323/1036/20.

144. Oliver, "The Comparison of the Abilities of Races," 161.

145. Quoting R. S. Woodworth, presidential address to the American Association for the Advancement of Science, 1910, in ibid, 164.

146. Ibid., 169.

147. Ibid.

148. Ibid., 171.

149. Oliver, "Mental Tests in the Study of the African," 42.

150. Oliver gives the ages in "The Comparison of the Abilities of Races [Part I]," 173.

151. Oliver, "The Adaptation of Intelligence Tests to Tropical Africa. [Part] 11," 9.

152. Oliver, "Adaptation of Intelligence Tests to Tropical Africa," 190–91; a copy of the final visual images used in the test can be found in BY/26/7, KNA. Instructions were given in English in those schools where English was the primary teaching language and in Swahili in those schools where Swahili was the language of instruction.

153. The schools were the Prince of Wales School (for European boys) and the Alliance High School, a Protestant school funded by the Kenyan government located in the Kikuyu Province and designed to train students in commerce, agriculture, and teaching. For a description of the school, including the curriculum, during the exact period when Oliver was conducting his tests, see Thurnwald, *Black and White in East Africa*, esp. 261–70.

154. Oliver, "Mental Tests in the Study of the African," 44.

155. Oliver, "The Comparison of the Abilities of Races," 174.

156. Oliver, "Mental Tests in the Study of the African," 45.

157. Ibid.

158. See, for instance, correspondence in CO 533/388/11 "Re-Organisation of African Education—Kenya, 1929," and CO 533/408/17 "Native Education Policy—Kenya, 1931," BNA.

159. Department proposal quoted in Williams, "The Provision of Hospitals for, and the Medical Training of, Africans in Kenya," 48.

160. Trowell, "The Medical Training of Africans," 341, 345.

161. Minute by Acheson, August 17, 1932 and minute by Stanton, August 12, 1932, CO 822/47/5, "East African Native—Brain Structure and Mental Capacity, 1932," BNA.

162. See the minutes by the assistant secretaries of state for East and West Africa respectively, J. E. W. Flood, December 8, 1933, CO 822/55/1, and A. Fiddian, August 24, 1932, CO 822/47/5.

163. R. V. Vernon, minute, October 24, 1929, CO 323/1036/20.

164. H. L. Gordon to Dr. Carlyle Johnstone, Acting Director of the Medical Services, July 8, 1934, BY/26/7, KNA.

165. For the committee's full remit, see Francis Hemming (secretary to the EAC) to L. B. Freeston (secretary to the East Africa Governors Conference), August 31, 1934, CO 822/61/14; and Report by Dr. Hemming on his meeting with Mr. Freeston, November 16, 1933, CO 822/55/1.

166. The prime minister served as president of the Economic Advisory Council. For the recurring doubts expressed by Hemming, secretary to the EAC and the prime minister, see minute by Stanton, October 18, 1933, CO 822/55/1; Hemming to Freeston, October 29, 1934, Hemming to Freeston, November 16, 1934, CO 822/61/14.

167. For the final proposal, see J. Byrne, Kenya governor, to Sir Philip Cunliffe-Lister, secretary of state for the colonies, July 5, 1934, BY/26/7, KNA. On Gordon's "press campaign," see dispatch from Colonial Office to Secretaries of Native Affairs of East Africa including Kenya, Uganda, Tanganyika, Nyasaland, Northern Rhodesia, and Somaliland, October 11, 1934, CO 822/61/14.

168. J. E. W. Flood to M. Moore, April 17, 1934, CO 8222/61/14.

169. Hemming to L.B. Freeston, October 29, 1934, CO 822/61/14.

170. Oldham to J. L. Gilks, director of Kenya Medical Services, October 8, 1932, BY/26/7, KNA.

171. J. H. O[ldham], unpublished draft letter to the *Times*, no date, MP, Africa I, #13.

172. Francis Hemming to Freeston, November 16, 1934, and minute by Freeston, November 8, 1934, CO 822/61/14.

173. James Sequeira to Dr. Paterson, February 5, 1934, BY/26/7; "Resolution on Research in Africa," 122; see also CO 822/72/8.

174. "Mental Quotient of the Natives," 2; also reported in *East African Medical Journal* 11 (1934/35): 101. Also see Dubow, *Scientific Racism in Modern South Africa*, 202; and Vaughan, *Curing Their Ills*, 110–11.

175. J. Byrne to Cunliffe-Lister, July 5, 1934, BY/26/7.

176. FO 141/592/5, "Sudan requested to co-operate with a Committee of the Economic Advisory Council to examine influences on mental development of East African people, 1934," BNA.

177. See the letters to the *Times* in 1934 on January 1, 2, 5, 8, 11, and 22, August 28 and 31, September 3, 6, and 19, and October 9. Questions regarding the status of the project and proposal were asked in Parliament on November 29 and 30, 1933, in the fall of 1934, and in the summer of 1935.

178. J. E. W. Flood, minute, September 19, 1936, CO 822/72/7, "East African Native— Brain Structure and Mental Capacity, 1936," PRO.

179. He served only from June through November when he lost his seat in Parliament, but this period was long enough for him to have a significant effect on the African survey's importance to the Colonial Office and also to officials in British Africa.

180. "Trusteeship, House of Commons, 9th July 1935," 153–54; the questioner was Sir Ernest Graham-Little, who had originally hosted Gordon's talk to the Empire Parliamentary Association in 1933.

181. Coupland to Malcolm Hailey, March 13, 1936, LP, GD40/17/128.

182. On Coupland's work, see CO 847/6/3, CO 847/11/2, and CO 847/14/2, "African Research: Professor Coupland's Scheme," for 1936, 1938, and 1939 respectively, BNA.

183. On Gordon's further work, see Gordon, "An Inquiry into the Correlation of Civilization and Mental Disorder in the Kenya Native"; and Gordon, "On Certification on Mental Disorder in Kenya." On the East African branch of the British Medical Association's endorsement of continued research along these lines, see "Resolution on Research in Africa," 122.

184. See, for instance, the proposals put forward to the Rhodes Trust by the Africa Co-Operation Society of Southern Rhodesia for an "Inter-Territorial Survey of British Africa Relating to Soil, Water, Rivers, Forests, Tsetse Fly, Malaria, Etc," in Rhodes Trust Archives, File 2790. This proposal was deferred and then rejected because it was seen to duplicate work already done by the African Survey. Also relevant are the debates concerning the establishment of the Rhodes-Livingstone Institute (RLI) in Northern Rhodesia/Zambia; the RLI was ultimately funded, in large part because it was seen to meet a strategic need for anthropological research in British territories.

185. Nadel, "The Application of Intelligence Tests in the Anthropological Field," 187–88; also see Nadel, "A Field Experiment in Racial Psychology." Nadel was appointed in 1937 as government anthropologist to the Anglo-Egyptian Sudan; before that, his fieldwork was in Nigeria.

186. Gustavo Tosti, quoted in Oldham, *Christianity and the Race Problem*, 61. Tosti was from Italy, and his argument appeared in an article on Durkheim and suicide that was published in the *American Journal of Sociology* in 1898.

187. Huxley, *Africa View*, 395.

188. Huxley, Haddon, and Carr-Saunders, *We Europeans*, 216, 92.

189. Haddon, "The Universal Races Congress."

190. Firth, *Human Types*, chap. 1, "Racial Traits and Mental Differences," 22.

191. Ibid., 10–11, 35.

192. Ibid., 26.

193. Ibid., 32.

194. Worthington, *Science in Africa*, 585; and Hailey, *An African Survey*, 37.

195. Hailey, *An African Survey*, 37. In one of his articles, Gordon introduced the term *bradyphysis* as a synonym for backwardness and defined it as "a non-progressive state showing deficiencies of social, moral, intellectual, and material self-development" in Kenyan tribes. Trowell also pointed out the tautology; see "The Medical Training of Africans," 344.

196. Hailey, *African Survey*, 38. Also see Mumford and Smith, "Racial Comparisons and Intelligence Testing." Dr. Mumford, a psychologist at the University of London's Institute of Education, was also an advisor to the Survey, and sections of this article correspond almost verbatim to the Survey's findings.

197. Worthington, *Science in Africa*, 586. Also see Smith, "Africa: What Do We Know of It?" 56.

198. Hailey, *African Survey*, 40.

199. Quoted in the *East African Medical Journal* 12 (1935): 262. Also see "Anthropologists in Congress," 398–403; this editorial included the cautionary note.

200. "Africa Section Resolutions," 62/188/1 and Resolution proposed for full Congress by K. A. H. Houghton, 62/188/18, Royal Anthropological Institute Archives, London.

201. The several hundred pages of summaries of colonial medical department research

confirms this; see Colonial Medical Research Committee, *Memorandum on Medical Research in the Colonies*; Colonial Medical Research Committee, *Medical Research in the Colonies, Protectorates, and Mandated Territories*; Colonial Medical Research Committee, *Medical Research in the Colonies, Protectorates, and Mandated Territories*, London School of Hygiene and Tropical Medicine Library.

202. Clark, "Race," 66–67.

203. Mayhew, "Education in the Colonies," 85.

204. Hailey, *African Survey*, 36.

205. Minute by C. White, December 12, 1938, CO 822/85/11.

206. Thomas, *Politics of the Womb*; Hunt, *A Colonial Lexicon*; White, *Children of the French Empire*; Summers, "Intimate Colonialism."

207. Dubow, *Scientific Racism in Modern South Africa*, 132–33; for the fascinating Egyptian story see El Shakry, *The Great Social Laboratory*.

208. In addition to Dubow's discussion of eugenics and "poor whitism," also see Bell, "American Philanthropy, the Carnegie Corporation and Poverty in South Africa."

209. See, for instance, discussion and memoranda in CO 847/20/6 "Public Announcements regarding the intentions of H.M. Government as to the future of Native Races in Africa, 1940," and CO 859/80/13 "Colour Discrimination in the Colonies General Policy of the Colonial Office, 1941–42," BNA.

210. He asked this during his inaugural address as Lord Rector of Glasgow University, November 16, 1900; quoted in Godard, *Racial Supremacy*, 4–5.

CHAPTER SIX

1. Philip Mitchell, Secretary for Native Affairs, Tanganyika, to Sir Harold Carmichael, May 2, 1934, File 21984, "African Research Survey," vol. I, Tanzania National Archives; B.Ma., "Anthropology," *Encyclopedia Britannica* (1938), in Malinowski Papers accompanied by lengthy draft comments, Encyclopedia 7, Temp 14.

2. On Richard Gregory and *Nature*, see Werskey, "*Nature* and Politics between the Wars." Gregory joined the General Committee of the African Research Survey in 1935.

3. All quotations from "Magic and Administration in Africa," 629–31; the piece could have been written by Richard Gregory himself, or by one of his regular contributors such as A. G. Church, who had firsthand experience in British Africa.

4. Waller, "Witchcraft and Colonial Law in Kenya." The verdict in the case was first reported in the *Times (London)* on February 6, 1932 under the heading "Murder of a Witch in East Africa."

5. In 1911 Ukamba Province had been the site of an unsettling prophet movement. For a wonderful overview of officials' reactions to various types of social movements, including the 1911 "mania," see Mahone, "The Psychology of Rebellion."

6. Details in CO 533/420/8 "Murder of a Witch, Death Sentence Passed on Sixty Wakamba Natives, 1932," BNA.

7. Telegram from Kenya governor, February 13, 1932, CO 533/420/8; and Rex v. Kumwaka Wa Mulumbi & 69 Others, judgment issued March 26, 1932, in *Law Reports of Kenya*, 137–39.

8. Many of the original letters were "destroyed under statute," but records of their protest were noted by Freeston, March 31, 1932, CO 533/420/8; individual protesters included the specialist in African linguistics, Alice Werner, and Kenya's former director of Public Works, William McGregor Ross.

9. Geoffrey Dawson to Cunliffe-Lister, March 29, 1932, CO 533/420/8.

10. W[illiam] B[lack] Stevenson to John Buchan, M.P., February 6, 1932, enclosed in Buchan to Malcolm Macdonald, February 9, 1932; and W. B. Stevenson, letter to *The Herald* February 6, 1932, all in CO 533/420/8. Also see Hutcheson and Stevenson, *Kikuyu: 1898–1923*; and Stevenson, *Four Months among African Missions.*

11. W. Hutcheson, MD to Hon Sec of Foreign Affairs (forwarded to Colonial Office), February 7, 1932, CO 533/420/8.

12. Mr. A. de V. Wade, "Memorandum on the Report of the Commission of Inquiry into the Administration of Justice in East Africa, in so far as that report affects the Administration of Justice in Kenya," October 12, 1933, CO 822/53 "Enquiry into the Administration of Justice in Criminal Matters as Affecting Natives," BNA.

13. Governor of Kenya to Sec of State for Colonies, April 1, 1932, CO 533/420/8.

14. On these points see Frank Melland to Frederick Lugard, April 23, 1932, File 4, Box 12, Lugard Papers; Lugard and Melland exchanged a number of letters on this subject between 1932 and 1934.

15. Melland, "A Shadow over Africa"; for his views on "native policy" and development, see Melland, "Native Policy in Africa."

16. Lugard, *Report by Sir F. D. Lugard on the Amalgamation of Northern and Southern Nigeria*, 6; and Lugard, "Witchcraft in Africa." The psychologist was Richard Thurnwald; for a brief report on Thurnwald's proposed research dated October 21, 1929, see File 8/8 "Office Correspondence and Memoranda 1930–1938," International African Institute Archives (IAI), London School of Economics, London.

17. The eight panels on Africa are discussed further below; resolutions reported in 62/188/1 and 62/188/2, "African Section Resolutions," Royal Anthropological Institute Archives, London.

18. "Licenses for Witches," *Daily Telegraph*, August 4, 1934, MS 176 Press Cuttings #1, RAI Archives; this is a summary of Frank Melland's paper.

19. See Hailey, *African Survey*, 1637.

20. For the "museum specimen" quotation, see *Higher Education in East Africa*, 7; also see "Future Policy in Africa, Carlton Hotel Discussion, 1939," CO 847/17/11, BNA, in which Arthur Mayhew stressed during the discussion, "You cannot keep the educated African within the sphere marked out for him by anthropologists and politically nervous administrators."

21. "Bureau of Anthropology"; "Anthropology and the Empire"; Haddon, "An Imperial Bureau of Anthropology"; and "Imperial Bureau of Anthropology." Also see chapter one for details of nineteenth-century discussions.

22. The French proposal failed because of tensions between administrators and professional anthropologists; see Sibeud, "The Elusive Bureau of Colonial Ethnology."

23. J. E. W. F[lood], minute on behalf of himself, Furse, and Amery, May 5, 1920; also see Flood's minute after he, Furse, and Sir H. Read met with Haddon and a Mr. Adams at Cambridge, July 31, 1920: "They seem to attach an altogether undue importance to ethnology, in fact, they both are crazy." Minutes in "Tropical African Services Course," CO 877/1, BNA. The terms *ethnology* and *anthropology*, despite their slightly different meanings to specialists, were often used interchangeably by officials.

24. On several "disastrous" interventions in Nigeria in the 1920s, see Lackner, "Social Anthropology and Indirect Rule." Also see Nigeria's former lieutenant governor F. H. Ruxton's endorsement of social anthropology that appeared in response to Malinowski's 1929 article "Practical Anthropology": Ruxton, "An Anthropological No-Man's-Land."

25. For comments on importance of anthropology, see *Report of the East Africa Commission*, 80.
26. *Report by the Hon. W. G. A. Ormsby-Gore on His Visit to West Africa in 1926*, 13.
27. See the outline of lectures for 1930–31, "Oxford University Tropical African Services Courses," Correspondence File "H," MP; in the original Chichewa is listed as "Chinyanja." See also *Report of a Committee on the System of Appointment in the Colonial Office and the Colonial Services*, 36; and Jeffries, *The Colonial Office and Its Civil Service*, 59.
28. Malinowski, "Memorandum on Colonial Research," December 1927, OP, Box 2, File 2.
29. See, for instance, Kuper, *Anthropology and Anthropologists*, esp. chaps. 1 and 2 on Malinowski and Radcliffe-Brown; and Stocking, *After Tylor*, esp. chaps. 6 and 7, on Malinowski and Radcliffe-Brown respectively.
30. Malinowski, "Memorandum for the Rockefeller Foundation Written for Mr. Embree in March 1926," OP, Box 2, File 2. He later complicated this view in an article that opened with a discussion of "The Curse of Science."
31. *Report of the Commission on Closer Union of the Dependencies in Eastern and Central Africa* 67.
32. *Report of the Joint Committee on Closer Union in East Africa* [156], 34.
33. B. Malinowski, "The Teaching of Practical Anthropology in Connection with Colonial Studies," 15 pp. [written in hand, "Memo drafted for Lionel Curtis & Co.," no date, but likely written between January and June 1929], Malinowski Papers, Africa I, #2, Mal 541. In another note, Malinowski mentions a "discussion between Mr. Oldham, Lionel Curtis, Philip Kerr and B. M. at Chatham House in June last [1929] [when] a certain type of Colonial Studies was outlined." "Re: Discussion with Philip Kerr," March 1930.
34. Malinowski, "Reflections on the Possibilities of the African Institute," March 2, 1930, MP, Africa I, #8, Temp 550; and Malinowski, "Practical Anthropology," 36.
35. Malinowski, "Practical Anthropology," 36–37.
36. All quotations from untitled and undated minutes of meeting that took place on March 18, 1930; present were Malinowski, Philip Kerr/Lord Lothian, Lord Lugard, Joseph Oldham, Mrs. [Sybil] Rolfe, Dr. Drummond Shiels, Professor Eric Walker, Professor Seligman, Julian Huxley, and Sir Basil Blackett. See A. T. K. Grant to Malinowski, March 28, 1930 and minutes [18 pp.], MP, Africa I, #9, Temp 589.
37. Malinowski, "Rationalization of Anthropology and Administration," 408.
38. Mitchell, "The Anthropologist and the Practical Man," 221.
39. Malinowski, "Rationalization of Anthropology and Administration," 418–19.
40. Ibid., 419.
41. Ibid., 415–16.
42. These views were common among members of the Frankfurt School, especially Walter Benjamin and Max Horkheimer, but were also shared by less radical critics of "modernity." See Whitehead, *Science and the Modern World*, esp. chap. 13, "Requisites for Social Progress"; and Horkheimer, "Notes on Science and the Crisis."
43. Malinowski, "Rationalization of Anthropology and Administration," 405–6.
44. Ibid., 406. The reference to "pre-logical beings" comes from the work of the French ethnographer Lucien Lévy-Bruhl who published *Les fonctions mentales dans les sociétés inférieures* (1910) and *La mentalité primitive* (1922); Malinowski rejected Lévy-Bruhl's arguments that non-Europeans were averse to "abstract thought and reason."

45. The topics in quotation marks are taken from the headings found in Malinowski, "Practical Anthropology."

46. See his description of anthropology's contribution to "modern humanism," circa 1924, quoted in Kuper, *Anthropology and Anthropologists*, 19.

47. On Young's objections, see Sir Hubert Young to Sir Cecil Bottomley, June 5, 1934, SEC1/1727, ZNA.

48. See the discussion of the African Survey and the comments by Malcolm Hailey on Evans-Pritchard's and Fortes's proposed research project in "Research into Problems of Political Organisation under Indirect Rule in Africa by Dr. E. E. Evans-Pritchard and Dr. M. Fortes, 1940," CO 847/20/8.

49. General Committee Report of Progress, October 31 1933, "Points for Chairman's Speech," GD40/17/123.

50. Hilda Matheson, "Update on African Research Survey to the General Committee," April 10, 1934, LP, GD40/17/124.

51. William Macmillan to Hilda Matheson, August 21, 1933, LP, GD40/17/122.

52. The executive committee also met with the British Association's Committee on the Human Geography in Tropical Africa, but this was an ad hoc body with few professional ambitions for the field of human geography.

53. Hilda Matheson to Lord Lothian, November 14, 1934, LP, GD40/17/125.

54. All quotations from Smith, "Africa: What Do We Know of It?" 1–2, 77–79, 81.

55. Matheson to Lothian, March 14, 1934, LP, GD40/17/124.

56. On Labouret's early involvement, see Hilda Matheson to Lord Lothian, November 23 and 27, 1933, LP, GD40/17/123; Westermann was consulted by Matheson just a few days before the anthropology meeting as well, Matheson to Lothian, March 9, 1934, LP, GD40/17/124.

57. Julian Huxley, as a member of the executive committee, had been meant to attend, but was unavailable; Lord Lugard, who was still chair of the IIALC's board, attended in his place.

58. Matheson to Lothian, March 14, 1934, GD40/17/124.

59. Firth's and Leakey's relationship to Malinowski is evident in Malinowski's own files in the LSE archives; on Bateson's role, see Stocking, *After Tylor*, 294 and 408.

60. "Memorandum" [no listed author or date, but accompanied by letter that said, "I enclose a memorandum which has been received unofficially from Mr. J. H. Driberg"], SEC1/1727, "African Research Survey," ZNA; also Sir Cecil Bottomley to Sir Hubert Young, March 27, 1934, same file.

61. Driberg Memorandum, SEC1/1727, ZNA.

62. Sir Cecil Bottomley to Sir Hubert Young, March 27, 1934, SEC1/1727, ZNA; these letters were sent to the governors of all British African territories.

63. Sir Cecil Bottomley to Governors, December 11, 1933, CO 847/2/1, BNA.

64. P. E. Mitchell to Sir Harold [Carmichael], May 2, 1934, File 21984, TNA, emphasis added.

65. He sent a copy of this letter to Mitchell accompanied by Bottomley's original letter and Driberg's memorandum.

66. Sir Harold Carmichael to Sir Cecil Bottomley, May 4, 1934, File 21984, TNA.

67. P. E. Mitchell to Sir Harold [Carmichael], May 2, 1934, File 21984, TNA; the last phrase in the sentence was added by Mitchell by hand.

68. All quotations from Sir Harold Carmichael to Sir Cecil Bottomley, May 4, 1934, File 21984, TNA.

69. Sir Harold Carmichael to Sir Cecil Bottomley, May 4, 1934, File 21984, TNA.

70. Mitchell deals with the question of equality in P. E. Mitchell to Joseph Oldham, March 21, 1929 (in which he discusses Malinowski's article on Practical Anthropology and Oldham's *Report on Closer Union*), OP, Box 4, Folder 2.

71. P. E. Mitchell to Malinowski, June 15, 1931, and "Civil Research in Tanganyika Territory," June 14, 1931, MP, File #62.

72. Philip Mitchell, "Introduction," xvii–xviii; also see a review of this book by Baker, "An Experiment in Applied Anthropology."

73. Minute to Northern Rhodesia Chief Secretary, April 25, 1934, SEC1/1727, ZNA.

74. Ogilvie, "Co-Operative Research in Geography."

75. E. H. Jalland, Acting Secretary for Native Affairs, to the Chief Secretary, "The African Research Survey," May 9, 1934, SEC1/1727, ZNA.

76. Sir Hubert Young to Sir Cecil Bottomley, June 5, 1934, SEC1/1727, ZNA.

77. Ibid. In a later context, during Hailey's tour of West Africa, some of these fears were confirmed by at least one of the governors who hosted Hailey. Sir Arthur Richards, then governor of the Gambia, wrote to Warrington Yorke at the Liverpool School of Tropical Medicine, "Sir Malcolm Hailey clearly intimated that he had a low opinion of African Governors. He came here from Accra and Sierra Leone and I apparently confirmed his worst fears." Richards to Warrington Yorke, May 19, 1936, TM13/130/12, University of Liverpool archives. My thanks to Marisa Chambers for sharing this letter with me.

78. Bottomley to Young, July 23, 1934, SEC1/1727, ZNA.

79. Minute by F. Stockdale, August 15, 1934, on conversation with Young, CO795/72/2, "The Livingstone Memorial Museum and Research Institute, 1934," BNA; also cited in Brown, "Anthropology and Colonial Rule," 177.

80. Lyn Schumaker has done a thorough historical and anthropological analysis of the Rhodes-Livingstone Institute; see Schumaker, *Africanizing Anthropology*.

81. This quotation comes from the document written by Malinowski, which discussed the joint potential of the IIALC and the proposed Institute of African Studies in Oxford: "Items of Expenditure Under the Million Dollar Interlocking Scheme for African Research," no date [March 1930], MP, Africa I, #8.

82. Malinowski, "Introduction," vii. See a similar tongue-in-cheek reference by Margery Perham to the British as "we savages" in Perham, "Future Relations of Black and White," 87.

83. "A Five-Year Plan of Research," 12.

84. For a close analysis of this relationship, see Berman and Lonsdale, "Custom, Modernity, and the Search for Kihooto."

85. Margery Perham, "Notes upon Research in Africa" [no date, but sent to Malinowski on June 10, 1931], MP, File #62.

86. Edwin Smith, quoted in "Anthropologists in Congress" [reporting on the first International Congress of Anthropological and Ethnological Sciences held in London July 30 to August 4 1934], 403.

87. On the complex effects of involving Africans in the ethnographic research process, see the three chapters on this subject in Tilley with Gordon, *Ordering Africa*.

88. Description in "Minutes of the Third Meeting of the Executive Council June 9–10, 1927, London," File 1/3 "Executive Council Minutes, 1925–1931," IAI Archives, LSE.

89. Smith, "The Story of the Institute," 15–16; the latter quotation is from an anonymous reviewer of a Swahili manuscript that Smith claimed "might be said of many

of these manuscripts." One of the first books to be published from this process was Pita Nwana's *Omenuko* (London: Atlantis Press, 1935).

90. Nyabongo, *Africa Answers Back*; and Nyabongo, "Religious Practices and Beliefs of Uganda." According to a reviewer, Nyabongo's brother was the ruler of the Toro kingdom in Uganda, but the title of "prince" was a loose translation of his status; see E. B. H., "Africa Answers Back," *Man* 36 (1936): 214–15. Also see Döring, "The Fissures of Fusion."

91. Smith, "Africa: What Do We Know of It?" 77.

92. Perham, *Ten Africans*, 13.

93. Huxley, "Travel and Politics in East Africa," 260.

94. *Nature*, December 5, 1936, 947–48, clipping in File 39/50 "Correspondence, Hunter, Monica," IAI Archives, LSE.

95. Audrey Richards's review in *London Spectator*, October 9, 1936; in File 39/50, "Correspondence, Hunter, Monica," IAI Archives.

96. All quotations from Mair, "The Study of Culture Contact as a Practical Problem," 417–18.

97. Perham, "Some Problems of Indirect Rule in Africa," 102.

98. Mair, "The Study of Culture Contact," 417–18; also see Mair, *Native Policies in Africa*, concluding chapter.

99. Richards, "Anthropological Problems in North-Eastern Rhodesia," 130.

100. Ibid., 141.

101. Perham, "Some Problems of Indirect Rule in Africa," 104.

102. Cecil Bottomley minute, July 10, 1931, following meeting with Oldham in which they discussed the help officials might be able to give to anthropologists, in CO 822/35/7, "International Institute of African Languages and Culture—Programme of Civil Research, 1931," BNA.

103. Isaac Schapera, "Review of Reaction to Conquest," *Bantu Studies* (1937): 53–60, on p. 53, clipping in File 39/50 "Correspondence, Hunter, Monica," IAI Archives.

104. Minute by G[erald] Creasy, October 19, 1935, recording his conversation with Jones the previous day, in CO 847/4/6, "Institute of African Languages and Cultures. Dr. Fortes' research work in the Gold Coast, 1935," BNA. Also see Fortes, "Culture Contact as a Dynamic Process."

105. Comments by acting attorney general (Mr. Jenkins) and acting chief secretary (Major Dutton) in Northern Rhodesia, *Legislative Council Debates*, 21–37, first two quotations on 30, third on 24.

106. Minute by Sir P[hilip] C[unliffe] L[ister], August 16, 1934, CO 795/72/2.

107. Minute by Frank Stockdale to Mr. Seel, August 15, 1934, CO 795/72/2.

108. Minute by A. T. Stanton, August 11, 1934, CO 795/72/2.

109. Hubert Young to Sir Philip [Cunliffe-Lister], September 14, 1934, CO 795/72/2.

110. P[hilip[C[unliffe] L[ister], August 23, 1934, CO 795/72/2, his emphasis.

111. Minute by Cunliffe-Lister, September 12, 1934, CO 795/72/2.

112. Philip Cunliffe-Lister to Hubert Young, September 24, 1934, CO 795/72/2.

113. Hilda Matheson to Lord Lothian, August 8, 1934, LP, GD40/17/125. The "Merle Davis project" resulted in *Modern Industry and the African*; it was funded by the Carnegie Corporation and the Phelps Stokes Fund.

114. A summary of comments made by Richards to Richard Brown; Brown, "Anthropology and Colonial Rule," 177.

115. Joseph Oldham paraphrased in Sir George Tomlinson to Sir Cecil Bottomley,

November 6, 1934, CO 795/72/2; also Lothian to Matheson, September 25, 1934, and Matheson to Lothian, October 25, 1934, LP, GD40/17/125.

116. Hubert Young to Cunliffe-Lister, September 29, 1934; and Hubert Young, "Livingstone Memorial Institute," CO 795/72/2.

117. Hubert Young to Cunliffe-Lister, September 29, 1934, CO 795/72/2.

118. Minute by J. B. Williams, October 3, 1934, CO 795/72/2.

119. E. B. Boyd to C. Bottomley, October 4, 1934, CO 795/72/2. Wilson ultimately was approved as director in 1937 and assumed the post in 1938.

120. Paraphrasing Cunliffe-Lister, Boyd wrote, "Now that he [the secretary of state] learns that he has written to the 'Anthropologist,' it is impossible to go back upon it and say we are not going to have it after all."

121. Tomlinson to Bottomley, November 6, 1934, CO 795/72/2. Even before Young had submitted his revised proposal, Cecil Bottomley had commented that his project seemed "closely akin to the objects of research of the International Institute of African Languages and Cultures." Bottomley to J. Maffey, September 7, 1934, CO 795/72/2.

122. Conversation with Oldham summarized by Tomlinson in Tomlinson to Bottomley, November 6, 1934, CO 795/72/2.

123. Tomlinson to Bottomley, November 6, 1934, CO 795/72/2.

124. Minute by Cunliffe-Lister, November 7, 1934, CO 795/72/2.

125. Cunliffe-Lister to Young, November 13, 1934, CO 795/72/2. This letter was written in consultation with Oldham and cites him as one of the individuals with whom the Colonial Office consulted in order to arrive at a decision; see Tomlinson to Bottomley, November 13, 1934, CO 795/72/2.

126. Hubert Young to Cunliffe-Lister, December 31, 1934, and Cunliffe-Lister to Young, telegram, January 15, 1935, CO 795/72/2.

127. Minute by Bottomley, April 9, 1937 [this comment was written in the margins beside another minute by Tomlinson, "I suppose the ecological side of the Institute which was mentioned in 1934 has now been dropped." April 8, 1937], CO 795/81/4, "The Livingstone Memorial Museum and Research Institute, 1936–37."

128. See Hailey to Lothian, August 16, 1934, Hailey to Lothian October 1, 1934, and Matheson to Hailey, January 24, 1935, LP, GD40/17/125.

129. Minute by S. E. V. Luke, July 2, 1935, summarizing conversation on May 21, 1935, between Cunliffe-Lister and Malcolm Hailey "for the record," CO 795/72/2.

130. Malinowski quoted by Reginald Coupland to Joseph Oldham, June 13, 1935, OP, Box 1, File 2.

131. Coupland to Oldham, June 13, 1935, ibid.

132. On Perham's early involvement in the project, see Oldham to Lothian, May 8, 1933, LP, GD40/17/121 (in which Oldham raises the question of whether Perham could serve as secretary to the project and the final report could be issued under the general committee to avoid prejudices due to her sex); also, "General Committee Report of Progress, October 1933," Matheson to Lothian November 23, 1933, and Perham to Matheson, January 7, 1934, LP, GD40/17/123; Matheson to Lothian, June 14, 1934, GD40/17/124; and Margery Perham, "A Study of Native Administration in British Africa" [a summary of her project proposal, July 1934], Lothian to Hailey, July 25, 1934, GD40/17/125.

133. While holding their fellowships, Perham pursued her comparative study of native administration and Mair undertook her work on culture contact in Uganda and

wrote *An African People in the Twentieth Century*. Mair dedicated her book, *Native Policies in Africa*, to Malinowski, while Perham acknowledged her debt to him in the introduction to her book *Native Administration in Nigeria*.

134. Perham was at Nuffield College, Oxford, and Mair was at the London School of Economics. On Mair's involvement in the Survey see Malcolm Hailey to Bronislaw Malinowski, July 1, 1936, B. Malinowski to Tracy Kittridge (assistant director of Social Sciences Division, Rockefeller Foundation), March 4, 1937, MP, File 15 "Students"; and Hilda Matheson to Lord Lothian, March 3, 1937, LP, GD40/17/129.

135. I also include a truncated discussion of this speech in my introduction to *Ordering Africa*.

136. Perham, "Some Problems of Indirect Rule," 93, her emphasis.

137. Ibid., 95, 100.

138. Ibid., 95–96.

139. Ibid., 97, 99, 100, 103–5.

140. Ibid., 95 and 104.

141. The Africans were Stella J. Thomas, a Nigerian of Sierra Leonean descent who received a law degree from Oxford University in the 1920s and in 1943 became the first woman in Nigeria appointed magistrate; Joseph Sackeyfio, from the Gold Coast; and Modjaben Dowuona, also from the Gold Coast, who according to Perham (p. 91) "became one of the first two African administrative officers in the Gold Coast and was later a member of the staff of the University of Accra."

142. Comments by William Macmillan, in Margery Perham, "Some Problems of Indirect Rule," 108.

143. Comments by S. J. Thomas, 110, in ibid.

144. Comments by M. Dowuona, 115, in ibid.

145. Comments by Shiels, 116–17, in ibid; Shiels's views on race and administration are discussed in chapter five.

146. Comments by S. J. Thomas, 110, in ibid.

147. Perham's reply, 115, in ibid.

148. Perham, "Some Problems of Indirect Rule," 96, 103.

149. Ibid., 101.

150. Mair, *Native Policies in Africa*, 4 and 6.

151. Ibid., 6.

152. Ibid., 3–6.

153. Ibid., 7.

154. Hailey to Malinowski, July 1, 1936, MP, File 15.

155. Matheson to Lothian, March 3, 1937, LP, GD40/17/129.

156. Malinowski to Hailey, August 21, 1936, MP, Correspondence Files.

157. Hailey, "Rough Outline of Part II. Of Chapter VI. Sociological Facts." July 17, 1936, MP, Correspondence Files.

158. Hailey paraphrased in minute by J. A. Calder, January 26, 1937, CO 795/81/4.

159. Minute by Ormsby-Gore, February 4, 1937, CO 795/81/4.

160. Hailey and Lugard were added as signatories between April 5 and May 20, 1937; see Young to Ormsby-Gore, April 5, 1937, and Hubert Young, "General Appeal," May 20, 1937, CO 795/81/4 and CO 795/88/12, respectively.

161. Hailey to Sir Hubert Young, June 17, 1937, File 2959, "Rhodes-Livingstone Institute," RTA.

162. Ibid.

163. Mair, "The Social Sciences in Africa South of the Sahara."

164. This view is articulated in Diana Jeater's opening remarks in her thought-provoking study, *Law, Language, and Science*, esp. 16–18.

165. *The Lancet* ran a series of articles on the "mind as a therapeutic agent" in 1909, a theme taken up later by a number of specialists including W. H. R. Rivers in his now-classic 1915–16 Fitzpatrick Lectures before the Royal College of Physicians; see Shaw, "The Influence of Mind as a Therapeutic Agent"; and Rivers, *Medicine, Magic, and Religion*. Also see the work of Frederick Mott on war veterans' psychic "shell-shock": Mott, "A Lecture on Body and Mind"; and Mott, "An Address on Body and Mind." On the growth of "psychic research" across the United Kingdom in this period, see Hazelgrove, *Spiritualism and British Society between the Wars*. For the text of a bill introduced in November 1930, which aimed to legalize investigations into psychical research that would otherwise be punishable through Britain's existing "witchcraft" legislation, see *Spiritualism and Psychical Research (Exemption) Bill*. For the first of a series of twelve lectures on "Motives and Mechanisms of the Mind," published between January and March of 1931, see [Howe,] "Motives and Mechanisms of the Mind: I."

166. MS 176 Press Cuttings #1, "1934 International Congress," RAI Archives.

167. This summary is derived from the archival records of the congress in the Royal Anthropological Institute; see *Congrès International des Sciences Anthropologiques et Ethnologiques Première Session, Londres, 1934 Guide* [32 pp.].

168. Minute on congress planning, October 16, 1933, CO 323/1221/6 "International Congress of Anthropological and Ethnological Sciences, London, 1934," BNA.

169. Comments by Charles Seligman in "International Congress: Preliminary Conference, Basel: 20–22 April 1933," 62/176/1, RAI Archives.

170. The congress section on "Anatomy and Physical Anthropology" passed the following resolution: "In view of the fallacious use of the term Aryan, increasingly prevalent at the present day, with the implied association of inherent mental aptitude and cultural achievements, this section urges the desirability for organized research in various countries, and considers that this research could best be carried out in connection with or under the direction of the International Institute for Intellectual Cooperation." 62/188/15, RAI Archives.

171. *Edinburgh Evening Dispatch*, July 31, 1934; also see "Prof. Sir Eliot Smith Denounces Aryan Fallacy of Science. The Common Ancestor of Humans and Apes. When Lines of Evolution Diverged." *Northern Evening Despatch*, July 31, 1934, MS 176 Press Cuttings #1, "1934 International Congress," RAI.

172. "African Sub-Section: Chair, Rev. Edwin Smith," annotated announcement notes, no date [likely January 1934], 62/177/9, RAI Archives.

173. These were six final panel titles, which were mentioned as topics in the earlier announcement.

174. Evans-Pritchard, "Oracle Magic of the Azande"; Evans-Pritchard, "Witchcraft (*Mangu*) among the Azande"; Evans-Pritchard, "Sorcery and Native Opinion"; Evans-Pritchard, "Mani, a Zande Secret Society"; Evans-Pritchard, "The Zande Corporation of Witchdoctors [Part I and Part II]"; and Evans-Pritchard, "The Intellectualist (English) Interpretation of Magic."

175. Evans-Pritchard, "The Zande Corporation of Witchdoctors," 291; this article was published in July, just a few months after the public controversy over the court case.

176. Lagae, *Les Azande ou Niam-Niam*.

177. *Africa* 4 (1931): 108.

178. De Graer, "L'art de guérir chez les Azande," 220–54, and 361–408; on De Graer's work on sleeping sickness, see Lyons, *The Colonial Disease*, 131.

179. On Evans-Pritchard's debt to De Graer, see Evans-Pritchard, "Zande Therapeutics"; also see Part IV, chapter 3 of *Witchcraft, Oracles, and Magic*; for his review of Lagae's work, see Evans-Pritchard, "Review of *Les Azande ou Niam-Niam*."

180. Dowuona quoted in *Daily Sketch*, August 4, 1934, in MS 176 Press Cuttings #1, "1934 International Congress," RAI.

181. Melland, "Northern Rhodesia: Retrospect and Prospect"; and H[obley], "Obituary: Frank Hulme Melland: 1879–3 February, 1939."

182. Hobley, *Bantu Beliefs and Magic*.

183. Melland, *In Witch-Bound Africa*, 191.

184. A complete history of Melland's role in this court case has yet to be written. See Fields, *Revival and Rebellion in Colonial Africa*; and Ranger, "The Mwana Lesa Movement of 1925."

185. Melland gave two lectures in this series; the first on "The Natural Resources of Africa," and the second on "Witchcraft in Africa," on February 17 and 24, 1932; see *JRAI* 62 (1932): v.

186. Frank Melland to Frederick Lugard, April 23, 1932, Box 12, File 4, Lugard Papers, Rhodes House.

187. Ibid.

188. Medical Officer, Ukamba Province, quoted in Native Affairs Department, *Annual Report, 1932* (Nairobi: Government Printer, 1932), 84.

189. Chesterman, "Witchcraft in Africa."

190. C. Chesterman, "The Curse and Cure of Witchcraft," July 17, 1932, BBC Archives; my thanks to Louise Weston of the BBC Archives for locating this address for me.

191. Frank Melland to Frederick Lugard, April 23, 1932, Box 12, File 4, Lugard Papers, Rhodes House.

192. Frank Melland to Granville Orde-Browne, April 13, 1934, Sir Granville Orde-Browne Papers, MSS Afr. s 1117, Rhodes House Library.

193. Melland, "The Misjudged Witch-Doctor," 15.

194. Granville Orde-Browne to Frank Melland, September 11, 1934; Orde-Browne was responding to the draft of Melland's paper, which was ultimately published in the special issue of *Africa*, edited by Dietrich Westermann in 1935.

195. W. Addison, letter to the editor, *Times (London)*, April 23, 1932; Addison does not identify the territory in which he worked, but I suspect it was in West Africa.

196. "Celebrations in Liverpool: Lord Olivier on British Policy in Africa," *Times (London)*, July 18, 1932, p. 14, col. B.

197. Frank Melland to Granville Orde-Browne, April 13, 1934,

198. Details from Melland to Orde-Browne, April 13, 1934; Orde-Browne to Melland, April 14, 1934; and Melland to Orde-Browne, April 16, 1934.

199. Orde-Browne to M. D. W. Jeffries, May 28, 1934; Jeffries, who had been based in Nigeria for twenty years, was also a critic of existing laws and explained in his reply (June 3, 1934) that his "views on the subject are not accepted in official quarters."

200. Orde-Browne to Melland, April 14, 1934.

201. Ibid., and Orde-Browne to Jeffries, June 6, 1934.

202. Melland to Orde-Browne, April 16, 1934.

203. F. Melland, comments in Strickland, "The Cooperative Movement in the East" [discussion at Chatham House on May 9, 1932], 830.

204. Melland, "The Problem of Africa: the Problem of Witchcraft," 17.

205. Melland to Orde-Browne June [indecipherable day], 1934.

206. A copy of Evans-Pritchard's paper, which is in the RAI Archives, is almost identical to his published paper: Evans-Pritchard, "Witchcraft," 418–19.

207. Dora Earthy, remarks during panel on "essentials of indigenous African culture," in *Congrès International des Sciences Anthropologiques et Ethnologiques: Compte-rendu*; also see Earthy, *Valenge Women*.

208. "African Section Resolutions," 62/188/1, RAI Archives.

209. Roberts, "Witchcraft and Colonial Legislation," 488–89.

210. Clifton Roberts, quoted in "Witchcraft Discussed," *East Africa* (August 16, 1934): 994–95. A similar argument about the incompatibility of English law with African witchcraft cases is made in Lugard, "How the Administrator Gains from Anthropology's New Insight into Tribal Life," newspaper clipping (no title), July 30, 1934, the day the International Congress of Anthropological Sciences opened, MS 176 Press Cuttings File #1, RAI Archives, London.

211. Granville Orde-Browne, quoted in "Witchcraft Discussed."

212. *Congrès International des Sciences Anthropologiques et Ethnologiques: Compte-rendu de la première session, Londres, 1934*, 229.

213. This part of Leakey's speech was not recorded in the official transcript, but it was reported in the *Morning Post*, August 4, 1934, MS 176 Press Cuttings File #1, RAI Archives.

214. Leakey and Smith quoted in "Witchcraft Discussed."

215. Dauncey Tongue, quoted in "The Secret of Black Witchcraft," *Sunday Dispatch*, August 5, 1934.

216. Modjaben Dowuona, paper summary in *Congrès International des Sciences Anthropologiques et Ethnologiques*, 226.

217. Dowuona, quoted in *Daily Sketch*, August 4, 1934.

218. Dowuona, paper summary in *Congrès International des Sciences Anthropologiques et Ethnologiques*, 226.

219. Jomo Kenyatta, quoted in "The Secret of Black Witchcraft," *Sunday Dispatch*, August 5, 1934.

220. J. Kenyatta, comment after panels on "Witchcraft and Colonial Legislation," in *Congrès International des Sciences Anthropologiques et Ethnologiques*, 230.

221. "African Section Resolutions," 62/188/2, RAI archives; this resolution is marked in hand "too late."

222. Editorial notes, "The African Explains Witchcraft," *Africa* 8 (1935): 504–59.

223. Smith, "Inzuikizi," 473.

224. Roberts, "Witchcraft and Colonial Legislation," 492.

225. Orde-Browne to Melland, September 11, 1934. See also Orde-Browne, "Witchcraft and British Colonial Law."

226. Orde-Brown, "Witchcraft and the Law," 53.

227. On Malinowski's role in commenting on and revising the section in the Survey on "Studies in Social Anthropology," see Matheson to Malinowski, November 26, 1937, and Malinowski to Matheson, January 1, 1938 (with enclosure, "Comments on Chapter 'Studies in Social Anthropology,' African Research Survey"), MP, Encyclopedia File #7, 14.

228. Hailey, *African Survey*, 31–32.

229. Ibid., 32.

230. Ibid., 40.

231. Ibid., 40–41.
232. Ibid., 45. In *Africa Emergent*, Macmillan wrote, "The particular case of the educated classes is a reminder that it is not enough 'to build on native foundations' as interpreted by anthropology. The definition of 'native' is too narrow where it fails to allow for the growth of this new class, which is an essential part of the Africa that is." Macmillan, *Africa Emergent*, 316.
233. Hailey, *African Survey*, 45–47.
234. Ibid., 47–48.
235. Ibid., 50–51.
236. Ibid., 46.
237. Ibid., 58.
238. Ibid.
239. Ibid., 1624.
240. Ibid., 1612.
241. William Ormsby-Gore to C. Parkinson, November 8, 1938, CO 323/1524/8, BNA.
242. Hailey, *African Survey*, 196.
243. Orde-Browne to Melland April 14, 1934.
244. CO 847/13/11 "Laws Relating to Witchcraft, 1938," and CO 847/19/9 "Laws Relating to Witchcraft, 1940," BNA.
245. Hailey, *African Survey*, 296.
246. "Witchcraft in Africa," *Manchester Guardian*, January 16, 1940, in CO 847/19/9.
247. Mr. Musgrave Thomas, Judge of High Court, Nyasaland to Sir Grattan Bushe, February 27, 1940, CO 847/19/9.
248. F[rederick] J. Pedler, minute, December 14, 1939, CO 847/13/11 "Laws Relating to Witchcraft—1938," BNA.
249. T. S. L. Fox-Pitt to Audrey Richards, April 12, 1939, File 1/15, Audrey Richards Papers, LSE.
250. Gluckman, "The Logic of African Science and Witchcraft," 62.
251. Audrey Richards, "The Pragmatic Value of Magic in Primitive Societies," talk before the Royal Institution in 1962, in File 1/15 "Magic and Witchcraft," Audrey Richards Papers, LSE.
252. See Reginald Coupland, confidential memorandum for Lord Hailey on "International Co-Operation in the Study of African Problems and the work of the International Institute of African Languages and Cultures," June 23, 1938, RIIA, File 16/26d; and the letters between Oldham, Coupland, and Lothian in 1938–39 in Box 204, IMC, File "IIALC" and subfile on "Relations with Hailey Survey."
253. D[aryll] F[orde], "Note on the Position and Prospects of the International African Institute—Confidential," April 27, 1944, in IMC, Box 204, File on "International Africa Institute."
254. McCulloch, *Psychiatry and the African Mind*; Sadowsky, *Imperial Bedlam*; Jackson, *Surfacing Up*; Keller, *Colonial Madness*; and Mahone, "Psychology of the Tropics."
255. I have not done a full count of these studies, but since the CSSRC existed until 1962 it seems quite likely that it funded several hundred studies in total if between 1945 and 1953 it had already funded 94.
256. Schumaker, *Africanizing Anthropology*; Mills, "British Anthropology at the End of Empire"; Mills, *Difficult Folk*; De l'Estoile, "L'Afrique comme laboratoire"; and Cooper, "Development, Modernization, and the Social Sciences in the Era of Decolonization."
257. Scientific Council for Africa South of the Sahara, *Research in the Social Sciences in Af-*

rica. The seventy categories were developed at the second meeting and are published in Scientific Council for Africa South of the Sahara and CCTA, *Social Sciences: Inter-African Conference*.

258. Tooth, *Report to the Colonial Social Science Research Council on the Use of Unadapted Tests of Intelligence*.

CHAPTER SEVEN

1. Elias, *Africa and the Development of International Law*; Fisch, "Africa as Terra Nullius," 347–75; Anghie, "Colonialism and the Birth of International Institutions"; and Louis, "African Origins of the Mandates Idea."

2. On the importance of the African territories to the founding of the schools of tropical medicine, see FO 2/890 "London and Liverpool Schools of Tropical Medicine, 1898–1904," British National Archives, London. On Africa's importance to social anthropology, see Moore, *Anthropology and Africa*; Schumaker, *Africanizing Anthropology*; and Tilley with Gordon, *Ordering Africa*. On the social sciences more generally, see Bates, Mudimbe, and O'Barr, *Africa and the Disciplines*.

3. Robertshaw, *A History of African Archaeology*.

4. Adams, *Against Extinction*; Van Heijnsbergen, *International Legal Protection of Wild Fauna and Flora*; Hayden, *The International Protection of Wild Life*. Historical treatments include Neumann, "The Post-War Conservation Boom in British Africa"; MacKenzie, *Empire of Nature*; Anderson and Grove, *Conservation in Africa*; and Carruthers, *The Kruger National Park*.

5. See Young, *The African Colonial State in Comparative Perspective*; Havinden and Meredith, *Colonialism and Development*; Cooper, *Decolonization and African Society*.

6. For a recent discussion of Africa's place in the "modern world," written in dialogue with Christopher Bayly's *Birth of the Modern World*, see Vaughan, "Africa and the Birth of the Modern World."

7. Dussel, "Beyond Eurocentrism," 4.

8. The authors who first began to take up these questions explicitly include Rabinow, *French Modern*; Wright, *The Politics of Design in French Colonial Urbanism*; and Stoler, *Race and the Education of Desire*. More recent contributions include Prakash, *Another Reason*; Stoler and Cooper, *Tensions of Empire*; and Sengoopta, *Imprint of the Raj*. A number of scholars have challenged some of the characterizations of colonialism in these arguments, but their objections rarely undermine the idea that colonialism had profound effects on European institutions and ideas; see, for instance, Zinoman, *The Colonial Bastille*. A similar challenge to the idea that colonial structures were all-pervasive or all-powerful can be found in Vaughan, *Curing Their Ills*.

9. Chakrabarty, *Provincializing Europe*.

10. See Cañizares-Esguerra, "New World, New Stars"; Cañizares-Esguerra also engages with the "laboratories of modernity" literature. Also see Werbner, "Vernacular Cosmopolitanism"; and Tilley, "Global Histories, Vernacular Science, and African Genealogies."

11. Appiah, "Cosmopolitan Patriots," 617.

12. Smuts, "South Africa in Science," 3–4; this speech was also excerpted in Smuts, "Science in South Africa," 245–49. Smuts began this lecture with an analysis of Alfred Wegener's hypothesis of continental drift.

13. His exact words were: "In the period that followed the first visit of the British Association [1905] we South Africanised Science in South Africa. Is it too much to hope that in the next we shall Africanise it?" (9).

14. Barack Obama's response to Mbeki's stance draws attention to an enduring theme of this book: the question of whether African and European epistemologies are competing or complementary. "There should not be a conflict or contradiction between traditional values and modern science. It's not an issue of Western science versus African science. It's just science." Obama quoted in Zeleny and Goering, "Obama Challenges South Africa to Face AIDS Crisis."

15. See, for instance, Leakey, *Kenya: Contrasts and Problems*, esp. chap. 8, "Science and the African."

16. Leakey, "Comparative Methods of Colonial Administration" [a talk delivered at a meeting at Chatham House December 10, 1930, which was marked "Not for publication"], Rhodes House Library. Also see Leakey, *White African*.

17. "The Cradle of Man; Kenya's Claim: An Expedition and Its Finds"; Leakey, "The Cradle of Man; More Evidence from Kenya: Elmenteita Finds"; "The Cradle of Modern Man; Evidence from Africa: Mr. Leakey on Finds in Kenya"; and Leakey, "East Africa Past and Present."

18. Raymond Dart, "The Present Position of Anthropology in South Africa" [Presidential Address to Section E, S.A. Association for Advancement of Science, 1925], quoted in Dubow, "Human Origins, Race Typology," 7.

19. Azikiwe, *Renascent Africa*, chap. 22, "Superstition or Super-Science?" Also quoted in Tilley, "Global Histories, Vernacular Science, and African Genealogies."

20. See Haas, "Introduction: Epistemic Communities and International Policy Coordination," 1–35.

21. Huxley, *The Conservation of Wild Life*, 54.

22. It was well known at the time of Johnston's 1885–86 expedition that German naturalists were also prominent in this work and continued it once Germany took possession of Tanganyika (German East Africa). See Johnston, *The Kilima-Njaro Expedition*; and Gregory, *The Great Rift Valley*.

23. Zimmer, "A Biological Hot Spot in Africa"; Burgess et al., "The Biological Importance of the Eastern Arc Mountains of Tanzania and Kenya" (in a special issue on "Conservation in Areas of High Population Density in Sub-Saharan Africa"); and Nyamweru, "The African Rift System," 24.

24. Summerfield, "Tectonics, Geology, and Long-Term Landscape Development," 1.

25. Caldwell, "The Social Repercussions of Colonial Rule," 462. For a bibliography of examples and a discussion of the methodological challenges, see Lorimer, Brass, and Van de Walle, "Demography," 271–303.

26. I am indebted to the late Archie Mafeje for this question, which he raised in the context of a critique of anthropology; Mafeje, "Anthropology and Independent Africans."

27. Prashad, *The Progress of Science in India*.

28. For an excellent example, see Houtondji, *Endogenous Knowledge*. For a different kind of example (on mathematics), see Verran, *Science and an African Logic*. Also see Thomas-Emeagwali, *African Systems of Science, Technology, and Art*; and Ayittey, *Indigenous African Institutions*. Most authors who contribute to these genres, however, have a limited grasp of the history of scientific debates in and on colonial Africa and sometimes make misleading generalizations about the effects of colonialism and the nature of science and knowledge.

29. See, for example, the oversimplified and ahistorical analysis offered by Washington, "Op-Ed: Why Africa Fears Western Medicine."

30. For a recent example of policy discussions that makes only passing mention of "in-

digenous knowledge," see the special issue edited by Juma on "Science and Innovation in Africa," *International Journal of Technology and Globalization* 2 (2006), especially Juma, "Reinventing Growth: Science, Technology and Innovation in Africa," 323–39.

31. King, "Governing Technology and Growth," 117.

32. Robert Adams, who starred as Kisenga, founded the Negro Repertory Arts Theatre in London in 1944; he also starred in films with Paul Robeson (*King Solomon's Mines*) and in the Colonial Film Unit's *An African in London* (1943). *Men of Two Worlds* starred Orlando (Alhandu) Martins (1899–1985), born in Lagos, Nigeria, and Eseza Makumbi of Uganda.

33. Ladipo Solanke, Secretary-General of WASU to Under-Secretary of State for the Colonies, September 11, 1943 in CO 875/17/6, "Two Cities Film Unit in Africa—'Men of Two Worlds,' 1943," BNA. Ladipo Solanke (1884–1958) was born in Nigeria and came to the United Kingdom in 1922 to study law; he was one of the founding members of WASU in 1925 and an active organizer of African anticolonial activity in London. See Adi, *West Africans in Britain*.

34. Solanke, *WASU* 1 (1926), quoted in Zachernuk, *Colonial Subjects*, 119.

35. Ladipo Solanke to Under-Secretary of State, July 27, 1943; WASU, "A (Proposed) Film Play Entitled: 'The Men of Two Worlds,' WASU Comments," July 27, 1943, CO 875/17/6.

36. Harley, *Native African Medicine*. Some of Harley's papers survive in Duke University's archives.

37. Noel Sabine, Public Relations Officer to Thorold Dickinson, Director, July 30, 1943, CO 875/17/6.

38. I am thinking here of the projects supported by the Colonial Research Council, the Colonial Social Science Research Council, and the Scientific Council for Africa South of the Sahara, which were all funded through grants from Britain's Colonial Development and Welfare Acts. See the seventy-five-page report published in 1954 by the Scientific Council for Africa South of the Sahara, reviewing research in human geography, demography, anthropology, psychology, prehistory, economics, political science, and medicine: Scientific Council for Africa South of the Sahara, *Research in the Social Sciences in Africa South of the Sahara*. These research priorities are themselves open to criticism; see, for instance, Zileza, *Manufacturing African Studies and Crises*.

39. See the work of J. C. Carothers, whose reports on psychiatry and psychology are often presented in African studies courses as examples of racist ideology: Carothers, *The African Mind in Health and Disease*; Carothers, *The Psychology of Mau Mau*. Interestingly, Carothers's viewpoints played a rather minor role in scientific networks in British colonial Africa outside Kenya. The forthcoming work of Sloan Mahone will help to place this research in its wider context; also see McCulloch, *Colonial Psychiatry and "the African Mind."*

40. Two edited collections shed considerable light on these decades: Gifford and Louis, *The Transfer of Power in Africa*; and Gifford and Louis, *Decolonization and African Independence*. Two recent studies are very useful: Cooper, *Africa since 1940*; and Nugent, *Africa since Independence*.

41. Shortly after the speech, Cripps was appointed chancellor of the Exchequer; see Cripps, "Colonies' Contribution to World Trade Stability," 7.

42. For instance, Gluckman, "Human Laboratory across the Zambesi," 38–49; Worthington, "Organization of Science in East Africa," 451–53.

43. Cripps, "Colonies' Contribution to World Trade Stability," 7.
44. The Royal Society meeting was a unique undertaking; for its reports, see *The Royal Society Empire Scientific Conference June–July 1946—Report.*
45. *British Commonwealth Scientific Official Conference,* 45 and 66; for the origins of these proposals, see some of the memoranda in CO 847/23/4, "Scientific Research in Africa, 1942," BNA.
46. For the 1949 conference, see FO 371/73774, "African Regional Scientific Conference, 1949"; for the CSA's origins, see CO 927/124/7 "Recommendations Arising from African Regional Scientific Conference. Establishment of a Scientific Council for Africa South of the Sahara, 1950," BNA.
47. Commonwealth Relations Office to UK High Commissioner in South Africa, E. Baring, October 21, 1949, FO 371/73774 "African Regional Scientific Conference, 1949," emphasis in original, BNA.
48. C. Eastwood to A. Mackay, Treasury, October 5, 1950, CO 927/126/3 "Scientific Council for Africa South of the Sahara. Appointment of a Secretary General. Dr. E. B. Worthington, 1950," BNA.
49. CSA, *Eleventh Meeting of the Scientific Council Cape Town, 1960,* 11. Many of these are described in Scientific Council for Africa South of the Sahara, *Inter-African Technical and Scientific Co-Operation, 1948–1955.*
50. On the Bukoba research, see Berry, *The Culwick Papers, 1934–1944;* on Nyasaland, see Committee on Nutrition in the Colonial Empire, *Nutrition in the Colonial Empire,* esp. 143–44 (the IIALC provided the anthropologist). Also see Berry and Petty, *The Nyasaland Survey Papers, 1938–1943;* and Culwick, "Nutrition Work in British African Colonies since 1939."
51. The ecological survey was approved in 1935 and begun in 1937, but was never published; its author, J. G. Myers, died in an auto accident in 1942. For a discussion of his results and the ensuing agricultural work, see Willimott and Anthony, "Agricultural Research and Development in the Southwest Sudan." See also Culwick, *A Dietary Survey among the Azande;* Reining, *The Zande Scheme.* Some of Reining's papers related to the Zande scheme are held in the National Anthropological Archives of the Smithsonian Institution.
52. Cripps, "Colonies' Contribution to World Trade Stability," 8.
53. See Mills, *Difficult Folk?*
54. Personal Interview, Colin Trapnell, November 12, 1998, Bristol, England. Even in retirement, Trapnell continued to work with geographers and naturalists in Zambia to keep the traditions of the Ecological Survey alive.
55. Worthington, *A Development Plan for Uganda;* the background to his selection can be found in BV/16/184 and BV/16/139, KNA, Nairobi. He was selected by the East African governors for this task in November 1944, shortly after assuming the role of scientific secretary of the Colonial Research Committee, an appointment that was meant to be for seven years (1946–52).
56. Trapnell was asked to synthesize his research for Northern Rhodesia's Ten-Year Development Plan, which appeared as an appendix; see Lewin, *Agricultural and Forestry Development Plans for 10 Years,* appendix I, 17–29; Trapnell interview November 12, 1998. Philips was based in Ghana between 1952 and 1960 and advised Kwame Nkrumah; Thomas Nash became director of the West African Trypanosomiasis Research Institute in Nigeria and helped design a number of "development schemes" related to public health and trypanosomiasis control following the Second World War; John Ford directed the East African Tsetse and Trypanosomiasis Research Or-

ganization in Uganda and served as an advisor to the Southern Rhodesian government on agricultural and trypanosomiasis projects.

57. *Colonial Development and Welfare—Despatch from the Secretary of State for the Colonies to the Colonial Governors; International Monetary Fund [and] International Bank for Reconstruction and Development Inaugural Meetings.*

58. Legal expertise rounded out this triumvirate for the World Bank and the IMF, but it was an implied rather than explicit part of their mandate.

59. See Amadae, *Rationalizing Capitalist Democracy.*

60. World Bank, *Knowledge for Development;* and Stone, *Banking on Knowledge.* Also see King and McGrath, *Knowledge for Development?* In February 2005, the World Bank sponsored its first Africa-wide workshop on "Knowledge for Development" in Cairo; it held follow-up meetings in Uganda and South Africa in 2006 and in Burkina Faso in 2007.

61. Charles Wilcocks, M.D. (1896–1977), "A Tropical Doctor in Africa and Europe: An Autobiography," unpublished manuscript, no date, 232 and 234, GC 55, "Charles Wilcocks," Wellcome Library for the History of Medicine, London.

62. Hobson, "Socialistic Imperialism," 54.

63. Scientific Council for Africa South of the Sahara, *Eighth Meeting of the Scientific Council,* 5–6, emphasis added.

64. E. B. Worthington, interview in East Grinsted, November 16–17, 1996. He reiterated this point during the three successive visits I made to his home between 1996 and 1998.

65. Norman Leys to Thomas Jones, September 10, 1941, quoted in Cell, "Lord Hailey and the Making of the African Survey," 505.

66. Tim O'Meara, respondent in Scheper-Hughes, "The Primacy of the Ethical," 427. This quotation should not be taken as the "last word" on issues of objectivity or relativism, but I find O'Meara's point of view refreshing and his cautions an important antidote to glib denunciations of science. Also see Reyna, "Literary Anthropology and the Case against Science."

APPENDIX

1. My gratitude to Funke Sangodeyi for her assistance counting the totals and crunching the numbers for each year. Her help was invaluable and her camaraderie made this a less arduous and *almost* enjoyable task.

2. Sir Ralph Furse, "The Great Experiment," *Corona* 3 (1951): 447–48.

BIBLIOGRAPHY

Abbreviations

BAAS	British Association for the Advancement of Science
BNA	British National Archives
CCTA	Commission for Technical Cooperation in Africa
CDA	Colonial Development Act
CDWA	Colonial Development and Welfare Act
CRC	Civil Research Committee
CSA	Scientific Council for Africa South of the Sahara
DNB	*Dictionary of National Biography*
EMB	Empire Marketing Board
IHB	International Health Board/Division, Rockefeller Foundation
IIALC	International Institute of African Languages and Cultures
JAI/JRAI	*Journal of the (Royal) Anthropological Institute*
JRAS	*Journal of the Royal African Society*
KNA	Kenya National Archives
LLP	Lord Lugard Papers
LNHO	League of Nations Health Organisation
LP	Lord Lothian Papers
MP	Bronislaw Malinowski Papers
OP	Joseph Oldham Papers
PRGS	*Proceedings of the Royal Geographical Society*
RF	Rockefeller Foundation Archives
RIIA	Royal Institute of International Affairs (Chatham House)
TNA	Tanzania National Archives
WP	Worthington Papers
ZNA	Zambia National Archives

Archives

African National Archives

KENYA NATIONAL ARCHIVES, NAIROBI
TANZANIA NATIONAL ARCHIVES, DAR ES SALAAM
ZAMBIA NATIONAL ARCHIVES, LUSAKA

United Kingdom

BODLEIAN LIBRARY, UNIVERSITY OF OXFORD
Charles Elton Papers
H. A. L. Fisher Manuscripts (FM)

BRITISH NATIONAL ARCHIVES, LONDON
CO 323
CO 533
CO 583
CO 691
CO 758
CO 795
CO 822
CO 847
CO 852
CO 859
CO 875
CO 879
CO 927
FO 2
FO 84
FO 371
FO 881
INF 1
PRO 30

LIBRARY OF COMMONWEALTH AND AFRICAN STUDIES AT RHODES HOUSE, UNIVERSITY OF OXFORD
Mary Kingsley Papers
Lord Lugard Papers (LLP)
Donald Wingfield Malcolm Papers
Joseph Oldham Papers (OP)
Sir Granville Orde-Browne Papers
Frederick Johnson Pedler Papers
Margery Perham Papers
Rhodes Trust Archives (RTA)
E. B. Worthington Papers (WP)

LONDON SCHOOL OF ECONOMICS ARCHIVES
Monica Hunter Papers
International African Institute (IIALC) Papers

Bronislaw Malinowski Papers
Audrey Richards Papers

LONDON SCHOOL OF HYGIENE AND TROPICAL MEDICINE ARCHIVES
Geoffrey Hale Papers

NATIONAL ARCHIVES OF SCOTLAND, EDINBURGH
Lord Lothian Papers (LP)

OXFORD UNIVERSITY PRESS OUT-OF-PRINT ARCHIVES, OXFORD

ROYAL INSTITUTE OF INTERNATIONAL AFFAIRS (CHATHAM HOUSE),
ARCHIVES, LONDON
African Research Survey Files

ROYAL BOTANIC GARDENS—KEW ARCHIVES, LONDON
Amani Institute Papers and Reports, 1920–28
British East African Protectorate 1876–1906
East Africa, Kilimanjaro Expedition
Nyasaland Botanic Station, 1878–1905

ROYAL GEOGRAPHICAL SOCIETY ARCHIVES, LONDON
Verney Lovett Cameron Papers (VLC)
Harry Johnston Papers

ROYAL ANTHROPOLOGICAL INSTITUTE ARCHIVES, LONDON
International Congress of Ethnological and Anthropological Sciences Papers

SCHOOL OF ORIENTAL AND AFRICAN STUDIES ARCHIVES, LONDON
International Missionary Council Papers

TRINITY COLLEGE, CAMBRIDGE UNIVERSITY
Sir James Frazer Papers

WELLCOME LIBRARY FOR THE HISTORY OF MEDICINE
Charles Wilcocks Papers

United States
ROCKEFELLER FOUNDATION ARCHIVES, TARRYTOWN, NEW YORK
CARNEGIE CORPORATION ARCHIVES, COLUMBIA UNIVERSITY, NEW
YORK CITY

Reports, Conference Proceedings, and HMSO Publications

A Note on Some of the Scientific Studies Undertaken by Members of the Colonial Medical Service during the Period 1930–47, with a Bibliography. London: HMSO, 1949.

African Regional Scientific Conference, Johannesburg, October 17 to 28, 1949, 2 vols. Pretoria: Government Printer, 1949.

British Commonwealth Scientific Official Conference—Report of Proceedings. Cmd. 6970. London: HMSO, 1946.

Colonial Development and Welfare: Despatch from the Secretary of State for the Colonies to the Colonial Governors, 12 November 1945. Cmd. 6713. London: HMSO, 1945.

Colonial Survey Committee. *The Surveys and Explorations of British Africa.* London: HMSO, 1906.

The Colonial Territories (1948–1949). Cmd. 7715. London: HMSO, 1949.

Committee on Nutrition in the Colonial Empire. *Nutrition in the Colonial Empire.* London: HMSO, 1939.

Conférence Géographique de Bruxelles. Brussels, 1876.

Conference on Co-Ordination of General Medical Research in East African Territories, Entebbe, 27th to 29th November 1933. Nairobi: Government Printer, 1934.

Conference on Co-Ordination of General Medical Research in East African Territories, Nairobi, 20th to 22nd January. Nairobi: Government Printer, 1936.

Conference on Co-Ordination of Tsetse and Trypanosomiasis (Animal and Human) Research in East Africa, Held at Entebbe, January 1936. Nairobi: Government Printer, 1936.

Conference on Co-Ordination of Tsetse and Trypanosomiasis Research and Control in East Africa. Nairobi: Government Printer, 1943.

Conference on Tsetse and Trypanosomiasis (Animal and Human) Research Entebbe, 22nd to 25th November 1933. Nairobi: Government Printer, 1934.

Congrès International des Sciences Anthropologiques et Ethnologiques: Compte-rendu de la première session, Londres, 1934. London: Royal Anthropological Institute, 1934.

Congrès International des Sciences Anthropologiques et Ethnologiques Première Session, Londres, 1934 Guide. N.P., 1934.

Congrès International des Sciences Géographiques Compte Rendu des Séances. 2 vols. (1 au 11 Aout 1875). Paris: E. Martinet, 1878.

Correspondence Regarding the Position of Indians in East Africa (Kenya and Uganda). Cmd. 1311. London: HMSO, 1921.

East African Agricultural Research Station, Amani First Annual Report, 1928–29. London: HMSO, 1930.

Elliott, W. R. "Report on the Forestry and Agriculture of Northern Nigeria." In F. D. Lugard, *Northern Nigeria: Report for 1904,* 129–34. Cd. 2684. London: HMSO, 1905.

First Interim Report of the Colonial Development Advisory Committee 1st August 1929 to 28th February 1930. London: HMSO, 1930.

Further Correspondence Relating to the Preservation of Wild Animals in Africa. Cd. 5775. London: HMSO, 1911.

Further Correspondence Relating to the Transvaal and Orange River Colony. Cd. 1895. London: HMSO, 1904.

Great Britain, *Yellow Fever Commission (West Africa).* London, 1911–13.

Higher Education in East Africa. London: HMSO, 1937.

Imperial Agricultural Research Conference, 1927: Report and Summary of Proceedings. London: HMSO, 1928.

International Monetary Fund [and] International Bank for Reconstruction and Development Inaugural Meetings Savannah, Georgia March 8th to 18th, 1946. Cmd. 6800. London: HMSO, 1946.

Memorandum Relating to Indians in Kenya. London: HMSO, 1923. Cmd. 1922.

Meteorological Office (London). *Hints to Meteorological Observers in Tropical Africa*. London: HMSO, 1907.

Minutes of Evidence Taken by the Departmental Committee on Sleeping Sickness. Cd. 7350. London: HMSO, 1914.

Native Affairs Department (Kenya). *Annual Report, 1932*. Nairobi: Government Printer, 1932.

Papers Relating to the Health and Progress of Native Populations in Certain Parts of the Empire. London: HMSO, 1931.

Pim, Sir Alan. *Financial and Economic Position of Basutoland*. London: HMSO, 1935.

Proceedings of Agricultural Research Conference Held at Amani Research Station, February 2nd to 6th, 1931. Nairobi: Government Printer, 1931.

Proceedings of a Conference of East African Soil Chemists Held at the Agricultural Research Station, Amani, Tanganyika Territory, May 21st–26th, 1932. Nairobi: Government Printer, 1932.

Proceedings of the First International Conference on Sleeping Sickness. Cd. 3778. London: HMSO, 1907.

Proceedings of the Second Conference of East African Agricultural and Soil Chemists Held at Zanzibar, August 3rd to 9th, 1934. Nairobi: Government Printer, 1935.

Protocols and General Act of West African Conference at Berlin, February 1885. London: HMSO, 1885.

Report by the Hon. W. G. A. Ormsby-Gore on His Visit to West Africa in 1926. Cmd. 2744. London: HMSO, 1926.

Report from the Select Committee on Colonization and Settlement (India). London: House of Commons, 1859.

"Report of Committee on Colonial Scientific and Research Services." In *Colonial Office Conference, 1927, Summary of Proceedings*, 23–40. London: HMSO, 1927.

Report of a Committee on Agricultural Research and Administration in the Non-Self-Governing Dependencies. Cmd. 2825. London: HMSO, 1927.

Report of a Committee on the System of Appointment in the Colonial Office and the Colonial Services. Cmd. 3554. London: HMSO, 1930.

Report of Lovat Committee on Agricultural Research. Cmd. 3049. London: HMSO, 1928.

Report of Lovat Committee on Agricultural Research and Administration. Cmd. 2825. London: HMSO, 1927.

Report of Lovat Committee on Veterinary Services. Cmd. 3261. London: HMSO, 1929.

Report of the Commission Appointed to Enquire into the Financial and Economic Position of Northern Rhodesia. London: HMSO, 1938.

Report of the Commission on Closer Union of the Dependencies in Eastern and Central Africa. Cmd. 3234. London: HMSO, 1929.

Report of the Committee on the Staffing of the Veterinary Departments in the Colonies and Protectorates. London: HMSO, 1920.

Report of the East Africa Commission. Cmd. 2387. London: HMSO, 1925.

Report of the East Africa Sub-Committee of the Tsetse Fly Committee of the Economic Advisory Council. London: HMSO, 1935.

Report of the East African Guaranteed Loan Committee, 1926–1929. Cmd. 3494. London: HMSO, 1929.

Report of the Inter-Departmental Committee on Sleeping Sickness. Cd. 7349. London: HMSO, 1914.

"Report of the International Conference of Representatives of the Health Services of Certain African Territories and British India Held at Cape Town, November 15th to 25th, 1932." *Quarterly Bulletin of the Health Organisation of the League of Nations* 2 (1933).

Report of the Joint Select Committee on Closer Union in East Africa. London: HMSO, 1931.

"Report of the Pan-African Health Conference, Johannesburg, November 20–30, 1935." *Quarterly Bulletin of the Health Organisation of the League of Nations* 5 (1936).

Report of the President and the Secretary as to an Educational Program in Africa. New York: Carnegie Corporation, 1927.

Report of the Proceedings of the Conference of Colonial Directors of Agriculture. London: HMSO, 1938.

Report of the Second International Conference on Sleeping Sickness—Paris, November 5–7, 1928. Geneva: League of Nations, 1928.

Report of the Sixth International Geographical Congress, London 1895. London: John Murray, 1896.

Report of the Uganda Development Commission, 1920. Entebbe: Uganda, 1920.

Report on the Departmental Committee to Enquire into the Colonial Medical Services. London: HMSO, 1920.

Report on the Second Imperial Entomological Conference, June 1925. London: HMSO, 1925. Cmd. 2490.

Reports of the Colonial Development Fund. London: HMSO, 1929–39. Cmd. 3540, 3876, 4079, 4316, 4634, 4916, 5202, 5537, 5789, 6062.

Reports of the President and Treasurer of Harvard College, 1925–26. Cambridge, MA: Harvard University Press, 1927.

Reports of the Sleeping Sickness Commission of the Royal Society, n. 16. London: HMSO, 1915.

The Royal Society Empire Scientific Conference June–July 1946—Report, 2 vols. London: Royal Society, 1948.

Scientific Council for Africa South of the Sahara. *Eighth Meeting of the Scientific Council.* London: CCTA, no date [1957].

———. *Eleventh Meeting of the Scientific Council Cape Town, 1960.* Bukavu: CCTA, 1960.

———. *Inter-African Technical and Scientific Co-Operation, 1948–1955.* Bukavu: CCTA and CSA, [1956].

———. *Research in the Social Sciences in Africa South of the Sahara.* Bukavu: CSA, 1954.

Scientific Council for Africa South of the Sahara and CCTA. *Social Sciences: Inter-African Conference.* Bukavu: [CCTA], 1955.

Simpson, W. J. *Report by Professor W. J. Simpson on Sanitary Matters in Various West African Colonies and the Outbreak of Plague in the Gold Coast.* Cd. 4718. London: HMSO, 1909.

Spiritualism and Psychical Research (Exemption) Bill. London: HMSO, 1930.

Staples, R. R., and W. K. Hudson. *An Ecological Survey of the Mountain Area of Basutoland.* London: HMSO, no date [1938/39].

Summary of First Colonial Office Conference. Cmd. 2883. London: HMSO, 1927.

Tanganyika Annual Medical Report of 1928. Dar es Salaam: Government Printer, 1929.

Transactions of the Third International Congress of Soil Science, 2 vols. London: Thomas Murby, 1935.

Tsetse Research Department Annual Report, 1928–1929. Dar es Salaam: Government Printer, 1930.

Primary and Secondary Sources

"A nos lectures" and "L'exploration moderne de l'Afrique." *L'Afrique Explorée et Civilisée* 1 (1879–1880): 3–4 and 5–16.

Adams, Cyrus. "Foundations of Economic Progress in Tropical Africa." *Bulletin of the American Geographical Society* 43 (1911): 753–66.

Adams, W. M., A. S. Goudie, and A. R. Orme, eds. *The Physical Geography of Africa.* Oxford: Oxford University Press, 1996.

Adams, William. *Against Extinction: The Story of Conservation.* London: Earthscan, 2004.

Adas, Michael. *Machines as the Measure of Men: Science, Technology, and Ideologies of Western Dominance.* Ithaca, NY: Cornell University Press, 1989.

Adi, Hakim. *West Africans in Britain, 1890–1960: Nationalism, Pan-Africanism, and Communism.* London: Lawrence and Wishart, 1998.

"The African Explains Witchcraft." *Africa* 8 (1935): 504–59.

"The African Society Inaugural Meeting." *Journal of the African Society* 1 (1901): i–xx.

Akpan, M. B. "Liberia and Ethiopia, 1880–1914: The Survival of Two African States." In *Africa under Colonial Domination, 1880–1935,* edited by Adu Boahen. Berkeley: Heinemann, 1985.

Akurang-Parry, Kwabena O. "'We Cast about for a Remedy': Chinese Labour and African Opposition in the Gold Coast, 1874–1914." *International Journal of African Historical Studies* 34 (2001): 365–84.

Alcock, Rutherford. "President's Address, November 13, 1876." *PRGS* 21 (1877): 1–26.

Allan, William. *The African Husbandman.* London: International African Institute, 2003 [1965].

Alter, Peter. *The Reluctant Patron: Science and the State in Britain, 1850–1920.* Oxford: Berg, 1987.

Amadae, Sonja. *Rationalizing Capitalist Democracy: The Cold War Origins of Rational Choice Liberalism.* Chicago: University of Chicago Press, 2003.

Anderson, David. "Depression, Dust Bowl, Demography, and Drought: The Colonial State and Soil Conservation in East Africa in the 1930s." *African Affairs* 83, no. 332 (1984): 321–43.

Anderson, David, and Richard Grove, eds. *Conservation in Africa: People, Policies, and Practice.* Cambridge: Cambridge University Press, 1987.

Anderson, David, and David Throup. "The Agrarian Economy of Central Province, Kenya, 1918 to 1939." In *The Economies of Africa and Asia in the Inter-War Depression,* edited by Ian Brown, 8–28. London: Routledge, 1989.

Anderson, Warwick. *Colonial Pathologies: American Tropical Medicine, Race, and Hygiene in the Philippines.* Durham, NC: Duke University Press, 2006.

———. "Natural Histories of Infectious Disease: Ecological Vision in Twentieth-Century Biomedical Science." *OSIRIS* 19 (2004): 39–61.

Anghie, Antony. "Colonialism and the Birth of International Institutions: Sovereignty, Economy, and the Mandate System of the League of Nations." *New York University Journal of International Law and Politics* 34 (2001–02): 513–633.

Anker, Peder. *Imperial Ecology: Environmental Order in the British Empire, 1895–1945.* Cambridge, MA: Harvard University Press, 2001.

———. "The Politics of Ecology in South Africa on the Radical Left." *Journal of the History of Biology* 37, no. 2 (2004): 303–31.

"The Annual General Meeting." *Journal of the Royal African Society* 16 (1917): 155–64.

"Anthropologists in Congress." *African Affairs* 33 (1934): 398–403.

"Anthropology and the Empire: Deputation to Mr. Asquith." *Man* 9 (1909): 85–87.

Appiah, Kwame Anthony. "Cosmopolitan Patriots." *Critical Inquiry* 23 (1997): 617–39.

Armytage, W. H. G. *Sir Richard Gregory: His Life and Work.* London: Macmillan, 1957.

Arndt, H. W. *Economic Development: The History of an Idea.* Chicago: University of Chicago Press, 1987.

Arnold, David. *Colonizing the Body: State Medicine and Epidemic Disease in Nineteenth-Century India.* Berkeley: University of California Press, 1993.

———. "Medicine and Colonialism." In *Companion Encyclopedia of the History of Medicine.* Vol. 1. Edited by W. F. Bynum and Roy Porter, 1393–416. London: Routledge, 1993.

———. *Science, Technology, and Medicine in Colonial India.* Cambridge: Cambridge University Press, 2000.

"Arrival of Lieutenant Cameron." *Geographical Magazine* 3 (April 1876): 104.

Asad, Talal, ed. *Anthropology and the Colonial Encounter.* London: Ithaca Press, 1973.

Austen, Ralph. *Northwest Tanzania under German and British Rule: Colonial Policy and Tribal Politics, 1889–1939.* New Haven, CT: Yale University Press, 1968.

Axelson, Eric. *Portugal and the Scramble for Africa.* Johannesburg: Witwatersrand University Press, 1967.

Ayittey, George. *Indigenous African Institutions.* Ardsley, NY: Transnational Publishers, 2006.

Azikiwe, Nnamdi. *Renascent Africa.* Accra: Published by the author, [1937].

Azzi, Girolamo. *Agricultural Ecology.* London: Constable, 1956.

Baader, Gerhard, Susan Lederer, Morris Low, Florian Schmaltz, and Alexander Schwerin. "Pathways to Human Experimentation, 1933–1945: Germany, Japan, and the United States." *OSIRIS* 20 (2005): 205–31.

Bacillus [pseud.]. "Colour Prejudice." *East and West* (1912): 657–66.

Baker, C. A. *Johnston's Administration: A History of the British Central Africa Administration 1891–1897.* Zomba, Malawi: Government Press, 1970.

Baker, G. "An Experiment in Applied Anthropology." *Africa* 8 (1935): 304–14.

Baldwin, Robert. *Economic Development and Export Growth: A Study of Northern Rhodesia, 1920–1960.* London: University of Cambridge Press, 1966.

Balfour, Andrew, and Henry Harold Scott. *Health Problems of the Empire: Past, Present, and Future.* London: Collins, 1924.

Balfour, Andrew, E. Van Campenhout, Gustave Martin, and A. G. Bagshawe. *Interim Report on Tuberculosis and Sleeping-Sickness in Equatorial Africa.* Geneva, 1923.

Ballantyne, Tony, ed. *Science, Empire, and the European Exploration of the Pacific.* Aldershot, UK: Ashgate, 2004.

Ballard, Charles. "The Repercussions of Rinderpest: Cattle Plague and Peasant Decline in Colonial Natal." *International Journal of African Historical Studies* 19 (1986): 421–50.

Banning, Emile. *Africa and the Brussels Geographical Conference.* Brussels, 1877.

Barkan, Elazar. *The Retreat of Scientific Racism: Changing Concepts of Race in Britain and the United States between the Two World Wars.* Cambridge: Cambridge University Press, 1992.

Barnes, John, and David Nicholson, eds. *The Empire at Bay: The Leo Amery Diaries, 1929–1945.* London: Hutchinson, 1988.

Barrera, Giulia. "Dangerous Liaisons: Colonial Concubinage in Eritrea, 1890–1941." Evanston, IL: Program of African Studies, 1996.

Barrett, R. E. "Notes on the Epidemiology of Sleeping Sickness with Special Reference to Conditions in the West Nile District of Uganda." *East African Medical Journal* 11 (1934–35): 20–28.

Basalla, G. "The Spread of Western Science." *Science* 156 (1967): 611–22.

Bates, Robert, V. Y. Mudimbe, and Jean O'Barr, eds. *Africa and the Disciplines: The Contributions of Research in Africa to the Social Sciences and Humanities.* Chicago: University of Chicago Press, 1993.

Baumslag, Naomi. *Murderous Medicine: Nazi Doctors, Human Experimentation, and Typhus.* Westport, CT: Praeger, 2005.

Bayly, C. A. *The Birth of the Modern World, 1780–1914.* Oxford: Blackwell, 2004.

———. *Empire and Information: Intelligence Gathering and Social Communication in India, 1780–1870.* Cambridge: Cambridge University Press, 1996.

Beinart, William. "Soil Erosion, Animals, and Pastures over the Longer Term: Environmental Destruction in Southern Africa." In *The Lie of the Land,* edited by M. Leach and R. Mearns, 54–72. Oxford: James Currey, 1996.

———. "Soil Erosion, Conservation, and Ideas about Development: A Southern African Exploration, 1900–60." *Journal of Southern African Studies* 11, no. 1 (1984): 52–83.

Bell, Heather. *Frontiers of Medicine in the Anglo-Egyptian Sudan, 1899–1940.* Oxford: Clarendon Press, 1999.

Bell, Morag. "American Philanthropy, the Carnegie Corporation and Poverty in South Africa." *Journal of Southern African Studies* 26 (2000): 481–504.

Bendikat, Elfi. "The Berlin Conference in the German, French, and British Press." In *Bismarck, Europe, and Africa,* edited by Stig Förster et al., 377–97. London: Oxford University Press, 1988.

Berman, Bruce, and John Lonsdale. "Custom, Modernity, and the Search for Kihooto: Kenyatta, Malinowski, and the Making of Facing Mount Kenya." In *Ordering Africa,* edited by Helen Tilley with Robert Gordon. Manchester: Manchester University Press, 2007.

Berry, Sara. "Hegemony on a Shoestring: Indirect Rule and Access to Agricultural Land." *Africa* 62 (1992): 327–55.

Berry, Veronica, ed. *The Culwick Papers, 1934–1944: Population, Food, and Health in Colonial Tanganyika.* London: Academy Books, 1994.

Berry, Veronica, and Celia Petty, eds. *The Nyasaland Survey Papers, 1938–1943: Agriculture, Food, and Health.* London: Academy Books, 1992.

Blyden, Edward Wilmott. "Africa and the Africans." *Fraser's Magazine* 18 (1878): 178–96.

———. *The African Society and Miss Mary H. Kingsley Articles Reprinted from "The Sierra Leone Weekly News" (March, April, May, and June, 1901).* London: John Scott, 1901.

———. "West Africa before Europe." *Journal of the Royal African Society* 2 (1903): 359–74.

———. *West Africa before Europe and Other Addresses Delivered in England in 1901 and 1903.* London: C. M. Phillips, 1905.

Boahen, A. Adu, ed. *Africa under Colonial Domination, 1880–1935: UNESCO General History of Africa.* Berkeley: University of California Press, 1985.

Bock, Gisela. "Sterilization and 'Medical' Massacres in National Socialist Germany: Ethics, Politics, and the Law." In *Medicine and Modernity: Public Health and Medical Care in Nineteenth- and Twentieth-Century Germany,* edited by Manfred Berg and Geoffrey Cocks, 149–72. Cambridge: Cambridge University Press, 1997.

Bocking, Stephen. *Ecologists and Environmental Politics: A History of Contemporary Ecology.* New Haven, CT: Yale University Press, 1997.

Bonneuil, Christophe. "Crafting and Disciplining the Tropics: Plant Science in the French Colonies." In *Science in the Twentieth Century,* edited by John Krige and Dominique Pestre, 77–96. Amsterdam: Harwood Academic Publishers, 1997.

———. "Mettre en ordre et discipliner les tropiques: Les sciences du végétal dans l'empire français, 1870–1940." 2 vols. PhD diss., University of Paris, 1997.

Bonneuil, Christophe, and Mina Kleiche. *Du jardin d'essais colonial à la station expérimentale, 1880–1930*. Paris: CIRAD, 1993.

"Botanic Station-Sierra Leone (with plan)." *Kew Bulletin* 130 (1897): 303–17.

"Botanical Enterprise in British Central Africa." *Kew Bulletin* 104 (1895): 186–91.

"Botanical Enterprise in West Africa." *Kew Bulletin* 130 (1897): 329–33.

"Botanical Station at Lagos." *Kew Bulletin* 18 (1888): 149–56.

Bourn, David, Robin Reid, David Rogers, Bill Snow, and William Wint. *Environmental Change and the Autonomous Control of Tsetse and Trypanosomosis in Sub-Saharan Africa*. Oxford: Environmental Research Group, 2001.

Bourne, Ray. *Aerial Survey in Relation to the Economic Development of New Countries, with Special Reference to an Investigation Carried Out in Northern Rhodesia*. Oxford: Clarendon Press, 1928.

———. "Some Ecological Conceptions." *Empire Forestry Journal* 13 (1934): 15–30.

Bowler, Peter. *Biology and Social Thought, 1850–1914*. Berkeley: Office for History of Science and Technology, 1993.

Brantley, Cynthia. "Kikuyu-Maasai Nutrition and Colonial Science: The Orr and Gilks Study in Late 1920s Kenya Revisited." *International Journal of African Historical Studies* 30 (1997): 49–86.

———. *Feeding Families: African Realities and British Ideas of Nutrition and Development in Early Colonial Africa*. Portsmouth, NH: Heinemann, 2002.

Bravo, Michael. "Ethnological Encounters." In *Cultures of Natural History*, edited by N. Jardine, J. A. Secord, and E. C. Spary, 338–57. Cambridge: Cambridge University Press, 1996.

Bridges, Roy C. "The First Conference of Experts on Africa." In *Experts in Africa*, edited by J. C. Stone, 12–28. Aberdeen: Aberdeen University African Studies Group, 1980.

———. "The R.G.S. and the African Exploration Fund, 1876–1880." *Geographical Journal* 129 (1963): 25–35.

British Association. *Notes and Queries on Anthropology for the Use of Travellers and Residents in Uncivilized Lands*. London: Edward Stanford, 1874.

"British Association for the Advancement of Science—Report—Geography Section." *Nature* 16 (September 6, 1877): 409.

British Medical Association. "Brain and Mind in East Africa." *British Medical Journal* (April 30, 1932): 812–13.

Broc, N. "Les Français face à l'inconnue saharienne: Géographes, explorateurs, ingénieurs (1830–1881)." *Annales de Géographie* 535 (1987): 302–38.

Brown, Ian, ed. *The Economies of Africa and Asia in the Inter-War Depression*. London: Routledge, 1989.

Brown, Karen. "Tropical Medicine and Animal Diseases: Onderstepoort and the Development of Veterinary Science in South Africa, 1908–1950." *Journal of Southern African Studies* 31 (2005): 513–29.

Brown, Richard. "Anthropology and Colonial Rule: The Case of Godfrey Wilson and the Rhodes-Livingstone Institute, Northern Rhodesia." In *Anthropology and the Colonial Encounter*, edited by Talal Asad, 173–97. London: Ithaca Press, 1973.

Brubaker, Rogers, and Frederick Cooper. "Beyond 'Identity.'" *Theory and Society* 29 (2001): 1–47.

Bruce, David. "Sleeping Sickness in Uganda: Duration of the Infectivity of the *Glossina palpalis* after the Removal of the Lake-Shore Population." In *Reports of the Sleeping Sickness Commission of the Royal Society* 10 (1910): 56–62.

Bulmer, Martin. "Philanthropic Foundations and the Development of the Social Sciences in the Early Twentieth Century: A Reply to Donald Fisher." *Sociology* 18 (1984): 572–79.

Bulmer, Martin, and Joan Bulmer. "Philanthropy and Social Science in the 1920s: Beardsley Ruml and the Laura Spelman Rockefeller Memorial, 1922–29." *Minerva* 19 (1981): 347–407.

"Bureau of Anthropology." *JRAI* 38 (1908): 489–92.

Burgess, N. D., et al. "The Biological Importance of the Eastern Arc Mountains of Tanzania and Kenya." *Biological Conservation* 134 (2007): 209–31.

Burkitt, Dr. "The Medical Aspect of Closer Settlement of Europeans in the Kenya Highlands." *East African Medical Journal* 6 (1929–30): 188–90.

Burnet, F. M. *Biological Aspects of Infectious Disease*. Cambridge: The University Press, 1940.

———. *Changing Patterns: An Atypical Autobiography*. New York: American Elsevier, 1968.

———. "Inapparent Virus Infections: With Special Reference to Australian Examples." *British Medical Journal* 1 (1936): 99–103.

———. *Virus as Organism: Evolutionary and Ecological Aspects of Some Human Virus Diseases*. Cambridge, MA: Harvard University Press, 1945.

Burnett, D. Graham. *Masters of All They Surveyed: Exploration, Geography, and a British El Dorado*. Chicago: University of Chicago Press, 2000.

Burt, Cyril. "The Native Brain—Intelligence Tests." *Times (London)*, December 15, 1933.

Bush, Barbara. *Imperialism, Race, and Resistance: Africa and Britain, 1919–1945*. London: Routledge, 1999.

Butler, J. R. M. *Lord Lothian (Philip Kerr), 1882–1940*. London: Macmillan, 1960.

Buxton, P. A. "Tsetse Flies of East Africa." *African Affairs* 37 (1938): 92–94.

Cain, P. J. "Economics and Empire: The Metropolitan Context." In *The Oxford History of the British Empire, The Nineteenth Century*, edited by Andrew Porter. London: Oxford University Press, 1999.

Cain, P. J., and A. G. Hopkins. *British Imperialism: Innovation and Expansion, 1688–1914*. Loondon: Longman, 1993.

Caldwell, J. C. "The Social Repercussions of Colonial Rule: Demographic Aspects." In *Africa under Colonial Domination 1880–1935*, edited by A. A. Boahen, 458–86. Berkeley: University of California Press, 1985.

Cameron, V. Lovett. *Across Africa*. New York: Harper and Brothers, 1877.

———. "Letter to the Editor." *Times (London)*, November 25, 1884, 4, col. A.

———. "Lieut. Cameron's Letters Detailing the Journey of the Livingstone East Coast Expedition from Lake Tanganyika to the West Coast of Africa." *PRGS* 20 (1875–76): 117–34.

———. "'On the Anthropology of Africa,' Delivered on May 23, 1876." *JAI* 6 (1877): 167–81.

———. "On Proposed Stations in Central Africa as Bases for Future Exploration." *Report of the Forty-Seventh Meeting of the British Association for the Advancement of Science* [August 1877], 141–42. London: John Murray, 1878.

Campbell, Chloe. *Race and Empire: Eugenics in Colonial Kenya*. Manchester: Manchester University Press, 2007.

Cañizares-Esguerra, Jorge. *Nature, Empire, and Nation: Explorations in the History of Science in the Iberian World*. Stanford, CA: Stanford University Press, 2006.

——. "New World, New Stars: Patriotic Astrology and the Invention of Indian and Creole Bodies in Colonial Spanish America, 1600–1650." *American Historical Review* 104 (1999): 33–68.

Carney, Michael. *Stoker: The Life of Hilda Matheson OBE, 1888–1940.* Llangynog, UK: Michael Carney, 1999.

Carothers, J. C. *The African Mind in Health and Disease.* Geneva: World Health Organization, 1953.

——. *The Psychology of Mau Mau.* Nairobi: Government Printer, 1954.

Carpenter, G. D. H. *A Naturalist on Lake Victoria with an Account of Sleeping Sickness and the Tse-Tse Fly.* New York: E. P. Dutton, 1920.

——. "Progress Report on Investigations into the Bionomics of *Glossina palpalis.*" In *Reports of the Sleeping Sickness Commission of the Royal Society,* 79–108 London: HMSO, 1912.

——. "Second Report on the Bionomics of *Glossina fuscipes (palpalis)* of Uganda." In *Reports of the Sleeping Sickness Commission of the Royal Society,* no. 14, 1–37. London: HMSO, 1913.

——. "Sleeping Sickness: A Lecture Delivered to the Uganda Branch of the British Medical Association, on 7th June, 1929." *Kenya and East African Medical Journal* 6 (1929–30): 131–48.

——. "Third, Fourth, and Fifth Reports on the Bionomics of *Glossina palpalis* on Lake Victoria." In *Reports of the Sleeping Sickness Commission of the Royal Society,* 3–101. London: HMSO, 1919.

Carrière, B. "Le Transsaharien: Histoire et géographie d'une enterprise inachevée." *Acta Geographica* 74 (1988): 23–38.

Carroll, Patrick. "Engineering Ireland: The Material Constitution of the Technoscientific State." PhD diss., University of California, San Diego, 1999.

Carruthers, Jane. *The Kruger National Park: A Social and Political History.* Natal: University of Natal Press, 1995.

Carswell, Grace. "Soil Conservation Policies in Colonial Kigezi, Uganda: Successful Implementation and an Absence of Resistance." In *Social History and African Environments,* edited by William Beinart and JoAnn McGregor, 131–54. Oxford: Currey Press, 2003.

Cassidy, John. "The Next Crusade: Paul Wolfowitz at the World Bank." *New Yorker* (April 9, 2007): 36–51.

"Celebrations in Liverpool: Lord Olivier on British Policy in Africa." *Times (London),* July 18, 1932, 14, col. B.

Cell, John. "Anglo-Indian Medical Theory and the Origins of Segregation in West Africa." *American Historical Review* 91 (1986): 307–35.

——. *By Kenya Possessed.* Chicago: University of Chicago Press, 1976.

——. *Hailey: A Study in British Imperialism, 1872–1969.* Cambridge: Cambridge University Press, 1992.

——. "Lord Hailey and the Making of the African Survey." *African Affairs* 88 (1989): 481–505.

"Central Africa." *Times (London),* May 8, 1875, 6.

"C. F. M. Swynnerton." *Times (London),* June 13, 1938, 16, col. C.

Chakrabarty, Dipesh. *Provincializing Europe: Postcolonial Thought and Historical Difference.* Princeton, NJ: Princeton University Press, 2000.

Chamberlin, J. E., and S. Gilman, eds. *Degeneration: The Dark Side of Progress.* New York: Columbia University Press, 1985.

Chambers, David Wade. "Period and Process in Colonial and National Science." In *Scientific Colonialism: A Cross-Cultural Comparison*, edited by Nathan Reingold and Marc Rothenberg, 297–321. Washington, DC: Smithsonian Institution Press, 1987.

Chambers, David Wade, and R. Gillispie. "Locality in the History of Science: Colonial Science, Technoscience, and Indigenous Knowledge." *OSIRIS* 15 (2000): 221–40.

Chesterman, Clement C. "Witchcraft in Africa." *Times (London)*, May 10, 1932, 12, col. D.

Christy, Cuthbert. "Sleeping Sickness." *Journal of the Royal African Society* 3 (1903): 1–11.

Church, A. G. *East Africa, a New Dominion: A Crucial Experiment in Tropical Development and Its Significance to the British Empire*. London: H. F. & G. Witherby, 1927.

"Church Congress: Question of Race Equality." *Times (London)*, October 8, 1925, 7, col. A.

Churchill, Winston. "The Development of Africa." *JRAS* 6 (1907): 291–96.

———. *My African Journey*. London: Norton, 1989 [1908].

"Cinquantième anniversaire de la Fondation de la Société Royale Belge de Géographie." *Bulletin de la Société Royale Belge de Géographie* 50 (1926): 173–87.

Cittadino, Eugene. *Nature as the Laboratory: Darwinian Plant Ecology in the German Empire, 1880–1900*. Cambridge: Cambridge University Press, 1990.

Clarence-Smith, Gervase. "The Effects of the Great Depression on Industrialisation in Equatorial and Central Africa." In *The Economies of Africa and Asia in the Inter-War Depression*, edited by Ian Brown, 170–202. London: Routledge, 1989.

Clark, Professor W. E. Le Gros. "Race." In *Oxford University Summer School on Colonial Administration 27 June–8 July 1938*, 65–67. Oxford: Oxford University Press, 1938.

Clements, Keith. *Faith on the Frontier: A Life of J. H. Oldham*. Edinburgh: T & T Clark, 1999.

Cohen, Jon. "Vaccine Theory of AIDS Origins Disputed at Royal Society." *Science* 289 (2000): 1850–51.

Cohen, William. "Malaria and French Imperialism." *Journal of African History* 24 (1983): 23–36.

Cohn, Bernard. *Colonialism and Its Forms of Knowledge: The British in India*. Princeton, NJ: Princeton University Press, 1996.

Collins, Peter. "The British Association as Public Apologist for Science, 1919–1946." In *Parliament of Science*, edited by Roy MacLeod and Peter Collins, 211–36. London: Science Reviews, 1981.

Colonial Medical Research Committee. *Medical Research in the Colonies, Protectorates, and Mandated Territories*. London: Colonial Office, 1929.

———. *Medical Research in the Colonies, Protectorates, and Mandated Territories*. London: Colonial Office, 1930.

———. *Memorandum on Medical Research in the Colonies*. London: Colonial Office, 1928.

Colson, Elizabeth. *The Plateau Tonga of Northern Rhodesia: Social and Religious Studies*. Manchester: Manchester University Press, 1962.

Comaroff, John. "Reflections on the Colonial State, in South Africa and Elsewhere: Factions, Fragments, Facts, and Fictions." *Bulletin of the Institute of Ethnology Academia Sinica* 83 (1997): 1–50.

Comaroff, John, and Jean Comaroff. "Medicine, Colonialism, and the Black Body." In *Ethnography and the Historical Imagination*, 215–34. Boulder, CO: Westview Press, 1992.

"Conference on Co-Ordination of Medical Research on East African Territories." *East African Medical Journal* 11 (1934): 73–75.

"The Congo." *Times (London)*, August 28, 1884, 2, col. A.

Constantine, Stephen. *The Making of British Colonial Development Policy, 1914–1940*. London: Frank Cass, 1984.

Conte, Chris. "Colonial Science and Ecological Change: Tanzania's Mlalo Basin, 1888–1946." *Environmental History* 4 (1999): 220–44.

Cooper, Frederick. *Africa since 1940: The Past of the Present*. Cambridge: Cambridge University Press, 2002.

———. *Decolonization and African Society: The Labor Question in French and British Africa*. Cambridge: Cambridge University Press, 1996.

———. "Development, Modernization, and the Social Sciences in the Era of Decolonization: The Examples of British and French Africa." *Revue d'Histoire des Sciences Humaines* 10 (2004): 9–38.

———. "Modernizing Bureaucrats, Backward Africans, and the Development Concept." In *International Development and the Social Sciences: Essays on the History and Politics of Knowledge*, edited by Frederick Cooper and Randall Packard. Berkeley: University of California Press, 1997.

Cooper, Frederick, and Ann Stoler, eds. *Tensions of Empire: Colonial Cultures in A Bourgeois World*. Berkeley: University of California Press, 1997.

Cooper, Frederick, and Randall Packard, eds. *International Development and the Social Sciences: Essays on the History and Politics of Knowledge*. Berkeley: University of California Press, 1997.

Coupland, Reginald. *The Exploitation of East Africa, 1856–1890*. London: Faber and Faber, 1939.

Cowen, John. "Race Prejudice." *Westminster Review* 173 (1910): 631–38.

Cowen, M. P., and R. W. Shenton. *Doctrines of Development*. London: Routledge, 1996.

"The Cradle of Man; Kenya's Claim: An Expedition and Its Finds." *Times (London)*, August 1, 1928, 15, col. G.

"The Cradle of Modern Man; Evidence from Africa: Mr. Leakey on Finds in Kenya." *Times (London)*, September 1930, 7, col. C.

Crane, Diana. "Transnational Networks in Basic Science." *International Organization* 25 (1971): 585–601.

Crawford, Elisabeth, Terry Shinn, and Sverker Sörlin. "The Nationalization and Denationalization of the Sciences: An Introductory Essay." In *Denationalizing Science*, edited by Elisabeth Crawford et al., 1–42. Dordrecht: Kluwer Academic Publishers, 1993.

Cripps, Stafford. "Colonies' Contribution to World Trade Stability." *Crown Colonist* (January 1948): 7–8.

Cueto, Marcos. "The Cycles of Eradication: The Rockefeller Foundation and Latin American Public Health, 1918–1940." In *International Health Organizations and Movements, 1918–1939*, edited by P. Weindling, 222–43. Cambridge: Cambridge University Press, 1995.

Culwick, G. M. *A Dietary Survey among the Azande of the South-Western Sudan*. Khartoum: Agricultural Publications Committee, 1950.

———. "Nutrition Work in British African Colonies since 1939." *Africa* 14 (1943): 24–26.

Cunningham, A., and Perry Williams, eds. *The Laboratory Revolution in Medicine*. Cambridge, Cambridge University Press, 1992.

Curtin, Philip. "The End of the 'White Man's Grave'?: Nineteenth-Century Mortality in West Africa." *Journal of Interdisciplinary History* 21 (1990): 63–88.

———. *The Image of Africa: British Ideas and Actions, 1780–1850.* Madison: University of Wisconsin Press, 1964.

———. "Medical Knowledge and Urban Planning in Tropical Africa." *American Historical Review* 90 (1985): 594–613.

Curtin, Philip, Steven Feierman, Leonard Thompson, and Jan Vansina. *African History from Earliest Times to Independence.* 2nd ed. London: Longman, 1995.

Cust, R. N. "The Ethics of African Geographical Explory [sic]." *Asiatic Quarterly Review* 4 n.s. (1892): 348–64.

———. "On the Attitude of the White Man to His Coloured Fellow-Creature All over the World" [1902]. In *Linguistic and Oriental Essays Written from the Year 1840 to 1903*, by Robert Cust, 20–24. London: Luzac, 1904.

———. "The Scramble for Africa." *Africa* (October 1884), reprinted with addendum (1896) in R. N. Cust, *Linguistic and Oriental Essays*, vol. 5, 449–53. London: Luzac, 1898.

Da Costa, B. F. Bruto. *Sleeping Sickness: A Record of Four-Years War against It in Principe, Portuguese West Africa.* Translated by J. A. Wyllie. London: Balliére, Tindall, and Cox, 1916.

———. *Sleeping Sickness in the Island of Principe: Sanitation, Statistics, Hospital Services, and Work of Official Conservancy Brigade.* Translated by J. A. Wyllie. London: Balliére, Tindall, and Cox, 1913.

Darwin, Charles. *The Descent of Man and Selection in Relation to Sex, Revised Edition* (2nd). New York: Rand, McNally, 1874.

Davis, Merle, ed. *Modern Industry and the African: An Enquiry into the Effects of the Copper Mines of Central Africa upon Native Society and the Work of Christian Missions.* London: Macmillan, 1933.

Davis, Michael. "Social Medicine as a Field for Social Research." *American Journal of Sociology* 44 (1938): 274–79.

De Boer, H. S. "A Survey of European Children of School-Going Age Attending Schools in Nairobi, Kenya Colony." *Kenya Medical Journal* 1 (1924–25): 265–69.

De Courcel, Geoffrey. "The Berlin Act of 26 February 1885." In *Bismarck, Europe, and Africa: The Berlin Africa Conference, 1884–1885 and the Onset of Partition*, edited by Stig Föster, Wolfgang Mommsen, and Ronald Robinson, 247–61. London: Oxford University Press and German Historical Institute, 1988.

De Graer, R. P. "L'art de guérir chez les Azande: Essai d'ethnographie médicale [Part I and Part II]." *Congo: Revue generale de la Colonie Belge* 10 (1929): 220–54 and 361–408.

De l'Estoile, Benoît. "L'Afrique comme laboratoire: Experiénces réformatrices et revolution anthropologique dans l'empire colonial britannique (1920–1950)." PhD diss., L'École des Hautes Études en Sciences Sociales, 2004.

De Raadt, Pieter. "The History of Sleeping Sickness." In *Protozoal Diseases*, edited by H. M. Gilles. London: Arnold, 1999.

Delagrange, M. "Rapport sur la colonisation, l'emigration et la main d'oeuvre dans les pays intertropicaux." *Congrès International des Sciences Géographiques (1 au 11 Aout 1875) Compte Rendu des Séances* (Paris: E. Martinet, 1878), vol. 1, 523–32.

Denham, Helen [now Tilley]. "The Cunning of Unreason and Nature's Revolt: Max Horkheimer and William Leiss on the Domination of Nature." *Environment and History* 3 (1997): 149–75.

Depping, Guillaume. "Sociétés de géographie: Le mouvement géographique." *Journal Officiel de la République Française* 290 (October 23, 1881): 5877–78.

"Development of School Children in Kenya." *The Lancet* 218 (1931): 721.

"*Diagnoses Africanae*, I." *Kew Bulletin* 85 (1894): 17–32.

Döpke, Wolfgang. "'Magomo's Maize': State and Peasants during the Depression in Colonial Zimbabwe." In *The Economies of Africa and Asia in the Inter-War Depression*, edited by Ian Brown, 29–58. London: Routledge, 1989.

Döring, Tobias. "The Fissures of Fusion: Akiki Nyabongo's *Africa Answers Back* (1936) and What It May Teach Us." In *Fusion of Cultures?* edited by Peter Stummer and Christopher Balme, 139–52. Amsterdam: Rodopi, 1996.

Dougall, James. "Characteristics of African Thought." *Africa* 5 (1932): 249–65.

Dover, Cedric. *Brown Phoenix*. London: The College Press, 1950.

———. *Half-Caste*. London: Martin Secker and Warburg, 1937.

Doyal, Leslie. *The Political Economy of Health*. London: Pluto Press, 1979.

"Dr. H. L. Gordon." *Eugenics Review* 39 (1947–48): 109.

Drayton, Richard. *Nature's Government: Science, Imperial Britain, and the "Improvement" of the World*. New Haven, CT: Yale University Press, 2000.

———. "Science, Medicine, and the British Empire." In *The Oxford History of the British Empire—Historiography, volume 5*, edited by Robin Winks, 264–76. London: Oxford University Press, 1999.

Driberg, Jack H. *The Savage as He Really Is*. London: Routledge, 1929.

Driver, Felix. *Geography Militant: Cultures of Exploration and Empire*. Oxford: Blackwell, 2001.

Dubos, René. *Man Adapting*. New Haven, CT: Yale University Press, 1965.

Dubow, Saul. *A Commonwealth of Knowledge: Science, Sensibility, and White South Africa, 1880–2000*. London: Oxford University Press, 2006.

———. "A Commonwealth of Science: The British Association in South Africa, 1905 and 1929." In *Science and Society in Southern Africa*, edited by Saul Dubow, 67–99. Manchester: Manchester University Press, 2000.

———. "Human Origins, Race Typology, and the Other Raymond Dart." *African Studies* 55 (1996): 1–30.

———. *Scientific Racism in Modern South Africa*. Cambridge: Cambridge University Press, 1995.

Dubow, Saul, ed. *Science and Society in Southern Africa*. Manchester: Manchester University Press, 2000.

Dudley, S. F. "Can Yellow Fever Spread into Asia? An Essay on the Ecology of Mosquito-Borne Disease." *Journal of Tropical Medicine and Hygiene* 37 (1934): 273–78.

———. "The Ecological Outlook on Epidemiology—President's Address." *Proceedings of the Royal Society of Medicine* 30 (1936): 57–70.

Duignan, Peter, and L. H. Gann, eds. *Colonialism in Africa 1870–1960*. Vol. 4, *The Economics of Colonialism*. Cambridge: Cambridge University Press, 1975.

Duke, H. Lyndhurst. "An Inquiry into an Outbreak of Human Trypanosomiasis in a '*Glossina morsitans*' Belt to the East of Mwanza, Tanganyika Territory." *Proceedings of the Royal Society of London Series B* 94 (1923): 250–65.

Dumett, R. E. "Joseph Chamberlain, Imperial Finance and Railway Policy in British West Africa in the Late Nineteenth Century." *English Historical Review* 90 (1975): 287–321.

Dunlap, Thomas. *Nature and the English Diaspora: Environment and History in the United States, Canada, Australia, and New Zealand*. Cambridge: Cambridge University Press, 1999.

Dunn, J. "'For God, Emperor, Country!': The Evolution of Ethiopia's Nineteenth-Century Army." *War in History* 7 (1994): 278–99.

Dussel, Enrique. "Beyond Eurocentrism: The World-System and the Limits of Modernity."

In *The Cultures of Globalization*, edited by Frederic Jameson and Masao Miyoshi, 3–31. Durham, NC: Duke University Press, 1998.

Earthy, Dora. *Valenge Women: The Social and Economic Life of Valenge Women in Portuguese East Africa—an Ethnographic Study*. London: IIALC, 1933.

Edge, P. Granville. "The Incidence and Distribution of Human Trypanosomiasis in British Tropical Africa." *Tropical Diseases Bulletin* 35 (1938): 3–18.

"Editorial." *Second Annual Bulletin—Agricultural Department of Northern Rhodesia*, 3–4. Livingstone: Government Printer, 1932.

Edney, Matthew. *Mapping an Empire: The Geographical Construction of British India, 1765–1843*. Chicago: University of Chicago Press, 1997.

El Shakry, Omnia. *The Great Social Laboratory: Subjects of Knowledge in Colonial and Postcolonial Egypt*. Stanford, CA: Stanford University Press, 2007.

Elias, T. O. *Africa and the Development of International Law*. Dordrecht: Martinus Nijhoff, 1988.

Elliot, G. F. Scott. "Suggestions for an Inquiry into the Resources of the Empire." *Geographical Journal* 27 (June 1906): 553–58.

Elton, Charles. *Animal Ecology*. London: Sidgwick and Jackson, 1927.

———. *The Ecology of Animals*. London: Methuen, 1933.

———. "Plague and the Regulation of Numbers." *Journal of Hygiene* 24 (1925): 138.

———. "The Study of Epidemic Diseases among Wild Animals." *Journal of Hygiene* 31 (October 1931): 435–56.

Eluwa, G. I. C. "Background to the Emergence of the National Congress of British West Africa." *African Studies Review* 14 (1971): 205–18.

Engels, Dagmar, and Shula Marks, eds. *Contesting Colonial Hegemony: State and Society in Africa and India*. London: British Academic Press, 1994.

Evans-Pritchard, Edward. "The Intellectualist (English) Interpretation of Magic." *Bulletin of the Faculty of Arts* [Cairo] 1 (1933): 1–21.

———. "Mani, a Zande Secret Society." *Sudan Notes and Records (SNR)* 14 (1931): 105–48.

———. "Oracle Magic of the Azande." *SNR* 11 (1928): 1–53.

———. "Review of *Les Azande ou Niam-Niam*." *SNR* 12 (1929): 261–65.

———. "Sorcery and Native Opinion." *Africa* 4 (1931): 22–55.

———. "Witchcraft." *Africa* 8 (1935): 417–42.

———. "Witchcraft (*Mangu*) among the Azande." *SNR* 12 (1929): 163–249.

———. *Witchcraft, Oracles, and Magic among the Azande*. Oxford: Oxford University Press, 1937.

———. "The Zande Corporation of Witchdoctors [Part I and Part II]." *JRAI* 62 (1932): 291–336; and vol. 63 (1933): 63–100.

———. "Zande Therapeutics." In *Essays Presented to C. G. Seligman*, edited by Raymond Firth, Bronislaw Malinowski, and Isaac Schapera, 49–61. London: Kegan Paul, 1934.

Fabian, Johannes. *Out of Our Minds: Reason and Madness in the Exploration of Central Africa*. Berkeley: University of California Press, 2000.

Fairhead, James, and Melissa Leach. *Misreading the African Landscape: Society and Ecology in a Forest-Savanna Mosaic*. Cambridge: Cambridge University Press, 1996.

———. "Rethinking Forest-Savanna Mosaic: Colonial Science and Its Relics in West Africa." In *The Lie of the Land*, edited by Melissa Leach and Robin Mearns, 118–19. Oxford: James Currey, 1996.

Farley, John. *Bilharzia: A History of Imperial Tropical Medicine*. Cambridge: Cambridge University Press, 1991.

———. "The International Health Division of the Rockefeller Foundation: The Russell Years, 1920–1934." In *International Health Organizations and Movements, 1918–1939*, edited by P. Weindling, 203–21. Cambridge: Cambridge University Press, 1995.

Farr, Sir William. "Inaugural Address [November 19, 1872]." *Journal of the Statistical Society of London* 35 (1872): 417–30.

Faulkner, O. T. "The Aims and Objects of the Agricultural Department in Nigeria." In *First Annual Bulletin of the Agricultural Department, 1st July, 1922*, 5–17. Lagos: Government Printer, 1922.

Faulkner, O. T., and J. R. Mackie. "The Introduction of Mixed Farming in Northern Nigeria." *Empire Journal of Experimental Agriculture* 4 (1936): 89–96.

———. *West African Agriculture*. Cambridge: Cambridge University Press, 1933.

Feierman, Steven. "Healing as Social Criticism in the Time of Colonial Conquest." *African Studies* 54 (1995): 73–88.

———. "Struggles for Control: The Social Roots of Health and Healing in Modern Africa." *African Studies Review* 28 (1985): 73–145.

Feinstein, Charles. *An Economic History of South Africa: Conquest, Discrimination, and Development*. Cambridge: Cambridge University Press, 2005.

Felkin, Robert W. "On the Geographical Distribution of Tropical Diseases in Africa." *Proceedings of the Royal Physical Society* (Edinburgh) 12 (1894): 415–87.

Ferguson, James. *Global Shadows: Africa in the Neoliberal World Order*. Durham, NC: Duke University Press, 2006.

Fields, Karen. *Revival and Rebellion in Colonial Africa*. Princeton, NJ: Princeton University Press, 1985.

Findlay, George. *Miscegenation: A Study of the Biological Sources of Inheritance of the South African Population*. Pretoria: The "Pretoria News" and Printing Works, 1936.

Findlen, Paula. *Possessing Nature: Museums, Collecting, and Scientific Culture in Early Modern Italy*. Berkeley: University of California Press, 1994.

Finot, Jean. *Les préjugé des races*. Paris: Felix Algan, 1905.

———. *Race Prejudice*. Translated by Florence Wade-Evans. London: Archibald Constable, 1906.

Firth, Raymond. *Human Types*. London: Thomas Nelson, 1938.

Fisch, J. "Africa as Terra Nullius: The Berlin Conference and International Law." In *Bismarck, Europe, and Africa*, edited by Stig Föster et al., 347–75. London: Oxford University Press and German Historical Institute, 1988.

Fisher, Donald. "The Rockefeller Foundation and the Development of Scientific Medicine in Great Britain." *Minerva* 16 (1978): 20–41.

———. "Rockefeller Philanthropy and the British Empire: The Creation of the London School of Hygiene and Tropical Medicine." *History of Education* 7 (1978): 129–43.

———. "The Role of Philanthropic Foundations in the Reproduction and Production of Hegemony: Rockefeller Foundations and the Social Sciences." *Sociology* 17 (1983): 206–33.

"A Five-Year Plan of Research." *Africa* 5 (1932): 1–13.

Ford, John. "African Trypanosomiasis: An Assessment of the Tsetse Fly Problem Today." In *African Environment: Problems and Perspectives*, edited by Paul Richards, 67–72. London: International African Institute, 1975.

———. "The Ecological Method." Review of J. W. Bews, *Life as a Whole*, 1937. *Journal of Animal Ecology* 7 (1938): 156–57.

———. *The Role of the Trypanosomiases in African Ecology: A Study of the Tsetse Fly Problem*. Oxford: Clarendon Press, 1971.

Ford, John, and R. de Z. Hall. "The History of Karagwe (Bukoba District)." *Tanganyika Notes and Records* 42 (1947): 3–27.

Ford, John, C. H. Hartley, et al. *Borneo Jungle: An Account of the Oxford Expedition to Sarawak.* London: Drummond, 1938.

Ford, John, E. F. Whiteside, and A. T. Culwick. "The Trypanosomiasis Problem." *East African Agricultural Journal* (1948): 187–94.

Forman, Charles. "Science for Empire: Britain's Development of the Empire through Scientific Research, 1895–1940." PhD diss., University of Wisconsin, 1941.

Fortes, M. "Culture Contact as a Dynamic Process: An Investigation in the Northern Territories of the Gold Coast." *Africa* 9 (1936): 24–55.

———. "A New Application of the Theory of Neogenesis to the Problem of Mental Testing. (Perceptual Tests of 'g')." PhD diss., University of London, 1930.

———. "Perceptual Tests of 'General Intelligence' for Inter Racial Use." *Transactions of the Royal Society of South Africa* 20 (1932): 281–99.

———. "Review of S. Porteus, *The Psychology of a Primitive People.*" *Man* 32 (1932): 98–100.

Fox, Robert, and Anna Guagnini. *Laboratories, Workshops, and Sites: Concepts and Practices of Research in Industrial Europe, 1800–1914.* Berkeley: Office for History of Science and Technology, University of California, 1999.

Frankel, Sally Herbert. *Capital Investment in Africa: Its Causes and Effects.* London: Oxford University Press, 1938.

Fraser, A. D., and H. L. Duke. "Duration of the Infectivity of the *Glossina palpalis* after the Removal of the Lake-Shore Population." *Reports of the Sleeping Sickness Commission of the Royal Society* 12 (1912): 63–74.

Frere, H. Bartle. "Memorandum of Instructions for the Livingstone East Coast Expedition; Given at Zanzibar." *PRGS* 17 (1872–73): 158–61.

Freshfield, Douglas. "The Place of Geography in Education." *PRGS* 8 (1886): 698–718.

Freyhofer, Horst. *The Nuremberg Medical Trial: The Holocaust and the Origin of the Nuremberg Medical Code.* New York: P. Lang, 2004.

Füredi, Frank. *The Silent War: Imperialism and the Changing Perception of Race.* London: Pluto Press, 1998.

Furse, Ralph. "The Great Experiment." *Corona* 3 (1951): 447–48.

Gallagher, Nancy E. *Egypt's Other Wars: Epidemics and the Politics of Public Health.* Syracuse, NY: Syracuse University Press, 1990.

———. *Medicine and Power in Tunisia, 1780–1900.* Cambridge: Cambridge University Press, 1983.

Galton, Francis. "Address Delivered to the Anniversary Meeting." *JAI* 16 (1887): 386–402.

———. *The Art of Travel, or Shifts and Contrivances Available in Wild Countries.* London: J. Murray, 1855.

———. "Opening Remarks by the President." *JAI* 15 (1886): 336–38.

———. "Presidential Address: Geography." *BAAS Report for 1872.* London, 1873, 198–203.

Gann, L. H., and Peter Duignan. *The Rulers of British Africa, 1870–1914.* London: Hoover Institution Publications, 1978.

"German Colonies in Tropical Africa." *Kew Bulletin* 96 (December 1894): 410–12.

"German Colonies in Tropical Africa and the Pacific." *Kew Bulletin* 117–18 (September–October 1896): 174–85.

Giblin, James. "Trypanosomiasis Control in African History: An Evaded Issue?" *Journal of African History* 31 (1990): 59–80.

Gifford, Prosser, and W. R. Louis, eds. *Decolonization and African Independence: The Transfers of Power, 1960–1980*. New Haven, CT: Yale University Press, 1988.

———. *The Transfer of Power in Africa: Decolonization, 1940–1960*. New Haven, CT: Yale University Press, 1982.

Gilfoyle, Daniel. "Veterinary Immunology as Colonial Science: Method and Quantification in the Investigation of Horse Sickness in South Africa, c. 1905–1945." *Journal of the History of Medicine and Allied Sciences* 61 (2006): 26–65.

———. "Veterinary Research and the African Rinderpest Epizootic: The Cape Colony, 1896–1898." *Journal of Southern African Studies* 29 (2003): 133–54.

Gill, Clifford Allchin. *The Genesis of Epidemics and the Natural History of Disease: An Introduction to the Science of Epidemiology Based upon the Study of Epidemics of Malaria, Influenza, and Plague*. London: Baillière, Tindall and Cox, 1928.

Gillman, Clement. "East African Vegetation Types." *Journal of Ecology* 26 (1936): 502–5.

———. "Review: *The Soils, Vegetation, and Agriculture of North-Eastern Rhodesia*." *East African Agricultural Journal* 10 (1944): 61–62.

Glasgow, J. P. "Shinyanga: A Review of the Work of Tsetse Research Laboratory." *East African Agricultural and Forestry Journal* 26 (1960): 22–34.

Glassman, Jonathan. "Slower Than a Massacre: The Multiple Sources of Racial Thought in Colonial Africa." *American Historical Association* 109 (2004): 720–54.

Gluckman, Max. "Human Laboratory across the Zambesi." *Libertas* 6 (1946): 38–49.

———. "The Logic of African Science and Witchcraft: An Appreciation of Evans-Pritchard's *Witchcraft Oracles and Magic among the Azande* of the Sudan." *Rhodes-Livingstone Journal* 1 (1944): 61–71.

———. "The Rhodes-Livingstone Institute and Museum." *Rhodes Livingstone Institute Journal* 1 (1944): 4–9.

Godard, John George. *Racial Supremacy: Being Studies in Imperialism*. London: Simpkin, Marshall, 1905.

"Gold Coast Botanical Station." *Kew Bulletin* 55 (1891): 169–75.

Goldberg, David Theo. *The Racial State*. Oxford: Blackwell, 2002.

Goldie, George Taubman. "Geographical Ideals [November 1906, address delivered to the Scottish Geographical Society]." *Geographical Journal* 29 (1907): 1–14.

Gooday, Graeme. "'Nature' in the Lab: Domestication and Discipline with the Microscope in Victorian Life Sciences." *British Journal for the History of Science* 24 (1991): 307–41.

Gordon, H. L. "Amentia in the East African." *The Eugenics Review* 25 (1934): 223–35.

———. "An Inquiry into the Correlation of Civilization and Mental Disorder in the Kenya Native." *East African Medical Journal* 12 (1935–36): 327–35.

———. "The Native Brain—Observations in Kenya—A Comparison with Europeans." *Times (London)*, December 8, 1933.

———. "A Note on Diagnosis of Amentia (Mental Deficiency) in Africans." *East African Medical Journal* 7 (1930–31): 208–14.

———. "On Certification on Mental Disorder in Kenya." *East African Medical Journal* 12 (1935–36): 358–65.

———. "Relation of Malaria to the Alleged Rarity of Neurosyphilis amongst 'Uncivilised' Races." *Kenya and East African Medical Journal* 6 (1929–30): 221–29.

Graham, Richard, ed. *The Idea of Race in Latin America, 1870–1940*. Austin: University of Texas Press, 1990.

Grandidier, Alfred. "Méthode pratiques à employer pour l'observation des longitudes en voyage." *Congrès International des Sciences Géographiques*, vol. 1, 598–99, 620–21. Paris: E. Martinet, 1878.

Gray, John. "Ismail Pasha and Sir Samuel Baker." *Uganda Journal* 25 (1961): 199–213.

Greenwood, Major. "The Epidemiological Point of View." *British Medical Journal* 2 (1919): 405–7.

Gregory, John Walter. *The Great Rift Valley: Being a Narrative of a Journey to Mount Kenya and Lake Baringo, with Some Account of the Geology, Natural History, Anthropology, and Future Prospects of British East Africa*. London: J. Murray, 1896.

Greig, Sir Robert. "Presidential Address: Agriculture and the Empire." *BAAS Report*. London: 1929.

Greswell, William. "Europe and Africa." *Blackwood's Magazine* 151 (1892): 843–52.

———. *Geography of Africa South of the Zambesi*. Oxford: Clarendon Press, 1892.

———. *Our South African Empire*. London: Chapman and Hall, 1885.

Grove, Richard. *Green Imperialism: Colonial Expansion and the Origins of Environmentalism, 1600–1860*. Cambridge: Cambridge University Press, 1995.

Grovogui, Siba. *Sovereigns, Quasi-Sovereigns, and Africans: Race and Self-Determination in International Law*. Minneapolis: University of Minnesota Press, 1996.

Gründer, H. "Christian Missionary Activities in Africa in the Age of Imperialism and the Berlin Conference of 1884–1885." In *Bismarck, Europe, and Africa*, edited by Stig Föster et al., 85–103. London: Oxford University Press, 1988.

Haas, Peter. "Introduction: Epistemic Communities and International Policy Coordination." *International Organization* 46 (1992): 1–35.

Hackett, L. W. *Malaria in Europe: An Ecological Study*. London: Oxford University Press, 1937.

Hacking, Ian. *Representing and Intervening: Introductory Topics in the Philosophy of Natural Science*. Cambridge: Cambridge University Press, 1983.

Haddon, A. C. "Anthropology, Its Position and Needs." *JAI* 33 (1903): 11–23.

———. "An Imperial Bureau of Anthropology." *Nature* 80 (March 18, 1909): 73–74.

———. "Presidential Address: Section H Anthropology." In *Report of the Seventy-Fifth Meeting of the British Association for the Advancement of Science—South Africa, 1905*, 511–25. London: John Murray, 1906.

———. "The Universal Races Congress." *Times (London)*, August 8, 1911, 6, col. E.

Hailey, Lord. *An African Survey: A Study of Problems Arising in Africa South of the Sahara*. London: Oxford University Press, 1938.

———. *An African Survey: A Study of the Problems Arising in Africa South of the Sahara*. 1956 rev. ed. London: Oxford University Press, 1957.

Haldane, J. B. S. "The Native Brain." *Times (London)*, December 19, 1933.

Haldane, R. B. "The Constitution of the Empire and the Development of Its Council." *Journal of the Society of Comparative Legislation* 4 n.s. (1902): 11–18.

Hallen, Barry. *African Philosophy: The Analytic Approach*. Trenton, NJ: Africa World Press, 2006.

Hancock, Keith. *Survey of British Commonwealth Affairs*. London: Oxford University Press, 1937.

Hansford, C. G. "Some Effects of the Development of the Cotton Industry on Native Agriculture in Uganda." *Empire Journal of Experimental Agriculture* 4 (1936): 81–88.

Hargreaves, John. "History: African and Contemporary." *African Research and Documentation* 1 (1973): 3–8.

Harley, George Way. *Native African Medicine with Special Reference to Its Practice in the Mano Tribe of Liberia*. Cambridge, MA: Harvard University Press, 1941.

Harries, Patrick. *Butterflies and Barbarians: Swiss Missionaries and Systems of Knowledge in Southeast Africa*. Oxford: James Currey, 2007.

Harris, Sheldon. *Factories of Death: Japanese Biological Warfare, 1932–1945 and the American Cover-Up*. New York: Routledge, 2002.

Harrison, E. *Soil Erosion—A Memorandum*. Dar es Salaam: Government Printer, 1937.

Hartley, B. J. "An Indigenous System of Soil Protection." *East African Agricultural Journal* 4 (1938): 63–66.

Havinden, Michael, and David Meredith. *Colonialism and Development: Britain and Its Tropical Colonies, 1850–1960*. London: Routledge, 1993.

Hayden, Sherman Strong. *The International Protection of Wild Life*. New York: Columbia University Press, 1942.

Hayes, T. R. "Agricultural Surveys in the Eastern Province, Uganda." *East African Agricultural Journal* 4 (1938): 211–17.

Haynes, Douglas. *Imperial Medicine: Patrick Manson and the Conquest of Tropical Disease*. Philadelphia: University of Pennsylvania Press, 2001.

Hazelgrove, Jenny. *Spiritualism and British Society between the Wars*. Manchester: Manchester University Press, 2000.

Headrick, Daniel. *The Tentacles of Progress: Technology Transfer in the Age of Imperialism, 1850–1940*. Oxford: Oxford University Press, 1988.

Hecht, Jennifer Michael. "The Solvency of Metaphysics: The Debate over Racial Science and Moral Philosophy in France, 1890–1919." *Isis* 90 (1999): 1–24.

Heffernan, Michael. "The Limits of Utopia: Henri Duveyrier and the Exploration of the Sahara in the Nineteenth Century." *Geographical Journal* 155 (1989): 342–52.

Helly, Dorothy. "British Attitudes towards Tropical Africa, 1860–1890." PhD diss., Radcliffe College, 1961.

———. "'Informed Opinion' on Tropical Africa in Great Britain, 1860–1890." *African Affairs* 68 (1969): 195–217.

"Henry Laing Gordon." *East African Medical Journal* (1947): 313–14.

Hetherington, Penelope. *British Paternalism and Africa, 1920–1940*. London: Frank Cass, 1978.

Hobley, C. W. *Bantu Beliefs and Magic with Particular Reference to the Kikuyu and Kamba Tribes of Kenya Colony*. London: Witherby, 1922.

———. "Obituary: Frank Hulme Melland: 1879–3 February, 1939." *Man* 39 (1939): 112.

Hobson, John. *Imperialism: A Study*. London: George Allen and Unwin, 1902.

———. "Socialistic Imperialism." *International Journal of Ethics* 12 (1901): 44–58.

Hochschild, Adam. *King Leopold's Ghost: A Story of Greed, Terror, and Heroism in Colonial Africa*. London: Pan Books, 2002 [1998].

Hodge, Joseph. *Triumph of the Expert: Agrarian Doctrines of Development and the Legacies of British Colonialism*. Athens: Ohio University Press, 2007.

Hoffman, Frederick. "The Race Traits and Tendencies among the American Negro." *Publications of the American Economic Association* 11 (1897): 1–329.

Hofmeyr, Jan. "Africa and Science: Presidential Address [22 July 1929]." In *Report of the British Association for the Advancement of Science—South Africa, 1929*. London: British Association, 1930.

Hogben, Lancelot. *Genetic Principles in Medicine and Social Science*. London: Williams and Norgate, 1931.

———. "Letter to the Editor." *Journal of the Royal African Society* 33 (1934): 432.

———. *Nature and Nurture*. London: Allen and Unwin, 1933.

Holden, Pat. "Doctors and Other Medical Personnel in the Public Health Services in

Africa, 1930–1965—Uganda, Tanganyika, Nigeria." Oxford: Oxford Development Records Project Report 17, 1984.

Hooper, Edward. *The River: A Journey Back to the Source of HIV and AIDS*. London: Allan Lane, 1999.

Hoppe, Kirk. *Lords of the Fly: Sleeping Sickness Control in British East Africa, 1900–1960*. Westport, CT: Praeger, 2003.

Horkheimer, Max. "Notes on Science and the Crisis." In *Critical Theory: Selected Essays*, edited by Matthew J. O'Connell, et al., translators. New York: Seabury Press, 1972 [1932].

Horton, Robin. "African Traditional Thought and Western Science, Part I. From Tradition to Science." *Africa* 37 (1967): 50–71.

———. "African Traditional Thought and Western Science, Part II. The 'Closed' and 'Open' Predicaments." *Africa* 37 (1967): 155–87.

Hosking, H. R. "The Improvement of Native Food Crop Production by Selection and Breeding in Uganda." *East African Agricultural Journal* 4 (1938): 84–88.

Houtondji, Paulin, ed. *Endogenous Knowledge: Research Trails*. Dakar: CODESRIA, 1997.

Howard, Albert. *An Agricultural Testament*. Oxford: Oxford University Press, 1943.

Howard, Albert, and Gabrielle Howard. *The Development of Indian Agriculture*. Oxford: Oxford University Press, 1929.

Howe, E. Graham. "Motives and Mechanisms of the Mind: I. Clearing the Ground." *The Lancet* 217 (January 1931): 36–41.

Howsin, Hilda. "Race and Colour Prejudice." *Imperial and Asiatic Quarterly Review* 31 (January–April 1911): 351–71.

Hunt, Nancy Rose. *A Colonial Lexicon: Of Birth Ritual, Medicalization, and Mobility in the Congo*. Durham, NC: Duke University Press, 1999.

———. "Colonial Medical Anthropology and the Making of the Central African Infertility Belt." In *Ordering Africa*, edited by Helen Tilley with Robert Gordon. Manchester: Manchester University Press, 2007.

Hutcheson, Chas. W., and W. B. Stevenson, eds. *Kikuyu: 1898–1923; Semi-Jubilee Book of the Church of Scotland Mission Kenya Colony*. Edinburgh: Blackwood, 1923.

Huxley, Julian. *Africa View*. London: Harper and Brothers, 1931.

———. "Appendix No. 36 Memorandum." *Joint Committee on Closer Union in East Africa*. Vol. 3, *Appendices*. London: HMSO, 1931.

———. *Biology and Its Place in Native Education in East Africa*. London: HMSO, 1930.

———. *The Conservation of Wild Life and Natural Habitats in Central and East Africa*. Paris: UNESCO, 1961.

———. *Essays of a Biologist*. London: Chatto and Windus, 1923.

———. *Essays in Popular Science*. London, 1937 [1926].

———. *Memories I & II*. New York: Penguin, 1970–73.

———. "The Native Brain: Size and Growth." *Times (London)*, December 18, 1933.

———. *Science and Social Needs*. London: Harper and Brothers, 1935.

———. "Travel and Politics in East Africa." *JRAS* 30 (July 1931): 245–61.

Huxley, Julian, and De Beer, G. R. *The Elements of Experimental Embryology*. Cambridge: Cambridge University Press, 1934.

Huxley, Julian, A. C. Haddon, and Alexander Carr-Saunders. *We Europeans: A Survey of Racial Problems*. London: Penguin Books, 1935.

Iliffe, John. *East African Doctors: A History of the Modern Profession*. Cambridge: Cambridge University Press, 1998.

"Imperial Bureau of Anthropology." *The Lancet* 173 (April 10, 1909): 1055–56.

"International Geographical Congress." *Times (London)*, August 1, 1895, 11, col. A.

International Society of Soil Science. *Fifty Years Progress in Soil Science*, special issue, *Geoderma* 12, no. 4 (1974).

Ireland, Montague George de Courcey, H. R. Hosking, and L. Loewenthal. *An Investigation into Health and Agriculture in Teso, Uganda*. Entebbe: Government Press, 1937.

Jackson, C. H. N. "Some New Methods in the Study of *Glossina morsitans*." *Proceedings of the Zoological Society, London* (1937): 811–96.

Jackson, Lynette. *Surfacing Up: Psychiatry and Social Order in Colonial Zimbabwe, 1908–1968*. Ithaca, NY: Cornell University Press, 2005.

Jagailloux, Serge. *La medicalisation de l'Égypt au XIX siècle, 1798–1918*. Paris: Éditions recherche sur les civilisations, 1986.

James, Frank A. J. L., ed. *The Development of the Laboratory: Essays on the Place of Experiment in Industrial Civilisation*. Basingstoke: Macmillan, 1989.

James, S. P. *Report on a Visit to Kenya and Uganda to Advise on Anti-Malarial Measures*. London: HMSO, 1929.

Jamison, Andrew. "National Political Cultures and the Exchange of Knowledge: The Case of Systems Ecology." In *Denationalizing Science*, edited by Elisabeth Crawford et al., 187–208. Dordrecht: Kluwer Academic Publishers, 1993

Janzen, John. *The Quest for Therapy: Medical Pluralism in Lower Zaire*. Berkeley: University of California Press, 1978.

———. "Toward a Historical Perspective on African Medicine and Health." *Beiträge zur Ethnomedizin, Ethnobotanik und Ethnozoologie* 8 (1983): 99–138.

Jeater, Diana. *Law, Language, and Science: The Invention of the "Native Mind" in Southern Rhodesia, 1890–1930*. Portsmouth, NH: Heinemann, 2007.

Jeffries, Charles. *The Colonial Empire and Its Civil Service*. Cambridge: Cambridge University Press, 1938.

Jelliffe, Smith Ely. "The Ecological Principle in Medicine." *Journal of Abnormal and Social Psychology* 32 (1937): 100–121.

Jezequel, Jean-Hervé. "Itinéraire lettres sous la colonisation: L'emergence d'une élite de la fonction publique en Afrique de l'ouest." PhD diss., École des Hautes Études en Sciences Sociales, Paris, 2002.

———. "Voices of Their Own? African Participation in the Production of Knowledge in French West Africa, 1910–1950." In *Ordering Africa*, edited by Helen Tilley with Robert Gordon, 145–73. Manchester: Manchester University Press, 2007.

Johnson, Douglas, and David Anderson, eds. *The Ecology of Survival: Case Studies from Northeast African History*. London: Lester Cook Academic Publishing, 1988.

Johnson, W. B. *Notes upon a Journey through Certain Belgian, French, and British African Dependencies to Observe General Medical Organisation and Methods of Trypanosomiasis Control*. Lagos: Government Printer, 1929.

Johnston, Harry H. *British Central Africa: An Attempt to Give Some Account of a Portion of the Territories under British Influence North of the Zambezi*. London: Methuen, 1897.

———. "British Interests in Eastern Equatorial Africa." *Scottish Geographical Magazine* 1 (1885): 144–56.

———. *Despatch from His Majesty's Special Commissioner in Uganda Relating to Travellers in the Protectorate*. Cd. 590. London: HMSO, 1901.

———. *The Kilima-Njaro Expedition: A Record of Scientific Exploration in Eastern Equatorial Africa*. London: Paul, Trench, 1886.

———. "The Preservation of the African Fauna and Its Relation to Tropical Diseases." *Nature* 88 (December 7, 1911): 178–79.

———. *Report by His Majesty's Special Commissioner on the Protectorate of Uganda.* Cd. 671. London: HMSO, 1901.

Johnston, Keith, ed. *Africa.* London: Edward Stanford, 1878.

Johnston, W. Ross. *Sovereignty and Protection: A Study of British Jurisdictional Imperialism in Late Nineteenth Century.* Durham, NC: Duke University Press, 1973.

Jones, G. Howard. *The Earth Goddess: A Study of Native Farming on the West African Coast.* London: Longmans, Green, 1936.

Juma, Calestous, ed. *Going for Growth: Science, Technology, and Innovation in Africa.* London: Smith Institute, 2005.

———. "Science and Innovation in Africa." *International Journal of Technology and Globalization* 2 (2006).

Kay, H. D. "John Boyd Orr." *Biographical Memoirs of Fellows of the Royal Society* 18 (1972): 43–81.

Keller, Evelyn Fox. *Secrets of Life, Secrets of Death: Essays on Language, Gender, and Science.* New York: Routledge, 1992.

Keller, Richard. *Colonial Madness: Psychiatry in French North Africa.* Chicago: University of Chicago Press, 2007.

Keltie, J. Scott. *Applied Geography: A Preliminary Sketch.* London: G. Philip and Son, 1890.

Keltie, J. Scott, and M. Epstein, eds. *The Statesman's Year Book for the Year 1918.* London: Macmillan, 1918.

Kendle, John Edward. *The Colonial and Imperial Conferences, 1887–1911: A Study in Imperial Organization.* London: Longmans, 1967.

Kennedy, Dane. *Islands of White: Settler Society and Culture in Kenya and Southern Rhodesia, 1890–1939.* Durham, NC: Duke University Press, 1987.

Kennedy, Paul M. "Review: The Theory and Practice of Imperialism." *Historical Journal* 20 (1977): 761–69.

Kerr, Philip. "Editorial Manifesto in the First Number, November 1910." Reprinted in *The Round Table* (1970): 381–83.

Kevles, Daniel. *In the Name of Eugenics: Genetics and the Uses of Human Heredity.* Berkeley: University of California Press, 1985.

King, Sir David. "Governing Technology and Growth." In *Going for Growth: Science, Technology, and Innovation in Africa*, ed. C. Juma. London: Smith Institute, 2005.

King, Kenneth, and Simon McGrath, eds. *Knowledge for Development? Comparing British, Japanese, Swedish, and World Bank Aid.* New York: Palgrave Macmillan, 2004.

Kingsley, Mary. *Travels in West Africa: Congo Français, Corisco, and Cameroons.* New York: Barnes and Noble, 1965 [1897].

———. *West African Studies.* 2nd ed. London: Macmillan, 1901 [1899].

Kirk-Greene, A. H. M. *On Crown Service: A History of HM Colonial and Overseas Civil Service, 1837–1997.* London: I. B. Tauris, 1999.

———. "The Thin White Line: The Size of the Colonial Service in Africa." *African Affairs* 79 (1980): 25–44.

Kirkpatrick, T. W. "East Africa and the Tsetse Fly: A Review." *East African Agricultural Journal* 3 (1937): 411–15.

———. *Studies on the Ecology of Coffee Plantations in East Africa I. The Climate and Eco-Climates of Coffee Plantations.* Amani: Government Printer, 1935.

———. *Studies on the Ecology of Coffee Plantations in East Africa II.* Amani: Government Printer, 1937.

Kish, George. "The Participants." In *Geography through a Century of International Congresses*, 35–49. Paris: UNESCO, 1972.

Kitson, Elisabeth. "A Study of the Negro Skull with Special Reference to the Crania from Kenya Colony." *Biometrika* 23 (1931): 271–314.

Kjekshus, Helge. *Ecology Control and Economic Development in East African History.* London: James Currey, 1996 [1977].

Kohler, Robert. *Landscapes and Labscapes: Exploring the Lab-Field Border in Biology.* Chicago: University of Chicago Press, 2002.

Kohn, George C., ed. *Encyclopedia of Plague and Pestilence.* New York: Facts on File, 1995. Rev. ed. 2001.

Koponen, Juhani. *Development for Exploitation: German Colonial Policies in Mainland Tanzania, 1884–1914.* Helsinki: Lit Verlag, 1994.

Koskenniemi, Martti. *The Gentle Civilizer of Nations: The Rise and Fall of International Law 1870–1960.* Cambridge: Cambridge University Press, 2001.

Koyaji, Ratanshaw. "Colour Prejudice in the British Colonies." *Indian Review* 14 (1913): 945–48.

Krupenikov, I. A. *History of Soil Science: From Its Inception to the Present.* Rotterdam: Brookfield, 1993.

Kubicek, Robert. *The Administration of Imperialism: Joseph Chamberlain at the Colonial Office.* Durham, NC: Duke University Press, 1969.

Kuczynski, Robert R. *The Cameroons and Togoland: A Demographic Study.* London: Oxford University Press, 1939.

Kuklick, Henrika. *The Savage Within: The Social History of British Anthropology, 1885–1945.* Cambridge: Cambridge University Press, 1991.

Kuklick, Henrika, and Robert Kohler, eds. *Science in the Field. OSIRIS* 11 (1996). Chicago: University of Chicago Press.

Kumar, Deepak. "Patterns of Colonial Science in India." *Indian Journal for the History of Science* 15 (1980): 105–13.

Kuper, Adam. *Anthropology and Anthropologists: The Modern British School.* Rev. ed. London: Routledge, 1989 [1973].

Lackner, Helen. "Social Anthropology and Indirect Rule: The Colonial Administration and Anthropology in Eastern Nigeria: 1920–1940." In *Anthropology and the Colonial Encounter,* edited by Talal Asad, 123–51. London: Ithaca Press, 1973.

Lagae, C. Robert. *Les Azande ou Niam-Niam: L'organisation d'Azande croyances religieuses et magiques, coutumes familiales.* Brussels: Vromant, 1926.

Lagemann, Ellen. *The Politics of Knowledge: The Carnegie Corporation, Philanthropy, and Public Policy.* Middletown, CT: Wesleyan University Press, 1989.

Laidlaw, Zoë. *Colonial Connections, 1815–1845: Patronage, the Information Revolution, and Colonial Government.* Manchester: Manchester University Press, 2005.

Lambrecht, Frank. "Aspects of the Evolution and Ecology of Tsetse Flies and Trypanosomiases in Prehistoric African Environment." *Journal of African History* 5 (1964): 1–24.

Lankester, Ray. *The Kingdom of Man.* New York: Holt, 1907.

Largent, Mark A. "Bionomics: Vernon Lyman Kellogg and the Defense of Darwinism." *Journal of the History of Biology* 32 (1999): 465–88.

Last, Murray. "The Importance of Knowing about Not Knowing: Observations from Hausaland." In *The Social Basis of Health and Healing in Africa,* edited by Steven Feierman and John Janzen, 393–406. Berkeley: University of California Press, 1992.

"Latest Intelligence." *Times (London),* November 19, 1884.

Latour, Bruno. *Science in Action.* Cambridge, MA: Harvard University Press, 1987.

Lavin, Deborah. *From Empire to International Commonwealth: A Biography of Lionel Curtis.* Oxford: Clarendon Press, 1994.

———. "Lionel Curtis and the Founding of Chatham House." In *Chatham House and British Foreign Policy, 1919–1945*, edited by in A. Bosco and C. Navari. London: Lothian Foundation Press, 1994.

Leach, Melissa, and Robin Mearns. "Environmental Change and Policy: Challenging Received Wisdom on the African Environment." In *The Lie of the Land*, 1–33. Oxford: James Currey, 1996.

Leach, Melissa, and Robin Mearns, eds. *The Lie of the Land: Challenging Received Wisdom on the African Environment*. Oxford: James Currey, 1996.

League of Nations. "Bibliography of the Technical Work of the Health Organisation of the League of Nations 1920–1945." *Bulletin of the Health Organisation* 11 (1945): 3–235.

———. *Report on the Principles and Methods of Anti-Malarial Measures in Europe*. Geneva: League of Nations, 1932.

———. *Report on the Therapeutics of Malaria*. Geneva: League of Nations, 1933.

Leakey, Harry. "Appendix No. 6, Memorandum from the Rev. Canon Harry Leakey, M. A. Representative for Native Interests in Kenya Legislative Council." In *Appendices of the Report of the Joint Committee on Closer Union in East Africa*, vol. 3, 20–25. London: HMSO, 1931.

Leakey, L. S. B. "The Cradle of Man; More Evidence from Kenya: Elmenteita Finds." *Times (London)*, March 7, 1929, 15, col. G.

———. "East Africa Past and Present." *Geographical Journal* 76 (1930): 494–500.

———. *Kenya: Contrasts and Problems*. London: Methuen, 1936.

———. "The Native Brain: Is Size a True Guide? Reasons for Stunted Growth." *Times (London)*, December 13, 1933.

———. *White African: An Early Autobiography*. London: Hodder and Stoughton, 1937.

Leakey, L. S. B., et al. "Coloured Medical Students." *Times (London)*, February 24, 1938, 8, col. C.

Lee, Christopher Joon-Hai. "Colonial Kinships: The British Dual Mandate, Anglo-African Status, and the Politics of Race and Ethnicity in Inter-War Nyasaland." PhD diss., Stanford University, 2003.

———. "The 'Native' Undefined: Colonial Categories, Anglo-African Status, and the Politics of Kinship in British Central Africa, 1929–38." *Journal of African History* 46 (2005): 455–78.

Lee, J. M. *Colonial Development and Good Government: A Study of the Ideas Expressed by the British Official Classes in Planning Decolonization, 1939–1964*. Oxford: Clarendon Press, 1967.

Leiss, William. *The Domination of Nature*. Montreal: McGill-Queen's University Press, 1994 [1972].

Lester, Robert. *Summary of Grants from the British Dominions and Colonies Fund, 1911–1935*. New York: Carnegie Corporation, n.d.

Lewin, C. J. *Agricultural and Forestry Development Plans for 10 Years*. Lusaka: Government Printer, 1945.

Lewis, Michael. *Inventing Global Ecology: Tracking the Biodiversity Ideal in India, 1945–1997*. Athens: Ohio University Press, 2004.

Li, S. J. "Natural History of Parasitic Disease: Patrick Manson's Philosophical Method." *Isis* 93 (2002): 115–29.

Lieven, Dominic. "Dilemmas of Empire, 1850–1918: Power, Territory, Identity." *Journal of Contemporary History* 34 (1999): 163–200.

"List of the Staffs of the Royal Gardens, Kew, and of Botanical Departments and Establishments at Home, and in India and the Colonies." *Kew Bulletin* (1898): appendix III, 55–62.

Livingston, Julie. *Debility and the Moral Imagination in Botswana*. Bloomington: Indiana University Press, 2005.

Livingstone, Burton E. "Environments." *Science* 80 (1934): 569–76.

Livingstone, David. "The Spaces of Knowledge: Contributions towards a Historical Geography of Science." *Environment and Planning D: Society and Space* 13 (1995): 5–34.

Lloyd, Trevor. "Africa and Hobson's Imperialism." *Past and Present* 55 (1972): 130–53.

Lonsdale, John, and Bruce Berman. "Coping with Contradictions: The Development of the Colonial State in Kenya, 1895–1914." *Journal of African History* 20 (1979): 487–505.

Lorimer, Frank, William Brass, and Etienne van de Walle. "Demography." In *The African World: A Survey of Social Research*, edited by Robert Lystad, 271–303. New York: Praeger, 1965.

Louis, William Roger. "African Origins of the Mandates Idea." *International Organization* 19 (1965): 20–36.

———. *Imperialism at Bay, 1941–1945: The United States and the Decolonization of the British Empire*. London: Oxford University Press, 1977.

Low, D. Anthony, and John Lonsdale. "Introduction: Towards the New Order, 1945–1963." In *History of East Africa, Volume III*, edited by D. A. Low and Alison Smith, 1–63. Oxford: Clarendon Press, 1976.

Lugard, Frederick. *Northern Nigeria: Report for 1901*. Cd. 1388. London: HMSO, 1903.

———. *Report by Sir F. D. Lugard on the Amalgamation of Northern and Southern Nigeria and Administration, 1912–1919*. Cmd. 468. London: HMSO, 1919.

———. "Witchcraft in Africa." *Times (London)*, April 20, 1932, 15, col. E.

Lyons, Maryinez. *The Colonial Disease: A Social History of Sleeping Sickness in Northern Zaire, 1900–1940*. Cambridge: Cambridge University Press, 1992.

———. "Diseases of Sub-Saharan Africa." In *The Cambridge World History of Human Disease*, edited by Kenneth Kiple. Cambridge: Cambridge University Press, 1993.

———. "The Power to Heal: African Medical Auxiliaries in Colonial Belgian Congo and Uganda." In *Contesting Colonial Hegemony*, edited by Dagmar Engels and Shula Marks, 202–23. London: British Academic Press, 1994.

Macalister, A. "Presidential Address." *JAI* 24 (1895): 452–69.

Mackenzie, Fiona. *Land, Ecology, and Resistance in Kenya, 1880–1952*. Edinburgh: Edinburgh University Press, 1998.

MacKenzie, John. *The Empire of Nature: Hunting, Conservation, and British Imperialism*. Manchester: University of Manchester Press, 1988.

———. "Experts and Amateurs: Tsetse, Nagana, and Sleeping Sickness in East and Central Africa." In *Imperialism and the Natural World*, 187–212. Manchester: Manchester University Press, 1990.

MacKinnon, M. "Education in Its Relation to the Physical and Mental Development of European Children in the Tropics." *Journal of Tropical Medicine and Hygiene* (May 1, 1923): 136–40.

MacLeod, Roy, ed. *Nature and Empire: Science and the Colonial Enterprise*. In *OSIRIS* vol. 15. Chicago: University of Chicago Press, 2000.

MacLeod, Roy, and Kay Andrews. "The Committee of Civil Research: Scientific Advice for Economic Development, 1925–1930." *Minerva* 7 (1969): 680–705.

MacLeod, Roy, and Kay MacLeod. "The Social Relations of Science and Technology 1914–

1939." In *The Fontana Economic History of Europe—The Twentieth Century*, edited by Carlo M. Cipolla, 301–63. London: Fontana Books, 1976.

MacLeod, Roy, and Philip Rehbock, eds. *Darwin's Laboratory: Evolutionary Theory and Natural History in the Pacific*. Honolulu: University of Hawai'i Press, 1994.

Macmillan, Margaret. *Peacemakers: Six Months That Changed the World*. London: John Murray, 2001.

Macmillan, W. M. *Africa Emergent: A Survey of Social, Political, and Economic Trends in British Africa*. London: Faber and Faber, 1938.

Madison, Mark. "Potatoes Made of Oil: Eugene and Howard Odum and the Origins of American Agroecology." *Environment and History* 3 (1997): 209–38.

Mafeje, Archie. "Anthropology and Independent Africans: Suicide or End of an Era?" *African Sociological Review* 2 (1998): 1–43.

"Magic and Administration in Africa." *Nature* 129 (April 30, 1932): 629–31.

Mahone, Sloan. "The Psychology of the Tropics: Conceptions of Danger and Lunacy in British East Africa." PhD diss., Oxford University, 2004.

———. "The Psychology of Rebellion: Colonial Medical Responses to Dissent in British East Africa." *Journal of African History* 46 (2006): 241–58.

Mair, Lucy. *An African People in the Twentieth Century*. London: Routledge, 1934.

———. *Native Policies in Africa*. London: Routledge, 1936.

———. "The Social Sciences in Africa South of the Sahara: The British Contribution." *Human Organization* 19 (1960): 98–107.

———. "The Study of Culture Contact as a Practical Problem." *Africa* 7 (1934): 415–22.

Malinowski, Bronislaw. *Culture as a Determinant of Behavior*. Cambridge, MA: Harvard University Press, 1936.

———. "Introduction." In *The Savage Hits Back*, by Julius Lips. New York: University Books, 1966 [1937].

———. "New and Old Anthropology." *Nature* 113 (1924): 299–301.

———. "Practical Anthropology." *Africa* 2 (1929): 22–38.

———. "The Rationalization of Anthropology and Administration." *Africa* 3 (1930): 405–29.

Mandaza, Ibbo. *Race, Colour, and Class in Southern Africa: A Study of the Coloured Question in the Context of the Colonial and White Settler Racial Ideology, and African Nationalism in Twentieth-Century Zimbabwe, Zambia, and Malawi*. Harare: Sapes Books, 1997.

Marks, Shula. "What Is Colonial about Colonial Medicine? And What Has Happened to Imperialism and Health?" *Social History of Medicine* 10 (1997): 207–19.

Martin, S. M. "The Long Depression: West African Export Producers and the World Economy, 1914–1945." In *The Economies of Africa and Asia in the Inter-War Depression*, edited by Ian Brown, 74–94. London: Routledge, 1989.

Mavhunga, Claperton. "Firearms Diffusion, Exotic and Indigenous Knowledge Systems in the Lowveld Frontier, South Eastern Zimbabwe, 1870–1920." *Comparative Technology Transfer and Society* 1 (2003): 201–31.

Maxon, Robert. "The Devonshire Declaration: The Myth of Missionary Intervention." *History in Africa* 18 (1991): 259–70.

———. *Struggle for Kenya: The Loss and Reassertion of Imperial Initiative, 1912–1923*. Rutherford, NJ: Farleigh Dickinson University Press, 1993.

Mayhew, Arthur. "Education in the Colonies." In *Oxford University Summer School on Colonial Administration 27 June–8 July 1938*, 83–86. Oxford: Oxford University Press, 1938.

McCann, James. *People of the Plow: An Agricultural History of Ethiopia, 1800–1990*. Madison: University of Wisconsin Press, 1995.

McClellan, James. *Colonialism and Science: Saint-Domingue in the Old Regime*. Baltimore: Johns Hopkins University Press, 1992.

McCulloch, Jock. *Colonial Psychiatry and "the African Mind."* Cambridge: Cambridge University Press, 1995.

McCulloch, W. E. "An Enquiry into the Dietaries of the Hausa and Town Fulani with Some Observations of the Effects on the National Health, with Recommendations Arising Therefrom." *West African Medical Journal* 3 (1929–30): 8–22; 36–47; 62–73.

McKay, D. V. "Colonialism and the French Geographical Movement." *Geographical Review* 33 (1943): 214–32.

McKinley, Edward. *The Lure of Africa: American Interests in Tropical Africa, 1919–1939*. New York: Bobbs-Merrill, 1974.

McPhee, Allan. *The Economic Revolution in British West Africa*. New York: Negro Universities Press, 1970 [1926].

Melland, Frank. *In Witch-Bound Africa: An Account of the Primitive Kaonde Tribe and Their Beliefs*. London: Seeley, 1923.

———. "The Misjudged Witch-Doctor." *The Listener* (January 1934): 14–16.

———. "Native Policy in Africa." *Times (London)*, July 5, 1932, 10, col. C.

———. "Northern Rhodesia: Retrospect and Prospect." *Journal of the Royal African Society* 29 (1930): 490–98.

———. "The Problem of Africa: The Problem of Witchcraft." *African Observer* 1 (1934): 13–18.

———. "A Shadow over Africa—the Terrors of Witchcraft—Law and Belief." *Times (London)*, April 13, 1932, 13, col. F.

Mendelsohn, J. Andrew. "From Eradication to Equilibrium: How Epidemics Became Complex after World War I." In *Greater Than the Parts: Holism in Biomedicine, 1920–1950*, edited by Christopher Lawrence and George Weisz, 303–31. New York: Oxford University Press, 1998.

"Mental Quotient of the Natives." *South African Medical Journal* 8 (1934): 2.

Miller, Charles. *The Lunatic Express: The Magnificent Saga of How the White Man Changed Africa—The Pioneers, Visionaries, and Politicians and Their Crazy Railway*. London: St. Giles Press, 1971.

Miller, David P., and Peter H. Reill, eds. *Visions of Empire: Voyages, Botany, and Representations of Nature*. Cambridge: Cambridge University Press, 1996.

Mills, David. "British Anthropology at the End of Empire: The Rise and Fall of the Colonial Social Science Research Council, 1944–1962." *Revue d'Histoire des Sciences Humaines* 6 (2002): 161–88.

———. *Difficult Folk?: A Political History of Social Anthropology*. Oxford: Berghahn Books, 2008.

Milne, G. "Normal Erosion as a Factor in Soil Profile Development." *Nature* 138 (September 1936): 548–49.

———. "A Soil Reconnaissance Journey through Parts of Tanganyika Territory December 1935 to February 1936." *Journal of Ecology* 35 (1947): 192–265.

Milne, G., et al. *Provisional Soil Map of East Africa—Amani Memoirs*. Tanganyika: Crown Agents for the Colonies, 1936.

———. "Provisional Soil Map of East Africa (Kenya, Uganda, Tanganyika, and Zanzibar)." *Transactions of the Third International Congress of Soil Science*, vol. 1, Commission Papers, 266–70. London: Thomas Murby, 1935.

————. "A Short Geographical Account of the Soils of East Africa." *Transactions of the Third International Congress of Soil Science*, vol. 1, Commission Papers, 270–74. London: Thomas Murby, 1935.

Miner, Horace. "Culture Change under Pressure: A Hausa Case." *Human Organization* 19 (1960): 164–67.

Mitchell, J. P. "The Uganda Medical School: Its Origins and Objectives." *Uganda Teacher's Journal* 1 (1939): 85–91.

Mitchell, Philip. "The Anthropologist and the Practical Man: A Reply and a Question." *Africa* 3 (April 1930): 217–23.

————. "Introduction." In *Anthropology in Action: An Experiment in the Iringa District of the Iringa Province Tanganyika Territory*, edited by G. Gordon Brown and A. Bruce Hutt. London: Oxford University Press and IIALC, 1935.

Mitman, Gregg. *The State of Nature: Ecology, Community, and American Social Thought, 1900–1950*. Chicago: University of Chicago Press, 1992.

Moffat, U. J. "Native Agriculture in the Abercorn District." In *Second Annual Bulletin— Agricultural Department of Northern Rhodesia*, 55–62. Livingstone: Government Printer, 1932.

Moore, Henrietta, and Megan Vaughan. *Cutting Down Trees: Gender, Nutrition, and Agricultural Change in the Northern Province of Zambia, 1890–1990*. London: James Currey Press, 1994.

Moore, Sally Falk. *Anthropology and Africa: Changing Perspectives on a Changing Scene*. Charlottesville: University of Virginia Press, 1994.

Moore, T. C. "Some Indigenous Crops of Northern Rhodesia." In *Second Annual Bulletin—Agricultural Department of Northern Rhodesia*, 35–50. Livingstone: Government Printer, 1932.

Morel, E. D. *Nigeria: Its People and Its Problems*. London: Smith, Elder, 1911.

Morrell, Jack. *Science at Oxford, 1914–1939: Transforming an Arts University*. Oxford: Clarendon Press, 1997.

Mott, Frederick. "An Address on Body and Mind: The Origin of Dualism." *The Lancet* 199 (January 1922): 1–5.

————. "A Lecture on Body and Mind." *The Lancet* 196 (August 1920): 383–87.

"Mr. Churchill on African Affairs." *Times (London)*, January 20, 1908, 3, col. B.

Müller-Hill, Benno. *Murderous Science: Elimination by Scientific Selection of Jews, Gypsies, and Others, Germany, 1933–1945*. London: Oxford University Press, 1988.

Mumford, W. B., and C. E. Smith. "Racial Comparisons and Intelligence Testing—Some Notes on Psychological and Anthropometrical Studies of African and European Differences: A Corrective to Some Popular Opinions." *African Affairs* 37 (1938): 46–57.

Munro, J. Forbes. *Africa and the International Economy, 1800–1960*. London: Dent and Sons, 1976.

Murphy, E. Jefferson. *Creative Philanthropy: Carnegie Corporation and Africa, 1953–1973*. New York: Teachers College Press, 1977.

Murray, Gilbert. "At Home in the Modern World." *Century Magazine* (May 1921): 30–38.

Mustafa, Raufu, and Kate Meager. "Agrarian Production, Public Policy, and the State in Kano Region, 1900–2000." Working Paper No. 35. Crekerne, UK: Drylands Research, 2000.

Mutwira, Roben. "Southern Rhodesia Wild Life Policy (1890–1953): A Question of Condoning Game Slaughter?" *Journal of Southern African Studies* 15 (1989): 250–62.

Nadel, Siegfried. "The Application of Intelligence Tests in the Anthropological Field." In

The Study of Society: Methods and Problems, edited by Sir Frederic Bartlett, M. Ginsberg, E. J. Lindgren, and R. H. Thouless, 184–98. London: Routledge, 1939.

———. "A Field Experiment in Racial Psychology." *British Journal of Psychology* 28 (1937): 195–211.

Nandy, Ashis. *The Savage Freud and Other Essays on Possible and Retrievable Selves*. Princeton, NJ: Princeton University Press, 1995.

Nash, T. A. M. *Africa's Bane: The Tsetse Fly*. London: Collins, 1969.

———. "The Anchau Settlement Scheme." *Farm and Forest* 2 (1941): 76–82.

———. *A Zoo without Bars: Life in the East Africa Bush, 1927–1932*. Tunbridge Wells, UK: Wayte Binding, 1984.

National Research Council, Board on Science and Technology for International Development. *Lost Crops of Africa*. 3 vols. Washington, DC: National Academy Press, 1996–.

"The Native Problem in Africa." *African Affairs* 28 (1928–29): 93–94.

Neumann, Roderick. "The Post-War Conservation Boom in British Africa." *Environmental History* 7 (2002): 22–47.

Northern Rhodesia. *The Agricultural Survey Commission Report, 1930–1932*. Livingstone: Government Printer, 1932.

———. *Annual Report for the Year 1927, Department of Agriculture*. Livingstone: Government Printer, 1928.

———. *Annual Report for the Year 1928, Department of Agriculture*. Livingstone: Government Printer, 1929.

———. *Annual Report for the Year 1929, Department of Agriculture*. Livingstone: Government Printer, 1930.

———. *Annual Report for the Year 1932, Department of Agriculture*. Livingstone: Government Printer, 1933.

———. *Annual Report for the Year 1940 Agriculture Department*. Livingstone: Government Printer, 1941.

———. *Legislative Council Debates Fourth Session of the Fifth Council, 26th June–8th July 1937*. Lusaka: Government Printer, 1937.

Nowell, Charles. "Portugal and the Partition of Africa." *Journal of Modern History*, 19 (1947): 1–17.

Nowell, W. "The Agricultural Research Station at Amani." *Journal of the African Society* 33 (1934): 1–20.

———. "The Work of the Amani Institute." *Annals of Applied Biology* 15 (1928): 120–24.

Nugent, Paul. *Africa since Independence*. New York: Palgrave Macmillan, 2004.

Nutman, F. J. "The Root-System of *Coffea arabica*, Parts 1 and 2." *Empire Journal of Experimental Agriculture* 1 (1933).

———. "The Root-System of *Coffea arabica*, Part 3." *Empire Journal of Experimental Agriculture* 2 (1934): 293–302.

Nyabongo, Akiki. *Africa Answers Back*. London: Routledge, 1936.

———. "Religious Practices and Beliefs of Uganda." PhD diss., Oxford University, 1939.

Nyamweru, Celia. "The African Rift System." In *The Physical Geography of Africa*, edited by W. M. Adams, A. S. Goudie, and A. R. Orme, 18–33. Oxford: Oxford University Press, 1996.

"Obituary: Robert Needham Cust." *Geographical Journal* 34 (1909): 685–86.

Ogilvie, Alan. "Co-Operative Research in Geography: With an African example." *Scottish Geographical Magazine* 50 (1934): 353–78.

———. "'Science in Africa' by E. B. Worthington: A Review." *Africa* 12 (April 1939): 233–38.

Oldham, Joseph. *Christianity and the Race Problem*. London: Student Christian Movement, 1924.

———. *White and Black in Africa: A Critical Examination of the Rhodes Lectures of General Smuts*. London: Longmans, 1930.

Oliver, R. A. C. "The Adaptation of Intelligence Tests to Tropical Africa." *Oversea Education* 4 (July 1933): 186–92.

———. "The Adaptation of Intelligence Tests to Tropical Africa, [Part] II." *Oversea Education* 5 (October 1933): 8–13.

———. "The Comparison of the Abilities of Races: With Special Reference to East Africa." *East African Medical Journal* 9 (1932–33): 160–75 and 193–204.

———. *General Intelligence Tests for Africans: Manual of Directions*. Nairobi: Kenya Colony, 1932.

———. "Mental Tests in the Study of the African." *Africa* 7 (1934): 40–46.

Olumwullah, Osaak. *Dis-Ease in the Colonial State: Medicine, Society, and Social Change among the AbaNyole of Western Kenya*. Westport, CT: Greenwood Press, 2002.

Orde-Browne, Granville. "Witchcraft and British Colonial Law." *Africa* 8 (1935): 481–87.

———. "Witchcraft and the Law." *African Observer* 6 (1937): 53–57.

Ormsby-Gore, William. "Agricultural and Veterinary Research in the Colonies, Protectorates, and Mandated Territories." In *Colonial Office Conference, 1927, Appendices to the Summary of Proceedings*, 43–62. London: HMSO, 1927.

———. "Opening Address." In *Report of the Second International Conference on Sleeping Sickness—Paris, November 5–7, 1928*, 14–19. Geneva: League of Nations, 1928.

Orr, John Boyd. *As I Recall: Memoirs of a Nobel Peace Prize Winner*. New York: Doubleday, 1967.

———. "Problems of African Native Diet: Foreword." *Africa* 9 (1936): 145–46.

Osborne, Michael. "Acclimatizing the World: A History of the Paradigmatic Colonial Science." *OSIRIS* 15 (2000): 135–51.

———. *Nature, the Exotic and the Science of French Colonialism*. Bloomington: Indiana University Press, 1994.

Owen, Nicholas. "Critics of Empire in Britain." In *The Oxford History of the British Empire—Twentieth Century, Volume 4*, edited by Judith Brown and William Roger Louis, 188–211. London: Oxford University Press, 1999.

Packard, Randall. "The Invention of the 'Tropical Worker': Medical Research and the Quest for Central African Labor on the South African Gold Mines, 1903–1936." *Journal of African History* 34 (1993): 271–92.

———. "Visions of Postwar Health and Development and Their Impact on Public Health Interventions in the Developing World." In *International Development and the Social Sciences: Essays on the History and Politics of Knowledge*, edited by Frederick Cooper and Randall Packard, 93–115. Berkeley: University of California Press, 1997.

———. *White Plague, Black Labor: Tuberculosis and the Political Economy of Health and Disease in South Africa*. Berkeley: University of California Press, 1989.

"La part des Suisses dans l'exploration et la civilization de l'Afrique." *L'Afrique Explorée et Civilisée* 4 (1883): 215–29.

Partington, John. *Building Cosmopolis: The Political Thought of H. G. Wells*. London: Ashgate, 2003.

Paterson, A. R. "The Provision of Medical and Sanitary Services for Natives in Rural Africa." *Transactions of the Royal Society of Tropical Medicine* 21 (1927–28): 439–62.

Patton, Adell. *Physicians, Colonial Racism, and Diaspora in West Africa*. Gainesville: University Press of Florida, 1996.

Pearce, Louise. *The Treatment of Human Trypanosomiasis with Tryparsamide: A Critical Review*. New York: Rockefeller Institute for Medical Research, 1930.

Pearce, R. D. *The Turning Point in Africa: British Colonial Policy, 1938–1948*. London: Frank Cass, 1982.

Pearson, Karl. *The Grammar of Science*. London: Black, 1900.

———. *The Life of Francis Galton*. Vol. 2. Cambridge: Cambridge University Press, 1924.

Pelis, Kimberly. *Charles Nicolle, Pasteur's Imperial Missionary: Typhus and Tunisia*. Rochester, NY: University of Rochester Press, 2006.

Perham, Margery. "Future Relations of Black and White" (*The Listener*, March 28, 1934). In *Colonial Sequence, 1930–1949: A Chronological Commentary upon British Colonial Policy Especially in Africa*, edited by Margery Perham. London: Methuen, 1967.

———. *Lugard*, 2 vols. London: Collins, 1956–60.

———. *Native Administration in Nigeria*. London: Oxford University Press, 1937.

———. "Some Problems of Indirect Rule in Africa." Address to Royal Society of Arts, March 24, 1934. In *Colonial Sequence, 1930–1949: A Chronological Commentary upon British Colonial Policy Especially in Africa*, edited by Margery Perham, 91–118. London: Methuen, 1967.

———. *Ten Africans*. London: Faber and Faber, 1936.

Philip, Kavita. "English Mud: Towards a Critical Cultural Studies of Colonial Science." *Cultural Studies* 12 (1998): 300–331.

Phillips, John. "The Application of Ecological Research Methods to the Tsetse (*Glossina* SPP.) Problem in Tanganyika Territory: A Preliminary Account." *Ecology* 11 (1930): 713–33.

———. "Ecological Investigation in South, Central, and East Africa: Outline of a Progressive Scheme." *Journal of Ecology* 19 (1931): 474–82.

———. "Who Dare Be So Bold?" In *Kwame Nkrumah and the Future of Africa*, 13–33. London: Faber and Faber, 1960.

Phoofolo, Pule. "Epidemics and Revolutions: The Rinderpest Epidemic in Late Nineteenth-Century Southern Africa." *Past and Present* 138 (1993): 112–43.

———. "Face to Face with Famine: The BaSotho and the Rinderpest, 1897–1899." *Journal of Southern African Studies* 29 (2003): 503–27.

"Physique and Health in Kenya." *The Lancet* 217 (1931): 1358–59.

Pick, D. *Faces of Degeneration: A European Disorder, c. 1848–1918*. Cambridge: Cambridge University Press, 1989.

Poskin, A. *L'Afrique equatoriale: Climatologie, nosologie, hygiène*. Brussels: Société Belge de Libraire, 1897.

Prakash, Gyan. *Another Reason: Science and the Imagination of Modern India*. Princeton, NJ: Princeton University Press, 1999.

Prashad, B., ed. *The Progress of Science in India during the Past Twenty-Five Years*. Calcutta: Indian Science Congress, 1938.

Prins, Gwyn. "But What Was the Disease? The Present State of Health and Healing in African Studies." *Past and Present* 124 (1989): 159–79.

Proctor, Robert. *Racial Hygiene: Medicine under the Nazis*. Cambridge, MA: Harvard University Press, 1988.

Pugach, Sara. "Of Conjunctions, Comportment, and Clothing: The Place of African Teaching Assistants in Berlin and Hamburg, 1889–1919." In *Ordering Africa*, edited by Helen Tilley with Robert Gordon, 119–44. Manchester: Manchester University Press, 2007.

Pyenson, Lewis. *Civilizing Mission: Exact Sciences and French Overseas Expansion, 1830–1940*. Baltimore: Johns Hopkins University Press, 1993.

——. *Empire of Reason: Exact Sciences in Indonesia, 1840–1940*. Leiden: Brill, 1989.

Rabinow, Paul. *French Modern: Norms and Forms of the Social Environment*. Cambridge, MA: MIT Press, 1989.

Raina, Dhruv. "Beyond the Diffusionist History of Colonial Science." *Social Epistemology* 12 (1998): 203–13.

Rajan, Ravi. *Modernizing Nature: Forestry and Imperial Eco-Development, 1800–1950*. Oxford: Oxford University Press, 2006.

Ranger, Terence. "The Mwana Lesa Movement of 1925." In *Themes in the Christian History of Central Africa*, edited by T. Ranger and J. Weller, 45–75. London: Heinemann, 1975.

——. "Plagues of Beast and Men: Prophetic Responses to Epidemic in Eastern and Southern Africa." In *Epidemics and Ideas: Essays on the Historical Perception of Pestilence*, edited by Terence Ranger and Paul Slack, 241–68. Cambridge: Cambridge University Press, 1992.

Ravenstein, E. G. "The Climate of Tropical Africa." *Times (London)*, November 20, 1891, 7, col. F.

——. *Climatological Observations at Colonial and Foreign Stations: Tropical Africa*. London: HMSO, 1904.

——. "Lands of the Globe Available for European Settlement." *PRGS* 13 (1891): 27–35.

——. "Statistics at the Paris Geographical Congress." *Journal of the Statistical Society of London* 38 (1875): 422–29.

——. "Tables of African Climatology." In *How to Live in Tropical Africa*, by J. Murray, 28. London: George Philip, 1895.

Rawlinson, H. C. "Address to the Royal Geographical Society." *PRGS* 20 (May 22, 1876): 377–448.

Rawson, Rawson W. "The Territorial Partition of the Coast of Africa." *PRGS* 6 n.s. (November 1884): 615–31.

Reed, Alfred C. "Environmental Medicine." *Science* 82 (1935): 447–52.

Reeves, Jesse Siddall. *The International Beginnings of the Congo Free State*. Baltimore: Johns Hopkins Press, 1894.

Reid, Donald M. "The Egyptian Geographical Society: From Foreign Layman's Society to Indigenous Professional Association." *Poetics Today* 14 (1993): 539–72.

Reining, Conrad. *The Zande Scheme: An Anthropological Case Study of Economic Development in Africa*. Evanston, IL: Northwestern University Press, 1966.

"Report of the Committee . . . Appointed to Inquire into the Climatological and Hydrographical Conditions of Tropical Africa." In *Report of the Sixty-Second Meeting of the British Association (Edinburgh, 1892)*. London: John Murray, 1893.

"Resolution on Research in Africa." *East African Medical Journal* 13 (1936–37): 122.

"Rex v. Kumwaka Wa Mulumbi & 69 Others." In *Law Reports of Kenya*, 137–39. Nairobi: Government Printer, 1933.

Reyna, S. P. "Literary Anthropology and the Case against Science." *Man* 29 (1994): 555–81.

Rich, Paul. *Race and Empire in British Politics*. Cambridge: Cambridge University Press, 1986.

Richards, Audrey. "Anthropological Problems in North-Eastern Rhodesia." *Africa* 5 (1932): 121–44.

———. *Hunger and Work in a Savage Tribe: A Functional Study of Nutrition among the Southern Bantu*. London: Routledge, 1932.

———. *Land, Labour, and Diet in Northern Rhodesia: An Economic Study of the Bemba Tribe*. London: Oxford University Press, 1939.

Richards, Paul. "'Alternative' Strategies for the African Environment: 'Folk Ecology' as a Basis for Community Oriented Agricultural Development." In *African Environment*, edited by Paul Richards, 102–17. London: International African Institute, 1975.

———. "Ecological Change and the Politics of African Land Use." *African Studies Review* 26 (1983): 1–72.

———. *Indigenous Agricultural Revolution: Ecology and Food Production in West Africa*. London: Unwin Hyman, 1985.

Richardson, Peter. *Chinese Mine Labour in the Transvaal*. London: Macmillan, 1982.

Rivers, W. H. R. *Medicine, Magic, and Religion*. London: K. Paul Trench, Trubner, 1924.

Robbins, Charles R. "Northern Rhodesia: An Experiment in the Classification of Land with the Use of Aerial Photographs." *Journal of Ecology* 22 (1934): 88–105.

Roberts, Andrew. *Salisbury: Victorian Titan*. London: Orion Books, 1999.

———. "The Imperial Mind, 1905–1940." In *The Colonial Moment in Africa: Essays on the Movement of Minds and Materials, 1900–1940*, edited by A. Roberts. Cambridge: Cambridge University Press, 1990.

Roberts, Clifton. *Tangled Justice: Some Reasons for a Change in Policy in Africa*. London: Macmillan, 1937.

———. "Witchcraft and Colonial Legislation." *Africa* 8 (1935): 488–94.

Roberts, W. J. "The Racial Interpretation of History and Politics." *International Journal of Ethics* 18 (1908): 475–92.

Robertshaw, Peter, ed. *A History of African Archaeology*. London: James Currey, 1990.

Robinson, Kenneth. "Experts, Colonialists, and Africanists, 1895–1960." In *Experts in Africa*, edited by J. C. Stone, 55–74. Aberdeen: Aberdeen University African Studies Group, 1980.

Robinson, Ronald. "The Moral Disarmament of African Empire, 1919–1947." *Journal of Imperial and Commonwealth History* 8 (1979–80): 86–104.

Robinson, Ronald, John Gallagher, with Alice Denny. *Africa and the Victorians: The Official Mind of Imperialism*. New York: St. Martin's Press, 1961.

Roemer, Milton I. "Internationalism in Medicine and Public Health." In *Companion Encyclopedia of the History of Medicine*, vol. 2, edited by W. F. Bynum and Roy Porter, 1417–35. London: Routledge, 1993.

Roeykens, Auguste. "Banning et la Conférence Géographique de Bruxelles en 1876." *Zaire* 8 (1954): 227–71.

———. *Léopold II et la Conférence Géographique de Bruxelles*. Brussels, 1955.

———. *Léopold II et l'Afrique 1855–1880: Essai de synthèse et de mise au point*. Gembloux: Editions J. Duculot, 1957.

Rohlfs, Gerard. "Projet de voyage en Afrique." *Congrès International des Sciences Géographiques*, vol. 1, 612–14. Paris: E. Martinet, 1878.

Roll-Hansen, Nils. "Studying Nature without Nature? Reflections on the Realism of So-Called Laboratory Studies." *Studies in History and Philosophy of Biological and Biomedical Sciences* 29 (1998): 165–87.

Ross, G. A. Park. "Insecticide as a Major Measure in Control of Malaria, Being an Account of the Methods and Organisation Put in Force in Natal and Zululand during the Past Six Years." In "Report of Pan-African Health Conference," 114–33.

Ross, Ronald. "The Progress of Tropical Medicine." *Journal of the Royal African Society* 4 (1905): 271–89.

Rounce, N. V. "The Ridge in Native Cultivation, with Special Reference to the Mwanza District." *East African Agricultural Journal* 4 (1939): 352–55.

Royce, Josiah. "Race Questions and Prejudices." *International Journal of Ethics* 16 (1906): 265–88.

"Rural Hygiene and Medical Services in Africa." In "Report of the Pan-African Health Conference, Johannesburg, Nov. 20–30, 1935." *Quarterly Bulletin of the Health Organisation* 5 (1936): 198–200.

Ruxton, F. H. "An Anthropological No-Man's-Land." *Africa* 3 (1930): 1–11.

Sadowsky, Jonathan. *Imperial Bedlam: Institutions of Madness in Colonial Southwest Nigeria.* Berkeley: University of California Press, 1999.

Sambon, Louis. "Sleeping Sickness in the Light of Recent Knowledge." *Journal of Tropical Medicine* 6 (1903): 201–9.

Sampson, H. C., and E. M. Crowther. "Crop Production and Soil Fertility Problems." In *The West Africa Commission, 1938–39—Technical Reports,* 53. London: Leverhulme Trust, 1943.

Sanderson, G. N. "The European Partition of Africa: Origins and Dynamics." In *The Cambridge History of Africa.* Vol. 6, *1870 to 1905,* edited by J. D. Fage and Roland Oliver, 99. Cambridge: Cambridge University Press, 1985.

Sanford, Henry Shelton. "Report on the Annual Meeting of the African International Association, in Brussels, in June, 1877." *Journal of the American Geographical Society of New York* 9 (1877): 103–8.

Sanger, Clyde. *Malcolm MacDonald: Bringing an End to Empire.* Montreal: McGill University Press, 1995.

Sanjek, Roger. "Anthropology's Hidden Colonialism: Assistants and Their Ethnographers." *Anthropology Today* 9 (1993): 13–18.

Saul, S. B. "The Economic Significance of 'Constructive Imperialism.'" *Journal of Economic History* 17 (1957): 173–92.

Saunders, T. "On the Present Aspects of Africa, with Reference to the Development of Trade with the Interior." *Journal of the Society of Arts* 22 (1874): 261.

Schapera, Isaac. "Review of Reaction to Conquest." *Bantu Studies* (1937): 53–60.

Scheper-Hughes, Nancy. "The Primacy of the Ethical: Propositions for a Militant Anthropology." *Current Anthropology* 36 (1995): 409–40.

Schirmer, H. "La géographie de l'Afrique en 1880 et en 1891." *Annales de Géographie* 1, no. 2 (1892): 185–96.

Schmidt, Ulf. *Justice at Nuremberg: Leo Alexander and the Nazi Doctors' Trial.* New York: Palgrave Macmillan, 2004.

Schneider, William. "Geographical Reform and Municipal Imperialism in France, 1870–80." In *Imperialism and the Natural World,* ed. John MacKenzie, 90–117. Manchester: Manchester University Press, 1990.

Schroeder-Gudehus, Brigitte. "Nationalism and Internationalism." In *Companion to the History of Modern Science,* edited by Robert Olby et al., 909–19. London: Routledge, 1990.

Schumaker, Lyn. *Africanizing Anthropology: Fieldwork, Networks, and the Making of Cultural Knowledge in Central Africa.* Durham, NC: Duke University Press, 2001.

Schweinfurth, G. "Discours prononcé au Caire a la séance d'inauguration le 2 Juin 1875 par le Dr. G. Schweinfurth." *Bulletin de la Société Royale de Géographie d'Égypte* 14 (1926): 113–27.

Scott, A. MacCallum. "A New Colour Bar." *Contemporary Review* 102 (1912): 221–27.

Scott, H. H. *A History of Tropical Medicine*, 2 vols. London: Arnold, 1939.

Scott, H. S. "A Note on the Educable Capacity of the African." *East African Medical Journal* 9 (1932–33): 99–110.

Scott, James C. *Seeing like a State: How Certain Schemes to Improve the Human Condition Have Failed*. New Haven, CT: Yale University Press, 1998.

"The Scramble for Africa." *Times (London)*, September 15, 1884, 8, col. A.

Searle, G. R. *Eugenics and Politics in Britain, 1900–1914*. Leyden: Noordhoff International, 1976.

Sears, Paul. "Ecology: A Subversive Subject." *BioScience* 14 (1964): 11–13.

"A Selected Bibliography on the Physical and Mental Abilities of the American Negro." *Journal of Negro Education* 3 (1934): 548–64.

Semmel, Bernard. *Imperialism and Social Reform: English Social-Imperial Thought, 1895–1914*. Cambridge: Cambridge University Press, 1960.

Shantz, H. L. "Agriculture in East Africa." In *Education in East Africa: A Study of East, Central, and South Africa by the Second African Education Commission under the Auspices of the Phelps-Stokes Fund*, edited by Thomas Jesse Jones, 353–401. New York: Phelps-Stokes Fund, 1924.

———. "Urundi, Territory and People." *Geographical Review* 12 (1922): 329–57.

Shantz, H. L., and C. F. Marbut. *The Vegetation and Soils of Africa*. New York: National Research Council and American Geographical Society, 1923.

Shapin, Steven. "Placing the View from Nowhere: Historical and Sociological Problems in the Location of Science." *Transactions of the Institute of British Geographers* 23 (1998): 5–12.

Shapiro, Karen. "Doctors or Medical Aids: The Debate over the Training of Black Medical Personnel for the Rural Black Population in South Africa in the 1920s and 1930s." *Journal of Southern African Studies* 13 (1987): 234–55.

Shaw, Claye. "The Influence of Mind as a Therapeutic Agent." *The Lancet* (November 1909): 1369–73.

Sheets-Pyenson, Susan. "Civilizing by Nature's Example: The Development of Colonial Museums of Natural History, 1850–1900." In *Scientific Colonialism: A Cross-Cultural Comparison*, edited by Nathan Reingold and Marc Rothenberg, 351–77. Washington, DC: Smithsonian Institution Press, 1987.

Shelford, Frederic. "The Late Mrs. J. R. Green and the African Society." *JRAS* 28 (1929): 413–14.

Shiels, T. Drummond. "The Native Brain." *Times (London)*, December 28, 1933.

Shimazu, Naoko. *Japan, Race, and Equality: The Racial Equality Proposal of 1919*. London: Routledge, 1998.

Showers, Kate. "Soil Erosion in the Kingdom of Lesotho: Origins and Colonial Response, 1830s–1950s." *Journal of Southern African Studies* 15 (1989): 263–85.

Sibeud, Emmanuelle. "The Elusive Bureau of Colonial Ethnology." In *Ordering Africa*, edited by Helen Tilley with Robert Gordon. Manchester: Manchester University Press, 2007.

Sidibé, M., J. F. Scheuring, D. Tembely, M. M. Sidibé, P. Hofman, and M. Frigg. "Baobab: Homegrown Vitamin C for Africa." *Agroforestry Today* 8 (1996): 13–15.

Silva White, Arthur. *The Development of Africa*. London: G. Philip and Son, 1890.

———. "On the Comparative Value of African Lands." *Scottish Geographical Magazine* 7 (1891): 191–95.

———. "The Position of Geography in the Cycle of the Sciences." *Geographical Journal* 2 (1893): 178–79.

———. "To What Extent Is Tropical Africa Suited for Development by the White Races, or under Their Superintendence." In *Report of the Sixth International Geographical Congress, London 1895*, 549–53.

Singh, Saint Nihal. "Asiatic Emigration: A World Question." *Hindustan Review* 29 (1914): 533–40.

Smith, Crosbie, and Jon Agar, eds. *Making Space for Science: Territorial Themes in the Shaping of Knowledge*. Basingstoke: Macmillan, 1998.

Smith, Edwin W. "Africa: What Do We Know of It?" *Journal of the Royal Anthropological Institute* 65 (1935): 1–81.

———. "Inzuikizi." *Africa* 8 (1935): 473–80.

———. "The Story of the Institute: A Survey of Seven Years." *Africa* 7 (1934): 1–27.

Smith, Gaddis. "The British Government and the Disposition of the German Colonies in Africa." In *Britain and Germany in Africa: Imperial Rivalry and Colonial Rule*, edited by Prosser Gifford and W. R. Louis, 275–99. New Haven, CT: Yale University Press.

Smith, H. C. "The Sukuma System of Grazing Rights—Known Locally as Kupela Iseso or Ngitiri." *East African Agricultural Journal* 4 (1938): 129–30.

Smith, Honor. "Medicine in Africa as I Have Seen It." *African Affairs* 54 (1955): 28–36.

Smith, Paul, ed. *Ecological Survey of Zambia: The Traverse Records of Colin Trapnell, 1932–43*, 3 vols. London: Board of Trustees of the Royal Botanic Gardens, Kew, 2001.

Smuts, J. C. "African Settlement. Part I." *Journal of the Royal African Society* 29 (January 1930): 109–31.

———. "British African Conference [Smuts Lecture]." *Times (London)*, November 4, 1929, 17, col. C.

———. "Science in South Africa." *Nature* 116 (1925): 245–49.

———. "South Africa in Science." *South African Journal of Science* 22 (1925): 1–19.

Soleillet, Paul. *Avenir de la France en Afrique*. Paris: Challamel Aîné, 1876.

Soloway, Richard. *Demography and Degeneration: Eugenics and the Declining Birthrate in Twentieth-Century Britain*. Chapel Hill: University of North Carolina Press, 1990.

Soper, Fred. "Rehabilitation of the Eradication Concept in Prevention of Communicable Diseases." *Public Health Reports* 80 (1965): 855–69.

Sörlin, Sverker. "National and International Aspects of Cross-Boundary Science: Scientific Travel in the 18th Century." In *Denationalizing Science*, edited by Elisabeth Crawford et al., 43–72. Dordrecht: Kluwer Academic Publishers, 1993.

Sparn, Enrique. "Cronología, diferenciación, numero de socios y distribución de las sociedades de geografía." *Boletín de la Academia Nacional de Ciencias* 32 (1932–35): 323–36.

Stengers, Jean. "Introduction." *La Conférence de Géographie de 1876*, vii–xxv. Brussels: Académie Royale des Sciences d'Outre Mer, 1976.

———. "King Leopold and Anglo-French Rivalry, 1882–1884." In *France and Britain in Africa: Imperial Rivalry and Colonial Rule*, edited by P. Gifford and R. Louis, 121–66. New Haven, CT: Yale University Press, 1971.

———. "Léopold II and the *Association Internationale du Congo*." In *Bismarck, Europe, and Africa*, edited by Stig Föster et al., 229–44. London: Oxford University Press, 1988.

———. "Leopold II et Brazza en 1882: Documents Inedits." *Revue Française d'Histoire Outre-Mer* 63 (1976): 105–36.

Stent, H. B. "Observations on the Fertilizer Effect of Wood Burning in the 'Chitemene'

System." In *Third Annual Bulletin—Agricultural Department of Northern Rhodesia*, 48–49. Livingstone: Government Printer, 1933.

Stepan, Nancy. *The Idea of Race in Science: Great Britain, 1800–1960*. London: Macmillan, 1982.

Stevenson, W. B. *Four Months among African Missions*. Edinburgh: Blackwood, 1927.

Stocking, George Jr. *After Tylor: British Social Anthropology, 1888–1951*. Madison: University of Wisconsin Press, 1995.

Stocking, Michael. "Soil Conservation Policy in Colonial Africa." *Agricultural History* 59 (1985): 148–61.

Stoler, Ann. *Race and the Education of Desire: Foucault's History of Sexuality and the Colonial Order of Things*. Durham, NC: Duke University Press, 1995.

Sengoopta, Chandak. *Imprint of the Raj: How Fingerprinting was Born in Colonial India*. London: Pan, 2004.

Stone, Diane, ed. *Banking on Knowledge: The Genesis of the Global Development Network*. London: Routledge, 2000.

Storey, William. "Guns, Race, and Skill in Nineteenth-Century Southern Africa." *Technology and Culture* 45 (2004): 687–711.

———. "Plants, Power, and Development: Founding the Imperial Department of Agriculture for the West Indies, 1880–1914." In *States of Knowledge: The Co-Production of Science and the Social Order*, edited by Sheila Jasanoff, 109–30. New York: Routledge, 2004.

Strickland, C. F. "The Cooperative Movement in the East." *International Affairs* 11 (1932): 812–32.

Strong, Richard. "The Importance of Ecology in Relation to Disease." *Science* 82 (1935): 307–17.

Summerfield, Michael. "Tectonics, Geology, and Long-Term Landscape Development." In *The Physical Geography of Africa*, edited by W. M. Adams, A. S. Goudie, and A. R. Orme, 1–17. Oxford: Oxford University Press, 1996.

Summers, Carol. "Intimate Colonialism: The Imperial Production of Reproduction in Uganda, 1907–1925." *Signs* 4 (1991): 787–807.

Swellengrebel, N. H. "Report on Investigation into Malaria in the Union of South Africa, 1930–31." *Journal of the Medical Association of South Africa* 5 (1931): 409–24; 443–56.

Swynnerton, Charles Francis Massy. "Entomological Aspects of an Outbreak of Sleeping Sickness near Mwanza, Tanganyika Territory." *Bulletin of Entomological Research* 13 (1923): 317–70.

———. "An Examination of the Tsetse Problem in North Mossurise, Portuguese East Africa." *Bulletin of Entomological Research* 11 (1921): 315–85.

———. "An Experiment in Control of Tsetse-Flies at Shinyanga, Tanganyika Territory." *Bulletin of Entomological Research* (1923): 313–37.

———. "Fauna in Relation to Flora." In *Report on Tanganyika Territory Covering the Period from the Conclusion of the Armistice to the End of 1920*. London: HMSO, 1921.

———. "Some Factors in the Replacement of the Ancient East African Forest by Wooded Pasture Land." *South African Journal for the Advancement of Science* 14 (1918): 493–518.

———. *The Tsetse Flies of East Africa, a First Study of Their Ecology, with a View to Their Control*. Transactions of the Royal Entomological Society of London No. 8. London: The Society, 1936.

Tanganyika Territory. *Tsetse Research Report, 1935–1938*. Dar es Salaam: Government Printer, 1939.

T[ansley], A. G. "Two Recent Ecological Papers." *New Phytologist* 5 (1906): 219–22.

Tansley, Arthur. "The Classification of Vegetation and the Concept of Development." *Journal of Ecology* 8 (1920): 118–49.

———. "The Use and Abuse of Vegetational Concepts and Terms." *Ecology* 16 (1935): 284–307.

Taylor, Peter J. "Technocratic Optimism, H. T. Odum, and the Partial Transformation of Ecological Metaphor after World War II." *Journal of the History of Biology* 21 (1988): 213–44.

Thackray, Arnold. "History of Science." In *A Guide to the Culture of Science, Technology, and Medicine*, edited by Paul Durbin. New York: Free Press, 1980.

Thimm, C. A., ed. *Bibliography of Trypanosomiasis Embracing Original Papers Published Prior to April, 1909*. London: Sleeping Sickness Bureau, 1909.

Thomas, Lynn. *Politics of the Womb: Women, Reproduction, and the State in Kenya*. Berkeley: University of California Press, 2003.

Thomas, William. "The Psychology of Race Prejudice." *American Journal of Sociology* 9 (1904): 593–611.

Thomas-Emeagwali, Gloria, ed. *African Systems of Science, Technology, and Art: The Nigerian Experience*. London: Frontline International, 1993.

Thomason, Arthur Landsborough. *Half a Century of Medical Research*. Vol. 2, *The Programme of the Medical Research Council*. London: HMSO, 1975.

Thomson, Joseph. "East Central Africa and Its Commercial Outlook." *Scottish Geographical Magazine* 2 (1886): 65–78.

Thornton, D., and N. V. Rounce. "Ukara Island and the Agricultural Practices of the Wakara." *Tanganyika Notes and Records* 1 (1936): 25–32.

Thornton, E., and A. J. Orenstein. "Co-Ordination of Health Work in Africa." In "Report of the Pan-African Health Conference, Johannesburg, November 20–30, 1935," 208–9.

Thurnwald, Richard. *Black and White in East Africa: The Fabric of a New Civilization*. London: Routledge, 1935.

Tierney, Patrick. *Darkness in El Dorado: How Scientists and Journalists Devastated the Amazon*. London: W. W. Norton, 2000.

Tignor, Robert. *W. Arthur Lewis and the Birth of Development Economics*. Princeton, NJ: Princeton University Press, 2006.

Tilley, Helen. "Global Histories, Vernacular Science, and African Genealogies." *Isis* 101 (2010): 110–19.

———. "Africa as a 'Living Laboratory': The African Research Survey and the British Colonial Empire—Consolidating Environmental, Medical, and Anthropological Debates, 1920–1940." PhD diss., Oxford University, 2001.

Tilley, Helen, with Robert Gordon, eds. *Ordering Africa: Anthropology, European Imperialism, and the Politics of Knowledge*. Manchester: Manchester University Press, 2007.

Tooth, Geoffrey. *Report to the Colonial Social Science Research Council on the Use of Unadapted Tests of Intelligence, Attainment, and Aptitude for the Selection of Candidates for Secondary and Technical Education in Nigeria and the Gold Coast*. London: Colonial Office Research Department, 1953.

Tothill, J. D., et al. *A Report on Nineteen Surveys Done in Small Agricultural Areas in Uganda*. Entebbe: Government Press, 1938.

Tothill, J. D., ed. *Agriculture in Uganda*. London: Oxford University Press, 1940.

Trapnell, Colin. "Ecological Methods in the Study of Native Agriculture in Northern Rhodesia." *Bulletin of Miscellaneous Information—Kew* 1 (1937): 1–10.

Trapnell, Colin, and J. N. Clothier. "Report of the Ecological Survey of 1933." *Annual Report for the Year 1933, Department of Agriculture*, 23–27. Livingstone: Government Printer, 1934.

———. "Report of the Ecological Survey of 1934." *Annual Report for the Year 1934, Department of Agriculture*, 26–31. Livingstone: Government Printer, 1935.

———. *The Soils, Vegetation, and Agricultural Systems of North-Western Rhodesia: Report of the Ecological Survey*. Lusaka: Government Printer, 1937.

Trigger, Bruce. "The History of African Archaeology in World Perspective." In *A History of African Archaeology*, edited by Peter Robertshaw, 309–19. London: James Currey, 1990.

Trowell, H. C. "The Medical Training of Africans." *East African Medical Journal* 11 (1934–35): 338–53.

"Trusteeship—House of Commons, 9 July 1935." *East African Medical Journal* 12 (1935–36): 153–54.

University of Oxford. *Scientific Results of the First Oxford University Expedition to Spitsbergen*. Vol. 1. London, 1925.

Uvarov, B. P. *Locust Research and Control, 1929–1950*. London: HMSO, 1951.

Vageler, Paul. *An Introduction to Tropical Soils*. London: Macmillan, 1933.

Vail, Leroy. "Ecology and History: The Example of Eastern Zambia." *Journal of Southern African Studies* 3 (1977): 129–55.

Van Heijnsbergen, P. *International Legal Protection of Wild Fauna and Flora*. Amsterdam: Ohmsha Press, 1997.

Vandewoude, Emile. "De Aardrijkskundige Conferentie (1976) [*sic*] vanuit het koninklijk Paleis gezien." In *La Conférence de Géographie de 1876: Recueil d'études*, 415–438. Brussels: Académie Royale des Sciences d'Outre Mer, 1976.

Vaughan, Megan. "Africa and the Birth of the Modern World." *Transactions of the Royal Historical Society* 16 (2006): 143–62.

———. *Curing Their Ills: Colonial Power and African Illness*. Cambridge: Polity Press, 1991.

———. "Healing and Curing: Issues in the Social History and Anthropology of Medicine in Africa." *Social History of Medicine* 7 (1994): 283–95.

Verran, Helen. *Science and an African Logic*. Chicago: University of Chicago Press, 2001.

Vincent, Joan. "Sovereignty, Legitimacy, and Power: Prolegomena to the Study of the Colonial State." In *State Formation and Political Legitimacy*, edited by Ronald Cohen and Judith Drick Tolands, 137–54. New Brunswick, NJ: Transactions Books, 1988.

Vint, F. W. "A Preliminary Note on the Cell Content of the Prefrontal Cortex of the East African Native." *East African Medical Journal* 9 (1932): 30–55.

Wagner, Michele. "Notes on the Shantz Collection, Tucson, Arizona." *History in Africa* 19 (1992): 445–49.

Wakefield, A. J. "Native Production of Coffee on Kilimanjaro." *Empire Journal of Experimental Agriculture* 4 (1936): 97–106.

Waller, Horace. "Letter to the Editor." *Times (London)*, January 7, 1887, 7, col. D.

Waller, Richard. "Witchcraft and Colonial Law in Kenya." *Past and Present* 180 (2003): 241–75.

Waller, Richard, and Kathy Homewood. "Elders and Experts: Contesting Veterinary Knowledge in a Pastoral Community." In *Western Medicine as Contested Knowledge*, edited by Andrew Cunningham and Bridie Andrews, 69–93. Manchester: University of Manchester Press, 1997.

Washington, Harriet. "Op-Ed: Why Africa Fears Western Medicine." *New York Times*, July 31, 2007.

Waters, Kenneth, and Albert Van Helden, eds. *Julian Huxley: Biologist and Statesman of Science*. Houston: Rice University Press, 1992.

Weindling, Paul. *Nazi Medicine and the Nuremberg Trials: From Medical War Crimes to Informed Consent*. New York: Palgrave Macmillan, 2004.

———. "Social Medicine at the League of Nations Health Organisation and the International Labour Office Compared." In *International Health Organisations and Movements, 1918–1939*, edited by Paul Weindling, 134–53. Cambridge: Cambridge University Press, 1995.

Weinmann, H. *Agricultural Research and Development in Southern Rhodesia, under the Rule of the British South Africa Company, 1890–1923*. Salisbury: University of Rhodesia, 1972.

Wells, H. G. *A Modern Utopia*. London: Chapman and Hall, 1905.

———. "Race Prejudice." *Independent* 62 (1907): 381–84.

Wells, H. G., Julian Huxley, and G. P. Wells. *The Science of Life*, 3 vols. London: Amalgamated Press, 1929–30.

Werbner, Pnina. "Vernacular Cosmopolitanism." *Theory, Culture, and Society* 23 (2006): 496–98.

Werskey, Gary. "*Nature* and Politics between the Wars." *Nature* 224 (1969): 462–72.

———. *The Visible College: A Collective Biography of British Scientists and Socialists in the 1930s*. London: Penguin, 1978.

"West African Botanic Stations." *Kew Bulletin* 84 (1893): 364–66.

West, Harry. "Inverting the Camel's Hump: Jorge Dias, His Wife, Their Interpreter, and I." In *Significant Others: Interpersonal and Professional Commitments in Anthropology*, edited by Richard Handler, 51–90. Madison: University of Wisconsin Press, 2004.

White, Luise. *Speaking with Vampires: Rumor and History in Colonial Africa*. Berkeley: University of California Press, 2000.

———. "Tsetse Visions: Narratives of Blood and Bugs in Colonial Northern Rhodesia, 1931–9." *Journal of African History* 36 (1995): 219–45.

White, Owen. *Children of the French Empire: Miscegenation and Colonial Society in French West Africa, 1895–1960*. Oxford: Clarendon Press, 1999.

Whitehead, A. N. *Science and the Modern World*. Cambridge: Cambridge University Press, 1926.

Wilcocks, Charles. *Tuberculosis in Tanganyika Territory: Final Report on Investigations Carried out between 1930 and 1936 under the Auspices of the Colonial Development Fund*. Dar es Salaam: Government Printer, 1938.

Wilcocks, Charles, J. F. Corson, and R. L. Sheppard. *A Survey of Recent Work on Trypanosomiasis and Tsetse Flies—Based on Reports and Papers Published during the Period 1932–1944*. London, 1946.

Wilkinson, Lise. "Epidemiology." In *Companion Encyclopedia of the History of Medicine*, vol. 2, edited by W. F. Bynum and Roy Porter, 1262–82. London: Routledge, 1993.

Williams, A. D. "The Provision of Hospitals for, and the Medical Training of, Africans in Kenya." *Kenya Medical Journal* 3 (1926): 46–51.

Williams, A. W. "The History of Mulago Hospital and the Makerere College Medical School." *East African Medical Journal* 29 (1952): 253–63.

Willimott, S. G., and K. R. M. Anthony. "Agricultural Research and Development in the Southwest Sudan." *Tropical Agriculture* 34 (1957): 239–48.

Winterbottom, J. M. "The Ecology of Man and Plants in Northern Rhodesia." *Rhodes-Livingstone Institute Journal* 3 (1945): 33–44.

Wise, M. J. "The Scott Keltie Report 1885 and the Teaching of Geography in Britain." *Geographical Journal* 152 (1986): 367–82.

"Witchcraft Discussed." *East Africa* (August 16, 1934): 994–95.

Worboys, Michael. "The British Association and Empire: Science and Social Imperialism, 1880–1940." In *The Parliament of Science: The British Association for the Advancement of Science*, edited by Roy MacLeod and Peter Collins, 170–87. London: Science Reviews, 1981.

———. "The Comparative History of Sleeping Sickness in East and Central Africa, 1900–1914." *History of Science* 32 (1994): 89–102.

———. "Manson, Ross, and Colonial Medical Policy: Tropical Medicine in London and Liverpool, 1899–1914." In *Disease, Medicine, and Empire: Perspectives on Western Medicine and the Experience of European Expansion*, edited by Roy MacLeod and Milton Lewis, 21–38. London: Routledge, 1988.

———. "Science and British Colonial Imperialism." PhD diss., University of Sussex, 1979.

———. "Tropical Diseases." In *Companion Encyclopedia of the History of Medicine*, vol. 1, edited by W. F. Bynum and Roy Porter, 512–36. London: Routledge, 1993.

"Work of the Rockefeller Foundation." *Science* 76 (1932): 438–40.

World Bank. *Knowledge for Development: World Development Report 1998/1999*. New York: Oxford University Press, 1999.

Worthington, Edgar Barton. *A Development Plan for Uganda*. Entebbe: Government Press, 1946.

———. *The Ecological Century: A Personal Appraisal*. Oxford: Clarendon Press, 1983.

———. "The Lakes of Kenya and Uganda." *Geographical Journal* 79 (1932): 275–97.

———. "On the Food and Nutrition of African Natives." *Africa* 9 (April 1936): 150–65.

———. "Organization of Science in East Africa." In *African Regional Scientific Conference, Johannesburg, October 17 to 28, 1949*. Vol. 2, *Statements and Communications*, 451–53. Pretoria: Government Printer, 1949.

———. *A Report on the Fisheries of Uganda*. London: Crown Colonies, 1932.

———. *A Report on the Fishing Survey of Lakes Albert and Kioga*. London: Crown Agents for the Colonies, 1929.

———. *Science in Africa: A Review of Scientific Research Relating to Tropical and Southern Africa*. London: Oxford University Press, 1938.

———. *Science in the Development of Africa: A Review of the Contribution of Physical and Biological Knowledge South of the Sahara*. London: CCTA, 1958.

———. "Scientific Results of the Cambridge Expedition to the East African Lakes, 1930–1. General Introduction and Station List." *Linnean Society Journal—Zoology* 38 (1932): 99–119.

———. *A Survey of Research and Scientific Services in East Africa, 1947–1956*. Nairobi: East Africa High Commission, n.d. [1952].

Worthington, Edgar Barton, and S[tella] Worthington. *Inland Waters of Africa: The Result of Two Expeditions to the Great Lakes of Kenya and Uganda, with Accounts of Their Biology, Native Tribes, and Development*. London: Macmillan, 1933.

Wright, Gwendolyn. *The Politics of Design in French Colonial Urbanism*. Chicago: University of Chicago Press, 1991.

———. "Tradition in the Service of Modernity: Architecture and Urbanism in French Colonial Policy, 1900–1930." In *Tensions of Empire: Colonial Cultures in a Bourgeois World*, edited by Frederick Cooper and Ann Laura Stoler, 322–45. Berkeley: University of California Press, 1997.

Wright, C. T. Hagberg. "German Methods of Development in Africa." *Journal of the Royal African Society* 1 (1901): 23–38.

Wright, John K. "The Field of the Geographical Society." In *Geography in the Twentieth Century: A Study of Growth, Fields, Techniques, Aims and Trends*, 3rd ed., edited by Griffith Taylor, 543–65. New York: Philosophical Library, 1957.

Wybergh, W. "Imperial Organisation and the Colour Question." *Contemporary Review* 91 (two parts) (1907): 695–705; 805–15.

Wylie, Diana. "Norman Leys and McGregor Ross: A Case Study in the Conscience of African Empire, 1900–1939." *Journal of Imperial and Commonwealth History* 5 (1976–77): 294–309.

Wyndham, H. "The African Labourer and West African Agriculture: Review." *International Affairs* 12 (1933): 824–25.

Yaalon, Dan, and Simon Berkowicz, eds. *History of Soil Science: International Perspectives*. Reiskirchen, Germany: Catena Verlag, 1997.

Yearwood, Peter. "Great Britain and the Repartition of Africa, 1914–1918." *Journal of Imperial and Commonwealth History* 18 (1990): 316–41.

Young, Crawford. *The African Colonial State in Comparative Perspective*. New Haven, CT: Yale University Press, 1994.

———. "Country Report: The African Colonial State Revisited." *Governance* 11 (1998): 101–20.

———. "Nationalism, Ethnicity, and Class in Africa: a Retrospective." *Cahiers d'Étude Africaines* 26 (1986): 421–95.

Zachariah, Benjamin. *Developing India: An Intellectual and Social History, 1930–1950*. London: Oxford University Press, 2005.

Zachernuk, Philip. *Colonial Subjects: An African Intelligentsia and Atlantic Ideas*. Charlottesville: University of Virginia Press, 2000.

Zeleny, Jeff, and Laurie Goering. "Obama Challenges South Africa to Face AIDS Crisis." *Chicago Tribune*, August 22, 2006.

Zeller, Diane L. "The Establishment of Western Medicine in Buganda." PhD diss., Columbia University, 1971.

Zileza, Paul Tiyambe. *Manufacturing African Studies and Crises*. Dakar: CODESRIA, 1997.

Zimmer, Carl. "A Biological Hot Spot in Africa: With New Species Still to Discover." *New York Times*, March 6, 2007.

Zinoman, Peter. *The Colonial Bastille: A History of Imprisonment in Vietnam*. Berkeley: University of California Press, 2001.

INDEX

Page numbers followed by t indicate a table.